MATHPOWER™10

WESTERN EDITION

MATHPOWER™ 10

WESTERN EDITION

MATHPOWER™ 10–12,
***Western Edition*, Authors**
George Knill, B.Sc., M.S.Ed.
Hamilton, Ontario

Stella Ablett, B.Sc.
Vancouver, British Columbia

Cynthia Ballheim, B.Sc., M.A.
Calgary, Alberta

John Carter, B.Sc., M.Sc.
Toronto, Ontario

Eileen Collins, B.A., M.Ed.
Hamilton, Ontario

Eleanor Conrad, B.Sc.
Pugwash, Nova Scotia

Russel Donnelly, B.Sc., M.A.
Calgary, Alberta

Michael Hamilton, B.Sc., M.Sc.
St. Catharines, Ontario

Rosemary Miller, B.A.
Hamilton, Ontario

Alan Sarna, B.A.
Vancouver, British Columbia

Harold Wardrop, B.Sc.
Surrey, British Columbia

MATHPOWER™ 10,
***Western Edition*, Consultants**
Pat Angel
Islington, Ontario

Rene Baxter
Saskatoon, Saskatchewan

Terry Clifford
Winnipeg, Manitoba

Shirley Dalrymple
Newmarket, Ontario

Joe Fisher
Middle Musquodoboit, Nova Scotia

Lorne Lindenberg
Edmonton, Alberta

Mark Mahovlich
Saanichton, British Columbia

Kanwal Neel
Richmond, British Columbia

McGraw-Hill
Ryerson

Toronto Montreal New York Auckland Bogotá Caracas
Lisbon London Madrid Mexico City Milan New Delhi
San Juan Singapore Sydney Tokyo

McGraw-Hill
Ryerson Limited
A Subsidiary of The **McGraw-Hill** Companies

MATHPOWER™ 10
Western Edition

Copyright © 1998, McGraw-Hill Ryerson Limited, a Subsidiary of The McGraw-Hill Companies. All rights reserved. No part of this publication may be reproduced or transmitted in any form or by any means, or stored in a data base or retrieval system, without the prior written permission of McGraw-Hill Ryerson Limited, or, in the case of photocopying or other reprographic copying, a licence from CANCOPY (Canadian Reprography Collective), 6 Adelaide Street East, Suite 900, Toronto, Ontario M5C 1H6.

Any request for photocopying, recording, or taping of this publication shall be directed in writing to CANCOPY.

ISBN 0-07-552596-8

http://www.mcgrawhill.ca

9 10 TRI 7 6 5

Printed and bound in Canada

Care has been taken to trace ownership of copyright material contained in this text. The publishers will gladly take any information that will enable them to rectify any reference or credit in subsequent printings.

Claris and ClarisWorks are registered trademarks of Claris Corporation.

Microsoft® is a registered trademark of Microsoft Corporation.

Canadian Cataloguing in Publication Data
Main entry under title:

Mathpower 10

Western ed.
Includes index.
ISBN 0-07-552596-8

1. Mathematics. 2. Mathematics - Problems, exercises, etc.
I. Knill, George, date. II Title: Mathpower ten.

QA107.M37648 1998   510   C97-930021-5

PUBLISHER: Melanie Myers
EDITORIAL CONSULTING: Michael J. Webb Consulting Inc.
ASSOCIATE EDITORS: Sheila Bassett, Mary Agnes Challoner, Maggie Cheverie, Janice Nixon
SENIOR SUPERVISING EDITOR: Carol Altilia
PERMISSIONS EDITORS: Jacqueline Donovan, Crystal Shortt
PRODUCTION COORDINATOR: Yolanda Pigden
ART DIRECTION: Wycliffe Smith Design Inc.
COVER DESIGN: Wycliffe Smith Design Inc., Dianna Little
INTERIOR DESIGN: Wycliffe Smith Design Inc.
ELECTRONIC PAGE MAKE-UP: Tom Dart/First Folio Resource Group, Inc.
COVER ILLUSTRATIONS: Clarence Porter
COVER IMAGE: Harald Sund/The Image Bank

COPIES OF THIS BOOK MAY BE OBTAINED BY CONTACTING:

McGraw-Hill Ryerson Ltd.

WEBSITE:
http://www.mcgrawhill.ca

E-MAIL:
Orders@mcgrawhill.ca

TOLL FREE FAX:
1-800-463-5885

TOLL FREE CALL:
1-800-565-5758

OR BY MAILING YOUR ORDER TO:

McGraw-Hill Ryerson
Order Department,
300 Water Street
Whitby, ON L1N 9B6

Please quote the ISBN and title when placing your order.

CONTENTS

CHAPTER 2

Number Patterns

CHAPTER 3

Polynomials

Using *MATHPOWER™ 10, Western Edition*

Each chapter contains several numbered sections.
In a typical numbered section, you find the following features.

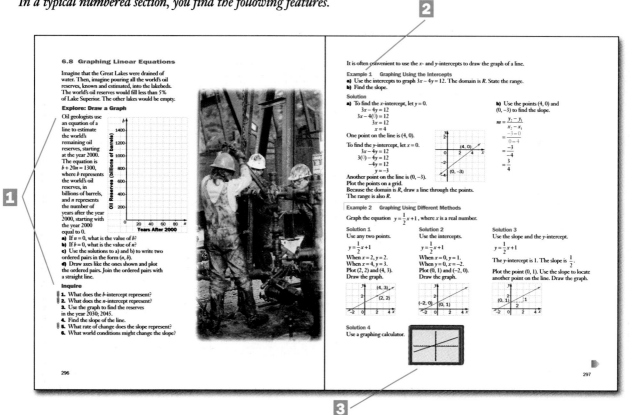

1 Explore and Inquire

You start with an exploration, followed by a set of inquire questions. The exploration and the inquire questions allow you to construct your own learning. Many explorations show how mathematics is applied in the world.

2 Examples

The examples show you how to use what you have learned.

3 Graphing Calculator Displays

These displays show you how technology can be used to solve problems.

4 Practice

By completing these questions, you practise what you have learned, so that you can stabilize your learning.

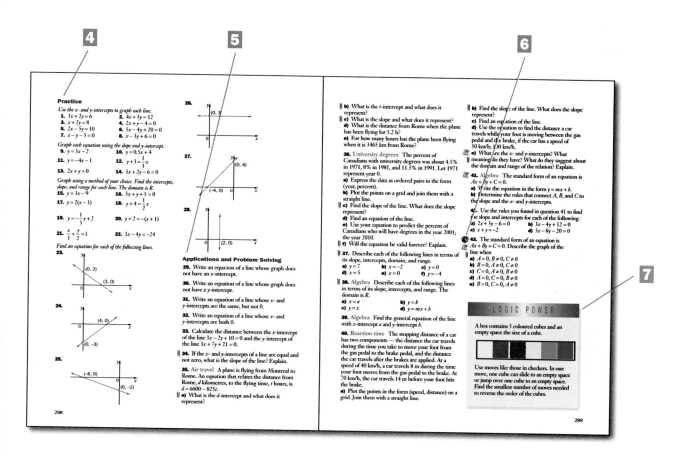

5 Applications and Problem Solving

These questions let you use what you have learned to solve problems, and to apply and extend what you have learned. The descriptors on many of the problems show connections to other disciplines, to other topics in mathematics, and to people's daily experiences.

6 Logos

The three logos indicate special kinds of problems.

When you see this logo, you will be asked to demonstrate an understanding of what you have learned by writing about it in a meaningful way.

This logo signals that you will need to think critically when you answer a question.

This logo indicates an opportunity to work with a classmate or in a larger group to solve a problem.

7 Power Problems

These problems are challenging and fun. They encourage you to reason mathematically.

Special Features of
MATHPOWER™ 10, Western Edition

Math Standard
There are 14 Math Standard pages before Chapter 1. By working through these pages, you will explore the mathematical concepts that citizens of the twenty-first century will need to understand.

Getting Started
A Getting Started section begins each chapter. This section reviews the mathematics that you will need to use in the chapter.

Mental Math
The Mental Math column in each Getting Started section includes a strategy for completing mental math calculations.

Problem Solving
The numerous ways in which problem solving is integrated throughout the book are described on pages xiv–xv.

Technology
Each chapter includes at least one Technology section. These sections allow you to explore the use of calculators and computers to solve problems. Graphing calculator displays are also integrated into many numbered sections. Most displays were generated using a TI-83 calculator. Exceptions are found in Chapters 3 and 4, and on page 25 of Chapter 1, where the more powerful TI-92 calculator was used.

Investigating Math
The explorations in the Investigating Math sections will actively involve you in learning mathematics, either individually or with your classmates.

Connecting Math and ...
Each chapter includes a Connecting Math section. In the explorations, you will apply mathematics to other subject areas, such as geography, physiology, English, and architecture.

Computer Data Bank
The Computer Data Bank sections are to be used in conjunction with the *MATHPOWER™ 10, Western Edition, Computer Data Bank*. In these sections, you will explore the power of a computer database program in solving problems. The explorations in these sections use ClarisWorks databases. For Microsoft Works users, alternative explorations are provided on blackline masters in the *MATHPOWER™ 10, Western Edition, Computer Data Bank Teacher's Resource*.

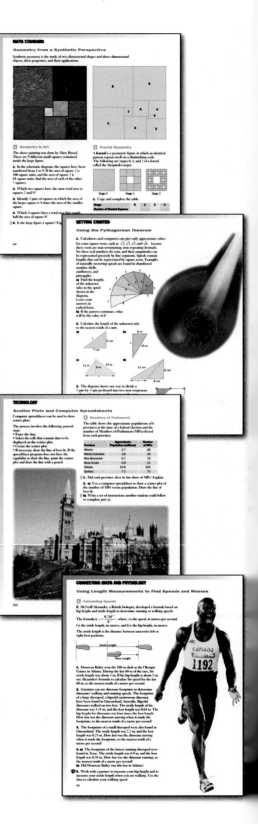

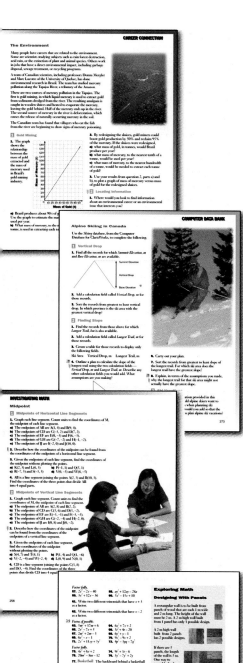

Career Connection

The explorations on the Career Connection pages will show you some applications of mathematics to the world of work.

Review/Chapter Check

Near the end of each chapter are sections headed Review and Chapter Check, which will allow you to test your progress. The questions in each Review are keyed to section numbers in the chapter, so that you can identify any sections that require further study.

Exploring Math

At the end of the Review section in each chapter, the Exploring Math column includes an enrichment activity designed as a problem solving challenge.

Cumulative Review

Chapter 4 and Chapter 8 end with cumulative reviews. The cumulative review at the end of Chapter 4 covers the work you did in Chapters 1–4. The first cumulative review at the end of Chapter 8 covers Chapters 5–8, and the second covers Chapters 1–8.

Data Bank

Problems that require the use of the Data Bank on page 404–413 are included in a Data Bank box at the end of the Problem Solving: Using the Strategies page in each chapter.

Answers

On pages 414–453, there are answers to most of the questions in this book.

Glossary

The illustrated glossary on pages 454–465 explains mathematical terms.

Indexes

The book includes three indexes — an applications index, a technology index, and a general index.

Problem Solving in *MATHPOWER*™ *10, Western Edition*

In whatever career you choose, and in other parts of your daily life, you will be required to solve problems. An important goal of mathematics education is to help you become a good problem solver.

George Polya was one of the world's best teachers of problem solving. The problem solving model he developed has been adapted for use in this book. The model is a guide. It will help you decide what to do when you "don't know what to do."

The problem solving model has the following four steps.

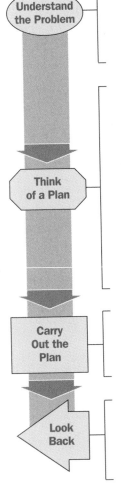

Read the problem and ask yourself these questions.
- What am I asked to find?
- Do I need an exact or approximate answer?
- What information am I given?
- What are the conditions or requirements?
- Is enough information given?
- Is there too much information?
- Have I solved a similar problem?

The main challenge in solving a problem is to devise a plan, or an outline of how to proceed. Organize the information and plan how to use it by deciding on a problem solving strategy. The following list of strategies, some of which you have used in previous grades, may help.

- Act out the problem.
- Look for a pattern.
- Work backward.
- Use a formula.
- Use logic.
- Draw and read graphs.
- Make an assumption.
- Guess and check.

- Use manipulatives.
- Solve a simpler problem.
- Use a diagram or flowchart.
- Sequence the operations.
- Use a data bank.
- Change your point of view.
- Use a table or spreadsheet.
- Identify extra information.

Estimate the answer to the problem. Choose the calculation method you will use to solve the problem. Then, carry out your plan, using paper and pencil, a calculator, a computer, or manipulatives. After solving the problem, write a final statement that gives the solution.

Check your calculations in each step of the solution. Then ask yourself these questions.
- Have I solved the problem I was asked to solve?
- Does the answer seem reasonable?
- Does the answer agree with my estimate?
- Is there another way to solve the problem?

Opportunities for you to develop your problem solving skills appear throughout this book.

In the first three chapters, there are nine numbered problem solving sections. Each section focuses on one strategy. The section provides an example of how the strategy can be used and includes problems that can be solved using the strategy.

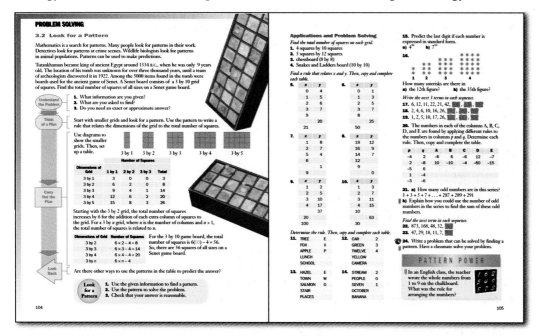

At the end of each chapter, you will find a section headed Problem Solving: Using the Strategies. Each of these sections includes a variety of problems that can be solved using different strategies. The section ends with problems contained in a Data Bank box. To solve the Data Bank problems, you can look up information in the Data Bank on pages 404–413, or you can use a data bank of your choice.

Every numbered section of the book includes the sub-heading Applications and Problem Solving. The problems under this sub-heading are related to that section and provide you with many opportunities to apply problem solving strategies.

Many numbered sections include Power Problems, which have been grouped into four types — Logic Power, Pattern Power, Number Power, and Word Power. These problems are challenging and fun.

Further problem solving opportunities are to be found in the Exploring Math columns. Each of these columns allows you to explore challenging mathematical ideas.

As described on pages xii–xiii, many special features in the book involve explorations. These features — including Math Standard, Technology, Investigating Math, Connecting Math, Computer Data Bank, and Career Connection sections — are filled with opportunities for you to refine your problem solving skills.

MATH STANDARD

Mathematics as Problem Solving

In any career you pursue, you will need to solve a wide variety of problems. This book includes many opportunities for you to improve your problem solving ability.

1 Solving Problems

1. How many different bracelets can be made using 5 identical blue beads and 2 identical yellow beads?

2. Find the number of squares that can be drawn on the grid shown, so that the following conditions are met:
• The vertices of each square are grid points.
• The red grid point lies inside each square, on one of its sides, or at one of its vertices.
One square has been drawn for you.

3. What is the smallest number of colours needed to paint a cube so that no two adjacent faces are painted the same colour?

4. On the diagram shown, line segments with lengths 1 to 5 can be found.

```
    1    1         3
 ├───┼───┼─────────┤
 A   B   C         D
```

AB = 1, AC = 2, CD = 3, BD = 4, and AD = 5.
Given the diagram below, what must the lengths of RS and ST be so that line segments with lengths from 1 to 9 can be found?

```
    1    1
 ├───┼───┼──────────────┼────────────┤
 P   Q   R              S            T
```

2 Comparing Strategies

1. The box is in the shape of a cube and has no top. Sixty-four small, identical cubes fill the box completely.
a) How many small cubes touch the bottom of the box?
b) How many small cubes touch the sides of the box?
c) How many small cubes do not touch the sides or the bottom of the box and are not visible from above the box?

2. Compare the problem solving strategies you used in question 1, parts b) and c), with your classmates' strategies.

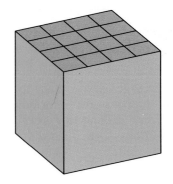

Mathematics as Communication

In any profession, it is important to communicate ideas, both orally and in writing. Using mathematical language can help you communicate ideas clearly.

1 Writing Instructions

Use a piece of paper with dimensions about 21 cm by 14 cm, which is about half the size of a standard sheet of paper. Draw four or five plane figures on the paper, marking their dimensions and their distances from each other or from the edges of the paper. Use circles, triangles, squares, rectangles, parallelograms, trapezoids, or kites. A sample arrangement of figures is shown.

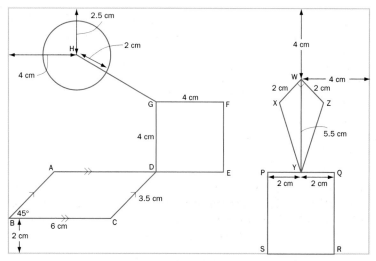

1. Write a set of instructions that you could give to a classmate to draw your design.

2. Read your instructions to 3 classmates and have each of them sketch your design, without looking at each other's sketches. Have your classmates compare their sketches with yours.

3. Are the sketches the same? If not, are there ways you could improve your instructions?

4. Repeat questions 1 to 3 but this time represent 3 or 4 three-dimensional shapes on the paper. Use cubes, rectangular prisms, square-based pyramids, or cones.

2 Describing Real Objects

Write a description of each of the following objects, so that someone who has never seen it will have a clear idea of how it looks. Compare your descriptions with a classmate's.

1. Newfoundland flag

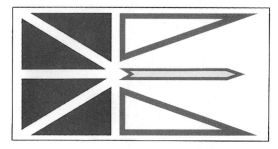

2. Canada Post mailbox

MATH STANDARD

Mathematics as Reasoning

To solve problems, you need to use logical reasoning to draw conclusions from given information. This textbook will help you increase your ability to reason logically.

1 Dividing Grids

Copy each grid. Show all the ways you could cut along grid lines to divide each grid into two congruent pieces with an X on each piece.

1.

2.
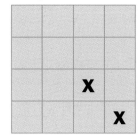

2 Arrangement of Sticks

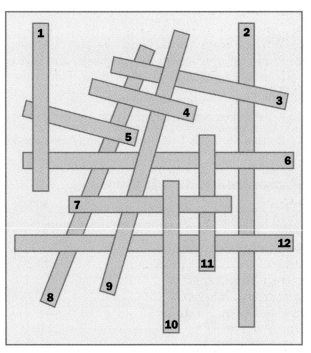

Twelve sticks, numbered 1 to 12, have been placed on a flat table. The sticks have different lengths. All the sticks have the same width. The thickness of each stick is the same as its width.

Some sticks rest completely on the table. Some sticks rest partly on the table and partly on other sticks. Some sticks rest completely on other sticks and do not touch the table. Some sticks have one end higher than the other.

1. List the numbers on the sticks that are level.

2. Which level stick is highest above the table?

3. List the numbers on the sticks that slope. State which end is higher on each sloping stick, the numbered end or the other end.

4. Does stick 7 touch stick 8?

5. Does stick 6 touch stick 9?

6. a) What is the greatest number of sticks in contact with any one stick?
b) Which stick(s) are in contact with the number of sticks from part a)?

7. a) What is the least number of sticks in contact with any one stick?
b) Which stick(s) are in contact with the number of sticks from part a)?

8. State the order in which you could remove the sticks from the table, so that each time you remove a stick, you do not move any other stick(s). Compare your order with a classmate's.

Mathematical Connections

There are connections between mathematics and many other disciplines, including the arts, the sciences, business, and so on.

1 Transforming Music

Transformations are often used in music.

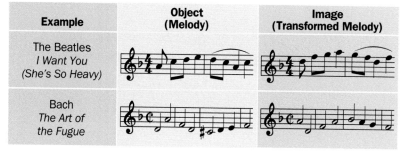

Example	Object (Melody)	Image (Transformed Melody)
The Beatles *I Want You* (She's So Heavy)		
Bach *The Art of the Fugue*		

1. a) What type of transformation is used in *I Want You*?
 b) Describe the transformation by stating how each note has moved.
 c) What is the musical name for this type of transformation?

2. a) What type of transformation is used in *The Art of the Fugue*?
 b) Describe the transformation by stating how each note has moved.
 c) What is the musical name for this type of transformation?

2 Water Sprinklers

The diagram shows a plan for connecting a water main to 10 sprinklers, named A to J, on a sports field. The point W is the connection to the water main. The lengths of the pipes, which are not necessarily straight, are shown in metres. To reduce the cost of the system, the plan should be modified to use the minimum length of pipe.

1. Copy the diagram and use it to decide which pipes to remove, so that water is supplied to all the sprinklers using the minimum length of pipe.

2. What is the minimum length of pipe needed?

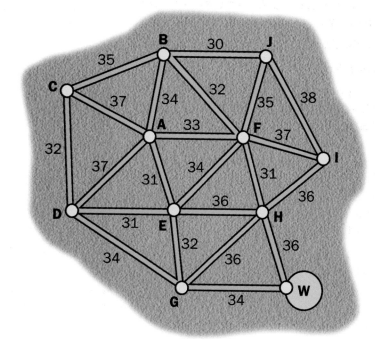

MATH STANDARD

Algebra

The language of much of mathematics is algebra.

1 Solving Equations

Solve each equation.

1. $x + 3 = -4$

2. $4t = 20$

3. $\dfrac{w}{3} = -2$

4. $2x + 5 = 13$

5. $6y - 11 = -65 + 3y$

6. $3(s - 4) = -11$

7. $\dfrac{n}{2} - \dfrac{n}{3} = 1$

8. $\dfrac{x+1}{2} = \dfrac{x-1}{3}$

9. In Winnipeg in July, the daily amount of bright sunshine averages 10 h. This is 2 h less than 3 times the amount in January. What is the daily amount of bright sunshine in Winnipeg in January?

10. In Prince Rupert, July averages 16 days of rain. This is 4 more than half the number of days of rain in October. On how many days does it rain in Prince Rupert in October?

2 Using Patterns

Each figure is made using toothpicks.

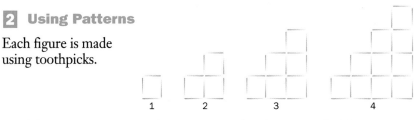

1. If the pattern continues, how many toothpicks are needed for
a) the 5th figure?   **b)** the 6th figure?

2. a) Write an expression for the number of toothpicks in terms of the figure number, n. Write the expression in the form $n(n + \blacksquare)$, where $\blacksquare$ is a whole number.
b) Use the expression from part a) to find the number of toothpicks in the 50th figure.

3. If the pattern continues, how many small squares are in
a) the 5th figure?   **b)** the 6th figure?

4. a) Write an expression for the number of small squares in terms of the figure number, n.
Write the expression in the form $\dfrac{n(n + \bullet)}{\blacktriangle}$, where $\bullet$ and $\blacktriangle$ are whole numbers.
b) Use the expression from part a) to find the number of small squares in the 50th figure.

3 Simplifying Expressions

Simplify.

1. $(4t^2 - 3t) + (2t^2 + 2t)$

2. $(3y^2 - y) - (2y^2 - 3y)$

3. $5x(-4x)$

4. $(-3xy^2)(4x^3y)$

5. $\dfrac{20s^2t^3}{-5st}$

6. $\dfrac{12xy^3z^4}{-3xy^2z^2}$

Functions

Functions are used to study how two things are related. Many applications of mathematics involve the relationships between variables. For example, the distance a greyhound runs in a certain length of time is related to its speed.
A way to find mathematical relationships is to use patterns.

1 T-Patterns

Each figure is made from squares of side 1 unit.

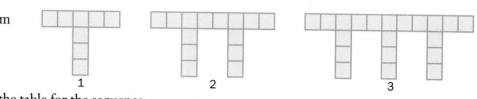

1. Copy and complete the table for the sequence.

Figure	1	2	3	4	5
Number of Squares					
Perimeter					

2. a) Write an expression for the number of squares in terms of the figure number, n.
b) How many squares are in the 75th figure? the 103rd figure?
c) What figure has 170 squares?

3. a) Write an expression for the perimeter in terms of the figure number.
b) What is the perimeter of the 68th figure? the 104th figure?
c) What figure has a perimeter of 270 units?

4. a) Write an expression for the perimeter in terms of the number of squares, s.
b) What is the perimeter of the figure with 92 squares? 314 squares?
c) How many squares are in the figure with a perimeter of 306 units? 858 units?

2 E-Patterns

Each figure is made from squares of side 1 unit.

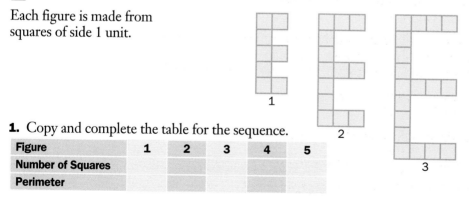

1. Copy and complete the table for the sequence.

Figure	1	2	3	4	5
Number of Squares					
Perimeter					

2. Write an expression for the number of squares in the nth figure in terms of n.

3. Write an expression for the perimeter of the nth figure in terms of n.

4. a) Write an expression for the perimeter in terms of the number of squares, s.
b) How does this expression compare with your answer to question 4a) in the T-patterns, above?

MATH STANDARD

Geometry from a Synthetic Perspective

Synthetic geometry is the study of two-dimensional shapes and three-dimensional objects, their properties, and their applications.

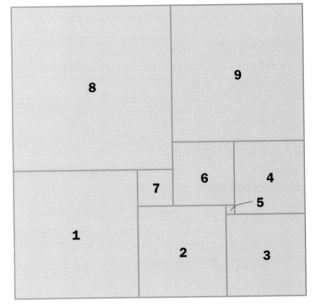

1 Geometry in Art

The above painting was done by Mary Russel. There are 9 different small squares contained inside the large figure.

1. In the schematic diagram, the squares have been numbered from 1 to 9. If the area of square 2 is 100 square units, and the area of square 3 is 81 square units, find the area of each of the other 7 squares.

2. Which two squares have the same total area as squares 2 and 9?

3. Identify 3 pairs of squares in which the area of the larger square is 4 times the area of the smaller square.

4. Which 4 squares have a total area that equals half the area of square 8?

5. Is the large figure a square? Explain.

2 Fractal Geometry

A **fractal** is a geometric figure in which an identical pattern repeats itself on a diminishing scale. The following are stages 0, 1, and 2 of a fractal called the Sierpinski carpet.

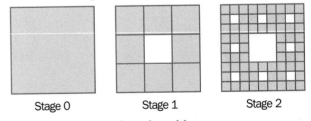

Stage 0   Stage 1   Stage 2

1. Copy and complete the table.

Stage	0	1	2	3
Number of Shaded Squares				

2. Describe the pattern in the table.

3. Write a rule for finding the number of shaded squares in the nth stage.

4. How many shaded squares are in the 5th stage? the 6th stage?

Geometry from an Algebraic Perspective

Transformations connect geometry and algebra to make geometry more dynamic. Understanding the connections between geometry and algebra can help you solve many problems.

1 Describing Flight

1. The diagram shows the transformations the space shuttle undergoes from when it leaves orbit until it lands. Describe the transformations.

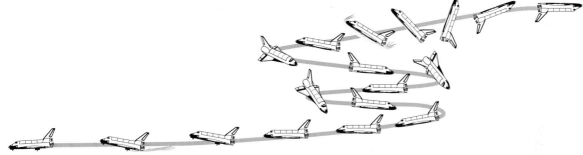

2. In an aerobatics display, two of the stunts an aircraft can do are a roll or a loop the loop. How are these two motions similar and how are they different?

2 Designing With Transformations

△ABC has vertices A(1, 4), B(0, 0) and C(5, 2).
1. Draw the image of △ABC, △A′B′C′, after a reflection in the y-axis.

2. Draw the image of △ABC, △A″B″C″, after a rotation of 180° clockwise about the origin.

3. Use the mapping $(x, y) \rightarrow (x + 1, y - 4)$ to draw the translation image of △ABC, △A‴B‴C‴.

4. a) Draw a figure of your choice in the first quadrant.
b) Use any combination of translations, rotations, and reflections to create a design.
c) Describe how you created the design.

3 Transforming Linear Relations

The graph shows five ordered pairs that satisfy the relation $x + y = 2$.
1. Copy the graph onto grid paper.

2. a) Translate the 5 points for the relation $x + y = 2$ using the mapping $(x, y) \rightarrow (x + 1, y + 3)$.
b) Write an equation for the translation image.

3. a) Reflect the 5 points for the relation $x + y = 2$ in the x-axis.
b) Write an equation for the reflection image.

4. a) Rotate the 5 points for the relation $x + y = 2$ 90° counterclockwise about the origin.
b) Write an equation for the rotation image.

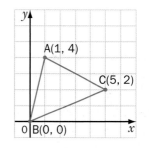

MATH STANDARD

Trigonometry

Trigonometry uses the ratios of the sides of a right triangle to find the measures of unknown sides and angles in the triangle. Using trigonometry to solve triangles is applied in such fields as navigation and surveying.

In right triangle ABC, the three primary ratios are

$$\sin A = \frac{\text{opposite}}{\text{hypotenuse}}$$

$$\cos A = \frac{\text{adjacent}}{\text{hypotenuse}}$$

$$\tan A = \frac{\text{opposite}}{\text{adjacent}}$$

1 Finding Sides and Angles

1. In each triangle, calculate x, to the nearest tenth of a metre.

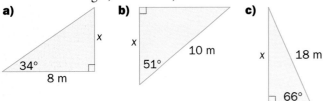

a) 34°, 8 m, x

b) x, 51°, 10 m

c) x, 18 m, 66°

2. In each triangle, calculate $\angle T$, to the nearest tenth of a degree.

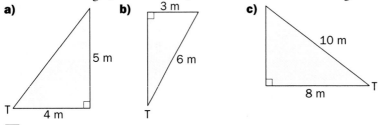

a) 5 m, T, 4 m

b) 3 m, 6 m, T

c) 10 m, 8 m, T

2 Problem Solving

1. The world's tallest and fastest roller coaster is at Six Flags Magic Mountain. At the start, riders are shot straight up a tower. They then begin the backward descent. The cars reach a speed of 160 km/h in 7 s during the descent. From a point 100 m from the foot of the tower, the angle of elevation of the top of the tower is 52°. Find the height of the tower, to the nearest metre.

2. Peggy's Cove Lighthouse, in Nova Scotia, is possibly the most photographed lighthouse in the world. The observation deck is about 20 m above sea level. From the observation deck, the angle of depression of a boat is 6°. How far is the boat from the lighthouse, to the nearest metre?

Statistics

Statistics is an attempt to make numbers "talk." Every day, we are bombarded with numbers by people who are trying to deliver a message, in advertising, politics, and other fields. Understanding statistics will help you interpret the true meaning of the message.

1 Goals Against Average

A hockey goaltender's goals against average (GAA) gives the average number of goals the goaltender allows in 60 min of play, or a complete game. The GAA is usually recorded to the nearest thousandth.
Suppose a goaltender plays 16 games and allows 26 goals.
GAA = 26 ÷ 16
   = 1.625
So, the goals against average is 1.625 goals/game.
Goaltenders do not always play for a complete game, so the number of minutes they play is recorded. If a goaltender allows 29 goals in 1650 min, the GAA is calculated by first finding the number of games played.
1650 ÷ 60 = 27.5
If 29 goals are allowed in the equivalent of 27.5 games, the GAA is 29 ÷ 27.5 ≐ 1.055
So, the goals against average is 1.055 goals/game, to the nearest thousandth.

1. The first time Dominik Hasek of the Buffalo Sabres had the best NHL goals against average, he played 3358 min and allowed 109 goals. What was his GAA?

2. The best ever goals against average was recorded by George Hainsworth when he played for the Montreal Canadiens. He played 2800 min and allowed 43 goals. What was his GAA?

2 Comparing Averages

1. In the first half of the season, Sharon played 1050 min and allowed 50 goals. Danica played 240 min and allowed 12 goals.
a) Calculate each goaltender's GAA, to the nearest thousandth, for the first half of the season.
b) Who had the better average?

2. In the second half of the season, Sharon played 115 min and allowed 5 goals. Danica played 904 min and allowed 41 goals.
a) Calculate each goaltender's GAA, to the nearest thousandth, for the second half of the season.
b) Who had the better average?

3. a) Calculate the GAA for each player for the whole season.
b) Who had the better average?

4. Who is the better goaltender? Explain.

MATH STANDARD

Probability

The world is full of uncertainty. Probability is the mathematics of chance. It can be used to predict what is likely to happen. Weather forecasters use probabilities when they predict the next day's weather. People can then plan for the next day without guessing at the weather.

1 Rolling Dice

A red die and a blue die are rolled. If the numbers rolled on the two dice are the same, the difference is zero. If the numbers rolled are not the same, the difference is found by subtracting the smaller number from the larger.

1. Copy and complete the table to show the possible outcomes. Three differences have been done for you.

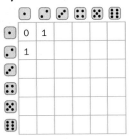

2. Find the probability of each of the following values for the difference.

a) 0   **b)** 1   **c)** 2   **d)** 3
e) 4   **f)** 5   **g)** 6   **h)** 2 or 3
i) greater than 3   **j)** less than 3
k) 0, 1, 2, 3, 4, or 5

2 Selecting Cubes

You have two bags. Bag A contains 3 red cubes and 3 green cubes. Bag B contains 3 red cubes and 3 green cubes. One cube is taken from bag A and another from bag B.

1. What is the probability that at least one of the cubes is red?

2. What would the answer to question 1 be if bag A contained 3 red cubes and 3 green cubes, but bag B contained
a) 2 red cubes and 2 green cubes?
b) 3 red cubes and 2 green cubes?

Discrete Mathematics

Discrete mathematics deals with sets of objects that can be counted. It has many applications, the most important being computer science.

1 Finding Possibilities

1. In how many different ways can the triangle be named using the letters A, B, and C?

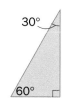

2. A baseball team has three different-coloured shirts — red, white, and blue. There are four different-coloured pants — grey, white, blue, and red. How many different combinations can they wear, if the combination of a red shirt and red pants is not allowed?

3. a) In how many different ways can you use zeros and ones to write a four-digit sequence? Sequences that begin with ones, such as 1111 and 1001, are allowed. Sequences that begin with zeros, such as 0101 and 0010, are not allowed.
b) Describe an application based on numbers that are sequences of zeros and ones.

2 Seating Arrangements

1. Three students are to sit in a row for the yearbook picture. In how many different ways can the students be seated?

2. One way for 3 people to sit at a circular table is as shown.

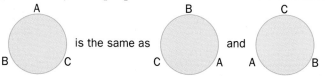

In all three cases, person A has person B on the right and person C on the left.
a) What is the other way the 3 people can sit at a circular table?
b) In how many different ways can 4 people sit at a circular table?

3 Making Words

The Pegasus language has only three letters, A, B, and C. A word in this language has 1, 2, 3, or 4 letters.
1. In this language, what is the maximum number of words with
a) 1 letter?　　**b)** 2 letters?　　**c)** 3 letters?　　**d)** 4 letters?

2. What is the maximum numbers of words in the Pegasus language?

MATH STANDARD

Investigating Limits

The beginnings of calculus go back to the ancient Greeks, who found areas using the "method of exhaustion." The following figures show how this method works for finding the area of a circle.

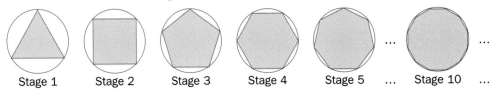

As the number of sides of the inscribed polygon increases, its area gets closer and closer to the area of the circle. We say the area of the circle is the **limit** of the areas of the inscribed polygons. The notion of a limit is a key idea in calculus.

1 Volume of a Pyramid

The sides of a the square-based pyramid are 24 cm. The height is 12 cm.

1. The pyramid has been divided into 6 rectangular prisms, as shown. Find the volume of each rectangular prism.

2. Find the sum of the volumes of the 6 rectangular prisms.

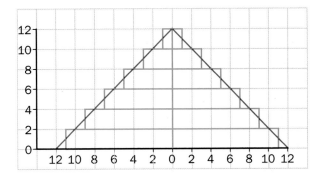

3. The same square-based pyramid has been divided into 12 rectangular prisms, as shown. Find the volume of each rectangular prism.

4. Find the sum of the volumes of the 12 rectangular prisms.

5. Find the volume of the pyramid using the formula $V = \frac{1}{3}$ (area of base)(height).

6. Compare your answers from questions 2 and 4 to the answer from question 5.

7. If you made the rectangular prisms thinner, would the sum of their volumes be closer to the volume found using the formula? Explain.

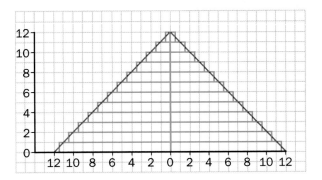

2 Other Limits

1. Draw diagrams to show how you could divide a circle into the following shapes to find an approximate value for its area.
a) rectangles   **b)** trapezoids

2. Describe how you could divide a cone into other shapes to find an approximate value for its volume.

Mathematical Structure

Mathematical structure involves the study of properties, such as $5 \times 7 = 7 \times 5$. These properties tie all the strands of mathematics together.

1 Geoboard Properties

The geoboard has 9 pegs. Suppose you stretch 3 rubber bands between two adjacent rows. Each rubber band must connect one row with the other. There are six possible ways of doing this.

These arrangements are now combined on the 3 by 3 board. The operation is called "snap" and is denoted by the symbol ■.

To find B ■ C, form B on the top two rows.

B ■ C is read as "B snap C."

Form C by stretching the same rubber bands to the third row.

The operation "snap" is performed by assuming the pegs from the middle row are removed. Any pegs that were connected via a peg in the middle row are now connected directly.

The result is E. So we say B ■ C = E.

1. Copy and complete the table.

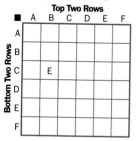

■	A	B	C	D	E	F
A						
B						
C		E				
D						
E						
F						

Top Two Rows / Bottom Two Rows

2. For real numbers, addition is commutative. $4 + 5 = 5 + 4$
For real numbers, multiplication is commutative. $4 \times 5 = 5 \times 4$
Is the operation "snap" commutative?

3. The identity element for addition is 0. $7 + 0 = 7$
The identity element for multiplication is 1. $7 \times 1 = 7$
What is the identity element for "snap"?

4. Addition is associative. $(2 + 3) + 4 = 2 + (3 + 4)$
Multiplication is associative. $(2 \times 3) \times 4 = 2 \times (3 \times 4)$
Is the operation "snap" associative?

CHAPTER 1

Real Numbers

O ur sun and the Earth belong to a spiral galaxy called the Milky Way. Using pictures from the Hubble Space Telescope, astronomers now estimate that there is an absolute minimum of 50 billion galaxies in the known universe. A typical galaxy contains 100 billion stars.

You might expect to see millions of stars under the perfect conditions that exist on remote deserts and mountains, where the night sky is almost black. However, the number of stars visible to the naked eye is actually quite small.

1. What is the minimum number of stars in the universe?

2. Counting one star per second, about how many years would you take to count the minimum number of stars in the universe?

3. It would take about 90 000 full moons packed together to fill the sky. Under perfect conditions, an area of sky the size of 30 full moons typically contains only one star that is visible to the naked eye. About how many stars can you see with the naked eye under perfect conditions?

4. About how many times more stars are there in the known universe than you can see with the naked eye under perfect conditions?

5. Viewed from a city, a visible star typically inhabits a patch of sky the size of 300 full moons. About how many stars can you see with the naked eye from a city?

GETTING STARTED

Using the Pythagorean Theorem

1. Calculators and computers can give only approximate values for some square roots, such as $\sqrt{2}, \sqrt{3}, \sqrt{5},$ and $\sqrt{6},$ because these roots are non-terminating, non-repeating decimals. Yet these real numbers do exist, and their magnitudes can be represented precisely by line segments. Spirals contain lengths that can be represented by square roots. Examples of naturally occurring spirals are found in chambered nautilus shells, sunflowers, and pineapples.

a) Find the lengths of the unknown sides in the spiral shown in the diagram. Leave your answers in radical form.

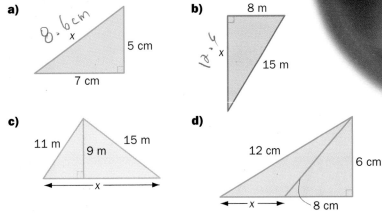

b) If the pattern continues, what will be the value of k?

2. Calculate the length of the unknown side, to the nearest tenth of a unit.

a)

8.6 cm
x
5 cm
7 cm

b)

8 m
x
12.4
15 m

c)

11 m, 15 m, 9 m
x

d)

12 cm
6 cm
x
8 cm

3. The diagram shows one way to divide a 3-pin-by-3-pin geoboard into two non-congruent polygonal regions, with vertices on the pins.

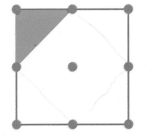

a) Sketch the five other different ways to divide the geoboard into two non-congruent polygonal regions, with vertices on the pins.

b) If the vertical and horizontal distance between two pins is 1 unit, calculate the perimeter of each polygonal region.

2

Warm Up

Expand.
1. $3(a + 4)$
2. $x(x - 3)$
3. $-5(y - 2)$
4. $-3(t + 6)$
5. $2x(x + 5)$
6. $4a(2 - a)$
7. $-2m(m + 6)$
8. $4y(2y + 1)$
9. $-5t(3t - 8)$
10. $-a(a - 1)$
11. $6(x^2 + 5x + 1)$
12. $2x(3x^2 - x - 2)$

Expand and simplify.
13. $2(x + 3) + 4(x + 5)$
14. $5(m - 2) - 3(m + 4)$
15. $3t(t - 4) - 5t(t + 5)$
16. $4a(a - 1) - (a + 5)$
17. $2x(x - 4) + 3x(1 - x)$
18. $y(2y + 1) - 4y(y - 3)$
19. $4t(2t - 3) - 6t(t - 7)$

Expand and simplify.
20. $(x + 4)(x + 3)$
21. $(a - 2)(a + 5)$
22. $(m - 3)(m - 6)$
23. $(y + 1)(y - 7)$
24. $(b - 4)(b - 4)$
25. $(c + 6)(c - 6)$
26. $(x + 5)^2$
27. $(y - 3)^2$
28. $(2x + 1)(x + 2)$
29. $(3a + 2)(2a + 5)$
30. $(m - 4)(4m + 1)$
31. $(4y + 3)(2y - 7)$
32. $(4b - 9)(2b - 3)$
33. $(3x - 5)(3x + 5)$
34. $(2x + 1)^2$
35. $(2x - 1)^2$

Write each number in scientific notation.
36. $45\ 000$
37. $0.000\ 23$
38. $8\ 000\ 000$
39. $0.000\ 000\ 6$

Write each number in standard form.
40. 5.7×10^8
41. 1.25×10^{12}
42. 3.8×10^{-7}
43. 4.08×10^{-11}

Write the number that is ten times as large as each number.
44. 5.6×10^7
45. 7.81×10^{-5}
46. 1.23×10^9
47. 3.4×10^{-8}

Write the number that is one-tenth as large as each number.
48. 8.2×10^8
49. 4.3×10^{-7}
50. 7.06×10^{10}
51. 9.84×10^{-9}

Write each number in scientific notation.
52. 70×10^6
53. 160×10^8
54. 10^7
55. 0.05×10^9
56. 30×10^{-8}
57. 0.07×10^{-7}
58. 90×10^4
59. 90×10^{-4}
60. 0.6×10^{-5}
61. 0.6×10^5

Mental Math

Evaluating Powers and Square Roots

Evaluate.
1. 2^4
2. 4^3
3. $(-4)^2$
4. -5^2
5. $2^2 \times 3^2$
6. $4^2 \div 2^3$
7. $(2^2)^3$
8. $2^3 + 3^2$

Evaluate.
9. $\sqrt{4} + \sqrt{9}$
10. $\sqrt{9} - \sqrt{4}$
11. $\sqrt{25} - \sqrt{16}$
12. $\sqrt{64} + \sqrt{25}$
13. $3\sqrt{49}$
14. $2\sqrt{36}$
15. $10\sqrt{4}$
16. $4\sqrt{121}$
17. $\dfrac{\sqrt{81}}{3}$
18. $\dfrac{\sqrt{100}}{2}$
19. $\dfrac{\sqrt{144}}{4}$
20. $\dfrac{\sqrt{64}}{2}$

Multiplying by Multiples of 5

To multiply 38×5, first multiply 38 by 10.
$38 \times 10 = 380$
Then, divide by 2.
$380 \div 2 = 190$
So, $38 \times 5 = 190$.

Estimate
$40 \times 5 = 200$

Calculate.
1. 26×5
2. 54×5
3. 73×5
4. 102×5
5. 898×5
6. 435×5
7. 5.2×5
8. 12.8×5
9. 36.4×5

To multiply 19×50, first multiply 19 by 100.
$19 \times 100 = 1900$
Then, divide by 2.
$1900 \div 2 = 950$
So, $19 \times 50 = 950$.

Estimate
$20 \times 50 = 1000$

Calculate.
10. 56×50
11. 61×50
12. 134×50
13. 256×50
14. 3.1×50
15. 10.2×50

16. Using a similar method to multiply 58 by 0.5 is easier than multiplying 58 by 5 or by 50. Explain why.

Calculate.
17. 88×0.5
18. 57×0.5
19. 132×0.5
20. 450×0.5
21. 9.6×0.5
22. 14.4×0.5

23. Explain why the rule for multiplying by 5 works.

3

TECHNOLOGY

Calculators and Repeating Decimals

Any repeating decimal can be written as a fraction
with a denominator made up of
• one or more 9s
or
• one or more 9s and one or more zeros

1 One Digit Repeating

1. Use a calculator to write each fraction as a repeating decimal.

a) $\dfrac{7}{9}$ **b)** $\dfrac{7}{90}$ **c)** $\dfrac{7}{900}$

2. Use the patterns to write the following repeating decimals as fractions
or mixed numbers in lowest terms.

a) $0.\overline{5}$ **b)** $0.\overline{6}$ **c)** $0.0\overline{8}$ **d)** $0.00\overline{2}$

e) $0.00\overline{1}$ **f)** $1.\overline{3}$ **g)** $2.0\overline{4}$ **h)** $7.00\overline{2}$

3. a) Write the repeating decimal $0.\overline{9}$ as a fraction. Explain your answer.

b) Another way to write the repeating decimal $0.\overline{9}$ is $0.\overline{9} = 3 \times 0.\overline{3}$.
Evaluate $3 \times 0.\overline{3}$ by first writing $0.\overline{3}$ as a fraction. Is the result the
same as the result from part a)?

2 Two Digits Repeating

1. Use a calculator to write each fraction as a repeating decimal.

a) $\dfrac{47}{99}$ **b)** $\dfrac{47}{990}$ **c)** $\dfrac{47}{9900}$

2. Use the patterns to write the following repeating decimals as fractions
or mixed numbers in lowest terms. Check your answers with a calculator.

a) $0.\overline{32}$ **b)** $0.\overline{63}$ **c)** $0.0\overline{45}$ **d)** $0.00\overline{56}$

e) $3.\overline{45}$ **f)** $6.\overline{66}$ **g)** $11.0\overline{35}$ **h)** $22.00\overline{78}$

3 Non-Repeating Parts

For repeating decimals with a non-repeating part, such as $0.1\overline{2}$,
write the fraction by breaking the decimal into two parts.

$$0.1\overline{2} = 0.1 + 0.0\overline{2}$$
$$= \frac{1}{10} + \frac{2}{90}$$
$$= \frac{9}{90} + \frac{2}{90}$$
$$= \frac{11}{90}$$

1. Write the following as fractions or mixed numbers in lowest terms.

a) $0.3\overline{4}$ **b)** $0.5\overline{6}$ **c)** $0.1\overline{23}$ **d)** $0.23\overline{4}$ **e)** $4.2\overline{5}$

1.1 The Real Number System

Canadian scientist Dr. Biruté Galdikas is the world's leading authority on orangutans. She carries out her research on these large apes in the rainforest of Tanjung Puting National Park on the island of Borneo. For part of the year, Dr. Galdikas teaches at Simon Fraser University.

The Swedish botanist Carolus Linnaeus (1707–1778) developed a system to classify living things, according to common characteristics. He arranged the classification categories as a series of nested sets. His sequence, from broadest to smallest category, is: kingdom, phylum, class, order, family, genus, and species. The Venn diagram shows the classification of orangutans.

Kingdom	Animal
Phylum	Chordate
Class	Mammal
Order	Primate
Family	Pongidae
Genus	Pongo
Species	Pongo Pygmaeus
	Orangutan

In mathematics, numbers are classified according to common characteristics. We have used the following sets of numbers.

Natural numbers $N = \{1, 2, 3, \ldots\}$
Whole numbers $W = \{0, 1, 2, 3, \ldots\}$
Integers $I = \{\ldots, -3, -2, -1, 0, 1, 2, 3, \ldots\}$
Rational numbers Q Any rational number can be expressed in the form $\frac{a}{b}$, where a and b are integers, and b does not equal 0.

Explore: Complete the Table

Express each rational number in the form $\frac{a}{b}$, in lowest terms.

Rational Number	5	$1\frac{1}{3}$	0.75	$0.\overline{6}$	−1.04	$-0.\overline{34}$	0	$-2\frac{1}{2}$	$\sqrt{25}$
Form $\frac{a}{b}$		$1\frac{3}{9}$	$\frac{3}{4}$	$\frac{6}{9}$	$-\frac{24}{25}$	$\frac{-34}{99}$		$-\frac{5}{2}$	

Inquire

1. Classify each rational number in the table as a repeating decimal or a terminating decimal.

2. Another name for a repeating decimal is a periodic decimal.
a) The digits that repeat are called the period. What is the period of each repeating decimal in the table?
b) The number of digits that repeat is called the length of the period. What is the length of the period of each repeating decimal in the table?

3. A terminating decimal can be considered as a type of repeating decimal. What repeats?

• Every rational number can be expressed as either a terminating or repeating decimal.

• Every terminating or repeating decimal represents a rational number.

Combining these two statements gives:
• A number is rational if, and only if, it can be expressed as a terminating or repeating decimal.

An **irrational number** is a number that cannot be expressed as a terminating or repeating decimal. Irrational numbers are non-terminating, non-repeating decimals.

They cannot be expressed in the form $\dfrac{a}{b}$, where a and b are integers and b does not

equal 0. The set of irrational numbers is named using the symbol $\overline{Q}$.

The following are examples of irrational numbers.

$$\pi = 3.141\,592\,653...$$
$$\sqrt{2} = 1.414\,213\,562...$$
$$-\sqrt{7} = -2.645\,751\,311...$$

The square roots of numbers like 2 and 7, which are not perfect squares, are irrational numbers.

Some irrational numbers have patterns, without being repeating decimals.
0.202 002 000 200 002 . . .
1.121 122 111 222 111 122 221 . . .
−0.100 200 300 400 500 . . .

The set of **real numbers**, R, is the set of all terminating decimals, repeating decimals, and non-terminating, non-repeating decimals. In other words, the set of rational numbers, Q, and the set of irrational numbers, $\overline{Q}$, together form the set of real numbers, R.

Just as all living things can be classified using a series of nested sets, so can sets of numbers.

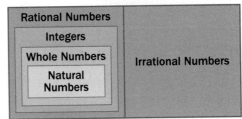

Example 1   Classifying Numbers

Name the sets of numbers to which each number belongs.

a) $\sqrt{81}$

b) $\sqrt{3.6}$

c) $-\sqrt{\dfrac{1}{4}}$

d) $\sqrt{0.36}$

Solution

a) $\sqrt{81} = 9$ natural, whole, integer, rational, real

b) $\sqrt{3.6} \doteq 1.897\,366\,596\ldots$ irrational, real

c) $-\sqrt{\dfrac{1}{4}} = -\dfrac{1}{2}$ rational, real

d) $\sqrt{0.36} = 0.6$ rational, real

Each real number corresponds to one point on a number line called the **real number line**.

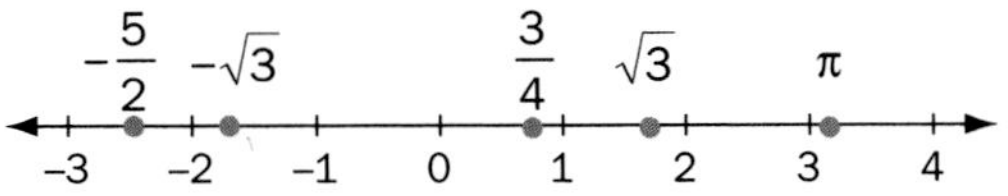

There is a one-to-one correspondence between the real numbers and all the points on the real number line. There are no "holes" or "gaps" in the real number line. It is complete.

Example 2   Graphing on a Real Number Line

Given that x is a real number, graph each of the following on a number line.

a) $x < 4$ **b)** $x \geq -3$ **c)** $-4 < x \leq 3$

Solution

a) $x < 4$

b) $x \geq -3$

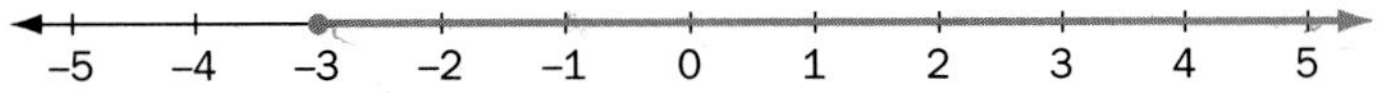

c) $-4 < x \leq 3$

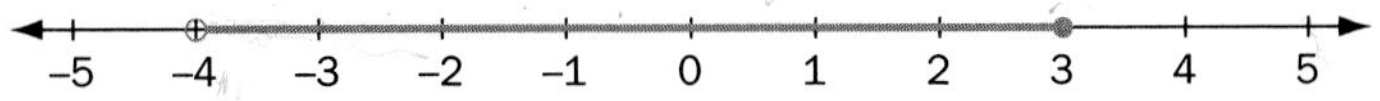

The **absolute value** of a real number is its distance from zero on a real number line. For a real number represented by x, the absolute value is written $|x|$, which means the positive value of x.

The diagram shows that the absolute value of -2 is 2.

$|-2| = 2$

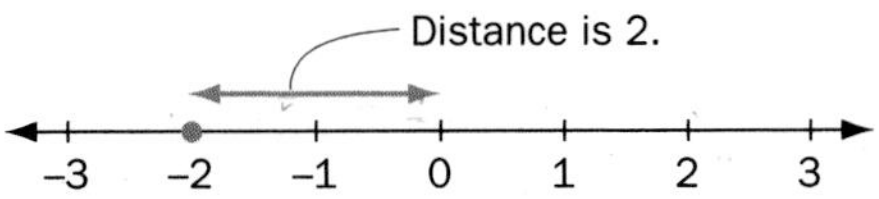

Example 3   Finding the Absolute Value

State the absolute value of each number.

a) $|-3|$ **b)** $|5-6|$ **c)** $|2|-|-5|$ **d)** $\left|-\dfrac{3}{4}\right|$

Solution

a) $|-3| = 3$

b) $|5-6| = |-1|$
$= 1$

c) $|2|-|-5| = 2-5$
$= -3$

d) $\left|-\dfrac{3}{4}\right| = \dfrac{3}{4}$

Practice

📖 *Identify each of the following as rational or irrational. Explain.*

1. 0.251 251 251 251 ... *rational*

2. 1.112 111 211 112 ... *irati*

3. −2.618 628 638 648 ... *irrational*

📖 *Identify each number as rational or irrational. Explain.*

4. $\dfrac{2}{5}$ **5.** 3.721   **6.** $3.\overline{7}$

7. $\sqrt{35}$ **8.** $-\sqrt{225}$ **9.** $\sqrt{0.29}$

10. $-\sqrt{\dfrac{25}{49}}$ **11.** $\sqrt{0.\overline{1}}$ **12.** $0.\overline{35}$

13. $\sqrt{\sqrt{16}}$ **14.** $2\dfrac{1}{3}$ **15.** π

Name the sets of numbers to which each number belongs.

16. $\sqrt{49}$ **17.** $\sqrt{0.04}$ **18.** $-\sqrt{15}$ **19.** $\sqrt{\dfrac{4}{9}}$

20. Graph the following on a number line.

$\sqrt{6}$ $-\sqrt{11}$ $\sqrt{16}$ 1.101 001 000 ...

Given that x is a real number, graph each of the following on a number line.

later   **21.** $x \geq 3$ **22.** $x \leq -3$ **23.** $x > -2$
24. $x < 2$ **25.** $x = -1$ **26.** $x \geq 0$

Given that x is a real number, graph each of the following on a number line.

later   **27.** $-3 < x \leq 2$ **28.** $-1 \leq x \leq 5$
29. $-4 < x < -1$ **30.** $2 \leq x \leq 7$

Evaluate.

31. $|-6|$ **32.** $|2-4|$ **33.** $|4|$

34. $\left|-\dfrac{2}{3}\right|$ **35.** $\left|\dfrac{1}{2}-\dfrac{3}{4}\right|$ **36.** $|1.2-1.5|$

37. $|-1|+|3|$ **38.** $|5|-|-2|$ **39.** $\underset{3\ -\ 4}{|-3|-|-4|}$

40. Graph the following on a number line.

$|-1|$ $|2|$ $\left|\sqrt{5}\right|$ $\left|-\sqrt{3}\right|$

Given that x is a real number, graph each of the following on a number line.

41. $x = |-5|$ **42.** $x \leq |-2|$

43. $x > |1|$ **44.** $|-1| < x < |-4|$

Express each of the following as a rational number in the form $\dfrac{a}{b}$.

45. $0.1 + 0.01 + 0.001 + \ldots$

46. $0.2 + 0.02 + 0.002 + \ldots$

47. $0.23 + 0.0023 + 0.000\,023 + \ldots$

48. Copy the diagram into your notebook.

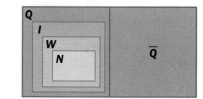

Place each of the following within the appropriate set(s) in the diagram.

$$\sqrt{36} \qquad -\sqrt{100} \qquad -\dfrac{3}{4}$$

$$0.06 \quad 1.101\,001\ldots \quad -\sqrt{10}$$

$$4.\overline{6} \qquad \sqrt[3]{8} \qquad \sqrt[4]{1}$$

Use x to write an inequality to describe each of the following graphs.

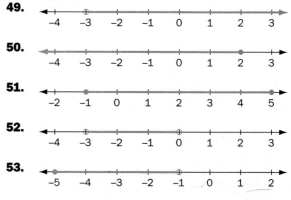

49.

50.

51.

52.

53.

Applications and Problem Solving

54. Arrange the following numbers in order from smallest to largest.

$2.\overline{35}, 2.3\overline{5}, 2.35$

🧩 **55.** Given that x is a real number, graph each of the following on a number line.
a) $x^2 > 9$ **b)** $x^2 \leq 4$

📖 **56.** Explain why you think the letter Q was chosen to represent rational numbers.

8

57. What is the absolute value of zero? Explain.

58. History Archimedes determined that the value of pi was somewhere between $\frac{22}{7}$ and $\frac{223}{71}$. In 480 A.D., Tsu Ch'ung-chi used $\frac{355}{113}$ as an approximate value of pi. Arrange these three numbers in order from smallest to largest.

59. Astronomy The magnitude, or brightness, of a star or planet is designated by a number — the smaller the number, the brighter the star or planet. Arrange the celestial bodies in order from brightest to dimmest.

Planet/Star	Magnitude	Planet/Star	Magnitude
Antares	1	Venus	−5
Sirius	−1.5	Altair	0.9
Polaris	2	Vega	0
Spica	1.2	Canopus	−0.9

60. Estimation At Columbia University in New York, pi has been calculated to 480 billion decimal places. If this number were printed using the size of type found in this book, how long would the number be?

61. What is the fiftieth digit after the decimal point for $\frac{1}{7}$?

62. State whether each of the following is always true, sometimes true, or never true. The sum of a rational number and an irrational number is
a) positive   **b)** irrational
c) a fraction   **d)** rational
e) negative   **f)** a whole number

63. Technology a) Because $\sqrt{4} = 2$, $\sqrt{4}$ is a rational number. Some scientific calculators give $\sqrt{3} = 1.732\,050\,808$. Does this display show that $\sqrt{3}$ is rational? Explain.
b) Find the product $1.732\,050\,808 \times 1.732\,050\,808$ without using the $\boxed{x^2}$ key. Explain the result.
c) Repeat the procedure for $\sqrt{2}$. Explain the result.

64. You are given 4 natural numbers A, B, C, and D, where A is less than B, B is less than C, and C is less than D.
a) Copy the diagram. Place the letters in the circles so that the sum of the two rational numbers is as large as possible.

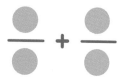

b) Place the letters in the circles so that the sum of the two rational numbers is as small as possible.

65. If you took the absolute value of every number in each of the following sets of numbers, which set of numbers would you obtain? Explain.
a) integers   **b)** negative integers
c) whole numbers   **d)** real numbers

66. Decide whether each of the following statements is always true, sometimes true, or never true. Explain.
a) An integer is a whole number.
b) A whole number is an integer.
c) The absolute value of an irrational number is a rational number.
d) A real number is an irrational number.
e) A repeating decimal is a real number.
f) The absolute value of a real number is positive.

LOGIC POWER

The figure is made from 24 toothpicks.

1. Show how you could remove 12 toothpicks to leave only 2 squares that intersect at 1 point.

2. Show how you could remove 8 toothpicks to leave only 2 squares that intersect at
a) 0 points
b) an infinite number of points

PROBLEM SOLVING

1.2 Guess and Check

One way to solve a problem is to guess at the answer and then check to see if it is correct. If it is not, you can keep guessing and checking until you get the right answer.

Jacques Villeneuve was the first Canadian to win the Indy 500 car race. The year he won, there were two other Canadians in the race, Paul Tracy and Scott Goodyear. The three of them drove a total of 538 laps of the track. Goodyear drove 64 laps more than Tracy, and Villeneuve drove 2 laps more than Goodyear. How many laps did each of them drive?

Understand the Problem
1. What information are you given?
2. What are you asked to find?
3. Do you need an exact or approximate answer?

Think of a Plan
Set up a table. Guess at the number of laps Tracy drove. Use this guess to write the number of laps driven by Goodyear and Villeneuve. If the total number of laps is not 538, make another guess at the number of laps driven by Tracy.

Carry Out the Plan

Guess				Check
Tracy	**Goodyear**	**Villeneuve**	**Total**	**Is the Total 538?**
100	164	166	430	Too Low
150	214	216	580	Too High
130	194	196	520	Too Low
135	199	201	535	Too Low
136	200	202	538	538 Checks!

Tracy drove 136 laps, Goodyear drove 200 laps, and Villeneuve drove 202 laps.

Look Back
Check the answer against the given information.
Since 200 – 136 = 64, Goodyear drove 64 laps more than Tracy.
Since 202 – 200 = 2, Villeneuve drove 2 laps more than Goodyear.
How could you set up a spreadsheet to solve the problem by guess and check?
Solve the problem by writing and solving an equation that represents the data.

Guess and Check
1. Guess an answer that fits one of the facts.
2. Check the answer against the other facts.
3. If necessary, adjust your guess and check again.

Applications and Problem Solving

Write your guess for each of questions 1–5. Compare each answer with a classmate's. Then, use your research skills to find the correct value.

1. How many kilometres long is the Trans-Canada Highway?

2. How many students are enrolled in your school?

3. How long would it take to drive from Halifax to Ottawa at a speed of 80 km/h?

4. How much does it cost to heat your school for a year?

5. Mount Logan is Canada's tallest mountain. How many metres tall is it?

6. In one game, the starting five players on the Westview basketball team scored a total of 93 points. Mary scored the fewest points. Sasha scored 5 more points than Mary. Paula scored 2 more points than Sasha. Amandi scored 2 more points than Paula. Heather scored 3 more points than Amandi. How many points did each player score?

7. Petra opened her math book to a place where the product of the page numbers was 14 042. What were the page numbers?

8. Use the numbers 1, 2, 3, 4, 5, and 6 once in each triangle. Place the numbers in the circles so that each side of triangle A adds to 10, each side of triangle B adds to 11, and each side of triangle C adds to 12.

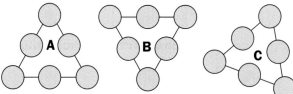

9. Kurt washed the wheels of the cars and motorcycles parked on the dealership lot. If there were 32 vehicles, and he washed 110 wheels, how many motorcycles were there?

10. The letters x, y, and z represent integers.
$$x - y - z = 2$$
$$x \div y \div z = 2$$
Find the values of x, y, and z that satisfy both equations. There are two solutions.

11. The whole number 5 can be written as the difference of the squares of two whole numbers as follows:
$$3^2 - 2^2 = 5$$
The whole number 4 can be written as the difference of squares of two whole numbers as follows:
$$2^2 - 0^2 = 4$$
There are five whole numbers less than 20 that cannot be written as the difference of squares of two whole numbers. What are they?

12. The number 9 has been placed in a circle. Place each of the other numbers from 1 to 10 in the other circles so that each of the five lines adds to the same multiple of 6.

13. The sum of two whole numbers is 119. When you subtract the smaller number from the larger number, the result is 45. What are the two numbers?

14. The number in each yellow square is found by adding the numbers in the two small green squares adjacent to it. Find the numbers in the green squares.

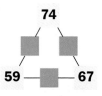

15. A unit fraction, such as $\frac{1}{7}$, is a fraction with 1 as the numerator.
  a) Find three different unit fractions that add to 1.
  b) Is it possible to have 2 different unit fractions that add to 1? Explain.

16. Write a problem that can be solved using the guess and check strategy. Have a classmate solve your problem.

1.3 Evaluating Irrational Numbers

The golden rectangle is a rectangle whose sides are in a ratio that is pleasing to the eye. The ratio of the length to the width of a golden rectangle is called the golden ratio. The golden ratio can be found in architecture, art, nature, and many other places. A French artist, Georges Seurat (1859–1891), used the golden ratio to compose many of his paintings. An example is *Le Pont de Courbevoie*, shown here.

Explore: Use the Diagram

The following steps can be used to construct a golden rectangle.

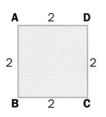 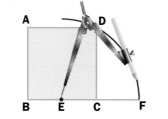

Start with square ABCD of side length 2 units.

Mark point E, the midpoint of BC. With centre E and radius ED, draw an arc to intersect BC extended at F.

Draw DG and FG to complete the golden rectangle ABFG.

Inquire

1. Calculate the length of ED. Leave your answer as a square root.

2. What is the length of BF?

3. Write an expression for the golden ratio.

4. Determine the approximate value of the golden ratio, to the nearest hundredth.

5. The Fibonacci sequence is 1, 1, 2, 3, 5, 8, 13, 21, ... What is the rule that determines this sequence?

6. The ratios of consecutive pairs of Fibonacci numbers eventually give a decimal approximation of the golden ratio.

$$\frac{1}{1} = 1, \ \frac{2}{1} = 2, \ \frac{3}{2} = 1.5, \ ...$$

What is the first pair of Fibonacci numbers that gives a value of the golden ratio correct to the nearest hundredth?

7. Describe how Georges Seurat used the golden ratio to compose *Le Pont de Courbevoie*.

12

Finding the square root of a number is the inverse operation to squaring. The square root operation is represented by the symbol $\sqrt{}$, which is called a **radical sign**. Numbers like $\sqrt{2}, \sqrt{5},$ and $\sqrt{10}$ are called **radicals**.

All positive numbers have two square roots. For example, the square roots of 16 are $+4$ and -4. However, when we write $\sqrt{16}$, we mean the **principal square root**. This is the positive root, $+4$.

Example 1   Distance to the Horizon

Because the Earth is curved, it is impossible to see beyond the horizon. The distance to the horizon depends on the observer's height above the ground. A spacecraft, S, is at an altitude, h, above the Earth's surface. The distance to the horizon is d. The radius of the Earth is r.

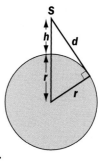

The formula for the distance to the horizon is
$$d = \sqrt{2rh + h^2}$$
Find the approximate distance to the horizon, to the nearest kilometre, for a spacecraft that is 200 km above the Earth. Use 6370 km as the radius of the Earth.

Solution

$$\begin{aligned} d &= \sqrt{2rh + h^2} \\ &= \sqrt{2(6370)(200) + 200^2} \\ &= \sqrt{2\,548\,000 + 40\,000} \\ &= \sqrt{2\,588\,000} \\ &\doteq 1608.7 \end{aligned}$$

√(2*6370*200+200
²)
1608.726204

The distance to the horizon is about 1609 km.

As with other arithmetic calculations, it is good practice to estimate the answers to problems involving square roots.

One way to estimate square roots is to use perfect squares whose square roots you know. To estimate $\sqrt{95}$, use the fact that 95 lies between the perfect squares 81 and 100. Since 95 is closer to 100, a good estimate for $\sqrt{95}$ is $\sqrt{100}$, or 10.

$81 < 95 < 100$

Another way to estimate the square root of a number is to divide the number into groups of 2 digits, starting at the decimal point.

Example 2   Estimating Square Roots Greater Than 1

Estimate. **a)** $\sqrt{3543}$ **b)** $\sqrt{945.2}$

Solution

For numbers greater than 1, estimate the square root of the group furthest to the left of the decimal point and add one zero for each other group.

a) $\sqrt{3543} = \sqrt{35\ 43}$
$6\ \ 0$
$\sqrt{3543} \doteq 60$

b) $\sqrt{945.2} = \sqrt{09\ 45.2}$
$3\ \ 0$
$\sqrt{945.2} \doteq 30$

Example 3   Estimating Square Roots Less Than 1

Estimate. **a)** $\sqrt{0.845}$ **b)** $\sqrt{0.000\ 575}$

Solution

For numbers less than 1, estimate the square root of the group closest to the decimal point.

a) $\sqrt{0.845} = \sqrt{0.84\ 50}$ **b)** $\sqrt{0.000\ 575} = \sqrt{0.00\ 05\ 75}$

$$\phantom{\sqrt{0.845} = } \begin{matrix}\downarrow\ \downarrow\\ 0.\ 9\end{matrix} \qquad \phantom{\sqrt{0.000\ 575} = }\begin{matrix}\downarrow\ \downarrow\ \ \downarrow\\ 0.\ 0\ \ 2\end{matrix}$$

$\sqrt{0.845} \doteq 0.9$ $\sqrt{0.000\ 575} \doteq 0.02$

Finding the cube root of a number is the inverse operation of cubing. To find the cube root of 8, you must find a number whose cube is 8.
Since $2^3 = 8$, $\sqrt[3]{8} = 2$.

The symbol $\sqrt[n]{}$ indicates an *n*th root.

$\sqrt[n]{x}$ stands for the principal *n*th root of *x*. If *n* is an odd

number and *x* is negative, as in $\sqrt[3]{-8}$, there is no positive root. In this example, the principal root is negative, namely, –2.

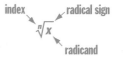

Example 4   Evaluating Cube Roots

Evaluate. Round to the nearest hundredth, if necessary.

a) $\sqrt[3]{-27}$

b) $\sqrt[3]{8000}$

c) $\sqrt[3]{150}$

Solution

a) $\sqrt[3]{-27} = -3$

b) $\sqrt[3]{8000} = 20$

c)

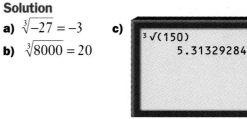

$\sqrt[3]{150} \doteq 5.31$

Practice

1. The following are approximations of $\sqrt{5}$.

2.2	(nearest tenth)
2.24	(nearest hundredth)
2.236	(nearest thousandth)
2.2361	(nearest ten thousandth)

Square each approximation and state whether the approximation is greater than or less than $\sqrt{5}$.

Estimate. Then, find an approximate value, to the nearest hundredth.

2. $\sqrt{44}$ **3.** $\sqrt{87}$ **4.** $\sqrt{792}$

5. $\sqrt{6680}$ **6.** $\sqrt{23\ 812}$ **7.** $\sqrt{82.55}$

8. $\sqrt{5.06}$ **9.** $\sqrt{0.22}$ **10.** $\sqrt{0.045}$

11. $\sqrt{0.0038}$ **12.** $\sqrt{0.000\ 41}$ **13.** $\sqrt{2.006}$

Estimate. Then, find an approximate value, to the nearest tenth.

14. $\sqrt{29} + \sqrt{72}$ **15.** $\sqrt{90} - \sqrt{56}$

16. $3\sqrt{19}$ **17.** $9\sqrt{189}$

18. $2\sqrt{11} + \sqrt{46}$ **19.** $7\sqrt{30} - 4\sqrt{40}$

20. $\sqrt{33} \times \sqrt{55}$ **21.** $2\sqrt{8} \times 5\sqrt{20}$

Estimate. Then, find an approximate value, to the nearest tenth.

22. $\dfrac{\sqrt{23}}{\sqrt{6}}$ **23.** $\dfrac{2\sqrt{40}}{\sqrt{27}}$

24. $\dfrac{3\sqrt{70}}{5\sqrt{11}}$ **25.** $\dfrac{\sqrt{61} + \sqrt{33}}{\sqrt{5}}$

26. $\dfrac{9}{\sqrt{7}+\sqrt{5}}$ **27.** $\dfrac{4}{\sqrt{6}-\sqrt{3}}$

28. Find decimal approximations of $\sqrt{12}$ and $2\sqrt{3}$, to four decimal places. What do your results suggest?

29. Find an integer equal to each of the following.
a) $\sqrt{16+9}$ **b)** $\sqrt{16}+\sqrt{9}$
c) $\sqrt{25-9}$ **d)** $\sqrt{25}-\sqrt{9}$
e) $\sqrt{25-16}$ **f)** $\sqrt{25}-\sqrt{16}$
g) $\sqrt{25\times9}$ **h)** $\sqrt{25}\times\sqrt{9}$
i) $\sqrt{25\times16}$ **j)** $\sqrt{25}\times\sqrt{16}$

30. Replace each ● with <, >, or = to make each statement true.
a) $\sqrt{9}+\sqrt{4}$ ● $\sqrt{13}$ **b)** $\sqrt{9}+\sqrt{4}$ ● $\sqrt{25}$
c) $\sqrt{9}-\sqrt{4}$ ● $\sqrt{5}$ **d)** $\sqrt{9}\times\sqrt{4}$ ● $\sqrt{36}$

Use decimal approximations to arrange the following in order from least to greatest.

31. $\sqrt{17},\ 3\sqrt{2},\ 2\sqrt{5}$

32. $8,\ 2\sqrt{15},\ 3\sqrt{7},\ \sqrt{62}$

33. $5\sqrt{3},\ \sqrt{74},\ 4\sqrt{5},\ 6\sqrt{2}$

Find an integer equal to each of the following.

34. $\sqrt[3]{8}+\sqrt[3]{27}$

35. $\sqrt[3]{-27}+\sqrt[3]{125}$

36. $2\left(\sqrt[3]{64}\right)+3\left(\sqrt[3]{-8}\right)$

37. $4\left(\sqrt[3]{-125}\right)-2\left(\sqrt[3]{-64}\right)$

38. $\sqrt[4]{16}+\sqrt[4]{81}$

Evaluate, to the nearest hundredth.

39. $\sqrt[3]{35}$ **40.** $\sqrt[3]{90}$

41. $2\left(\sqrt[3]{22}\right)$ **42.** $4\left(\sqrt[3]{-30}\right)$

43. $\sqrt[3]{56}+\sqrt[3]{88}$ **44.** $3\left(\sqrt[3]{85}\right)+5\left(\sqrt[3]{22}\right)$

Applications and Problem Solving

45. Between what two consecutive integers is each of the roots?
a) $\sqrt{19}$ **b)** $\sqrt{30}$ **c)** $\sqrt{200}$ **d)** $-\sqrt{10}$

46. Estimation Li estimated the perimeter of a triangle with sides $\sqrt{5}$ m, $\sqrt{8}$ m, and $\sqrt{10}$ m to be approximately 6.2 m.
a) Use estimates to determine if her estimate was reasonable.
b) Check the accuracy of her estimate with a calculator.

47. Evaluate. Round to the nearest hundredth, if necessary.
a) $\sqrt{\sqrt{16}}$ **b)** $\sqrt{\sqrt{81}}$
c) $\sqrt{\sqrt{64}}$ **d)** $\sqrt{\sqrt{0.0625}}$

48. Geography The city of Saskatoon has an area of 4749 km^2.
a) If the city had the shape of a square, what would be its side length, to the nearest tenth of a kilometre?
b) If the city had the shape of a circle, what would be its diameter, to the nearest tenth of a kilometre?

49. Space travel A spacecraft is 100 km above the surface of the moon. The moon has a radius of 1740 km. Use the formula $d=\sqrt{2rh+h^2}$, where h is the height of the satellite and r is the radius of the moon, to find the distance to the horizon. Round your answer to the nearest kilometre.

50. Solar energy Solar cells are used on spacecraft to convert sunlight into electrical energy. Each square centimetre of panel produces 0.01 W (one hundredth of a watt) of electrical power.
a) To deliver 5 W of power, what does the area of a square solar panel need to be?
b) What is the length of each side of the square panel, to the nearest tenth of a centimetre?
c) A circular solar panel delivers 4 W. What is the radius of the panel, to the nearest tenth of a centimetre?

51. Weather Meteorologists use the formula $D^3=830t^2$ to describe violent storms, such as tornadoes and hurricanes. D is the diameter of the storm in kilometres and t is the number of hours it will last.
a) If a storm has a diameter of 30 km, how long is it expected to last?
b) A typical hurricane lasts for about 18 h. What is the diameter of the hurricane?

15

52. Physics On the Earth's surface, the time, T seconds, a pendulum takes to swing back and forth once is given approximately by the formula $T = 2\sqrt{l}$, where l is the length of the pendulum in metres.
a) How long does a 2-m pendulum take to swing back and forth once, to the nearest tenth of a second?
b) The pendulum in a clock in Tokyo has a length of about 22.5 m. How long does the pendulum take to swing back and forth once, to the nearest tenth of a second?
c) How long is a pendulum that swings back and forth once a second?
d) If a pendulum were suspended from the top of Vancouver's Royal Centre Tower, and the bottom of the pendulum almost touched the ground, the pendulum would take about 23.7 s to swing back and forth once. How tall is the Royal Centre Tower, to the nearest metre?

53. Measurement A box is in the shape of a cube with 1-m sides. An ant wants to travel from A to G. The ant does not have to travel on edges.
a) If the ant travels from A to C and then to G, what is the total distance, to the nearest tenth of a metre?
b) To find the shortest distance the ant can travel from A to G, assume the bottom of the cube has been cut out and the sides flattened. Find the shortest distance the ant can travel from A to G, to the nearest tenth of a metre. At what point does the ant cross edge BC or edge DC?
c) Suppose the box is not a cube but has the dimensions shown. Find the shortest distance the ant can travel from A to G, to the nearest tenth of a metre.

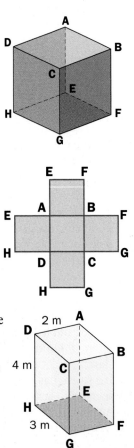

54. Mining The nickel Canada produces in a year has a volume of about 21 000 m³.
a) If this volume of nickel were made into a single cube, what would be the length of each edge, to the nearest tenth of a metre?
b) How does this volume of nickel compare with the volume of your school gymnasium?

55. Technology Why do calculators give only approximations of square roots of numbers that are not perfect squares?

56. Sports Suppose you drop a ball onto a hard surface from a height of 2 m, measured to the bottom of the ball. This distance is called h_{down}. The height the ball rises after hitting the hard surface, measured to the top of the ball, is called h_{up}. The formula to calculate the coefficient of restitution of the ball is

$$COR = \frac{\sqrt{h_{up}}}{\sqrt{h_{down}}}$$

a) The coefficient of restitution for an official basketball must lie between 0.76 and 0.8. A basketball dropped from 2 m bounced up 1.1 m. Is it an official basketball?
b) The coefficient of restitution for an official baseball must lie between 0.528 and 0.563. A baseball dropped from 5 m bounced up 1.5 m. Is it an official baseball?
c) The coefficient of restitution for a soccer ball and a volleyball must lie between 0.76 and 0.8. How many of the soccer balls and volleyballs in your school meet this requirement?

57. Measurement Many credit cards, bank cards, and membership cards approximate the shape of a golden rectangle. Measure membership cards and other types of cards that you have. Which ones approximate a golden rectangle?

PATTERN POWER

Find the value represented by ■.

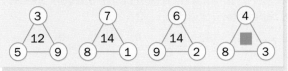

16

1.4 Simplifying Radicals

The approximate speed of a car prior to an accident can be found using the length of the tire marks left by the car after the brakes have been applied. The formula $s = \sqrt{169d}$ gives the speed, s, in kilometres per hour, where d is the length of the tire marks in metres. Radical expressions like $\sqrt{169d}$ can be simplified.

Explore: Complete the Table

Copy the table. Complete it by replacing each ▢ with a whole number.

$\sqrt{4 \times 9} = \sqrt{36} = 6$	$\sqrt{4} \times \sqrt{9} = 2 \times 3 = 6$
$\sqrt{9 \times 16} = \sqrt{144} = 12$	$\sqrt{9} \times \sqrt{16} = 3 \times 4 = 12$
$\sqrt{25 \times 4} = \sqrt{100} = 10$	$\sqrt{25} \times \sqrt{4} = 5 \times 2 = 10$
$\sqrt{\dfrac{36}{4}} = \sqrt{9} = 3$	$\dfrac{\sqrt{36}}{\sqrt{4}} = \dfrac{6}{2} = 3$
$\sqrt{\dfrac{100}{25}} = \sqrt{4} = 2$	$\dfrac{\sqrt{100}}{\sqrt{25}} = \dfrac{10}{5} = 2$
$\sqrt{\dfrac{144}{9}} = \sqrt{16} = 4$	$\dfrac{\sqrt{144}}{\sqrt{9}} = \dfrac{12}{3} = 4$

Inquire

1. Compare the two results in each of the first three rows of the table.

2. Copy and complete the statement: The square root of the product of two numbers equals the product of the square roots of the numbers.

3. Compare the two results in each of the last three rows of the table.

4. Copy and complete the statement: The square root of the quotient of two numbers equals the quotient of the square roots of the numbers.

5. Write the expression $\sqrt{169d}$ in the form $\sqrt{169}\,\sqrt{d}$, where ▢ represents a whole number.

6. Determine the speed of a car that leaves tire marks of each of the following lengths. Round the speed to the nearest kilometre per hour, if necessary.

a) 64 m   **b)** 100 m   **c)** 15 m

The following properties are used to simplify radicals.

- $\sqrt{ab} = \sqrt{a} \times \sqrt{b}, a \geq 0, b \geq 0$

- $\sqrt{\dfrac{a}{b}} = \dfrac{\sqrt{a}}{\sqrt{b}}, a \geq 0, b > 0$

Example 1   Simplifying Radicals

Simplify. **a)** $\sqrt{75}$ **b)** $\dfrac{\sqrt{48}}{\sqrt{6}}$ **c)** $\sqrt{\dfrac{2}{9}}$

Solution

a) $\sqrt{75} = \sqrt{25} \times \sqrt{3}$
$\phantom{\sqrt{75}} = 5\sqrt{3}$

b) $\dfrac{\sqrt{48}}{\sqrt{6}} = \sqrt{\dfrac{48}{6}}$
$\phantom{\dfrac{\sqrt{48}}{\sqrt{6}}} = \sqrt{8}$
$\phantom{\dfrac{\sqrt{48}}{\sqrt{6}}} = \sqrt{4} \times \sqrt{2}$
$\phantom{\dfrac{\sqrt{48}}{\sqrt{6}}} = 2\sqrt{2}$

c) $\sqrt{\dfrac{2}{9}} = \dfrac{\sqrt{2}}{\sqrt{9}}$
$\phantom{\sqrt{\dfrac{2}{9}}} = \dfrac{\sqrt{2}}{3}$ or $\dfrac{1}{3}\sqrt{2}$

Numbers like $\sqrt{75}$ and $\sqrt{\dfrac{2}{9}}$ are called **entire radicals**.

Numbers like $5\sqrt{3}$ and $\dfrac{1}{3}\sqrt{2}$ are called **mixed radicals**.

Example 2   Ordering Radicals

Without using a calculator, arrange the following in order from least to greatest.

$2\sqrt{13}$, $3\sqrt{6}$, $4\sqrt{5}$, $5\sqrt{2}$

Solution

Write the mixed radicals as entire radicals.

$2\sqrt{13} = \sqrt{4} \times \sqrt{13}$
$\phantom{2\sqrt{13}} = \sqrt{52}$

$3\sqrt{6} = \sqrt{9} \times \sqrt{6}$
$\phantom{3\sqrt{6}} = \sqrt{54}$

$4\sqrt{5} = \sqrt{16} \times \sqrt{5}$
$\phantom{4\sqrt{5}} = \sqrt{80}$

$5\sqrt{2} = \sqrt{25} \times \sqrt{2}$
$\phantom{5\sqrt{2}} = \sqrt{50}$

The order from least to greatest is

$\sqrt{50}$, $\sqrt{52}$, $\sqrt{54}$, $\sqrt{80}$
or $5\sqrt{2}$, $2\sqrt{13}$, $3\sqrt{6}$, $4\sqrt{5}$

Example 3   Multiplying Radicals

Simplify. **a)** $9\sqrt{2} \times 4\sqrt{7}$ **b)** $2\sqrt{3} \times 5\sqrt{6}$

Solution

a) $9\sqrt{2} \times 4\sqrt{7} = 9 \times 4 \times \sqrt{2} \times \sqrt{7}$
$\phantom{9\sqrt{2} \times 4\sqrt{7}} = 36\sqrt{14}$

b) $2\sqrt{3} \times 5\sqrt{6} = 2 \times 5 \times \sqrt{3} \times \sqrt{6}$
$\phantom{2\sqrt{3} \times 5\sqrt{6}} = 10\sqrt{18}$
$\phantom{2\sqrt{3} \times 5\sqrt{6}} = 10 \times \sqrt{9} \times \sqrt{2}$
$\phantom{2\sqrt{3} \times 5\sqrt{6}} = 30\sqrt{2}$

To eliminate a radical from the denominator of a fraction, use the idea of equivalent fractions.

$\dfrac{3}{4} = \dfrac{3}{4} \times \dfrac{2}{2}$
$\phantom{\dfrac{3}{4}} = \dfrac{6}{8}$

Multiplying by $\dfrac{2}{2}$ is the same as multiplying by 1.

Example 4   Fractions With Radicals in the Denominator

Simplify $\dfrac{1}{\sqrt{2}}$.

Solution
Multiply the numerator and denominator by $\sqrt{2}$.
This is the same as multiplying the fraction by 1.

$$\frac{1}{\sqrt{2}} = \frac{1}{\sqrt{2}} \times \frac{\sqrt{2}}{\sqrt{2}}$$

$$= \frac{1 \times \sqrt{2}}{\sqrt{2} \times \sqrt{2}}$$

$$= \frac{\sqrt{2}}{2}$$

The process shown in Example 4 is called **rationalizing the denominator**.
The denominator has been changed from an irrational number to a rational number.
It has been agreed that a radical is in simplest form when

• the radicand has no perfect square factors other than 1 $\sqrt{8} = 2\sqrt{2}$

• the radicand does not contain a fraction $\sqrt{\dfrac{1}{4}} = \dfrac{1}{2}$

• no radical appears in the denominator of a fraction $\dfrac{1}{\sqrt{3}} = \dfrac{\sqrt{3}}{3}$

Practice

Simplify.

1. $\sqrt{12}$ **2.** $\sqrt{20}$ **3.** $\sqrt{45}$

4. $\sqrt{50}$ **5.** $\sqrt{24}$ **6.** $\sqrt{63}$

7. $\sqrt{200}$ **8.** $\sqrt{32}$ **9.** $\sqrt{44}$

10. $\sqrt{60}$ **11.** $\sqrt{18}$ **12.** $\sqrt{54}$

13. $\sqrt{128}$ **14.** $\sqrt{90}$ **15.** $\sqrt{125}$

Simplify.

16. $\dfrac{\sqrt{14}}{\sqrt{7}}$ **17.** $\dfrac{\sqrt{10}}{\sqrt{2}}$ **18.** $\dfrac{\sqrt{60}}{\sqrt{3}}$

19. $\dfrac{\sqrt{40}}{\sqrt{5}}$ **20.** $\dfrac{\sqrt{33}}{\sqrt{3}}$ **21.** $\dfrac{\sqrt{7}}{\sqrt{4}}$

22. $\dfrac{3\sqrt{8}}{\sqrt{2}}$ **23.** $\dfrac{27\sqrt{15}}{3\sqrt{5}}$ **24.** $\dfrac{12\sqrt{75}}{4\sqrt{3}}$

Write as an entire radical.

25. $2\sqrt{3}$ **26.** $4\sqrt{2}$ **27.** $3\sqrt{10}$

28. $3\sqrt{5}$ **29.** $2\sqrt{7}$ **30.** $6\sqrt{8}$

Without using a calculator, arrange the following in order from least to greatest.

31. $3\sqrt{5}, 2\sqrt{11}, 4\sqrt{3}, 5\sqrt{2}$

32. $6\sqrt{2}, 3\sqrt{7}, 8, 2\sqrt{15}$

33. $5\sqrt{5}, 4\sqrt{7}, 3\sqrt{14}, 2\sqrt{30}$

Simplify.

34. $\sqrt{2} \times \sqrt{10}$ **35.** $\sqrt{3} \times \sqrt{6}$

36. $\sqrt{15} \times \sqrt{5}$ **37.** $\sqrt{7} \times \sqrt{11}$

38. $4\sqrt{3} \times \sqrt{7}$ **39.** $3\sqrt{6} \times 3\sqrt{6}$

40. $2\sqrt{2} \times 3\sqrt{6}$ **41.** $2\sqrt{5} \times 3\sqrt{10}$

42. $3\sqrt{3} \times 4\sqrt{15}$ **43.** $4\sqrt{7} \times 2\sqrt{14}$

44. $\sqrt{6} \times \sqrt{3} \times \sqrt{2}$ **45.** $2\sqrt{7} \times 3\sqrt{1} \times \sqrt{7}$

Simplify.

46. $\sqrt{\dfrac{1}{3}}$ **47.** $\sqrt{\dfrac{3}{7}}$ **48.** $\sqrt{\dfrac{5}{6}}$

49. $\dfrac{5\sqrt{5}}{2\sqrt{3}}$ **50.** $\dfrac{2\sqrt{2}}{\sqrt{18}}$ **51.** $\dfrac{2}{\sqrt{3}}$

52. $\dfrac{4\sqrt{2}}{\sqrt{8}}$ **53.** $\dfrac{4\sqrt{7}}{2\sqrt{14}}$ **54.** $\dfrac{3\sqrt{6}}{4\sqrt{10}}$?

Applications and Problem Solving

55. Express each of the following as an integer.

a) $\sqrt{5^2}$ **b)** $\left(\sqrt{5}\right)^2$ **c)** $\sqrt{(-5)^2}$

d) $-\sqrt{5^2}$ **e)** $-\left(\sqrt{5}\right)^2$ **f)** $-\sqrt{(-5)^2}$

56. Measurement Express the exact area of the triangle in simplest radical form.

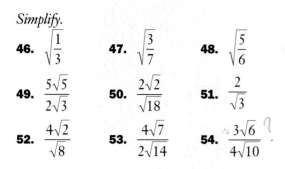

57. Measurement A square has an area of 675 cm². Express the side length in simplest radical form.

58. Canadian flag The ratio of the length to the width of a Canadian flag is always 2:1.
a) Determine the length of the diagonal of each of the following Canadian flags. Write each answer in simplest radical form.

b) Describe the relationship between the length of the diagonal and either dimension of the flag.
c) Use the relationship from part b) to predict the length of the diagonal of a 150 cm by 75 cm Canadian flag. Leave your answer in simplest radical form.

59. Simplify.

a) $\sqrt[3]{16}$ **b)** $\sqrt[3]{32}$ **c)** $\sqrt[3]{54}$

d) $\sqrt[3]{81}$ **e)** $\sqrt[4]{32}$ **f)** $\sqrt[4]{64}$

60. Board games There are many variations on the game of chess. Most are played on square boards that consist of a number of small squares. However, some of the variations do not use the 64 squares you are familiar with.

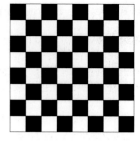

a) If each small square on a Grand Chess board is 2 cm by 2 cm, each diagonal of the whole board measures $\sqrt{800}$ cm. How many small squares are on the board?
b) A Japanese variation of chess is called Chu Shogi. If each small square on a Chu Shogi board measures 3 cm by 3 cm, each diagonal of the whole board measures $\sqrt{2592}$ cm. How many small squares are on the board?

61. The expression $\dfrac{1}{\sqrt[3]{2}}$ can be simplified to $\dfrac{\sqrt[3]{4}}{2}$. Show how.

62. Equations Solve. Express each answer in simplest radical form.

a) $x\sqrt{2} = \sqrt{14}$ **b)** $5x = \sqrt{50}$

c) $\dfrac{x}{\sqrt{3}} = \sqrt{6}$ **d)** $\dfrac{\sqrt{30}}{x} = \sqrt{5}$

63. a) For the property $\sqrt{ab} = \sqrt{a} \times \sqrt{b}$, explain the restrictions $a \geq 0$, $b \geq 0$.

b) For the property $\sqrt{\dfrac{a}{b}} = \dfrac{\sqrt{a}}{\sqrt{b}}$, the restrictions are $a \geq 0$, $b > 0$. Why is the second restriction not $b \geq 0$?

LOGIC POWER

Copy the diagram. Show two different ways to divide the shape along the lines into four congruent figures.

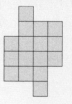

1.5 Operations With Radicals

Solar cells are attached to the surfaces of satellites.
They convert the energy of sunlight to electrical
energy. Solar cells are made in various shapes to cover
most of the surface area of satellites.

Explore: Use the Diagram

The scale drawing shows
6 solar cells, 3 triangles
and 3 rectangles, attached
to form one triangular
solar panel. The
dimensions shown are in
centimetres. Calculate the
lengths of AB, BC, and
CD. Write your answers
as mixed radicals in
simplest form.

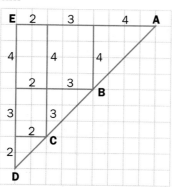

Inquire

1. The three mixed radicals in simplest form that represent the
lengths of AB, BC, and CD are called like radicals. Explain why.

2. Use the large right triangle ADE to write an expression for
the length of AD. Write your answer as a mixed radical in
simplest form.

3. Describe how the length of AD is related to the lengths
of AB, BC, and CD.

4. Compare the radical expressions you wrote for the lengths
of AB, BC, CD, and AD. Then, write a rule for adding like
radicals.

5. Simplify.
a) $3\sqrt{5} + 6\sqrt{5}$ **b)** $4\sqrt{3} + 5\sqrt{3} + \sqrt{3}$

Example 1   Adding and Subtracting Radicals

Simplify.
a) $\sqrt{12} + \sqrt{18} - \sqrt{27} + \sqrt{8}$
b) $4\sqrt{3} + 3\sqrt{20} - \sqrt{12} + 6\sqrt{45}$

Solution

Simplify radicals and combine like radicals.

a)
$$\sqrt{12} + \sqrt{18} - \sqrt{27} + \sqrt{8}$$
$$= \sqrt{4} \times \sqrt{3} + \sqrt{9} \times \sqrt{2} - \sqrt{9} \times \sqrt{3} + \sqrt{4} \times \sqrt{2}$$
$$= 2\sqrt{3} + 3\sqrt{2} - 3\sqrt{3} + 2\sqrt{2}$$
$$= -\sqrt{3} + 5\sqrt{2}$$

b)
$$4\sqrt{3} + 3\sqrt{20} - \sqrt{12} + 6\sqrt{45}$$
$$= 4\sqrt{3} + 3 \times 2\sqrt{5} - 2\sqrt{3} + 6 \times 3\sqrt{5}$$
$$= 4\sqrt{3} + 6\sqrt{5} - 2\sqrt{3} + 18\sqrt{5}$$
$$= 2\sqrt{3} + 24\sqrt{5}$$

Example 2   Multiplying a Radical by a Binomial

Expand and simplify $3\sqrt{2}(2\sqrt{6}+\sqrt{10})$.

Solution

Use the distributive property.

$$
\begin{aligned}
3\sqrt{2}(2\sqrt{6}+\sqrt{10}) &= 3\sqrt{2}(2\sqrt{6}+\sqrt{10}) \\
&= 3\sqrt{2}\times 2\sqrt{6}+3\sqrt{2}\times\sqrt{10} \\
&= 6\sqrt{12}+3\sqrt{20} \\
&= 6\times 2\sqrt{3}+3\times 2\sqrt{5} \\
&= 12\sqrt{3}+6\sqrt{5}
\end{aligned}
$$

Example 3   Binomial Multiplication

Simplify $(3\sqrt{2}+4\sqrt{5})(4\sqrt{2}-3\sqrt{5})$.

Solution

Multiply each term in the first binomial by each term in the second binomial.

$$
\begin{aligned}
(3\sqrt{2}+4\sqrt{5})(4\sqrt{2}-3\sqrt{5}) &= (3\sqrt{2}+4\sqrt{5})(4\sqrt{2}-3\sqrt{5}) \\
&= 12\sqrt{4}-9\sqrt{10}+16\sqrt{10}-12\sqrt{25} \\
&= 24-9\sqrt{10}+16\sqrt{10}-60 \\
&= -36+7\sqrt{10}
\end{aligned}
$$

Recall that FOIL means First, Outside, Inside, Last.

Binomials of the form $a\sqrt{b}+c\sqrt{d}$ and $a\sqrt{b}-c\sqrt{d}$, where a, b, c, and d are rational numbers, are **conjugates** of each other. The product of conjugates is always a rational number.

Example 4   Multiplying Conjugate Binomials

Simplify $(\sqrt{7}+2\sqrt{3})(\sqrt{7}-2\sqrt{3})$.

Solution

$$
\begin{aligned}
(\sqrt{7}+2\sqrt{3})(\sqrt{7}-2\sqrt{3}) &= (\sqrt{7}+2\sqrt{3})(\sqrt{7}-2\sqrt{3}) \\
&= \sqrt{49}-2\sqrt{21}+2\sqrt{21}-4\sqrt{9} \\
&= 7-12 \\
&= -5
\end{aligned}
$$

Conjugate binomials can be used to simplify a fraction with a binomial radical in the denominator.

Example 5   Rationalizing Binomial Denominators

Simplify $\dfrac{2}{\sqrt{6}-\sqrt{3}}$.

Solution

Multiply the numerator and the denominator by the conjugate of $\sqrt{6}-\sqrt{3}$, which is $\sqrt{6}+\sqrt{3}$.

$$
\begin{aligned}
\frac{2}{\sqrt{6}-\sqrt{3}} &= \frac{2}{\sqrt{6}-\sqrt{3}}\times\frac{\sqrt{6}+\sqrt{3}}{\sqrt{6}+\sqrt{3}} \\
&= \frac{2(\sqrt{6}+\sqrt{3})}{\sqrt{36}-\sqrt{9}} \\
&= \frac{2\sqrt{6}+2\sqrt{3}}{6-3} \\
&= \frac{2\sqrt{6}+2\sqrt{3}}{3}
\end{aligned}
$$

Practice

Simplify.

1. $2\sqrt{5} + 3\sqrt{5} + 6\sqrt{5}$
2. $4\sqrt{3} + 2\sqrt{3} - \sqrt{3}$
3. $6\sqrt{2} - \sqrt{2} + 7\sqrt{2} - 3\sqrt{2}$
4. $5\sqrt{7} + 3\sqrt{7} - 2\sqrt{7}$
5. $8\sqrt{10} - 2\sqrt{10} - 7\sqrt{10}$
6. $\sqrt{2} - 3\sqrt{2} - 9\sqrt{2} + 11\sqrt{2}$
7. $\sqrt{5} + \sqrt{5} + \sqrt{5} + \sqrt{5}$

Simplify.

8. $5\sqrt{3} + 2\sqrt{6} + 3\sqrt{3}$
9. $8\sqrt{5} - 3\sqrt{7} + 7\sqrt{7} - 4\sqrt{5}$
10. $2\sqrt{2} + 3\sqrt{10} + 5\sqrt{2} - 4\sqrt{10}$
11. $7\sqrt{6} - 4\sqrt{13} - \sqrt{13} + \sqrt{6}$
12. $9\sqrt{11} - \sqrt{11} + 6\sqrt{14} - 3\sqrt{14} - 2\sqrt{11}$
13. $12\sqrt{7} + 9 - 3\sqrt{7} + 4$
14. $8 + 7\sqrt{11} - 9 - 9\sqrt{11}$

Simplify.

15. $\sqrt{12} + \sqrt{27}$
16. $\sqrt{20} + \sqrt{45}$
17. $\sqrt{18} - \sqrt{8}$
18. $\sqrt{50} + \sqrt{98} - \sqrt{2}$
19. $\sqrt{75} + \sqrt{48} + \sqrt{27}$
20. $\sqrt{54} + \sqrt{24} + \sqrt{72} - \sqrt{32}$
21. $\sqrt{28} - \sqrt{27} + \sqrt{63} + \sqrt{300}$

Simplify.

22. $8\sqrt{7} + 2\sqrt{28}$
23. $3\sqrt{50} - 2\sqrt{32}$
24. $5\sqrt{27} + 4\sqrt{48}$
25. $3\sqrt{8} + \sqrt{18} + 3\sqrt{2}$
26. $\sqrt{5} + 2\sqrt{45} - 3\sqrt{20}$
27. $4\sqrt{3} + 3\sqrt{20} - 2\sqrt{12} + \sqrt{45}$
28. $3\sqrt{48} - 4\sqrt{8} + 4\sqrt{27} - 2\sqrt{72}$

Expand and simplify.

29. $\sqrt{2}(\sqrt{10} + 4)$
30. $\sqrt{3}(\sqrt{6} - 1)$

Multipling congogate binomiots

31. $\sqrt{6}(\sqrt{2} + \sqrt{6})$
32. $2\sqrt{2}(3\sqrt{6} - \sqrt{3})$
33. $\sqrt{2}(\sqrt{3} + 4)$
34. $3\sqrt{2}(2\sqrt{6} + \sqrt{10})$
35. $(\sqrt{5} + \sqrt{6})(\sqrt{5} + 3\sqrt{6})$
36. $(2\sqrt{3} - 1)(3\sqrt{3} + 2)$
37. $(4\sqrt{7} - 3\sqrt{2})(2\sqrt{7} + 5\sqrt{2})$
38. $(3\sqrt{3} + 1)^2$
39. $(2\sqrt{2} - \sqrt{5})^2$
40. $(2 + \sqrt{3})(2 - \sqrt{3})$
41. $(\sqrt{6} - \sqrt{2})(\sqrt{6} + \sqrt{2})$
42. $(2\sqrt{7} + 3\sqrt{5})(2\sqrt{7} - 3\sqrt{5})$

Simplify. Rationalizing Binomial Denominator

43. $\dfrac{1}{\sqrt{2} + 2}$
44. $\dfrac{3}{\sqrt{5} - 1}$
45. $\dfrac{\sqrt{2}}{\sqrt{6} - 3}$
46. $\dfrac{2}{\sqrt{6} + \sqrt{3}}$
47. $\dfrac{3}{\sqrt{5} - \sqrt{2}}$
48. $\dfrac{\sqrt{3}}{\sqrt{3} + \sqrt{2}}$
49. $\dfrac{2\sqrt{6}}{2\sqrt{6} + 1}$
50. $\dfrac{\sqrt{2} - 1}{\sqrt{2} + 1}$
51. $\dfrac{\sqrt{2} + \sqrt{5}}{\sqrt{6} - \sqrt{10}}$
52. $\dfrac{2\sqrt{7} + \sqrt{5}}{3\sqrt{7} - 2\sqrt{5}}$

Applications and Problem Solving

Simplify.

53. $\sqrt[3]{16} + \sqrt[3]{54}$
54. $\sqrt[3]{24} + \sqrt[3]{81}$
55. $2(\sqrt[3]{32}) + 5(\sqrt[3]{108})$
56. $\sqrt[3]{54} + 5(\sqrt[3]{16})$
57. $\sqrt[3]{16} - \sqrt[3]{54}$
58. $\sqrt[3]{108} - \sqrt[3]{32}$
59. $2(\sqrt[3]{40}) - \sqrt[3]{5}$
60. $5(\sqrt[3]{48}) - 2(\sqrt[3]{162})$

61. Without using a calculator, arrange the following expressions in order from greatest to least.

$$\sqrt{3}(\sqrt{3} + 1), (\sqrt{3} + 1)(\sqrt{3} - 1), (1 - \sqrt{3})^2, (\sqrt{3} + 1)^2$$

62. a) Without using a calculator, decide which of the following radical expressions does not equal any of the others.

$$\frac{60}{\sqrt{450}} \qquad 6\sqrt{2} - 4\sqrt{2} \qquad \frac{4}{\sqrt{2}}$$

$$6\sqrt{8} + \sqrt{8} - 5\sqrt{8} \qquad \frac{8}{\sqrt{18}} + \frac{4}{\sqrt{18}}$$

b) How is the radical expression you identified in part a) related to each of the others?

63. Nature Many aspects of nature, including the number of pairs of rabbits in a family and the number of branches on a tree, can be described using the Fibonacci sequence. This sequence is 1, 1, 2, 3, 5, 8, ...
The expression for the nth term of the Fibonacci sequence is called Binet's formula. The formula is

$$F_n = \frac{1}{\sqrt{5}} \left(\frac{1 + \sqrt{5}}{2} \right)^n - \frac{1}{\sqrt{5}} \left(\frac{1 - \sqrt{5}}{2} \right)^n$$

Use Binet's formula to find F_2.

64. Measurement
Write and simplify an expression for
a) the area of the rectangle
b) the perimeter of the rectangle

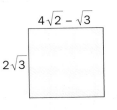
$4\sqrt{2} - \sqrt{3}$
$2\sqrt{3}$

65. Measurement
Write and simplify an expression for the area of the square.

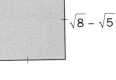
$\sqrt{8} - \sqrt{5}$

66. Measurement
Express the perimeter of the quadrilateral in simplest radical form.

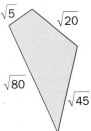
$\sqrt{5}$
$\sqrt{20}$
$\sqrt{80}$
$\sqrt{45}$

67. Measurement
Express the volume of the rectangular prism in simplest radical form.

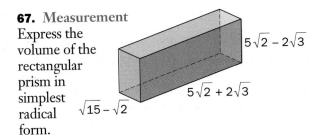
$5\sqrt{2} - 2\sqrt{3}$
$5\sqrt{2} + 2\sqrt{3}$
$\sqrt{15} - \sqrt{2}$

68. Measurement If a rectangle has an area of 4 square units and a width of $\sqrt{7} - \sqrt{5}$ units, what is its length in simplest radical form?

69. Measurement
Express the ratio of the area of the larger circle to the area of the smaller circle in simplest radical form.

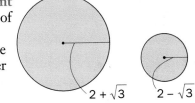
$2 + \sqrt{3}$
$2 - \sqrt{3}$

70. Coordinate geometry State the perimeter of each of the following triangles in simplest radical form.

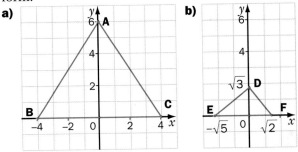

71. Equation Is the statement $\sqrt{a + b} = \sqrt{a} + \sqrt{b}$ always true, sometimes true, or never true? Explain.

LOGIC POWER

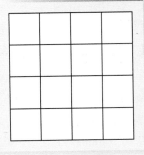

Copy the diagram. Place 4 As, 3 Bs, 3Cs, 3 Ds, and 3 Es in the squares so that the same letter does not appear more than once in any row, column, or diagonal.

24

Radical Expressions and Graphing Calculators

Complete the following on a graphing calculator that has the capability to simplify radical expressions and perform operations on them. If you do not have this type of calculator, work with paper and pencil, and use a graphing or scientific calculator to check your work. Do this by evaluating the given expression and your answer to check that the values are the same.

1 Simplifying Radicals

Simplify.

1. $\sqrt{425}$ **2.** $\sqrt{294}$ **3.** $\sqrt{507}$

4. $\sqrt{8} \times \sqrt{10}$ **5.** $2\sqrt{21} \times \sqrt{35}$ **6.** $3\sqrt{15} \times 4\sqrt{20}$

7. $\dfrac{3}{\sqrt{27}}$ **8.** $\dfrac{5}{\sqrt{112}}$ **9.** $\dfrac{-4}{\sqrt{32}}$

10. $\dfrac{\sqrt{192}}{\sqrt{6}}$ **11.** $\dfrac{\sqrt{75}}{\sqrt{162}}$ **12.** $\dfrac{\sqrt{88}}{\sqrt{33}}$

2 Operations With Radical Expressions

Simplify.

1. $\sqrt{20} + \sqrt{45}$ **2.** $\sqrt{72} + \sqrt{98} + \sqrt{242}$

3. $\sqrt{63} - \sqrt{28}$ **4.** $\sqrt{54} - \sqrt{96}$

5. $5\sqrt{27} + 3\sqrt{12}$ **6.** $7\sqrt{90} - 6\sqrt{40}$

7. $2\sqrt{125} + 4\sqrt{5} - 3\sqrt{80}$ **8.** $3\sqrt{99} - 6\sqrt{44} - 2\sqrt{11}$

Expand and simplify.

9. $\sqrt{5}(\sqrt{10} + \sqrt{15})$ **10.** $\sqrt{6}(\sqrt{18} - \sqrt{3})$

11. $2\sqrt{3}(\sqrt{27} + 5\sqrt{24})$ **12.** $4\sqrt{7}(3\sqrt{21} - 2\sqrt{14})$

13. $(2\sqrt{5} - 3\sqrt{3})(2\sqrt{5} + 3\sqrt{3})$

14. $(4\sqrt{11} + 5\sqrt{2})^2$

15. $(3\sqrt{6} - 5\sqrt{10})^2$

16. $(4\sqrt{3} - 3\sqrt{2})(5\sqrt{2} - 2\sqrt{3})$

Simplify.

17. $\dfrac{2}{\sqrt{3} + \sqrt{2}}$ **18.** $\dfrac{\sqrt{5}}{\sqrt{10} - \sqrt{5}}$

19. $\dfrac{3\sqrt{3}}{2\sqrt{6} - 3\sqrt{2}}$ **20.** $\dfrac{4 - \sqrt{10}}{7 - 2\sqrt{10}}$

21. $\dfrac{\sqrt{6} + 2\sqrt{3}}{\sqrt{6} - 2\sqrt{3}}$ **22.** $\dfrac{3\sqrt{7} - 2\sqrt{2}}{3\sqrt{2} - 2\sqrt{7}}$

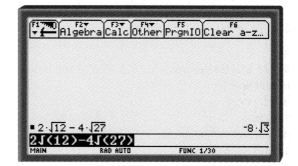

3 Problem Solving

1. The area of a triangle is 12 square units, and its base length is $4 + \sqrt{2}$ units. Write a radical expression in simplest form for the height of the triangle.

2. A chessboard has a 2-cm wide border around the playing area. The diagonal of each small square on the board measures 4 cm.

a) Write and simplify an expression that represents the area of the whole board, including the border.

b) Evaluate the expression from part a), to the nearest square centimetre.

Special Right Triangles

1 The 45°–45°–90° Triangle

An isosceles right triangle has two equal legs and two 45° angles. It is known as a 45°–45°–90° triangle.

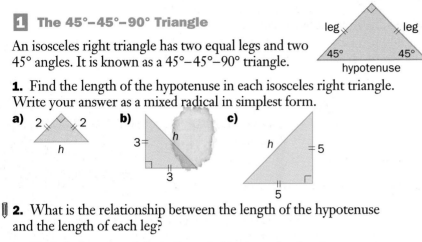

1. Find the length of the hypotenuse in each isosceles right triangle. Write your answer as a mixed radical in simplest form.

a)   2 ⟋⟍ 2
      h

b)   3 ⟍ h
         3

c)   h ⟍ 5
         5

2. What is the relationship between the length of the hypotenuse and the length of each leg?

3. Find the lengths of the indicated sides in each triangle.

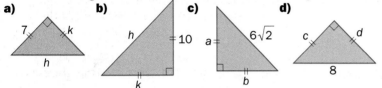

a) 7, k, h   **b)** h, k, 10   **c)** $6\sqrt{2}$, a, b   **d)** c, d, 8

4. If the length of each leg is 1, what is the length of the hypotenuse?

5. If the length of each leg is *s*, what is the length of the hypotenuse in terms of *s*?

2 The 30°–60°–90° Triangle

An altitude of an equilateral triangle bisects a 60° angle and a side of the triangle. A 30°–60°–90° triangle can be made by drawing an altitude of an equilateral triangle and removing one of the small right triangles.

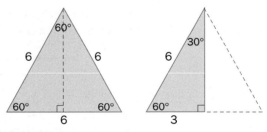

1. Find the lengths of the indicated sides in each triangle. Write radical answers as mixed radicals in simplest form.

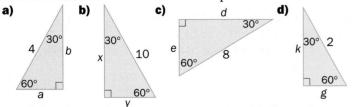

a) 4, 30°, b, 60°, a   **b)** 30°, x, 10, 60°, y   **c)** d, 30°, e, 60°, 8   **d)** 30°, 2, k, 60°, g

2. What is the relationship between the length of the hypotenuse and the length of the side opposite the 30° angle?

3. What is the relationship between the length of the side opposite the 60° angle and the length of the side opposite the 30° angle?

4. Find the lengths of the indicated sides in each triangle.

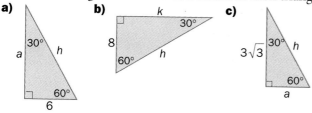

a) **b)** **c)**

5. If the length of the hypotenuse is 1, what are the lengths of the other two sides?

6. If the length of the hypotenuse is s, what are the lengths of the other two sides in terms of s?

3 The Equilateral Triangle

For each of the following, express the answer in simplest radical form.

1. Given equilateral triangle DEF, find
a) the altitude, a
b) the area, A

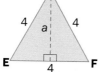

2. Given equilateral triangle XYZ, find
a) the altitude, a
b) the area, A

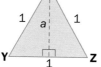

3. Given equilateral triangle PQR, express
a) the altitude, a, in terms of s
b) the area, A, in terms of s

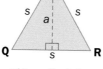

4. Use your formula from question 3b) to find the area of each of the following triangles.

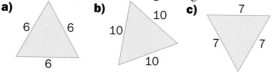

a) **b)** **c)**

4 Problem Solving

1. An equilateral triangle has an area of 36 cm². Calculate the side length and the height of the triangle, to the nearest tenth of a centimetre.

2. A regular hexagon has an area of 600 cm². Calculate the length of each side, to the nearest tenth of a centimetre.

3. A 30°–60°–90° triangle has an area of 20 cm². Calculate the length of each side, to the nearest tenth of a centimetre.

4. The equilateral triangle is inscribed in a circle. The circle has an area of 64π. State the exact area of the triangle as a mixed radical in simplest form.

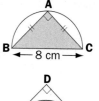

5. a) Find the area of △ABC, in square centimetres.

b) Triangle DEF is similar to △ABC. Find the area of △DEF, in square centimetres.

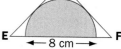

27

PROBLEM SOLVING

1.6 Solve Fermi Problems

"About how many buckets of water would it take to empty Loch Ness and find the monster?" Problems like this one that involve large numbers and give approximate answers are known as *Fermi problems*. They are named after the great Italian physicist Enrico Fermi (1901–1954), who liked to pose and solve them.
Solving Fermi problems will give you a greater appreciation of arithmetic and will help you improve your estimation skills.

About how many Canadian $1 coins are needed to cover a basketball court?

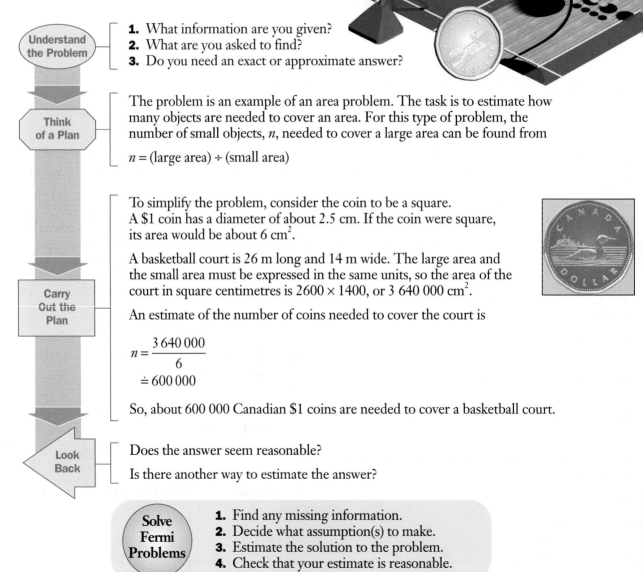

Understand the Problem

1. What information are you given?
2. What are you asked to find?
3. Do you need an exact or approximate answer?

Think of a Plan

The problem is an example of an area problem. The task is to estimate how many objects are needed to cover an area. For this type of problem, the number of small objects, n, needed to cover a large area can be found from

$n = $ (large area) ÷ (small area)

Carry Out the Plan

To simplify the problem, consider the coin to be a square.
A $1 coin has a diameter of about 2.5 cm. If the coin were square, its area would be about 6 cm^2.

A basketball court is 26 m long and 14 m wide. The large area and the small area must be expressed in the same units, so the area of the court in square centimetres is 2600×1400, or 3 640 000 cm^2.

An estimate of the number of coins needed to cover the court is

$$n = \frac{3\,640\,000}{6}$$
$$\doteq 600\,000$$

So, about 600 000 Canadian $1 coins are needed to cover a basketball court.

Look Back

Does the answer seem reasonable?

Is there another way to estimate the answer?

Solve Fermi Problems

1. Find any missing information.
2. Decide what assumption(s) to make.
3. Estimate the solution to the problem.
4. Check that your estimate is reasonable.

To solve a Fermi problem that involves volume, you may find it convenient to assume that an object approximates a cube or other simple shape.

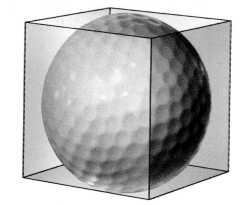

Suppose you want to estimate how many golf balls are needed to fill a 10 m by 8 m by 5 m room. A golf ball has a diameter of about 4 cm. If a golf ball were a cube, it would have a volume of about 64 cm^3.

The volume of the room in cubic centimetres is $1000 \times 800 \times 500$, or 400 000 000 cm^3.

For this type of problem, the number of small objects, n, needed to fill a large object can be found from

$n =$ (large volume) $\div$ (small volume)

$= \dfrac{400\,000\,000}{64}$

$\doteq 6\,000\,000$

So, the number of golf balls needed to fill the room is about 6 000 000.

Applications and Problem Solving

For each of the following problems, state any assumptions you make.

1. About how many Canadian $2 coins would cover the floor of your classroom?

2. Estimate the number of CDs, in their cases, that would cover the floor of the gymnasium in your school.

3. About how many students could stand on a football field?

4. About how many volleyballs would fill your principal's office?

5. About how many pieces of popped popcorn would fill your classroom?

6. About how many table tennis balls would fill a school bus?

7. About how many blades of grass are there on a soccer field?

8. About how many tennis balls would fill a minivan?

9. About how many math textbooks, lying end to end, would reach from St. John's to Vancouver along the Trans-Canada Highway?

10. Loch Ness contains about 9×10^{12} L of water. About how many buckets of water would it take to empty Loch Ness?

11. The Grand Canyon in Arizona is about 440 km long, and averages 1.5 km deep and 15 km wide. If you built a cubic apartment in the Grand Canyon for each person on Earth, what would be the dimensions of each apartment? Would each apartment be larger than your classroom?

12. About how many cars are there in your city or town?

13. You have a new pencil and a pencil sharpener. What is the length of the longest line you can draw?

14. About how many telephones are there within 1 km of your school?

15. About how many cats are kept as pets in Canada?

16. About how many kilometres can you walk in one pair of sneakers?

1.7 Reviewing the Exponent Laws

An **order of magnitude** is an approximate
size of a quantity, expressed as a power of 10.

Explore: Use the Table

The table shows some speeds in metres per second,
expressed to the nearest order of magnitude.

Entity	Speed (m/s)
Light (in space)	10^8
Sound (in air)	10^2
Horse (galloping)	10^1
Human (walking)	10^0
Garden Snail	10^{-3}

Express 10^0 metres per second in standard form. How does
the result compare with normal human walking speed?

Inquire

1. Use division to determine, to the nearest order of
magnitude, how many times faster
a) light is than sound   **b)** a horse is than a snail

2. Write the rule you used to divide two powers of 10.

3. To the nearest order of magnitude, the Concorde can fly 10^6
times faster than a snail can travel. Use multiplication to express
the speed of the Concorde in metres per second, to the nearest
order of magnitude.

4. Write the rule you used to multiply two powers of 10.

The following summary shows the exponent laws for integral
exponents.

Exponent Law for Multiplication

$$3^2 \times 3^4 = (3 \times 3)(3 \times 3 \times 3 \times 3)$$
$$= 3 \times 3 \times 3 \times 3 \times 3 \times 3$$
$$= 3^6$$

$$a^m \times a^n = \underbrace{(a \times a \times \ldots \times a)}_{m \text{ factors}}\underbrace{(a \times a \times \ldots \times a)}_{n \text{ factors}}$$
$$= \underbrace{a \times a \times a \times \ldots \times a}_{m + n \text{ factors}}$$
$$= a^{m+n}$$

Exponent Law for Division

$$\frac{6^5}{6^2} = \frac{6 \times 6 \times 6 \times 6 \times 6}{6 \times 6}$$
$$= 6 \times 6 \times 6$$
$$= 6^3$$

$$\frac{a^m}{a^n} = \frac{\overbrace{a \times a \times a \times \ldots \times a}^{m \text{ factors}}}{\underbrace{a \times a \times \ldots \times a}_{n \text{ factors}}}, a \neq 0$$
$$= \underbrace{a \times a \times a \times \ldots \times a}_{m - n \text{ factors}}$$
$$= a^{m-n}$$

Power Law

$(5^2)^3 = (5 \times 5)^3$
$\quad = (5 \times 5)(5 \times 5)(5 \times 5)$
$\quad = 5 \times 5 \times 5 \times 5 \times 5 \times 5$
$\quad = 5^6$

$$(a^m)^n = \underbrace{(a \times a \times \ldots \times a)}_{m \text{ factors}}^n$$

$$= \underbrace{(a \times a \times \ldots \times a)}_{m \text{ factors}} \times \underbrace{(a \times a \times \ldots \times a)}_{m \text{ factors}} \times \ldots \times \underbrace{(a \times a \times \ldots \times a)}_{m \text{ factors}}$$

$$\underbrace{\qquad\qquad\qquad\qquad\qquad\qquad\qquad\qquad}_{n \text{ times}}$$

$$= \underbrace{a \times a \times a \times \ldots \times a}_{mn \text{ factors}}$$

$$= a^{mn}$$

Power of a Product

$(5 \times 2)^3 = (5 \times 2) \times (5 \times 2) \times (5 \times 2)$
$\quad = 5 \times 5 \times 5 \times 2 \times 2 \times 2$
$\quad = 5^3 \times 2^3$

$$(ab)^m = \underbrace{(ab) \times (ab) \times \ldots \times (ab)}_{m \text{ factors}}$$

$$= \underbrace{(a \times a \times \ldots \times a)}_{m \text{ factors}} \times \underbrace{(b \times b \times \ldots \times b)}_{m \text{ factors}}$$

$$= a^m b^m$$

Power of a Quotient

$$\left(\frac{2}{5}\right)^3 = \left(\frac{2}{5}\right) \times \left(\frac{2}{5}\right) \times \left(\frac{2}{5}\right)$$

$$= \frac{2 \times 2 \times 2}{5 \times 5 \times 5}$$

$$= \frac{2^3}{5^3}$$

$$\left(\frac{a}{b}\right)^m = \underbrace{\left(\frac{a}{b}\right) \times \left(\frac{a}{b}\right) \times \ldots \times \left(\frac{a}{b}\right)}_{m \text{ factors}}$$

$$= \frac{\overbrace{a \times a \times \ldots \times a}^{m \text{ factors}}}{\underbrace{b \times b \times \ldots \times b}_{m \text{ factors}}}$$

$$= \frac{a^m}{b^m}, b \neq 0$$

A **power** is an expression in the form a^m. To simplify an expression with powers, rewrite the expression without negative exponents or brackets.

Example 1   Simplifying Expressions With Powers

Simplify. **a)** $(3a^2b)(-2a^3b^2)$ **b)** $(m^3)^4$ **c)** $(-4p^3q^2)^3$

Solution

a) $(3a^2b)(-2a^3b^2) = 3 \times (-2) \times a^2 \times a^3 \times b \times b^2$
$\qquad\qquad\qquad\quad = -6a^5b^3$

b) $(m^3)^4 = m^{3 \times 4}$
$\qquad\quad = m^{12}$

c) $(-4p^3q^2)^3 = (-4)^3 \times (p^3)^3 \times (q^2)^3$
$\qquad\qquad\quad = -64p^9q^6$

Example 2 Simplifying a Power of a Quotient

Simplify $\left(\dfrac{6x^5 y^3}{8y^4}\right)^2$.

Solution 1
Use the power of a quotient law first.

$$\left(\frac{6x^5 y^3}{8y^4}\right)^2 = \frac{(6)^2(x^5)^2(y^3)^2}{(8)^2(y^4)^2}$$

$$= \frac{36x^{10}y^6}{64y^8}$$

$$= \frac{9x^{10}}{16y^2}$$

Solution 2
Simplify the fraction first.

$$\left(\frac{6x^5 y^3}{8y^4}\right)^2 = \left(\frac{3x^5}{4y}\right)^2$$

$$= \frac{(3)^2(x^5)^2}{(4)^2(y)^2}$$

$$= \frac{9x^{10}}{16y^2}$$

The following summarizes the rules for zero and negative exponents.

Zero Exponent

$$\frac{2^3}{2^3} = 2^{3-3} = 2^0$$

but $\dfrac{2^3}{2^3} = 1$

so $2^0 = 1$

$$\frac{a^m}{a^m} = a^{m-m} = a^0$$

but $\dfrac{a^m}{a^m} = 1$

so, if $a \neq 0$, $a^0 = 1$

Negative Exponents

$2^3 \times 2^{-3} = 2^{3+(-3)} = 2^0$

so $2^3 \times 2^{-3} = 1$

$$\frac{2^3 \times 2^{-3}}{2^3} = \frac{1}{2^3} \quad \leftarrow \text{Divide both sides by } 2^3.$$

$$2^{-3} = \frac{1}{2^3}$$

$a^m \times a^{-m} = a^{m+(-m)} = a^0$

so $a^m \times a^{-m} = 1$

$$\frac{a^m \times a^{-m}}{a^m} = \frac{1}{a^m} \quad \leftarrow \text{Divide both sides by } a^m.$$

so, if $a \neq 0$, $a^{-m} = \dfrac{1}{a^m}$

Example 3 Expressions With Zero and Negative Exponents

Simplify. **a)** $\left(\dfrac{3}{4}\right)^{-2}$ **b)** $\dfrac{(-6)^0}{2^{-3}}$ **c)** $\dfrac{2^{-4}+2^{-6}}{2^{-3}}$

Solution

a)
$$\left(\frac{3}{4}\right)^{-2} = \frac{1}{\left(\dfrac{3}{4}\right)^2}$$

$$= \frac{1}{\dfrac{9}{16}}$$

$$= \frac{16}{9}$$

b)
$$\frac{(-6)^0}{2^{-3}} = \frac{1}{2^{-3}}$$

$$= \frac{1}{\dfrac{1}{2^3}}$$

$$= \frac{1}{\dfrac{1}{8}}$$

$$= 8$$

c)
$$\frac{2^{-4}+2^{-6}}{2^{-3}} = \frac{\dfrac{1}{2^4}+\dfrac{1}{2^6}}{\dfrac{1}{2^3}}$$

$$= \frac{2^2+1}{2^6} \times \frac{2^3}{1}$$

$$= \frac{2^2+1}{2^3}$$

$$= \frac{5}{8}$$

Practice

Express as a power of 2.

1. $2^4 \times 2^3$ **2.** $2^6 \div 2^2$ **3.** $(2^4)^3$

4. 2×2^7 **5.** $2^3 \times 2^m$ **6.** $2^7 \div 2^y$

7. $2^x \div 2^4$ **8.** $(2^x)^y$ **9.** $2^{-3} \times 2^4$

10. $2^{-2} \div 2^{-5}$ **11.** $(2^3)^{-1}$ **12.** $2^{-4} \times 2^0$

Evaluate.

13. 3^{-2} **14.** 5^0 **15.** 2^{-3}

16. $(-2)^{-4}$ **17.** $(2^{-1})^2$ **18.** $-(-3)^0$

19. $\dfrac{1}{5^{-2}}$ **20.** $\dfrac{1}{(-4)^{-1}}$ **21.** $-(2^3)^{-2}$

Simplify.

22. $a^4 \times a^3$ **23.** $(m^6)(m^2)$

24. $b^5 \times b^6 \times b$ **25.** $a \times b^2 \times a^4$

26. $(x^3)(y)(y^4)(x^5)$ **27.** $(x^3)(x^{-5})$

28. $m^{-4} \times m^{-5}$ **29.** $y^{-1} \times y^{-3} \times y^2$

30. $a^5 \times a^0$ **31.** $(a^{-3})(b^{-2})(a^2)$

Simplify.

32. $x^6 \div x^3$ **33.** $m^7 \div m$ **34.** $t^4 \div t^{-2}$

35. $y^{-5} \div y^{-3}$ **36.** $m^4 \div m^0$ **37.** $t^0 \div t^{-5}$

Simplify.

38. $(x^3)^2$ **39.** $(a^2b^3)^4$ **40.** $(x^2)^{-1}$

41. $(t^4)^0$ **42.** $(a^{-1}b^2)^{-2}$ **43.** $(x^2y^3)^{-3}$

Simplify.

44. $\left(\dfrac{x}{2}\right)^3$ **45.** $\left(\dfrac{a}{b}\right)^4$ **46.** $\left(\dfrac{x^2}{y^3}\right)^5$

47. $\left(\dfrac{x}{3}\right)^{-1}$ **48.** $\left(\dfrac{m^{-3}}{n}\right)^0$ **49.** $\left(\dfrac{a^{-2}}{b^{-3}}\right)^{-2}$

Simplify.

50. $5m^4 \times 3m^2$ **51.** $(4ab^4)(-5a^3b^2)$

52. $5a(-2ab^2)(-3b^3)$ **53.** $(-6m^3n^2)(-4mn^5)$

54. $(7x^2)(6x^{-2})$ **55.** $(3x^{-3}y^2)(-2x^2y^{-3})$

56. $(-6a^{-1}b^2)(-a^{-3}b^{-4})$ **57.** $(-10x^4) \div (-2x)$

58. $\dfrac{45a^2b^4}{9ab^2}$ **59.** $\dfrac{(4m^2n^4)(7m^3n)}{14mn^5}$

60. $\dfrac{3ab^3 \times 10a^4b^2}{15a^2b^6}$ **61.** $\dfrac{4a^4b^3}{a^5b^6} \times \dfrac{-a^3}{-(b^2)}$

62. $(35x^5) \div (5x^{-3})$ **63.** $(-6m^{-4}n^2) \div (2m^{-1}n^{-6})$

64. $\dfrac{-54a^5b^{-7}}{-6a^{-2}b^{-3}}$ **65.** $\dfrac{(-2x^{-3}y)(-12x^{-4}y^{-2})}{6xy^{-3}}$

Simplify.

66. $(2m^3)^2$ **67.** $(-4x^2)^3$ **68.** $(-3m^3n^2)^2$

69. $(5c^{-3}d^3)^{-2}$ **70.** $(2a^{-3}b^{-2})^{-3}$ **71.** $(-3x^3y^{-2})^{-4}$

72. $\left(\dfrac{4x}{3y}\right)^2$ **73.** $\left(\dfrac{-2a^2}{3y^3}\right)^3$ **74.** $\left(\dfrac{3a}{-b^4}\right)^4$

75. $\left(\dfrac{2m^2}{n^3}\right)^{-2}$ **76.** $\left(\dfrac{6ab^3}{2ab}\right)^3$ **77.** $\left(\dfrac{4x^{-3}y^4}{8x^2y^{-2}}\right)^{-2}$

Evaluate.

78. $\dfrac{6}{x^0 + y^0}$ **79.** $4^{-1} + 2^{-3}$

80. $\dfrac{3^{-3} + 3^{-4}}{3^{-5}}$ **81.** $\dfrac{(6^4 + 4^6)^0}{3^{-1}}$

Applications and Problem Solving

82. History The Burgess Shale in British Columbia's Yoho National Park contains one of the world's best fossil collections. The fossils are about 5.4×10^8 years old. This is about 4.5×10^4 times older than the first known human settlement in British Columbia. About how many years ago did humans first settle in British Columbia?

83. Chemistry A piece of wood burns completely in one second at 600°C. The time the wood takes to burn is doubled for every 10°C drop in temperature and halved for every 10°C increase in temperature. In how many seconds would the wood burn at
a) 500°C?   **b)** 650°C?

84. Without evaluating the expressions, determine which is bigger, 20^{100} or 400^{40}.

85. Evaluate.

a) $\dfrac{6^1 + 6^{-1}}{6^1 - 6^{-1}}$ **b)** $\dfrac{5^{-4} - 5^{-6}}{5^{-3} + 5^{-5}}$

c) $2^{-n}(2^n - 2^{1+n})$ **d)** $3\left(3^{2x} - \dfrac{1}{3^{-2x}}\right)$

86. Equations Determine the value of x.
a) $x^2 \times x^3 = 32$ **b)** $x^5 \div x^2 = 64$

c) $x^{-1} \times x^{-3} = \dfrac{1}{81}$ **d)** $x^2 \div x^5 = \dfrac{1}{125}$

87. Equation For which values of x is the following equation true? Explain.
$x^{-4} \div x^{-4} = 1$

1.8 Rational Exponents

Most of the power used to move a ship is needed to push along the bow wave that builds up in front of the ship. Ships are designed to use as little power as possible.

Ship designers test models of a ship before the real ship is built. When a model is being tested, in special water tanks, it must move at a speed to make a bow wave of the same height, relative to the model, as the real bow wave would be to the real ship.

To calculate the speed to use when testing a model the following formula is used.

$$S_m = \frac{S_r \times L_m^{\frac{1}{2}}}{L_r^{\frac{1}{2}}}$$

where S_m is the speed of the model in metres per second, S_r is the speed of the real ship in metres per second, L_m is the length of the model in metres, and L_r is the length of the real ship in metres. This formula involves powers with fractional exponents.

Explore: Complete the Statements

Using the power law for exponents, 9 can be written

as $\left(9^{\frac{1}{2}}\right)^2$, because $\left(9^{\frac{1}{2}}\right)^2 = 9^{\frac{1}{2} \times 2} = 9^1$ or 9.

Copy and complete the following statements.
The first one has been partially completed.

NRC

Я⅃G

1. $9 = \left(9^{\frac{1}{2}}\right)^2$
   but $9 = (3)^2$
   so $\left(9^{\frac{1}{2}}\right)^2 = (3)^2$
   and $9^{\frac{1}{2}} = 3$

2. $25 = \left(25^{\frac{1}{2}}\right)^2$
   but $25 = (5)^2$
   so $\left(25^{\frac{1}{2}}\right)^2 = (5)^2$
   and $25^{\frac{1}{2}} = 5$

3. $8 = \left(8^{\frac{1}{3}}\right)^3$
   but $8 = (2)^3$
   so $\left(8^{\frac{1}{3}}\right)^3 = (2)^3$
   and $8^{\frac{1}{3}} = 2$

4. $16 = \left(16^{\frac{1}{4}}\right)^4$
   but $16 = (2)^4$
   so $\left(16^{\frac{1}{4}}\right)^4 = (2)^4$
   and $16^{\frac{1}{4}} = 2$

Inquire

1. What does the exponent mean in $25^{\frac{1}{2}}$? $\sqrt{25}$

2. What does the exponent mean in

a) $8^{\frac{1}{3}}$? $\sqrt[3]{8}$ **b)** $16^{\frac{1}{4}}$? $\sqrt[4]{16}$

3. Evaluate.

a) $36^{\frac{1}{2}}$ 6^{-2} **b)** $27^{\frac{1}{3}}$ 3^{-3} **c)** $81^{\frac{1}{4}}$ 3^{-4} **d)** $100^{\frac{1}{2}}$ 10^{-2}

4. A ship is to be built 100 m long and able to travel at 15 m/s. The model of the ship is 4 m long. At what speed should the model be tested?

In the power law for exponents, $(a^m)^n = a^{mn}$, substituting $m = \dfrac{1}{n}$ gives

$$\left(a^{\frac{1}{n}}\right)^n = a^{\frac{1}{n} \times n} = a^1 \text{ or } a$$

If $a \geq 0$, we can take the nth root of both sides of the equation $\left(a^{\frac{1}{n}}\right)^n = a,$

which gives $a^{\frac{1}{n}} = \sqrt[n]{a}$.

This result suggests the following definition.

$a^{\frac{1}{n}} = \sqrt[n]{a}$, where n is a natural number

- If n is an even number, then we must have $a \geq 0$. Suppose that n is even and a is negative. For example, if $n = 2$ and $a = -4$, then $(-4)^{\frac{1}{2}}$ becomes $\sqrt{-4}$. There is no real square root of -4.
- If n is an odd number, then a can be any real number. For example, if $n = 3$ and $a = -8$, then $(-8)^{\frac{1}{3}}$ becomes $\sqrt[3]{-8}$, which is -2.

Note how brackets are used with fractional exponents. The expression $\sqrt{-4}$ has no meaning, but $-\sqrt{4} = -2$. Similarly, $(-4)^{\frac{1}{2}}$ becomes $\sqrt{-4}$, which has no meaning. But $-4^{\frac{1}{2}}$ becomes $-\left(4^{\frac{1}{2}}\right) = -\sqrt{4} = -2$.

Example 1 Exponents in the Form $\dfrac{1}{n}$

Evaluate. **a)** $49^{\frac{1}{2}}$ **b)** $(-27)^{\frac{1}{3}}$ **c)** $(-8)^{-\frac{1}{3}}$

Solution

a) $49^{\frac{1}{2}} = \sqrt{49}$
$= 7$

b) $(-27)^{\frac{1}{3}} = \sqrt[3]{-27}$
$= -3$

c) $(-8)^{-\frac{1}{3}} = \dfrac{1}{(-8)^{\frac{1}{3}}}$
$= \dfrac{1}{\sqrt[3]{-8}}$
$= -\dfrac{1}{2}$

The following suggests how to evaluate an expression with a fractional exponent in which the numerator is not 1, such as $4^{\frac{3}{2}}$. The power law $(a^m)^n = a^{mn}$ is used.

Method 1

$$4^{\frac{3}{2}} = \left(4^{\frac{1}{2}}\right)^3$$
$$= (\sqrt{4})^3$$
$$= (2)^3$$
$$= 8$$

Method 2

$$4^{\frac{3}{2}} = (4^3)^{\frac{1}{2}}$$
$$= \sqrt{4^3}$$
$$= \sqrt{64}$$
$$= 8$$

Notice that $(\sqrt{4})^3$ and $\sqrt{4^3}$ have the same value.

This result suggests the following definition for rational exponents.

$$a^{\frac{m}{n}} = \sqrt[n]{a^m} = \left(\sqrt[n]{a}\right)^m, \text{ where } m \text{ and } n \text{ are natural numbers}$$

If n is an even number, then $a \geq 0$.
If n is an odd number, then a can be any real number.

To calculate $a^{\frac{m}{n}}$

• take the nth root of a, then raise the result to the mth power

$$9^{\frac{3}{2}} = (\sqrt{9})^3$$
$$= 3^3$$
$$= 27$$

or

• raise a to the mth power, then take the nth root

$$9^{\frac{3}{2}} = \sqrt{9^3}$$
$$= \sqrt{729}$$
$$= 27$$

It is common practice to take the nth root first.

Example 2   Exponents in the Form $\dfrac{m}{n}$

Evaluate. **a)** $(-8)^{\frac{4}{3}}$ **b)** $9^{-2.5}$ **c)** $\left(\dfrac{25}{4}\right)^{-\frac{3}{2}}$

Solution

a) $(-8)^{\frac{4}{3}} = \left(\sqrt[3]{-8}\right)^4$
$$= (-2)^4$$
$$= 16$$

b) $9^{-2.5} = 9^{-\frac{5}{2}}$
$$= \frac{1}{9^{\frac{5}{2}}}$$
$$= \frac{1}{(\sqrt{9})^5}$$
$$= \frac{1}{3^5}$$
$$= \frac{1}{243}$$

c) $\left(\dfrac{25}{4}\right)^{-\frac{3}{2}} = \dfrac{1}{\left(\dfrac{25}{4}\right)^{\frac{3}{2}}}$
$$= \frac{1}{\dfrac{(\sqrt{25})^3}{(\sqrt{4})^3}}$$
$$= \frac{1}{\dfrac{125}{8}}$$
$$= \frac{8}{125}$$

Example 3   Using a Calculator

Use a calculator to evaluate the following to the nearest hundredth.

a) $2^{3.5}$ **b)** $7^{\frac{2}{3}}$

Solution

a)

2^3.5
11.3137085

Estimate
$2^3 = 8$
$2^4 = 16$
$2^{3.5} \doteq 12$

$2^{3.5} \doteq 11.31$

b)

7^(2/3)
3.65930571

Estimate
$7^{\frac{2}{3}} \doteq 8^{\frac{2}{3}}$
$\doteq 2^2$
$\doteq 4$

$7^{\frac{2}{3}} \doteq 3.66$

Practice

Write in radical form.

1. $2^{\frac{1}{3}}$ **2.** $37^{\frac{3}{2}}$ **3.** $x^{\frac{1}{2}}$ **4.** $a^{\frac{3}{5}}$

5. $6^{\frac{4}{3}}$ **6.** $6^{\frac{3}{4}}$ **7.** $7^{-\frac{1}{2}}$ **8.** $9^{-\frac{1}{5}}$

9. $x^{-\frac{3}{7}}$ **10.** $b^{-\frac{6}{5}}$ **11.** $(3x)^{\frac{1}{2}}$ **12.** $3x^{\frac{1}{2}}$

Write using exponents.

13. $\sqrt{7}$ **14.** $\sqrt{34}$ **15.** $\sqrt[3]{-11}$

16. $\sqrt[5]{a^2}$ **17.** $\sqrt[3]{6^4}$ **18.** $(\sqrt[3]{b})^4$

19. $\dfrac{1}{\sqrt{x}}$ **20.** $\dfrac{1}{\sqrt[3]{a}}$ **21.** $\dfrac{1}{(\sqrt[5]{x})^4}$

22. $\sqrt[3]{2b^3}$ **23.** $\sqrt{3x^5}$ **24.** $\sqrt[4]{5t^3}$

Evaluate.

25. $4^{\frac{1}{2}}$ **26.** $125^{\frac{1}{3}}$ **27.** $16^{-\frac{1}{4}}$

28. $(-32)^{\frac{1}{5}}$ **29.** $25^{0.5}$ **30.** $(-27)^{-\frac{1}{3}}$

31. $(64)^{-\frac{1}{6}}$ **32.** $0.04^{\frac{1}{2}}$ **33.** $81^{0.25}$

34. $0.001^{\frac{1}{3}}$ **35.** $\left(\dfrac{4}{9}\right)^{\frac{1}{2}}$ **36.** $\left(\dfrac{-27}{-8}\right)^{\frac{1}{3}}$

Evaluate.

37. $8^{\frac{2}{3}}$ **38.** $4^{\frac{3}{2}}$ **39.** $9^{2.5}$

40. $81^{\frac{3}{4}}$ **41.** $16^{-\frac{3}{4}}$ **42.** $(-32)^{\frac{2}{5}}$

43. $(-8)^{-\frac{5}{3}}$ **44.** $(-27)^{-\frac{2}{3}}$ **45.** $1^{\frac{5}{3}}$

46. $(-1)^{-\frac{8}{5}}$ **47.** $\left(\dfrac{100}{9}\right)^{\frac{3}{2}}$ **48.** $\left(\dfrac{27}{8}\right)^{-\frac{2}{3}}$

Evaluate, if possible.

49. $(-9)^{\frac{1}{2}}$ **50.** $100\,000^{\frac{3}{5}}$ **51.** $\left(\dfrac{27}{8}\right)^{\frac{2}{3}}$

52. $3^{\frac{1}{2}} \times 3^{\frac{1}{2}}$ **53.** $-9^{\frac{1}{2}}$ **54.** $(2^5)^{0.4}$

55. $-8^{\frac{5}{3}}$ **56.** $4^{\frac{3}{2}} \div 16^{\frac{1}{4}}$ **57.** $(-1)^{-\frac{3}{2}}$

58. $(\sqrt[3]{5^2})(\sqrt[3]{5})$ **59.** $\left(\dfrac{36}{121}\right)^{-\frac{1}{2}}$ **60.** $81^{0.75}$

61. $(-0.0016)^{\frac{1}{4}}$ **62.** $\dfrac{(0.027)^{-\frac{2}{3}}}{(0.25)^{-\frac{1}{2}}}$ **63.** $(625^{-1})^{-\frac{1}{4}}$

64. $9^{\frac{3}{7}} \times 3^{\frac{1}{7}}$ **65.** $\left[(\sqrt{125})^4\right]^{\frac{1}{6}}$ **66.** $\sqrt[3]{\sqrt{64}}$

67. $\sqrt{\sqrt[3]{729}}$ **68.** $\dfrac{(0.09)^{\frac{1}{2}}}{(0.008)^{\frac{1}{3}} \times 2^{-3}}$

Write an equivalent expression using exponents.

69. $\sqrt{\sqrt{x^4}}$ **70.** $\sqrt[3]{\sqrt{x^6}}$ **71.** $\sqrt{\sqrt{3x^6}}$

72. $\sqrt{\sqrt[3]{8x^7}}$ **73.** $\sqrt{\sqrt{81x^8}}$ **74.** $\left(x^{\frac{2}{3}} y^{\frac{1}{3}}\right)^3$

75. $\left(a^{\frac{1}{3}} b^{\frac{1}{4}}\right)^{12}$ **76.** $\sqrt[3]{-27x}$ **77.** $(81a^8b^4)^{\frac{1}{4}}$

78. $\left(27x^6 y^{-9}\right)^{\frac{2}{3}}$ **79.** $\left(\sqrt{x^3}\right)\left(\sqrt[3]{x}\right)$

80. $\left(\sqrt[3]{x^2}\right)\left(\sqrt[4]{x^3}\right)$ **81.** $\left(\sqrt[5]{x^3}\right)\left(\sqrt[3]{x^2}\right)$

82. $\left(\sqrt[3]{a^2 b^4}\right)^2$ **83.** $\left(\sqrt[4]{a^3 b^5}\right)^{\frac{1}{2}}$

Estimate. Then, find an approximation of each to the nearest hundredth.

84. $6^{0.4}$ **85.** $3^{2.8}$ **86.** $4^{-1.2}$

87. $5^{\frac{1}{3}}$ **88.** $7^{-\frac{3}{5}}$ **89.** $10^{\frac{3}{7}}$

Applications and Problem Solving

90. Ship building The design of a new ship calls for the ship to be 300 m long and travel at 12 m/s. To test the design, a model 15 m long is used. At what speed should the model be tested, to the nearest tenth of a metre per second?

91. Music The frequency of a note on a piano is measured in vibrations per second, or hertz (Hz). The frequency of each of the other notes in the octave above middle C is a multiple of the frequency of middle C. Copy and complete the table. Round each frequency to the nearest whole number of hertz.

Note	Multiple of C	Frequency (Hz)
C	1	262
C#	$\sqrt[12]{2}$	
D	$\left(\sqrt[12]{2}\right)^2$	
D#	$\left(\sqrt[12]{2}\right)^3$	
E	$\left(\sqrt[12]{2}\right)^4$	
F	$\left(\sqrt[12]{2}\right)^5$	
F#	$\left(\sqrt[12]{2}\right)^6$	
G	$\left(\sqrt[12]{2}\right)^7$	
G#	$\left(\sqrt[12]{2}\right)^8$	
A	$\left(\sqrt[12]{2}\right)^9$	
A#	$\left(\sqrt[12]{2}\right)^{10}$	
B	$\left(\sqrt[12]{2}\right)^{11}$	
C	$\left(\sqrt[12]{2}\right)^{12}$	

92. Equations Evaluate x, where x is a natural number.
a) $2^x = 32$ **b)** $3^{x+1} = 81$ **c)** $(-1)^x = 1$
d) $6^{x-2} = 36$ **e)** $2^{2x} = 16$ **f)** $(-1)^x = -1$

93. Measurement a) The diagrams show 2 squares with whole-number areas that can be made on a 4-pin-by-4-pin geoboard.

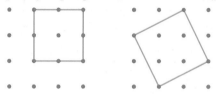

If the shortest distance between 2 pins is 1 unit, what is the area of each square?
b) Draw the 3 other different-sized squares with whole-number areas that can be made on the same geoboard.
c) Of the 5 different-sized squares, which ones do not have whole-number side lengths? Express their side lengths using fractional exponents.
d) Draw the 8 different-sized squares with whole-number areas that can be made on a 5-pin-by-5-pin geoboard. For the squares that do not have whole-number side lengths, express the side lengths using fractional exponents.
e) Repeat part d) for a 6-pin-by-6-pin geoboard, and state how many different-sized squares can be made.

NUMBER POWER

Copy the diagram.

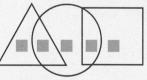

Identify 5 consecutive whole numbers and arrange them in the 5 small squares so that
• the sum of the numbers in the triangle is 28
• the sum of the numbers in the circle is 41
• the sum of the numbers in the rectangle is 30
• the sum of all 5 numbers is 70

List the 5 numbers in order from left to right in the diagram.

Proving That $\sqrt{2}$ Is Not Rational

1 Perfect Squares and Prime Factors

The prime factorization of 12 is $2 \times 2 \times 3$. There are three numbers in this prime factorization.

1. Write the prime factorization of each of the following natural numbers. How many numbers are there in each prime factorization?
a) 7   **b)** 9   **c)** 10   **d)** 18   **e)** 8   **f)** 16

2. Write the prime factorization of the square of each of the numbers in question 1. How many numbers are there in each prime factorization?

3. Use the word "even" or the word "odd" to complete the following statement.
The square of a natural number has an ▨ number of numbers in its prime factorization.

4. a) Suppose that there are n numbers in the prime factorization of a natural number, x. How many numbers are there in the prime factorization of x^2? Explain.
b) How does your finding in part a) prove that your statement in question 3 is true?

2 Proof by Contradiction

The method of proof by contradiction involves:
• assuming the opposite of what you want to prove
• showing that the assumption leads to a contradiction of known facts

The following is a proof that $\sqrt{2}$ is not rational. Write a reason to explain each step preceded by an asterisk.
To prove that $\sqrt{2}$ is not rational, assume that $\sqrt{2}$ is rational.

* If $\sqrt{2}$ is rational, it can be written as a fraction $\frac{a}{b}$.
Since $\sqrt{2} = \frac{a}{b}$,

* squaring both sides gives $2 = \frac{a^2}{b^2}$.

* Then $2b^2 = a^2$.
* There is an even number of numbers in the prime factorization of a^2.
* There is an odd number of numbers in the prime factorization of $2b^2$.
* A number with an even number of numbers in its prime factorization cannot equal a number with an odd number of numbers in its prime factorization.
The original assumption that $\sqrt{2}$ is rational is false.

Therefore, $\sqrt{2}$ is not rational.

39

PROBLEM SOLVING

1.9 Use a Data Bank

A data bank is a collection of information organized so that the information is easy to retrieve. A data bank can give you the information you need to solve a problem.

Amelia Earhart (1898–1937) was one of the world's leading aviators. Her solo flights in small planes broke a number of speed, altitude, and distance records. Her flight from Harbour Grace, Newfoundland, to Culmore, Ireland, was the second solo flight across the Atlantic. She was the first person to fly solo across the Pacific from Hawaii to California.

On her flight across the Pacific, Amelia Earhart took off from Honolulu, Hawaii, on January 11, 1935, at 17:16 local time. She landed in Oakland, California, at 13:31 local time on January 12. She flew a distance of 3875 km. Find her average speed, to the nearest kilometre per hour.

Understand the Problem
1. What information are you given?
2. What are you asked to find?
3. Do you need an exact or approximate answer?

Think of a Plan
The average speed can be found by dividing the distance flown by the time taken. The time taken must first be calculated. Honolulu, Hawaii, and Oakland, California, are in different time zones, so a time zone map is needed. Refer to the Data Bank on pages 404 to 405.

Carry Out the Plan
Honolulu is in time zone −10, so Honolulu is 10 h behind Greenwich Mean Time (GMT). When Amelia Earhart left Honolulu on January 11, the time in Greenwich was 17:16 + 10, or 03:16 GMT on January 12. California is in time zone −8, so it is 8 h behind GMT. The time when she landed was 13:31 + 8 or 21:31 GMT on January 12. From 03:16 GMT on January 12 to 21:31 GMT on January 12 is 18 h 15 min or 18.25 h.

$$\text{Average speed} = \frac{\text{distance flown}}{\text{time taken}}$$
$$= \frac{3875}{18.25}$$
$$\doteq 212 \text{ km/h}$$

Estimate
$$4000 \div 20 = 200$$

Her average speed was 212 km/h, to the nearest kilometre per hour.

Look Back
Does the answer seem reasonable?

Could you calculate the flying time without working in GMT?

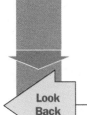

Use a Data Bank
1. Locate the information you need.
2. Solve the problem.
3. Check that your answer is reasonable.

Applications and Problem Solving

Use the Data Bank on pages 404 to 413 to solve the following problems.

1. Charles Lindbergh was the first person to fly solo across the Atlantic. He left New York City at 07:52 local time on May 20, 1927, and landed in Paris at 23:21 local time on May 21. He flew 5830 km.
a) Calculate the time he took to fly from New York City to Paris.
b) Find his average speed, to the nearest kilometre per hour.

2. One year, the Great North American Race for antique cars started in Ottawa and travelled through or near Toronto, Buffalo, Syracuse, Philadelphia, Washington, D.C., Charleston, Lexington, Nashville, Little Rock, Dallas, Austin, San Antonio, Laredo, and Monterrey, ending in Mexico City. The race took 15 days. There were 108 teams in the race, 12 of them from Canada. Each car had a driver and a navigator. Only two navigational aids were allowed — a clock, and a speedometer without the odometer. An odometer keeps track of the distance travelled by a car. Without an odometer, teams could only monitor the speed and the time. Teams were told what speeds to maintain and for how long. At checkpoints along the route, points could be lost for arriving either late or early. As one navigator stated about the race, "It's all math."
a) Use the map of highway driving distances in the Data Bank to calculate the length of the race.
b) The cars actually travelled a distance of about 7700 km. Give possible reasons why this value does not agree with your answer from part a).

3. The outside air temperature is −17°C, and the wind is gusting at between 15 km/h and 35 km/h.
a) What is the lowest wind chill temperature?
b) What is the highest wind chill temperature?

4. Estimate the humidex reading on a day when the air temperature is 29°C and the relative humidity is 73%.

5. On Monday at 21:00 local time, a plane leaves Vancouver for Sydney, Australia. The plane cruises at 890 km/h and stops for 1 h in Hawaii to refuel. At about what local time and on what day does the plane land in Sydney?

6. You are a travel agent planning a trip for 24 archaeology students from the University of Paris. The students will fly to Toronto and spend two days visiting the dinosaur exhibit at the Royal Ontario Museum. They will then fly to Calgary and take a 150-km bus trip to the Royal Tyrrell Museum in Drumheller, Alberta. They will spend three days at the museum studying the 40 dinosaur skeletons that make up the world's largest exhibit of complete dinosaurs. Next, the students will travel 125 km by bus to Dinosaur Provincial Park, where they will spend two days. Finally, they will take a 250-km bus trip back to Calgary and fly back to Paris. Plan an itinerary for the trip, leaving Paris on June 1. Use local times to describe all departures, arrivals, and times spent at the different locations. Assume that plane trips will average 850 km/h and bus trips will average 50 km/h.

7. a) Draw a diagram to show the relative diameters of the planets in our solar system.
b) State the scale of your diagram.

8. Route 66 was a popular two-lane highway that started in Chicago and ended in Los Angeles. Much of Route 66 still exists, but most travellers now prefer to use the superhighways that run parallel to it. The cities on Route 66 include Chicago, Springfield (Illinois), St. Louis, Springfield (Missouri), Tulsa, Oklahoma City, Amarillo, Tucumcari, Albuquerque, Gallup, Flagstaff, Needles, Barstow, and Los Angeles. A newspaper reporter, writing a story about Route 66, leaves Chicago on May 10 at 08:00. Assume that the reporter drives at 80 km/h for a maximum of 6 h/day and sleeps each night at one of the places on the list. Write an itinerary for the reporter.

9. Which planet takes about 350 times as long as Mercury to orbit the sun?

10. Which is the sunniest province in Canada? Explain.

11. Write a problem that requires the use of a data bank. Have a classmate solve your problem.

COMPUTER DATA BANK

Using the Databases

Answer the following to familiarize yourself with the *MATHPOWER*™ *10, Western Edition, Computer Data Bank*. Use the *Skiing, Movies, Olympics, Nutrition,* and *Insurance* databases included in the Computer Data Bank for ClarisWorks.

1 Ski Trails

1. Find all the records for which *Novice Trails, %, Intermediate Trails, %,* and *Expert Trails, %* are available. How many records are displayed?

2. Select, by matching, all the records where the percent of intermediate trails is greater than the percent of expert trails. Hide the other records. How many records are displayed? What fraction is this of the records displayed in question 1? Would you have predicted the fraction? Explain.

2 Movie Costs

1. Find all the records for which *Cost, $ millions* is available.

2. Create a table for those records to display only the following fields.

Movie   Length, min   Cost, $ millions

3. How could you calculate cost per minute, in dollars? Add a calculation field called *Cost per Minute, $*, rounding the costs to the nearest dollar.

4. Sort the records from greatest to least cost per minute. Which movie cost the most per minute? the least?

3 Winning Times

1. Find all the records for *100-m Dash, Women*.

2. Sort those records from fastest to slowest winning time.

3. What is the median winning time? Explain how you know.

4. What is the mode winning time? How frequently does it occur?

5. Create a summary field called *Mean Winning Time, s*, rounded to 2 decimal places. What is the mean winning time?

4 Food Graphs

1. Select the records of four foods you enjoy eating that contain at least three of these components — water, protein, carbohydrates, and fat. Hide the other records.

2. How could you calculate the percent, by mass, of each component in a food? Add three calculation fields called *Protein, %, Carbohydrates, %,* and *Fat, %*, rounding the percents in each to 1 decimal place.

3. Create a table for the four records to display only the following fields.

Food   Water, %   Protein, %
Carbohydrates, %   Fat, %

4. Copy and paste these data onto a spreadsheet, and create a graph for each food to show the percents. Compare your graphs with a classmate's.

5. What type of graph did you use? Explain why.

5 Collision Coverage Premiums

1. Find all the records for collision coverage in Alberta for drivers less than 25 years old.

2. Create two summary fields called *Total Number of Vehicles Insured* and *Total Premiums, $*. What are the totals?

3. Create another summary field called *Mean Premium, $*, rounded to the nearest dollar, using the values from question 2. What is the mean premium?

Astronomy

Do you often think about conditions that exist far beyond the Earth? Do you wonder how far it is to the edge of the universe, how many stars are in the sky, or how long it has taken light to reach us from the stars? If you do, a career in astronomy may interest you.

Most astronomers work for government agencies, universities, or observatories. Astronomers study many aspects of the universe. Most astronomers then specialize in an area that interests them, such as the study of supernovas.

A supernova is the explosive destruction of a massive star, which is a star with a mass four or more times the mass of the sun. The explosions are so powerful that a supernova can give out as much energy in a few seconds as the sun does in millions of years.

Canadian astronomer Ian Shelton discovered a supernova from an observatory in Chile in 1987. This supernova, now called Supernova Shelton, was bright enough to be visible to the naked eye.

The first discovery of a supernova from a Canadian observatory was made from the Burke-Gaffney Observatory at St. Mary's University in Halifax, Nova Scotia, in 1995. The supernova, named 1995-F, was discovered by David Lane, a community college student and amateur astronomer, and Paul Gray, observatory technician and president of the Halifax branch of the Royal Astronomical Society of Canada.

1 Comparing Distances

1. Light travels through space at a speed of about 3×10^5 km/s. How far does light travel in
a) 5 s? **b)** 30 s? **c)** 5 min?

2. a) A light-year is the distance that light travels through space in a year. Calculate this distance in kilometres. Express your answer in scientific notation, and round the decimal part of the number to the first decimal place.
b) The sun is about 1.5×10^8 km from the Earth. Express this distance in light-years.
c) Supernova 1995-F is about 7×10^7 light-years from Earth. Express this distance in kilometres.
d) An astronomical unit is defined as the distance of the Earth from the sun. Express the distance from the Earth to supernova 1995-F in astronomical units.
e) How many years did it take light from the explosion of 1995-F to reach the Earth? Explain.

2 Comparing Sizes

The diameter of the sun is about 1.4×10^6 km. The diameter of the Earth is about 12 760 km.
1. About how many Earths could fit along the diameter of the sun?

2. After a massive star explodes, a neutron star as little as 20 km across may remain. About how many neutron stars could fit
a) along the diameter of the Earth?
b) along the diameter of the sun?

3 Locating Information

Where would you look to find information on
1. the education needed for a career in astronomy?

2. how to become an amateur astronomer?

CONNECTING MATH AND PHYSIOLOGY

Using Length Measurements to Find Speeds and Masses

1 Calculating Speeds

R. McNeill Alexander, a British biologist, developed a formula based on hip height and stride length to determine running or walking speeds.

The formula is $s = \dfrac{0.78 l^{1.67}}{h^{1.17}}$, where s is the speed, in metres per second, l is the stride length, in metres, and h is the hip height, in metres.

The stride length is the distance between successive left or right foot positions.

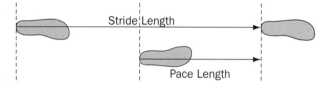

Stride Length

Pace Length

1. Donovan Bailey won the 100-m dash at the Olympic Games in Atlanta. During the last 60 m of the race, his stride length was about 5 m. If his hip height is about 1 m, use Alexander's formula to calculate his speed for the last 60 m, to the nearest tenth of a metre per second.

2. Scientists can use dinosaur footprints to determine dinosaurs' walking and running speeds. The footprints of a large theropod, a bipedal carnivorous dinosaur, have been found in Queensland, Australia. Bipedal dinosaurs walked on two feet. The stride length of the dinosaur was 3.31 m, and the foot length was 0.64 m. The hip height for dinosaurs was four times the foot length. How fast was the dinosaur moving when it made the footprints, to the nearest tenth of a metre per second?

3. The footprints of a small theropod were also found in Queensland. The stride length was 2.2 m, and the foot length was 0.25 m. How fast was the dinosaur moving when it made the footprints, to the nearest tenth of a metre per second?

4. a) The footprints of the fastest running theropod were found in Texas. The stride length was 6.9 m, and the foot length was 0.38 m. How fast was the dinosaur running, to the nearest tenth of a metre per second?
b) Did Donovan Bailey run this fast in Atlanta?

5. Work with a partner to measure your hip height and to measure your stride length when you are walking. Use the data to calculate your walking speed.

2 Calculating Masses

Dale Russell, a Canadian dinosaur specialist, worked with two other zoologists to develop a formula to estimate the masses of quadruped dinosaurs, which walked on four feet.

The researchers measured the circumferences of the humerus and femur bones, where they were smallest, and added the circumferences together. The sum was substituted into the formula

$$M = 0.000\,084\,(C_h + C_f)^{2.73}$$

where M is the mass of the dinosaur in kilograms, C_h is the circumference of the humerus, in millimetres, and C_f is the circumference of the femur, in millimetres.

1. The circumferences of the humerus and the femur of a quadruped Brachiosaurus were 650 mm and 720 mm. What was the mass of the Brachiosaurus?

2. The formula for the mass, in kilograms, of bipedal dinosaurs is $M = 0.000\,16(C_f)^{2.73}$, where C_f is the circumference of the femur, in millimetres. If a femur of a bipedal Anatosaurus had a circumference of 510 mm, what was its mass?

3. a) Why are the circumferences of bones, not the lengths of bones, used in the formulas?
b) Why are the smallest circumferences of bones used?

45

Review

1.1 *Name the sets of numbers to which each number belongs.*

1. $\sqrt{25}$ **2.** $\sqrt{0.09}$ **3.** $-\sqrt{7}$ **4.** $\sqrt{\dfrac{4}{9}}$

Evaluate.

5. $|-7|$ **6.** $\left|\dfrac{4}{5}-\dfrac{9}{10}\right|$ **7.** $|1.5-2.5|$

Given that x is a real number, graph each of the following on a number line.

8. $x \geq 2$ **9.** $-4 < x \leq 3$ **10.** $|-1| \leq x < |6|$

1.3 *Estimate. Then, find an approximate value, to the nearest hundredth.*

11. $\sqrt{44.56}$ **12.** $\sqrt{41\,980}$ **13.** $\sqrt{0.0008}$

Estimate. Then, find an approximate value, to the nearest thousandth.

14. $\sqrt{55} - \sqrt{22}$ **15.** $10\sqrt{12} + \sqrt{61}$

16. $\dfrac{\sqrt{56}}{\sqrt{5}}$ **17.** $\dfrac{\sqrt{44} + \sqrt{10}}{\sqrt{28}}$

Evaluate, to the nearest thousandth.

18. $\sqrt[3]{70} + \sqrt[3]{90}$ **19.** $3\sqrt[3]{16} - \sqrt[3]{128}$

20. Geography Prince Rupert has an area of 53.56 km². Quesnel has an area of 20.56 km².
a) How many times greater is the area of Prince Rupert than the area of Quesnel, to the nearest thousandth?
b) If each city was a square, what would be the side length of each, to the nearest thousandth of a kilometre?
c) How many times greater would the side length be for Prince Rupert than for Quesnel, to the nearest thousandth?
d) How are your answers to parts a) and c) related? Explain.

1.4 *Simplify.*

21. $\sqrt{18}$ **22.** $\sqrt{32}$ **23.** $\sqrt{500}$

Simplify.

24. $\dfrac{\sqrt{40}}{\sqrt{8}}$ **25.** $\dfrac{\sqrt{70}}{\sqrt{10}}$ **26.** $\dfrac{6\sqrt{30}}{\sqrt{5}}$ **27.** $\dfrac{\sqrt{200}}{\sqrt{2}}$

Write as an entire radical.

28. $2\sqrt{11}$ **29.** $4\sqrt{7}$ **30.** $6\sqrt{3}$

Simplify.

31. $\sqrt{10} \times \sqrt{6}$ **32.** $3\sqrt{5} \times 2\sqrt{10}$

Simplify.

33. $\dfrac{1}{\sqrt{5}}$ **34.** $\dfrac{\sqrt{2}}{\sqrt{3}}$ **35.** $\dfrac{4}{3\sqrt{2}}$ **36.** $\dfrac{\sqrt{3}}{4\sqrt{10}}$

37. Without using a calculator, arrange the following in order from least to greatest.

$$5\sqrt{2},\ 2\sqrt{10},\ 4\sqrt{3},\ 3\sqrt{5}$$

38. Measurement A rectangle has side lengths of 2$\sqrt{15}$ and 3$\sqrt{5}$. Express the area of the rectangle in simplest radical form.

1.5 *Simplify.*

39. $3\sqrt{2} + 7\sqrt{2} - 5\sqrt{2}$

40. $7\sqrt{3} - 2\sqrt{6} + 5\sqrt{6} - 3\sqrt{3}$

Simplify.

41. $\sqrt{45} + \sqrt{80}$ **42.** $\sqrt{12} - \sqrt{27}$

43. $\sqrt{18} - \sqrt{50} + \sqrt{32}$

44. $2\sqrt{20} - 3\sqrt{125} + 3\sqrt{80}$

Expand and simplify.

45. $\sqrt{3}(\sqrt{2} + 5)$ **46.** $\sqrt{2}(\sqrt{10} - \sqrt{6})$

47. $(4\sqrt{2} + \sqrt{5})(\sqrt{2} - 3\sqrt{5})$

48. $(2\sqrt{3} + \sqrt{5})^2$ **49.** $(\sqrt{7} - \sqrt{3})(\sqrt{7} + \sqrt{3})$

Simplify.

50. $\dfrac{2}{\sqrt{3} - 1}$ **51.** $\dfrac{4}{\sqrt{5} + \sqrt{2}}$

52. $\dfrac{2\sqrt{3}}{\sqrt{2} - 5}$ **53.** $\dfrac{2\sqrt{7} - \sqrt{3}}{3\sqrt{7} + 2\sqrt{3}}$

Simplify.

54. $\sqrt[3]{16} - \sqrt[3]{54}$ **55.** $3\sqrt[3]{81} + 5\sqrt[3]{24}$

56. Measurement A square has a side length of $4 - \sqrt{5}$. Write and simplify an expression for the area of the square.

1.7 *Evaluate.*

57. 5^{-2} **58.** 6^0 **59.** 3^{-3}

60. $(-3)^{-4}$ **61.** $(5^{-1})^2$ **62.** $\dfrac{1}{(-3)^{-1}}$

Simplify.

63. $m^2 \times m^5$ **64.** $y^{-3} \times y^{-2}$ **65.** $t^7 \div t^4$

66. $m^{-7} \div m^{-2}$ **67.** $(x^2 y^3)^4$ **68.** $(y^3)^0$

69. $(x^{-2} y^3)^{-2}$ **70.** $\left(\dfrac{m^3}{n^2}\right)^4$ **71.** $\left(\dfrac{x^{-3}}{y^{-2}}\right)^{-2}$

Simplify.

72. $(-2x^2 y^3)(-5x^3 y^4)$ **73.** $(-18a^3 b^2) \div (-2a^2 b)$

74. $3m^{-2} \times 4m^6$ **75.** $10x^{-2} \div (-2x^{-3})$

76. $(-2a^5 b^3)^2$ **77.** $(-3m^{-3} n^{-1})^{-3}$

78. $\left(\dfrac{3m^2}{2n^3}\right)^3$ **79.** $\left(\dfrac{-2x^{-3}}{3y^{-4}}\right)^{-2}$

80. $\dfrac{(3x^3 y)(6xy^4)}{-9xy^2}$ **81.** $\dfrac{3ab^4}{2a^3 b^2} \times \dfrac{12a^5 b}{15a^4 b}$

82. $\dfrac{(-2s^{-2} t)(5s^{-3} t^2)}{4s^2 t^{-3}}$ **83.** $\left(\dfrac{6a^{-2} b^{-3}}{2a^2 b^{-1}}\right)^{-2}$

Write in radical form.

84. $6^{\frac{1}{2}}$ **85.** $5^{-\frac{1}{2}}$ **86.** $7^{\frac{3}{5}}$ **87.** $10^{-\frac{4}{3}}$

Write using exponents.

88. $\sqrt[3]{-8}$ **89.** $\left(\sqrt[3]{m}\right)^5$ **90.** $\sqrt[3]{x^2}$ **91.** $\sqrt{\sqrt[5]{4a^4}}$

Evaluate.

92. $25^{\frac{1}{2}}$ **93.** $\left(\dfrac{1}{27}\right)^{\frac{1}{3}}$ **94.** $49^{-\frac{1}{2}}$

95. $1^{-\frac{1}{4}}$ **96.** $0.09^{0.5}$ **97.** $(-8)^{-\frac{1}{3}}$

98. $0.008^{-\frac{1}{3}}$ **99.** $27^{\frac{2}{3}}$ **100.** $-16^{-\frac{3}{4}}$

101. $\left(\dfrac{81}{16}\right)^{\frac{5}{4}}$ **102.** $\left(\dfrac{1}{9}\right)^{2.5}$ **103.** $\left(\dfrac{27}{125}\right)^{-\frac{2}{3}}$

104. $(-32)^{\frac{4}{5}}$ **105.** $(-8^{-1})^{-\frac{1}{3}}$ **106.** $\sqrt{\sqrt{16}}$

Simplify. Express each answer using exponents, if necessary.

107. $\sqrt{\sqrt[3]{y^4}}$ **108.** $\sqrt{\sqrt{81m^8}}$ **109.** $\sqrt[3]{-8x}$

110. $(\sqrt{x^3})(\sqrt{x})$ **111.** $(\sqrt[3]{-64})x$ **112.** $\sqrt[3]{-64x}$

113. Measurement The height and the base of a triangle each measure $2^{\frac{3}{2}}$. What is the area of the triangle?

Exploring Math

Crossing a Desert in a Jeep

1. A desert is 800 km wide. Terry must drive a jeep across it on a road that joins A and B.

The jeep can carry a total of 50 L of gasoline in its regular tank and in extra cans strapped to the side of the jeep. The jeep can travel 10 km on one litre of gasoline.

The only supply of gasoline is at the starting point, A. Since the desert is 800 km wide, the jeep would need 80 L of gasoline to cross it directly from A to B. Since the jeep can carry only 50 L of gasoline, Terry must place supplies of gasoline at points on the road. For example, Terry could leave point A with 50 L of gasoline and drive 200 km to C. The jeep would use 20 L of gasoline to get to C.

Terry could leave 10 L of gasoline at C, return to A using the last 20 L, and then refuel.

a) For Terry to use the minimum amount of gasoline to cross the desert, where along the road, and in what amounts, should Terry leave supplies of gasoline?

b) What is the minimum amount of gasoline needed?

2. Suppose the desert was 1000 km wide.

a) Where along the road, and in what amounts, should Terry leave gasoline to use the minimum amount of fuel to cross the desert?

b) What is the minimum amount of gasoline needed?

Chapter Check

1. Arrange the following in order from greatest to least.

$$|-1| \quad |2| \quad \left|\frac{3}{2}-\frac{9}{4}\right| \quad |-3|-|-4| \quad |2-5|$$

Given that x is a real number, graph each of the following on a number line.

2. $x \le 5$ **3.** $x > |-3|$

4. $-2 \le x < 4$ **5.** $0 \le x \le 6$

Simplify.

6. $\sqrt{50}$ **7.** $\sqrt{44}$ **8.** $\sqrt{80}$

Write as an entire radical.

9. $5\sqrt{2}$ **10.** $3\sqrt{5}$ **11.** $4\sqrt{100}$

Simplify.

12. $\sqrt{7} \times \sqrt{5}$ **13.** $2\sqrt{3} \times \sqrt{6}$

14. $\dfrac{\sqrt{30}}{\sqrt{6}}$ **15.** $5\sqrt{10} \times 3\sqrt{2}$

16. Without using a calculator, arrange the following in order from least to greatest.

$$3\sqrt{6}, \ 5\sqrt{2}, \ 2\sqrt{15}, \ 4\sqrt{3}$$

Estimate. Then, find an approximate value, to the nearest hundredth.

17. $\sqrt{75}$ **18.** $\sqrt{6030}$

19. $\sqrt{0.77}$ **20.** $\sqrt{0.0045}$

Estimate. Then, find an approximate value, to the nearest hundredth.

21. $2\sqrt{90} - \sqrt{30}$ **22.** $\dfrac{\sqrt{40}}{\sqrt{12}}$

23. Use decimal approximations to arrange the following in order from least to greatest.

$$6\sqrt{3}, \ 3\sqrt{10}, \ 7\sqrt{2}, \ \sqrt{93}$$

Simplify.

24. $\sqrt{48} - \sqrt{27} + \sqrt{12}$

25. $3\sqrt{40} + 5\sqrt{28} - \sqrt{63} - 2\sqrt{90}$

Expand and simplify.

26. $\sqrt{6}\left(3\sqrt{2} + 2\sqrt{8}\right)$ **27.** $\left(2 - \sqrt{3}\right)\left(1 + 3\sqrt{3}\right)$

28. $\left(3\sqrt{2} - 2\right)^2$ **29.** $\left(4 + \sqrt{5}\right)\left(4 - \sqrt{5}\right)$

Simplify.

30. $\dfrac{2}{\sqrt{7}}$ **31.** $\dfrac{3}{\sqrt{3} - 4}$ **32.** $\dfrac{5}{\sqrt{6} + \sqrt{3}}$

Simplify.

33. $2\sqrt[3]{24} + \sqrt[3]{81}$ **34.** $3\sqrt[3]{16} - \sqrt[3]{54}$

Evaluate.

35. 5^0 **36.** $(-2)^3$ **37.** 6^{-2}

38. -4^{-2} **39.** $\dfrac{1}{6^{-1}}$ **40.** $\dfrac{1}{(-5)^2}$

Simplify.

41. $y^{-2} \times y^6$ **42.** $a^6 \div a^{-4}$ **43.** $(y^4)^3$

44. $(5x^{-2}y^4)^{-2}$ **45.** $\left(\dfrac{m}{n}\right)^5$ **46.** $\left(\dfrac{s^{-2}}{t^3}\right)^{-3}$

Simplify.

47. $3a^3 \times 4a^2$ **48.** $30a^4b^2 \div (-5ab)$

49. $(-2m^{-3})(3m^2)$ **50.** $8x^{-4} \div (-2x^{-1})$

51. $(-3a^2b^5)^2$ **52.** $(2x^4y^{-2})^{-3}$

53. $\dfrac{10m^2n^{-2} \times 2m^{-1}n^4}{-4mn^{-3}}$ **54.** $\dfrac{(-4s^{-2}t^{-3})^{-1}}{-s^2t^{-1}}$

Write in radical form.

55. $13^{\frac{1}{2}}$ **56.** $6^{\frac{2}{3}}$ **57.** $3^{-\frac{5}{2}}$

Write using exponents.

58. $\sqrt[3]{5^2}$ **59.** $\sqrt{x^5}$ **60.** $\sqrt{\sqrt[4]{3t^7}}$

Evaluate.

61. $36^{\frac{1}{2}}$ **62.** $4^{-\frac{1}{2}}$ **63.** $-100^{-\frac{3}{2}}$ **64.** $32^{\frac{2}{5}}$

65. $81^{-\frac{3}{4}}$ **66.** $\sqrt{\sqrt{81}}$ **67.** $\left(\dfrac{8}{27}\right)^{\frac{1}{3}}$ **68.** $\left(\dfrac{8}{-27}\right)^{-\frac{2}{3}}$

Simplify. Express each answer using exponents.

69. $\sqrt{\sqrt[3]{a^4}}$ **70.** $\sqrt[4]{16x^2}$ **71.** $\sqrt[3]{-27a^6}$

72. Measurement A rectangle has side lengths of $3 + \sqrt{2}$ and $3 - \sqrt{2}$. Evaluate the area of the rectangle.

73. Board game The Japanese game of Tai Shogi is an expanded version of chess. Like chess, it is played on a square board covered with small squares. If each small square has a side length of 2 cm, the diagonal of the whole board measures $\sqrt{5000}$ cm. How many squares are on the board?

Using the Strategies

1. In how many different ways can you place the numbers 1, 2, 3, 4, 5, 6, 7, 8, and 9 into groups so that the sum of the numbers in each group is 15?

2. Copy the diagram.

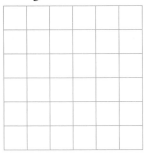

Place 6 circles in 6 different small squares, so that there is no more than one circle in each row, column, or diagonal.

3. Austin, Greenberg, Hill, Lo, and Pearson are the last names of five people who work on a cruise ship. Their jobs are captain, first officer, navigator, chef, and engineer. Each has a favourite port. The ports are Nassau, San Juan, Havana, Miami, and Puerto Plata. From the clues determine each person's job on the ship and each person's favourite port.
• One of Lo and Hill likes Nassau, and the other is the navigator.
• The first officer is not Pearson or the one who likes San Juan.
• The one who likes Miami is not Lo or Pearson, and none of these three is the chef.
• Of the one who likes Miami and the one who likes Havana, one is called Greenberg, and the other is the engineer.
• The one who likes San Juan is not named Austin or Pearson, and none of these three is the captain or the chef.

4. Find the product.

$$\left(1-\frac{1}{2}\right)\left(1-\frac{1}{3}\right)\left(1-\frac{1}{4}\right)\dots\left(1-\frac{1}{57}\right)$$

5. About how many cars could you park on the grounds surrounding your school?

6. Draw a square. Rotate the square 45° about one corner and draw the new square. Then, rotate the square another 45° in the same direction about the same corner and draw the next square. Continue the procedure until the square is returned to its original position.
a) How many squares are in the final diagram?
b) How many triangles are there?

7. The diagram shows one way to divide the grid into two congruent pieces using line segments that connect grid points. Find the other 12 ways.

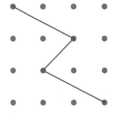

D A T A   B A N K

1. Sarah and Samantha Lennon flew from Winnipeg to Chicago. In Chicago, they rented a car and planned to visit the cities Atlanta, St. Louis, Cincinnati, Philadelphia, New York, Pittsburgh, New Orleans, and Houston, but not necessarily in that order. From the last city they visited, they planned to drive to Montreal and fly to Winnipeg.
a) Plan the driving route from Chicago to Montreal, passing through the eight cities, so that the driving distance is as short as possible. What is the total driving distance?
b) Including the flying distances, what is the total distance?
c) If Sarah and Samantha averaged 80 km/h driving and 800 km/h flying, how many hours did they spend travelling?

2. When a spectacular chinook hit Pincher Creek, Alberta, the air temperature changed from −19°C to +3°C in an hour, and the wind speed increased from calm to 68 km/h.
a) State the increase in the air temperature.
b) Estimate the change in the wind chill temperature.

CHAPTER 2

Number Patterns

Each year, the members of the Baseball Writers' Association select one pitcher from the American League and one from the National League to receive the Cy Young Award. Pat Hentgen was the first Toronto Blue Jay to win the award.

The table shows the results of the voting for the American League in the year that Pat Hentgen won. Different numbers of points are awarded for first, second, and third place votes.

1. How many points are given for each first, second, and third place vote?

2. Assume that a first place vote must get more points than a second, and a second place vote more points than a third. Find two ways in which points could be allocated to each vote so that Andy Pettite would win.

3. What do you think is the best way to decide the winner? Explain why.

4. Votes were cast by two baseball writers in each American League city. How many writers voted?

Player	Club	Votes			Total Points
		1st	2nd	3rd	
Pat Hentgen	Toronto	16	9	3	110
Andy Pettite	New York	11	16	1	104
Mariano Rivera	New York	1	1	10	18
Charles Nagy	Cleveland		1	9	12
Mike Mussina	Baltimore		1	2	5
Alex Fernandez	Chicago			1	1
Roberto Hernandez	Chicago			1	1
Ken Hill	Texas			1	1

Finding Patterns

1 **Patterns in Numbers and Letters**

1. Describe the pattern and write the next three terms.
a) 4, 7, 10, 13, ...
b) 3, 6, 12, 24, ...
c) 1, 2, 4, 7, 11, 16, ...
d) 4, 12, 6, 18, 9, ...
e) 2, 3, 9, 10, 30, ...
f) 3, 4, 7, 11, 18, 29, ...
g) 9, 27, 45, 63, ...

2. Describe the pattern and write the next three terms.
a) A, D, G, J, ...
b) A, D, B, E, C, F, D, ...
c) A, C, D, F, G, I, J, ...
d) 1, Z, 4, Y, 8, X, 13, ...

3. Describe the pattern in the numbers and draw the next two diagrams.

a)

12	2
11	4

27	7
16	9

48	14
23	16

75	23
32	25

b)

1	4
1	6
1	2
1	5

1	8
2	0
1	6
1	9

2	2
2	4
2	0
2	3

2	6
2	8
2	4
2	7

4. Determine the pattern and find the missing number.

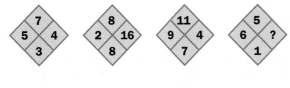

2 **Patterns in Diagrams**

1.

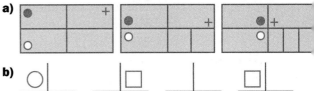

a) Draw the next three squares in the sequence.
b) Describe the pattern in words.
c) Draw the 75th square.
d) Draw the 153rd square.

2. Draw the next two diagrams in each sequence.

a)

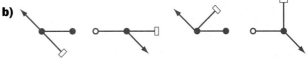

b)

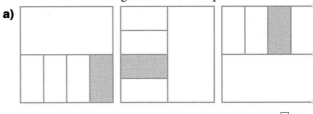

3. Draw the next diagram in each sequence.

a)

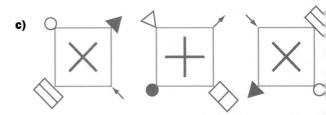

b)

c)

Exploring Sequences

1 Trapezoids

The trapezoids are made from toothpicks.   Diagram   1   2   3

1. How many toothpicks are in the fourth diagram? the fifth diagram?
2. Describe the pattern in words.
3. Write an expression that represents the number of toothpicks needed for the *n*th diagram in terms of *n*.
4. Using the expression from question 3, how many toothpicks are there in the 50th diagram? the 80th diagram?

2 I-Shapes

The I-shapes are made from asterisks.   Diagram   1   2   3

1. How many asterisks are in the fourth diagram? the fifth diagram?
2. Describe the pattern in words.
3. Write an expression that represents the number of asterisks in the *n*th I-shape in terms of *n*.
4. Using the expression from question 3, how many asterisks are there in the 65th I-shape? the 100th I-shape?

3 Triangles

Diagram 1 has   If you count only the small triangles,   Diagram 3 has
1 triangle.   Diagram 2 has 4 triangles.   16 small triangles.

Diagram 1   Diagram 2   Diagram 3

1. How many small triangles are there in the fourth diagram? the fifth diagram?
2. Describe the pattern in words.
3. How many small triangles are there in the 10th diagram?
4. Write an expression that represents the number of small triangles in the *n*th diagram in terms of *n*.

4 Generating a Sequence

A sequence can be generated as follows.
• Start with any positive integer.
• If the last number is odd, triple it and add 1 to find the next number.
• If the last number is even, divide it by 2 to find the next number.

1. Starting with the number 7 gives the sequence 7, 22, 11, 34, 17, ...
Continue the sequence until you find a repeating pattern of 3 numbers. What are they?

2. Repeat question 1, starting with the number 25.

3. Repeat question 1, starting with any other positive integer. Do you always get the same result?

2.3 Sequences

A number sequence is a set of numbers arranged in an order, with a first term, t_1, a second term, t_2, a third term, t_3, and so on. A sequence may stop at some number or may continue indefinitely. The following are examples of sequences.

5, 7, 9, 11, 13, ...
2, 6, 18, 54, ...
80, 40, 20, 10, ...
28, 57, 18, 94, 55, ...

A sequence does not need to follow a pattern. The sequence 28, 57, 18, 94, 55 was found using the random number generator on a graphing calculator. The sequences studied in this chapter do follow a pattern.

```
rand
        .2857189455
```

Explore: Use the Data

Spirals can be found in many places. The most sensational spirals are the galaxies in the universe. Examples of two common spirals are shown.

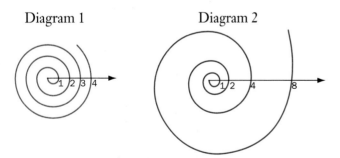

Diagram 1   Diagram 2

The spiral in diagram 1 was defined by Archimedes and is called an **Archimedean spiral**. The spiral in diagram 2 is called an **exponential spiral**. The distances of successive loops from the centre of each spiral are marked on the diagrams.

For each spiral, the distances from the centre form a sequence.

In the first sequence	In the second sequence
$t_1 = 1$	$t_1 = 1$
$t_2 = 2$	$t_2 = 2$
$t_3 = 3$	$t_3 = 4$
$t_4 = 4$	$t_4 = 8$

Inquire

1. How are the terms in the first sequence related?

2. Find t_5, t_6, and t_7 for the first sequence.

3. How are the terms in the second sequence related?

4. Find t_5, t_6, and t_7 for the second sequence.

5. A roll of tape is an example of an Archimedean spiral. State two other examples.

6. Our galaxy is an example of an exponential spiral. Use your research skills to find another example.

7. Explain why spirals in which the distances from the centre follow such patterns as 1, 2, 4, 8, ... or 1, 3, 9, 27, ... are called exponential spirals.

Sometimes a pattern can lead to a general rule for finding the terms of a sequence. The rule is called the general term, t_n.

For the sequence 1, 3, 5, 7, ...
$t_n = 2n - 1$
$t_1 = 2(1) - 1 = 1$
$t_2 = 2(2) - 1 = 3$
$t_3 = 2(3) - 1 = 5$
$t_4 = 2(4) - 1 = 7$

For the sequence 1, 3, 9, 27, ...
$t_n = 3^{n-1}$
$t_1 = 3^{1-1} = 3^0 = 1$
$t_2 = 3^{2-1} = 3^1 = 3$
$t_3 = 3^{3-1} = 3^2 = 9$
$t_4 = 3^{4-1} = 3^3 = 27$

In both sequences, the value of a term depends on the value of n.

Example 1   Generating a Sequence From a General Term
Given the general term, find the first 5 terms of the sequence, and graph t_n versus n.
a) $t_n = 3n - 2$ **b)** $t_n = n^2 + 1$

Solution 1
a) $t_n = 3n - 2$
$t_1 = 3(1) - 2 = 1$
$t_2 = 3(2) - 2 = 4$
$t_3 = 3(3) - 2 = 7$
$t_4 = 3(4) - 2 = 10$
$t_5 = 3(5) - 2 = 13$

b) $t_n = n^2 + 1$
$t_1 = (1)^2 + 1 = 2$
$t_2 = (2)^2 + 1 = 5$
$t_3 = (3)^2 + 1 = 10$
$t_4 = (4)^2 + 1 = 17$
$t_5 = (5)^2 + 1 = 26$

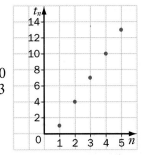

Solution 2

Use a graphing calculator.

a)

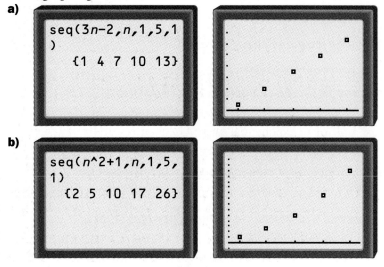

b)

It is sometimes convenient to describe the pattern of a sequence in terms of getting from one term to the next, rather than stating a general term. The pattern can be written as a **recursion formula**, which shows how to find each term from the term(s) before it.

Example 2   Generating a Sequence Using a Recursion Formula

Write the first 5 terms of the sequence determined by the following recursion formula.

$t_1 = 2$

$t_n = t_{n-1} + 3$

Solution

From the first line of the recursion formula, the first term is 2.
From the second line of the recursion formula,

$$t_2 = t_1 + 3 \qquad t_3 = t_2 + 3$$
$$= 2 + 3 \qquad\quad = 5 + 3$$
$$= 5 \qquad\qquad\ = 8$$
$$t_4 = t_3 + 3 \qquad t_5 = t_4 + 3$$
$$= 8 + 3 \qquad\quad = 11 + 3$$
$$= 11 \qquad\qquad = 14$$

```
2
                    2
Ans+3
                    5
                    8
                   11
                   14
```

So, the first 5 terms of the sequence are 2, 5, 8, 11, and 14.

Practice

Given the general term, state the first 5 terms of the sequence and graph the solution.

1. $t_n = 3n$ **2.** $t_n = 2n + 4$

3. $t_n = 5 - 2n$ **4.** $t_n = 10 - n$

5. $t_n = 2^n$ **6.** $t_n = n^2 - 1$

Find a general term that determines each sequence. Then, list the next 3 terms.

7. 5, 10, 15, 20, ... **8.** 2, 3, 4, 5, ...

9. 6, 5, 4, 3, ... **10.** 1, 4, 9, 16, ...

11. 2, 4, 6, 8, 10, ... **12.** −3, −6, −9, −12, ...

13. −1, 0, 1, 2, 3, ... **14.** $x, 2x, 3x, 4x, ...$

15. $1, 1 + d, 1 + 2d, 1 + 3d, ...$

List the first 4 terms of the sequence determined by each of the following.

16. $t_n = 3(n - 1)$

17. $t_n = (n - 1)^2$

18. $t_n = \dfrac{1}{n}$

19. $t_n = \dfrac{n + 1}{n}$

20. $t_n = (n + 1)(n - 1)$

21. $t_n = (-1)^n$

22. $t_n = 2^{n - 1}$

23. $t_n = 2^n - 1$

24. $t_n = \dfrac{n - 1}{n + 1}$

25. $t_n = (-1)^{n-1}$

26. $t_n = \dfrac{1}{3^n}$

27. $t_n = \dfrac{1}{2^n} + 1$

Find the indicated terms.

28. $t_n = 2n + 7$; t_6 and t_{15}

29. $t_n = 8n - 5$; t_9 and t_{11}

30. $t_n = 12 + 5n$; t_3 and t_{10}

31. $t_n = 9 - 4n$; t_4 and t_8

32. $t_n = n^2 + 4$; t_3 and t_7

33. $t_n = \dfrac{n - 2}{2}$; t_4 and t_{12}

34. $t_n = (n - 3)^2$; t_2 and t_{15}

Write the first 5 terms determined by each recursion formula.

35. $t_1 = 4$; $t_n = t_{n-1} + 3$

36. $t_1 = -1$; $t_n = 2t_{n-1} + 5$

37. $t_1 = 3$; $t_n = t_{n-1} - 2$

38. $t_1 = 6$; $t_n = t_{n-1} + 2n$

39. $t_1 = 5$; $t_{n+1} = t_n - 1$

40. $t_1 = 1$, $t_2 = 1$; $t_{n+2} = t_{n+1} + t_n$

Applications and Problem Solving

List the first 4 terms of each sequence. Then, find the general term for each.

41.

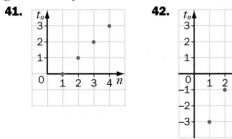

42.

43.

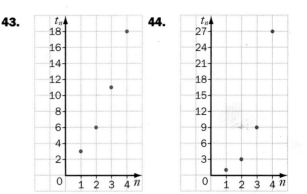

44.

45. Whooping cranes A Canadian zoo started with 16 whooping cranes and planned to hatch 20 chicks a year. A few birds were kept for breeding and the rest released into the wild. At the end of last year, a total of 136 cranes had spent some time in the zoo. If the pattern continues, how many cranes will have spent time at the zoo at the end of this year? at the end of next year?

46. Gold production Canada ranks fifth in the world in the production of gold. By the end of 1992, one of Canada's mines had produced 165 t of gold since it opened.
a) If the annual production averaged 4.5 t, what was the total production at the end of 1993? 1994?
b) What should be the total production by the end of 2010? 2020?

47. Hairstyles Human scalp hair grows at a rate of about 0.25 cm/week. Tania has decided to wear her hair longer. It is now 10 cm long.
a) How long will her hair be in 1 week? 2 weeks? 6 weeks?
b) How many weeks will it take for Tania's hair to grow from 10 cm to 15 cm in length?

48. Fitness Joshua exercised by cycling for 30 min then taking a brisk 15-min walk. The bike ride used 1000 kJ of energy. The walk used energy at a rate of 20 kJ/min.
a) What was the total amount of energy used for cycling and walking after Joshua had walked for 1 min? 5 min? 15 min?
b) Write the general term that determines the sequence.

49. Stadium seating In one section of a stadium, there are 30 seats in the first row, 32 in the second row, 34 in the third row, and so on. There are 60 rows.
a) Write the first 5 terms of the sequence.
b) Write the general term that determines the sequence.
c) How many seats are there in the 60th row?

50. Fibonacci sequence The terms of the Fibonacci sequence can be defined by the following recursion formula.
$$t_1 = t_2 = 1$$
$$t_n = t_{n-1} + t_{n-2}$$
Write the first 8 terms of the Fibonacci sequence.

51. Astronomy The Earth is about 150 000 000 km from the sun. Astronomers call this distance one astronomical unit (1 AU). The distances of other planets from the sun can be expressed in astronomical units. For example, if a planet were twice as far from the sun as the Earth, the planet's distance from the sun would be 2 AU. In 1776, the astronomer Johann Bode found a sequence that he thought could determine each planet's distance from the sun in astronomical units.

Planet	Bode's Distance (AU)	Actual Distance (AU)
Mercury	$\frac{(0+4)}{10}$	0.387
Venus	$\frac{(3+4)}{10}$	0.723
Earth	$\frac{(6+4)}{10}$	1
Mars	$\frac{(12+4)}{10}$	1.524
Minor Planets	$\frac{(24+4)}{10}$	2.3 to 3.3
Jupiter		5.203
Saturn		9.555
Uranus		19.22
Neptune		30.11
Pluto		39.84

a) Describe Bode's sequence in words.
b) Copy Bode's sequence and continue the sequence for the last 5 planets.
c) Calculate each planet's distance from the sun using Bode's sequence.
d) Compare the results from part c) with the actual distances. Name the first planet for which Bode's distance is not close to the actual distance.

52. a) Write 3 difference sequences of your own.
b) Describe each sequence in words.
c) Write the general term for each sequence.

53. Car depreciation A car dealership has determined that a car costing $60 000.00 new depreciates 20% in value each year.
a) What is the value of the car at the end of the first year? the second year?
b) Write a formula to determine the value of the car at the end of the nth year.
c) After how many years will the car be worth about $10 000.00?

54. A sequence begins 1, 2, 4, ... Can you predict the fourth term? Explain.

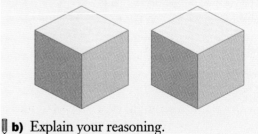

2.4 Use a Table or Spreadsheet

Much of the information you see in newspapers and magazines is displayed in tables. A table or an organized list is an efficient way to arrange information and solve problems.

You are a member of an archaeological team working at a site in North Africa. The site, S, is 80 km north of a road that runs east-west. The closest point on the road is 100 km from a town, T, where you buy supplies. Your all-terrain vehicle can average 20 km/h across the sand between the site and the road. Your vehicle can travel 40 km/h on the road. Where should you get on the road so that the travel time to the town is as short as possible?

Understand the Problem

1. What information are you given?
2. What are you asked to find?

Think of a Plan

If you drive straight to the road at point A, you will take 4 h to reach the road. It will then take you 2.5 h on the road to get to the town. The total time is 6.5 h.
Suppose you aim for a point closer to the town and get on the road there. Will the total time to the town be shorter?

Suppose you aim for point B, 10 km from A. The distance SB can be calculated using the Pythagorean Theorem.

$$SB^2 = SA^2 + AB^2$$
$$= 80^2 + 10^2$$
$$= 6400 + 100$$
$$= 6500$$
$$SB = \sqrt{6500} \text{ or } 80.62, \text{ to the nearest hundredth}$$

Your travel time on the sand for 80.62 km from S to B is 80.62 ÷ 20 or 4.03 h.
Your travel time on the road is 90 ÷ 40 or 2.25 h.
The total time is 4.03 + 2.25 or 6.28 h.

Carry Out the Plan

Complete a table for the other points 10 km apart on the road.

Aim Point	Sand Distance (km)	Sand Time (h)	Road Distance (km)	Road Time (h)	Total Time (h)
A	80	4	100	2.5	6.5
B	80.62	4.03	90	2.25	6.28
C	82.46	4.12	80	2	6.12
D	85.44	4.27	70	1.75	6.02
E	89.44	4.47	60	1.5	5.97
F	94.34	4.72	50	1.25	5.97
G	100	5	40	1	6

Look Back

Beyond point F, the total time is getting longer. The shortest time is 5.97 h, if you drive to point E or point F.

Does the answer seem reasonable?

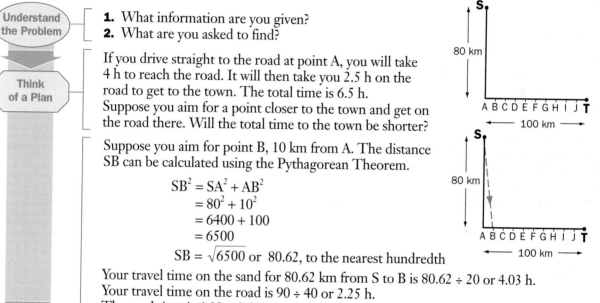

Another way to solve a problem of this type is to use a spreadsheet.

Set up a spreadsheet to calculate the shortest time.

	A	B	C	D	E	F	G
1	Aim	Distance From	Sand	Sand	Road	Road	Total
2	Point	Point A	Distance	Time	Distance	Time	Time
3		(km)	(km)	(h)	(km)	(h)	(h)
4	A	0	@sqrt(80^2+B4^2)	C4/20	100 – B4	E4/40	D4+F4
5	B	+B4+10					
6	C						
7	D						
8	E						
9	F						
10	G						

Use a Table or Spreadsheet
1. Organize the given information in a table or set up a spreadsheet.
2. Complete the table with the results of your calculations or use the spreadsheet to complete the calculations.
3. Find the answer from the table or spreadsheet.
4. Check that your answer is reasonable.

Applications and Problem Solving

1. As part of a survival course, Sandra has to run from point P to point Q. Point P is in the bush, 4 km from a road. The nearest point on the road is 10 km from point Q.

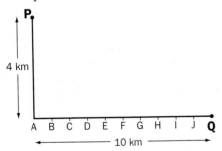

Sandra estimates that she can average 6 km/h through the bush and 10 km/h on the road. At what point should she meet the road in order to reach point Q in the shortest possible time? Round distances to the nearest hundredth of a kilometre, and times to the nearest hundredth of an hour.

2. What is the total number of digits in all the whole numbers from 1 to 1999?

3. How many whole numbers from 1 to 500 begin or end in a 2?

4. In Canadian football, a team can score points in the following ways.

Single: 1 point
Safety: 2 points
Field Goal: 3 points
Touchdown: 6 points
Touchdown plus Convert: 7 points
Touchdown plus Conversion: 8 points

Make an organized list to determine the number of different ways a team could score 21 points without kicking a single.

5. A rectangle has an area of 72 m². The width and length are whole numbers of metres.
a) What are the possible dimensions of the rectangle?
b) Which dimensions give the greatest perimeter?
c) Which dimensions give the smallest perimeter?

6. Two whole numbers have a product of 240. The sum of the two numbers is an odd number. What are the possible pairs of whole numbers?

7. How many isosceles triangles, with side lengths that are whole numbers of centimetres, have a perimeter of 20 cm?

70

8. A vending machine sells cold drinks. Each drink costs $1.00. The machine accepts only exact change. How many combinations of coins must the machine be programmed to accept if it accepts only
a) dollars and quarters?
b) dollars, quarters, and dimes?
c) dollars, quarters, dimes, and nickels?

9. The table shows part of the statistics for a hockey league after each team has played each of the other teams once. A win is worth 2 points, a tie 1 point, and a loss 0 points. Copy and complete the table.

Team	Games Played	Wins	Losses	Ties	Points
Aces		1			3
Bears			1	1	
Lions					
Pintos				2	2

10. The manager of a stock-car-racing facility can sell 2000 tickets each evening when the price per ticket is $32. A researcher has informed the manager that, for every $1 decrease in the ticket price, 100 more tickets will be sold. Find the ticket price that will give the greatest revenue from ticket sales.

11. Palindromes are words or numbers that read the same forward and backward. The word *noon* and the number 3223 are examples of palindromes. How many palindromes between 100 and 1000 are perfect squares?

12. The regions of the dart target have point values of 2, 4, and 6, as shown. How many different point totals are possible, if 3 darts are thrown and each dart lands on the target?

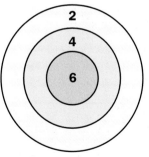

13. The sum of the digits in the number 629 is 6 + 2 + 9 or 17. How many whole numbers between 100 and 1000 have digits with a sum of 20?

14. Polyominoes are made from squares joined along entire sides. If polyominoes are made from squares with side lengths of 1 cm, how many different polyominoes can have a perimeter of 12 cm?

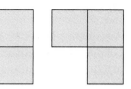

15. Céline and Ann-Marie ran around the same track in the same direction from the same starting point. Céline began running 6 min before Ann-Marie. Céline ran at 10 km/h, and Ann-Marie ran at 12 km/h. They ran until they had covered the same distance.
a) How far did they run?
b) For how many minutes did Céline run?

16. Suzanne and Marcus are both plumbers. Suzanne charges $55 for a service call, plus $35/h. Marcus charges $43 for a service call, plus $38/h. One day, they each made two service calls. If they earned the same amount of money that day, for how many hours did each plumber work?

17. Write a problem that can be solved using a table or organized list, or a spreadsheet. Have a classmate solve your problem.

NUMBER POWER

In a magic square, the sum of the numbers in each row, column, and diagonal is the same. The following are two special magic squares, which have been partially completed.

	twenty-two	
twenty-eight	fifteen	
twelve		twenty-five

4		8
	7	3
	5	

a) Copy and complete each square.
b) Lee Sallow, who discovered the two squares, called them alphamagic partners. How are the squares related?

2.5 Arithmetic Sequences

A sequence like 3, 5, 7, 9, ..., where the difference between consecutive terms is a constant, is called an **arithmetic sequence**. In an arithmetic sequence, the first term, t_1, is denoted by the letter a. Each term after the first is found by adding a constant, called the **common difference**, d, to the preceding term.

Explore: Complete the Table

In places, the walls of the Grand Canyon are over 1.5 km high. An object dropped from this height falls 5 m in the first second, 15 m in the second, 25 m in the third, 35 m in the fourth, and so on. The numbers 5, 15, 25, and 35 form an arithmetic sequence.

0 s	⎞
1 s	⎠ 5 m
2 s	⎠ 15 m
3 s	⎠ 25 m
4 s	⎠ 35 m

Copy and complete the table for this sequence.

	Term				
	t_1	t_2	t_3	t_4	t_5
Sequence	5	15	25	35	45
Sequence Expressed Using 5 and 10	5	5 + 1(10)			
Sequence Expressed Using a and d	a	a + 1(d)	a + 2(d)	a + 3(d)	a + 4(d)

Inquire

1. What are the values of a and d for this sequence?

2. When you write an expression for a term using the letters a and d, you are writing a formula for the term. What is the formula for t_5? t_8? t_9?

3. Evaluate t_8 and t_9.

4. Before the object hits the ground, terms after t_9 are all equal in value to t_9. Explain why.

The general arithmetic sequence is
$a, a + d, a + 2d, a + 3d, ...$
where a is the first term and d is the common difference.
$t_1 = a$
$t_2 = a + d$
$t_3 = a + 2d$
.
.
.
$t_n = a + (n - 1)d$, where n is a positive integer

Example 1 Finding Terms

Find the general term, t_n, and t_{19} for the arithmetic sequence 8, 12, 16, ...

Solution
For the given sequence, $a = 8$ and $d = 4$.

$$t_n = a + (n - 1)d$$

Substitute known values: $\quad = 8 + (n - 1)4$

Expand: $\quad = 8 + 4n - 4$

Simplify: $\quad = 4n + 4$

Three ways to find t_{19} are as follows.

Method 1	*Method 2*	*Method 3*
$t_n = a + (n - 1)d$	$t_n = 4n + 4$	Using a graphing calculator.
$t_{19} = a + (19 - 1)d$	$t_{19} = 4(19) + 4$	
$\quad = a + 18d$	$\quad = 76 + 4$	
$\quad = 8 + 18(4)$	$\quad = 80$	
$\quad = 8 + 72$		
$\quad = 80$		

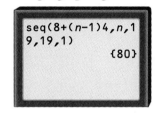

```
seq(8+(n-1)4,n,1
9,19,1)
                {80}
```

So, $t_n = 4n + 4$ and $t_{19} = 80$.

Example 2 Finding the Number of Terms

How many terms are there in the following sequence?
$-3, 2, 7, ..., 152$

Solution
For the given sequence, $a = -3$, $d = 5$, and $t_n = 152$.

Substitute the known values in the formula for the general term and solve for n.

$$t_n = a + (n - 1)d$$

Substitute known values: $\quad 152 = -3 + (n - 1)5$

Expand: $\quad 152 = -3 + 5n - 5$

Simplify: $\quad 152 = 5n - 8$

Add 8 to both sides: $\quad 152 + 8 = 5n - 8 + 8$

$$160 = 5n$$

Divide both sides by 5: $\quad \dfrac{160}{5} = \dfrac{5n}{5}$

$$32 = n$$

The sequence has 32 terms.

The terms between two non-consecutive terms of an arithmetic sequence are called **arithmetic means**.
For example, in the sequence
$$7, 10, 13, 16, 19, 22$$
10, 13, 16, and 19 are the four arithmetic means between 7 and 22.

Example 3   Arithmetic Means

a) Find the three arithmetic means between 22 and 74.

b) Graph t_n versus n for the sequence.

Solution

a) The sequence has 5 terms.

$t_1 = a = 22$ and $t_5 = 74$

$$
\begin{array}{ccccc}
22 & ? & ? & ? & 74 \\
t_1 & t_2 & t_3 & t_4 & t_5
\end{array}
$$

	$t_n = a + (n-1)d$
Substitute 5 for n:	$t_5 = a + (5-1)d$
Simplify:	$t_5 = a + 4d$
Substitute for t_5 and a:	$74 = 22 + 4d$
Subtract 22 from both sides:	$74 - 22 = 22 + 4d - 22$
	$52 = 4d$
Divide both sides by 4:	$\dfrac{52}{4} = \dfrac{4d}{4}$
	$13 = d$

The arithmetic means are $22 + 13 = 35$
$$35 + 13 = 48$$
$$48 + 13 = 61$$

b) The ordered pairs in the form (n, t_n) for the sequence are $(1, 22), (2, 35), (3, 48), (4, 61)$, and $(5, 74)$. Graph t_n versus n with pencil and paper or a graphing calculator.

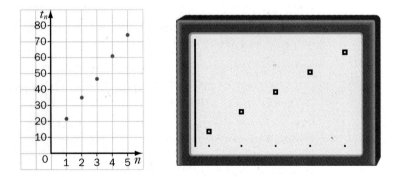

Practice

Find the next 3 terms of each arithmetic sequence.

1. $3, 7, 11, \dots$

2. $33, 27, 21, \dots$

3. $-23, -18, -13, \dots$

4. $25, 18, 11, \dots$

5. $5.8, 7.2, 8.6, \dots$

6. $\dfrac{3}{4}, \dfrac{5}{4}, \dfrac{7}{4}, \dots$

Determine which of the following sequences are arithmetic. If a sequence is arithmetic, write the values of a and d.

7. $5, 9, 13, 17, \dots$

8. $1, 6, 10, 15, 19, \dots$

9. $2, 4, 8, 16, 32, \dots$

10. $-1, -4, -7, -10, \dots$

11. $1, -1, 1, -1, 1, \dots$

12. $y, y^2, y^3, y^4, \dots$

13. $x, 2x, 3x, 4x, \dots$

14. $c, c + 2d, c + 4d, c + 6d, \dots$

Given the values of a and d, write the first 5 terms of each arithmetic sequence.

15. $a = 7, d = 2$
16. $a = 3, d = 4$
17. $a = -4, d = 6$
18. $a = 2, d = -3$
19. $a = -5, d = -8$
20. $a = 8, d = x$
21. $t_1 = 6, d = y + 1$
22. $t_1 = 3m, d = 1 - m$

Find the indicated terms for each arithmetic sequence.

23. 6, 8, 10, ...; t_{10} and t_{34}
24. 12, 16, 20, ...; t_{18} and t_{41}
25. 9, 16, 23, ...; t_9 and t_{100}
26. $-10, -7, -4, ...; t_{11}$ and t_{22}
27. $-4, -9, -14, ...; t_{18}$ and t_{66}
28. 5, $-1, -7, ...; t_n$ and t_{14}
29. 7, 10, 13, ...; t_n and t_{30}
30. 10, 8, 6, ...; t_n and t_{22}
31. $x, x + 4, x + 8, ...; t_n$ and t_{14}

Find the number of terms in each of the following arithmetic sequences.

32. 10, 15, 20, ..., 250
33. 1, 4, 7, ..., 121
34. 40, 38, 36, ..., -30
35. $-11, -7, -3, ..., 153$
36. $-2, -8, -14, ..., -206$
37. $x + 2, x + 9, x + 16, ..., x + 303$

Complete each arithmetic sequence by finding the arithmetic means.

38. 1, ■, 25
39. 14, ■, ■, 32
40. -3, ■, ■, -60
41. -1.5, ■, ■, ■, 4.5
42. 2, ■, ■, ■, ■, 107
43. $m + 40$, ■, ■, ■, $m + 4$

44. a) Find 4 arithmetic means between 30 and 70.
b) Graph t_n versus n for the sequence.

45. a) Find 5 arithmetic means between 11 and -7.
b) Graph t_n versus n for the sequence.

Applications and Problem Solving

46. The graph of an arithmetic sequence is shown.
a) What are the first 5 terms of the sequence?
b) What is t_{50}? t_{200}?

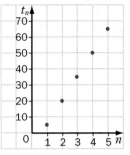

47. Copy and complete each arithmetic sequence. Graph t_n versus n for each sequence.
a) ■, ■, 14, ■, 26
b) ■, 3, ■, ■, -18

48. Olympic Games The modern summer Olympic Games were first held in Athens, Greece, in 1896. The games were to be held every 4 years, so the years of the games form an arithmetic sequence.
a) What are the values of a and d for this sequence?
b) In what years were the games cancelled and why?
c) What are the term numbers for the years the games were cancelled?
d) What is the term number for the next summer games?

49. Multiples How many multiples of 5 are there from 15 to 450, inclusive?

50. The 18th term of an arithmetic sequence is 262. The common difference is 15. What is the first term of the sequence?

51. Driving Portage la Prairie is 52 km west of Winnipeg by road. If you drove west from Portage la Prairie at 80 km/h, how far would you be by road from Winnipeg after
a) 1 h?   **b)** 2 h?   **c)** t hours?

52. Comets Comets approach the Earth at regular intervals. For example, Halley's Comet reaches its closest point to the Earth every 76 years. Comet Finlay is expected to reach its closest point to the Earth in 2009, 2037, and three times between these years. In which years between 2009 and 2037 will Comet Finlay reach its closest point to the Earth?

75

53. Electrician Amber works as an electrician. She charges $60 for each service call, plus an hourly rate. If she charges $420 for an 8-h service call
a) what is her hourly rate?
b) how much would she charge for a 5-h service call?

54. Salary Franco is the manager of a health club. He earns a salary of $25 000 a year, plus $200 for every membership he sells. What will he earn in a year if he sells 71 memberships?
72 memberships? 73 memberships?

55. Displaying merchandise Boxes are stacked in a store display in the shape of a triangle. The numbers of boxes in the rows form an arithmetic sequence. There are 41 boxes in the 3rd row from the bottom. There are 23 boxes in the 12th row from the bottom.
a) How many boxes are there in the first (bottom) row?
b) What is the general term for the sequence?
c) What is the maximum possible number of rows of boxes?

56. Wind sprints On the first day of practice, the soccer team ran eight 40-m wind sprints. On each day after the first, the number of wind sprints was increased by 2 from the day before.
a) What are the values of a and d for this sequence?
b) Write the general term for the sequence.
c) How many wind sprints did the team run on the 22nd day of practice? How many metres was this?

57. Pattern The U-shapes are made from asterisks.

Diagram   1   2   3

a) How many asterisks are in the 4th diagram?
b) What is the general term for the sequence in the numbers of asterisks?
c) How many asterisks are in the 25th diagram?
d) Which diagram contains 139 asterisks?

58. Scuba diving The pressure that a scuba diver experiences is the sum of the pressure of the atmosphere and the pressure of the water. The increase in pressure with depth under water follows an arithmetic sequence. If a diver experiences a pressure of 150 kPa (kilopascals) at a depth of 5 m and 180 kPa at a depth of 8 m,
a) what is the pressure at a depth of 6 m? 12 m?
b) what is the atmospheric pressure at the surface of the water that day?

59. The third term of an arithmetic sequence is 24 and the ninth term is 54. What is
a) the first term?   **b)** the general term?

60. The eighth term of an arithmetic sequence is 5.3 and the fourteenth term is 8.3. What is the fifth term?

61. Measurement The side lengths in a right triangle form an arithmetic sequence with a common difference of 2. What are the side lengths?

62. The sum of the first two terms of an arithmetic sequence is 16. The sum of the second and third terms is 28. What are the first three terms of the sequence?

63. Patterns a) How does the sum of the first and fourth terms of an arithmetic sequence compare with the sum of the second and third terms? Explain.
b) Find two other pairs of terms whose sums compare in the same way as the two pairs of terms in part a).

64. The first four terms of an arithmetic sequence are 4, 13, 22, and 31. Which of the following is a term of the sequence?
316   317   318   319   320

65. Algebra The first term of an arithmetic sequence is represented by $3x + 2y$. The third term is represented by $7x$. Write the expression that represents the second term.

66. Algebra Determine the value of x that makes each sequence arithmetic.
a) 2, 8, 14, $4x$, ...
b) 1, 3, 5, $2x - 1$, ...
c) $x - 2, x + 2, 5, 9$, ...
d) $x - 4, 6, x$, ...
e) $x + 8, 2x + 8, -x$, ...

Meteorology

In the spring of 1997, the thaw of a record winter snowfall caused the Red River to rise to unprecedented levels. About 29 000 Manitobans were forced to abandon their homes as the flooded river formed a 2000-km^2 lake stretching from the U.S. border to the outskirts of Winnipeg.

Predicting the weather and its effects is the work of meteorologists, most of whom are employed by the Atmospheric Environment Service (AES) of the federal government. Meteorologists prepare weather forecasts to meet the needs of various users, including the news media, and the aviation and shipping industries.

Satellite pictures, surface observations, and information from weather balloons are used in preparing forecasts. Regional computers provide up-to-date information on fluctuations, and computers at the national analysis centre in Montreal make long-range predictions.

Several Canadian universities offer meteorology programs. Meteorology students study mathematics and physics, as well as atmospheric science.

1 Precipitation in Canada

Each year about 5.5 trillion tonnes of precipitation, in the form of rain, snow, or hail, fall on Canada.

1. a) Estimate the mass of precipitation that falls on your province in a year. What are your assumptions?
b) Do you think that your estimate is reasonable? Explain.

2. About 36% of Canada's annual precipitation falls as snow. This is about one septillion or 1×10^{24} snowflakes.
a) What mass of snow, in tonnes, falls on Canada in a year?
b) What is the average mass of one snowflake, to the nearest thousandth of a milligram?

2 Greatest Precipitation

The greatest amount of precipitation to fall on Canada in a 24-h period was recorded at Ucluelet, British Columbia, where about 489.6 mm of rain fell.

1. If the rain fell at a steady rate, how much rain fell each hour?

2. a) Write the first three terms of an arithmetic sequence to represent the total amount of rain that fell by the end of hour one, the end of hour two, and the end of hour three.
b) What is the relationship between a and d for this sequence? Explain why.
c) What is t_{18} for this sequence?

3. If an arithmetic sequence represented the total amount of rain that fell by the start, not the end, of each hour, how would your answers to question 2 parts a) and c) change?

PROBLEM SOLVING

2.6 Solve a Simpler Problem

Many problems can be solved more easily if they are broken down into smaller problems.

Some Canadian cities, including Vancouver and Edmonton, have neighbourhoods in which the streets run east-west and north-south. Tony lives in Vancouver. His home is 3 blocks west and three blocks south of his school. In how many different ways can he walk to school, if he walks only east and north?

Understand the Problem

1. What information are you given?
2. What are you asked to find?
3. Do you need an exact or approximate answer?

Think of a Plan

Find how many possible routes there are for parts of the walk from his home to his school. Use the results to solve the complete problem.

Carry Out the Plan

Draw a grid that shows the positions of Tony's home, H, and his school, S. There is only 1 way for Tony to get to A, C, and F. Mark this information on the grid. Similarly, there is only 1 way for him to get to B, E, and K.

Since there is 1 way to A and 1 way to B, there are 2 ways to D.
Since there is 1 way to E and 2 ways to D, there are 3 ways to J.
Since there is 1 way to K and 3 ways to J, there are 4 ways to N.

Complete the grid to find the total number of ways in which Tony can walk from his home to his school.

The completed grid shows that Tony can walk to school in 20 different ways, if he walks only east and north.

Look Back

How could you check the answer?
Is there another way to solve the problem?

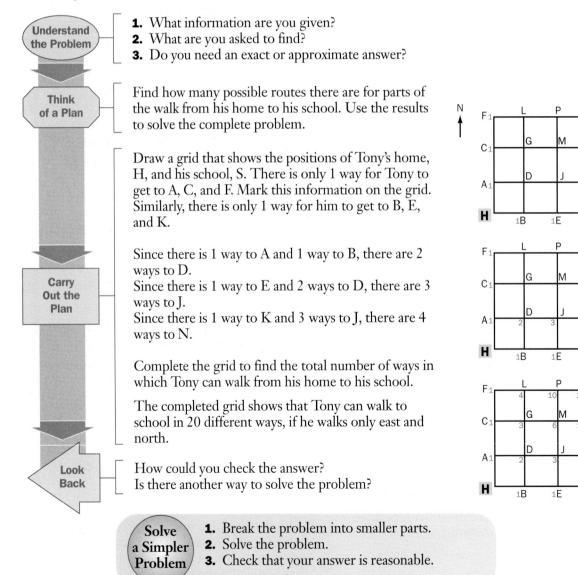

Solve a Simpler Problem

1. Break the problem into smaller parts.
2. Solve the problem.
3. Check that your answer is reasonable.

Applications and Problem Solving

1. Quadrilateral ABCD has vertices A(2, 3), B(8, 2), C(10, 7), and D(6, 9). Find the total area of quadrilateral ABCD. To make the problem simpler, the quadrilateral can be divided into four right triangles and a rectangle, as shown.

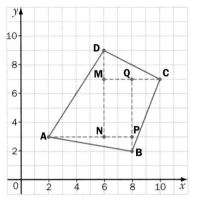

2. Quadrilateral WXYZ has vertices W(2, 3), X(3, −1), Y(1, −3), and Z(−4, 2). Find the area of the quadrilateral.

3. Pentagon DEFGH has vertices D(2, 3), E(7, 2), F(11, 5), G(9, 10), and H(4, 8). Find the area of the pentagon.

4. The streets in a town run north-south and east-west, as shown. How many ways are there for a taxi to get from A to B if the taxi travels only north and east?

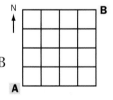

5. A town is separated into two parts by a bridge at M. How many routes are there for a bus to get from B to A if the bus travels only south and west?

6. One way to find the sum of the numbers from 1 to 10 is to use pairs of numbers.

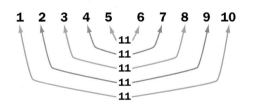

There are 5 pairs of numbers that add to 11. Since $11 \times 5 = 55$, the sum of the numbers is 55. Use this method to find the sum of
a) the whole numbers from 1 to 200
b) the whole numbers from 1 to 1000
c) the whole numbers from 1 to 1 000 000
d) the first twenty odd numbers

7. A vertical line through the number 2 divides the 2 into a maximum of 4 parts. Two vertical lines divide the 2 into a maximum of 7 parts. How many parts will 100 vertical lines produce?

8. A vertical line through the number 3 divides the 3 into a maximum of 5 parts. Two vertical lines divide the 3 into a maximum of 9 parts. How many parts will 100 vertical lines produce?

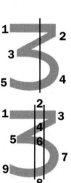

9. The first figure is made from one P-shape. The perimeter of the figure is 16 units. The second figure is made from two P-shapes. The perimeter of the figure is 26 units. The third figure is made from three P-shapes, and so on. Find the perimeter of the figure made from 75 P-shapes.

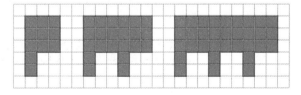
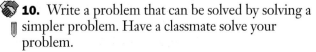

10. Write a problem that can be solved by solving a simpler problem. Have a classmate solve your problem.

2.7 Arithmetic Series

The following is an arithmetic sequence. 1, 3, 5, 7, 9, ...
A **series** is the sum of the terms of a sequence.

The series that corresponds to the sequence above is $1 + 3 + 5 + 7 + 9 + ...$
For this series, S_5 means the sum of the first 5 terms, so $S_5 = 25$.

Explore: Solve a Simpler Problem

In the version of pool known as "rotation," the aim is to pocket the balls
in numerical order from 1 to 15. Players receive points equal to the numbers
on the pocketed balls. The number of points on the table at the start
of the game is the sum of the numbers from 1 to 15.

To find this sum, consider
a simpler problem, the series
$1 + 2 + 3 + 4 + 5$.
a) What is the sum of
this series?
b) What is the average
of all the terms?
c) How many terms are there?

Inquire

1. How is the sum of the series related to
the number of terms and the average of all the
terms?

2. What is the average of the first and last terms?

3. How is the average of the first and last terms related
to the average of all the terms?

4. How is the sum of the series related to the number of terms
and the average of the first and last terms?

5. Use the result from question 4 to find the sum of the pool ball numbers.

When the mathematician Karl Gauss was eight years old, he used the following method to
find the sum of the natural numbers from 1 to 100.

Let S_{100} represent the sum of the first 100 natural numbers. Write out the series in order and
then in reverse order.

$$S_{100} = \quad 1 + \quad 2 + \quad 3 + ... + \ 98 + \ 99 + 100$$
$$S_{100} = 100 + \ 99 + \ 98 + ... + \quad 3 + \quad 2 + \quad 1$$

Add:
$$2S_{100} = 101 + 101 + 101 + ... + 101 + 101 + 101$$
$$= 100(101)$$

Divide both sides by 2:
$$\frac{2S_{100}}{2} = \frac{100(101)}{2}$$
$$S_{100} = 5050$$

The same method can be used to derive the formula for the sum of the general arithmetic series.

The general arithmetic sequence
$a, (a + d), (a + 2d), ..., (t_n - d), t_n$
has n terms, with the first term a and the last term t_n.
The corresponding series is

$$S_n = a + (a + d) + (a + 2d) + ... + (t_n - d) + t_n$$

Reverse: $\quad S_n = t_n + (t_n - d) + (t_n - 2d) + ... + (a + d) + a$

Add: $\quad 2S_n = (a + t_n) + (a + t_n) + (a + t_n) + ... + (a + t_n) + (a + t_n)$

$$2S_n = n(a + t_n)$$

Divide both sides by 2: $\quad S_n = \dfrac{n}{2}(a + t_n) \qquad (1)$

The (1) shows that we are naming the formula as "formula one."

Since $t_n = a + (n - 1)d$, substitute for t_n in formula (1).

$$S_n = \frac{n}{2}[a + a + (n - 1)d]$$

$$S_n = \frac{n}{2}[2a + (n - 1)d] \qquad (2)$$

Example 1 Sum of a Series Given First Terms

Find the sum of the first 60 terms of each series.
a) $5 + 8 + 11 + ...$
b) $-6 - 8 - 10 - ...$

Solution

a) $a = 5, d = 3$, and $n = 60$

$$S_n = \frac{n}{2}[2a + (n - 1)d]$$

$$= \frac{60}{2}[2(5) + (60 - 1)3]$$

$$= 30(10 + 177)$$

$$= 30(187)$$

$$= 5610$$

Estimate

$$30 \times 200 = 6000$$

b) $a = -6, d = -2$, and $n = 60$

$$S_n = \frac{n}{2}[2a + (n - 1)d]$$

$$= \frac{60}{2}[2(-6) + (60 - 1)(-2)]$$

$$= 30(-12 - 118)$$

$$= 30(-130)$$

$$= -3900$$

Example 2   Sum of a Series Given First and Last Terms

Find the sum of the arithmetic series.

$5 + 9 + 13 + \ldots + 201$

Solution

To use either formula (1), or formula (2), the number of terms is needed.
For the given series, $a = 5$, $d = 4$, and $t_n = 201$.

$$t_n = a + (n-1)d$$

Substitute known values: $\quad 201 = 5 + (n-1)4$

Expand: $\quad 201 = 5 + 4n - 4$

Simplify: $\quad 201 = 4n + 1$

Subtract 1 from both sides: $\quad 201 - 1 = 4n + 1 - 1$

$$200 = 4n$$

Divide both sides by 4: $\quad \dfrac{200}{4} = \dfrac{4n}{4}$

$$50 = n$$

Using formula (1)

$$S_n = \frac{n}{2}(a + t_n)$$

$$S_{50} = \frac{50}{2}(5 + 201)$$

$$= 25(206)$$

$$= 5150$$

Using formula (2)

$$S_n = \frac{n}{2}[2a + (n-1)d]$$

$$S_{50} = \frac{50}{2}[2(5) + (50-1)4]$$

$$= 25(10 + 196)$$

$$= 25(206)$$

$$= 5150$$

Estimate

$25 \times 200 = 5000$

Examples 1 and 2 could be solved using a graphing calculator.

For Example 2,

For Example 1, a graphing calculator could be used to evaluate the 60th term, then find the sum of the series.

Example 1a)   *Example 1b)*

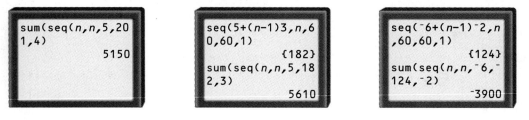

```
sum(seq(n,n,5,20
1,4)
              5150
```

```
seq(5+(n-1)3,n,6
0,60,1)
             {182}
sum(seq(n,n,5,18
2,3)
             5610
```

```
seq(-6+(n-1)-2,n
,60,60,1)
             {124}
sum(seq(n,n,-6,-
124,-2)
            -3900
```

Practice

Find the sum of the first 100 terms of each arithmetic series.

1. $1 + 5 + 9 + \ldots$

2. $2 + 5 + 8 + \ldots$

3. $10 + 8 + 6 + \ldots$

4. $0 - 2 - 4 - \ldots$

Find the indicated sum for each arithmetic series.

5. S_{10} for $2 + 4 + 6 + \ldots$

6. S_{20} for $10 + 15 + 20 + \ldots$

7. S_{30} for $-2 + 4 + 10 + \ldots$

8. S_{18} for $40 + 38 + 36 + \ldots$

9. S_{15} for $80 + 76 + 72 + \ldots$

Find the sum of each arithmetic series.
10. $4 + 6 + 8 + ... + 200$
11. $3 + 7 + 11 + ... + 479$
12. $100 + 90 + 80 + ... - 50$
13. $-8 - 5 - 2 + ... + 139$
14. $18 + 12 + 6 + ... - 216$
15. $-7 - 11 - 15 - ... - 171$

Given the first and last terms, find the sum of each arithmetic series.
16. $a = 6, t_6 = 11$
17. $a = 3, t_7 = 21$
18. $a = 5, t_{10} = -13$
19. $a = -4, t_{22} = -46$
20. $a = 4.5, t_{11} = 19.5$
21. $a = 20, t_{31} = 110$

Applications and Problem Solving

22. Find the sum of
a) the first 50 positive integers
b) the first 100 odd positive integers
c) the first 75 positive multiples of 3

23. Theatre seats A theatre has 30 seats in the front row, 31 seats in the second row, 32 seats in the third row, and so on. If there are 20 rows of seats, how many seats are there?

24. Salary increases Michelle is a marine biologist. She accepted a job that pays $46 850 in the first year and $56 650 in the eighth year. Her salary is an arithmetic sequence with eight terms.
a) What raise can Michelle expect each year?
b) What will her salary be in the sixth year?
c) In what year will her salary first exceed $50 000?
d) What is the total amount that Michelle will earn in the eight years?

25. Measurement The perimeter of a triangle is 30 units. The side lengths form an arithmetic sequence. If each side length is a whole number, what are the possible sets of side lengths?

26. Falling object An object dropped from the the same height as the Alberta Stock Exchange building in Calgary takes about 5 s to hit the ground. The object falls 4.9 m in the first second, 14.7 m in the second second, 24.5 m in the third second, and so on. How tall is the building?

27. Store display
The top 3 layers of boxes in a store display are arranged as shown. If the pattern continues, and there are 12 layers in the display, what is the total number of boxes in the display?

28. Selling sweaters A clothing store ordered 300 sweaters and sold 20 of them at $100 each in the first week. In the second week, the selling price was lowered by $10, and 40 sweaters were sold. In the third week, the selling price was lowered by another $10, and 60 sweaters were sold. If the pattern continued,
a) how many weeks did it take to sell all the sweaters?
b) what was the the selling price the final week?

29. The first 8 terms of an arithmetic series have a sum of 148. The common difference is 3. What are the first three terms of the series?

30. Measurement The interior angles of a hexagon are in an arithmetic sequence. The largest angle is 130°. What are the other angles?

31. Determine the first and last terms of an arithmetic series with 50 terms, a common difference of 6, and a sum of 7850.

32. The first three terms of an arithmetic series have a sum of 24 and a product of 312. What is the fourth term of the series?

33. Algebra Find the sum of each series.
a) $4x + 6x + 8x + ... + 22x$
b) $(x + y) + (x + 2y) + (x + 3y) + ... + (x + 10y)$

N U M B E R   P O W E R

Tia was born in the 20th century. She will celebrate her nth birthday in the year n^2. In what year was she born?

INVESTIGATING MATH

Geometric Sequences

In the sequence 3, 6, 12, 24, ..., each term after the first is found by multiplying the preceding term by 2. Therefore, the ratio of consecutive terms is a constant.

$$\frac{6}{3} = 2 \qquad \frac{12}{6} = 2 \qquad \frac{24}{12} = 2$$

```
3
              3
Ans*2
              6
             12
             24
             48
```

This type of sequence is called a **geometric sequence**. The ratio of consecutive terms is called the **common ratio**. For the sequence 3, 6, 12, 24, ..., the common ratio is 2.

1 Writing Terms

State the common ratio and write the next 3 terms.

1. 1, 3, 9, 27, ...
2. 5, 10, 20, 40, ...
3. 2, −8, 32, −128, ...
4. 7, −7, 7, −7, ...
5. 64, 32, 16, 8, ...
6. 900, 90, 9, ...
7. 800, −400, 200, −100, ...
8. 0.2, 1, 5, ...

Given the first term, t_1, and the common ratio, r, write the first 5 terms of each geometric sequence.

9. $t_1 = 4$; $r = 3$
10. $t_1 = 8$; $r = 2.5$
11. $t_1 = 1000$; $r = 0.2$
12. $t_1 = -3$; $r = 3$
13. $t_1 = 2$; $r = -4$
14. $t_1 = -80$; $r = -0.5$

2 Finding Geometric Means

The terms between two non-consecutive terms of a geometric sequence are called **geometric means** between the two terms.
The numbers 6 and 18 are two geometric means between 2 and 54, because

$$\frac{6}{2} = 3 \qquad \frac{18}{6} = 3 \qquad \frac{54}{18} = 3$$

The number 8 is a geometric mean between 4 and 16, because

$$\frac{8}{4} = 2 \qquad \frac{16}{8} = 2$$

The number −8 is also a geometric mean between 4 and 16, because

$$\frac{-8}{4} = -2 \qquad \frac{16}{-8} = -2$$

Insert two numbers between the given terms, so that the four numbers form a geometric sequence.
1. 3, 81
2. 1, 64
3. 6, 48
4. −625, −5

Complete each sequence by inserting geometric means. Give all possible answers.
5. 4, ▨, 100
6. 90, ▨, 10
7. −4, ▨, ▨, ▨, −64
8. 80, ▨, ▨, ▨, 5
9. 216, ▨, ▨, 1
10. 3, ▨, ▨, ▨, ▨, 96

40 2010
36 6

84

1. An art collector bought a painting for $10 000. The painting increased in value by 20% every year for three years. What was the value of the painting at the end of each of the three years?

2. An $80 000 sports car decreases in value by 10% each year. Find the value of the car at the end of each of the next four years, if it is new now.

3. The value of a rare stamp is expected to follow a geometric sequence from year to year. The stamp is now worth $800 and is expected to be worth $1250 two years from now.
How much is it expected to be worth
a) one year from now?
b) three years from now?

4. A photocopier was set to reduce the dimensions of a drawing to 70% of their original value. If the reduction was repeated until the final dimensions were about 24% of the original dimensions, how many reductions were carried out?

1. Use the following geometric sequences to describe how a geometric mean between two terms is related to the two terms.

1, 2, 4	3, 6, 12	−8, −4, −2
5, −10, 20	−18, 6, −2	0.5, 5, 50

2. Does the amount of an investment made at a fixed rate of compound interest follow an arithmetic sequence or a geometric sequence? Explain and give an example.

3. Given the sequence 5, ■, ■, ■, ■, 160,
a) complete the sequence assuming that it is arithmetic
b) complete the sequence assuming that it is geometric
c) graph t_n versus n for both sequences on the same set of axes
d) compare the two graphs

23.

Stamp reproduced courtesy of Canada Post Corporation.

TECHNOLOGY

Using a Spreadsheet

A word processor is used to manipulate text data to produce everything from letters to novels. In contrast to a word processor, an electronic spreadsheet manipulates numerical data. Almost any job that uses rows and columns of numbers can be performed using an electronic spreadsheet. Using a spreadsheet to analyze data has become a routine business procedure.

1 Interest Calculated Annually

Claudia has negotiated an $80 000 farm mortgage. The annual interest rate is 7%. The table shows the fixed annual payment for the 8-year mortgage.

1. a) Set up a spreadsheet to calculate the opening balance, the interest charged in each year, and the closing balance at the end of each year.
b) If the interest rate increases at the start of year 5, what payment options might the farmer have?
c) If the interest rate decreases at the start of year 5, what payment options might the farmer have?

Year	Opening Balance ($)	Annual Interest Rate (%)	Interest ($)	Annual Payment ($)	Closing Balance ($)
1	80 000.00	7		13 397.42	
2		7		13 397.42	
3		7		13 397.42	
4		7		13 397.42	
5		7		13 397.42	
6		7		13 397.42	
7		7		13 397.42	
8		7		13 397.42	

2. For an annual interest rate of 6.5%, the annual payment for the loan would be $13 138.98. Modify the spreadsheet to show the opening balance, the interest charged in each year, and the closing balance at the end of each year.

2 Interest Calculated Monthly

Pierre negotiated a $10 000.00, 1-year home renovation loan. The annual interest rate was 6%, compounded monthly. The fixed monthly payments were $860.66. Set up a spreadsheet, with column headings similar to the farm mortgage, to find the opening balance, the interest charged each month, and the closing balance at the end of each month for 1 year.

Readability

1 **Measuring Readability**

Readability statistics are one way to determine how well a piece of writing communicates with a reader. The Flesch-Kincaid formula puts a reading grade level on a piece of writing. The formula for determining the reading grade level is

Grade Level = 0.39 × (average number of words per sentence)
          + 11.8 × (average number of syllables per word)
          − 15.59

The steps used to calculate the reading level are as follows.

1. Select a 100-word passage at random. Start counting at the first word of the first complete sentence.

2. Count the number of syllables and divide by 100 to find the average number of syllables per word, to the nearest hundredth. (You know there are at least 100 syllables in a 100-word passage, so just count the extra syllables in words with more than one syllable and add 100). Use the following rules for counting syllables.
• Count one syllable for each vowel sound.
Track, coil, jet, and rough each have one syllable.
Turnip, thorough, and science each have two syllables.
• Do not count a silent e as a syllable.
Rope has one syllable.
Wildlife has two syllables.
• When endings such as -y, -le, and -ed are sounded at the end of a word, count them as a syllable.
Baby, rattle, and dropped each have two syllables.
• Count abbreviations, such as "Corp." and "Blvd.," as one-syllable words.

3. Find the number of sentences, to the nearest hundredth. For example, suppose that 95 of the 100 words take up 10 sentences, and the remaining 5 words form part of a 12-word sentence. The total number of sentences is $10 + \dfrac{5}{12} \doteq 10.42$.

4. Divide 100 by the number of sentences to find the average number of words per sentence, to the nearest hundredth.

5. Repeat steps 1 to 4 for two other 100-word passages. Average the three values for the number of syllables per word, to the nearest hundredth. Also, average the three values for the number of words per sentence, to the nearest hundredth.

6. Substitute the values from step 5 into the formula.

The Flesch Reading Ease score is on a scale from 0 to 100. The lower the score, the more difficult the writing is to read. The formula to calculate the score is

Score = 206.84 − 1.02 × (average number of words per sentence)
  − 0.85 × (average number of syllables per 100 words)

The table shows how to interpret the score.

Score	Reading Difficulty
90 – 100	Very easy (grade 4)
80 – 90	Easy (grade 5)
70 – 80	Fairly easy (grade 6)
60 – 70	Standard (grades 7, 8)
50 – 60	Fairly difficult (some high school)
30 – 50	Difficult (high school – college)
0 – 30	Difficult (college and up)

The following are three excerpts from the book *Deadly Appearances*, a Joanne Kilbourn mystery written by Gail Bowen. Bowen is a Canadian mystery writer, who teaches at the University of Regina. Use these excerpts to determine the reading grade level and the reading ease score for the book.

(Page 1) There was, and still is, something surreal about that moment: the famous face looming up out of nowhere. He was pulling himself up the portable metal staircase that was propped against the back of the truck bed. His body appeared in stages over the metal floor: head, shoulders and arms, torso, belly, legs, feet. He seemed huge. He was climbing those steps as if his life depended on it, and his face was shiny and red with exertion. The heat on the floor of the stage was unbearable. I could smell it. I remember thinking very clearly, a big man like that could die in this heat, then I turned and scrambled toward Andy.

(Page 174) A tiny young woman in a trench coat came in carrying a medical bag. She went not to Eve, but to me. She slid her fingers around my wrist, positioned her face close to mine.

"Shock," she said, still holding my wrist in her hand. Then there was a swab and a pinprick sensation at the crease of my elbow, and I felt warm and weary. "You'll be all right now. You're Joanne Kilbourn, aren't you? Well, Joanne, someone will get you some tea. Plenty of sugar," she said over her shoulder. "Hang in there, Joanne," and then, smooth as silk, she moved along.

(Page 201) The next morning I woke up in my own bed in the house on Eastlake Avenue. The room was full of light, and as I lay there, I could hear in the distance the mournful cries of geese flying south. I got out of bed, opened the window and curled up in the window seat to watch. The air that came into the room was fresh and cold and smelled of the north. I hugged my knees for warmth and looked out. There were no clouds. The sky was clear, hard blue. It was a flawless October day.

Suddenly the air was black with geese, hundreds, then thousands of them.

2 Writing Instructions About Numbers

1. Using words only, write a set of instructions so that a classmate can complete the steps required for each of the following.

a) Calculate $3 \times 6 \div 2 \times 7$.

b) Calculate $2 + 3 \div 5$.

c) Use a calculator to find a 12% commission on a $3500 sale.

d) Use a calculator to find the reciprocal of the square of a number.

e) Evaluate the quotient when a three-digit number with identical digits is divided by the sum of the digits.

2. Have a classmate check your instructions to ensure that they are clear and correct, and that no steps are missing.

3. a) Rewrite your instructions so that, if you measured the reading grade level, it would go down. Explain your reasoning.

b) Rewrite your instructions so that, if you measured the Flesch Reading Ease score, it would go down. Explain your reasoning.

Review

2.1 *Find the amount of each investment.*
1. $10 000 invested for 4 years at an annual interest rate of 6% compounded annually

2. $6000 invested for 2 years at an annual interest rate of 8% compounded quarterly

Calculate the price, including taxes,
a) *if the PST is calculated only on the price*
b) *if the PST is calculated on the price plus the GST*
3. $999 TV; 7% GST; 9% PST
4. $279.95 CD player; 7% GST; 8% PST

5. Weather The table shows average values for the snowfall and rainfall in four cities in April, and an estimate of the total precipitation.

City	Rainfall (mm)	Snowfall (cm)	Total Precipitation (mm)
Edmonton	8.8	13.2	22.0
Regina	14.3	10.9	25.2
Vancouver	59.3	0.3	59.6
Winnipeg	27.1	11.3	38.4

a) How is the total precipitation estimated from the rainfall and the snowfall?
b) The total precipitation is an estimate of how much rain would have fallen if all the precipitation was in the form of rain. How many centimetres of snowfall are assumed to be equivalent to 1 mm of rainfall?

6. Repaying a loan The table shows data for the repayment of a small loan. The column headings represent the Month (M), Opening Balance (OB), Annual Interest Rate (IR), Interest Charged (IC), Monthly Payment (MP), and Closing Balance (CB).

M	OB	IR (%)	IC	MP	CB
1	$2000.00	8	$13.33	$508.36	$1504.97
2	$1504.97	8	$10.03	$508.36	$1006.66
3	$1006.66	8	$6.71	$508.36	$ 505.01
4	$ 505.01	8	$3.37	$508.38	$ 0.00

a) What is the total interest paid on the loan?

b) If extra payments were allowed, how much would the extra payment be at the end of month 2 so that the loan would be paid off at the end of month 3?
c) Are the rows in the table related recursively? Explain.

2.3 *Given the general term, state the first 5 terms of each sequence. Then, graph t_n versus n.*
7. $t_n = 2n + 1$ **8.** $t_n = n^2 - 3$
9. $t_n = 7 - 2n$ **10.** $t_n = 3^n - 1$

Given the general term, find the tenth term of each sequence.
11. $t_n = \dfrac{3n + 2}{n - 2}$

12. $t_n = 2n^2 - 5$

Write the first 5 terms determined by each recursion formula.
13. $t_1 = -5; t_n = t_{n-1} - 3$
14. $t_1 = 2; t_n = t_{n-1} - n$
15. $t_1 = 2; t_2 = 2; t_n = t_{n-1} + 2t_{n-2}$

16. Fingernails A human fingernail grows about 0.6 mm a week. If the visible part of a fingernail is 15 mm long, how long will the nail be in
a) 1 week?   **b)** 2 weeks?   **c)** 4 weeks?

2.5 *Find the next 3 terms of each arithmetic sequence.*
17. 9, 15, 21, ...
18. 6, 1, −4, ...

Write the first 4 terms of each arithmetic sequence.
19. $a = 3, d = 5$
20. $a = -5, d = 2$

Find the indicated term for each arithmetic sequence.
21. 3, 5, 7, ...; t_8
22. −2, −6, −10, ...; t_{25}
23. 4, 11, 18, ...; t_n

Find the number of terms in each arithmetic sequence.
24. 4, 9, 14, ..., 169
25. 19, 11, 3, ..., −229

26. a) Insert two numbers between −1 and 26, so that the four numbers form an arithmetic sequence.
b) Graph t_n versus n for the sequence.

27. Find three arithmetic means between 9 and 65.

28. Pattern The rectangular shapes are made from asterisks.

```
                              ✳ ✳ ✳
                  ✳ ✳ ✳       ✳   ✳
      ✳ ✳ ✳       ✳   ✳       ✳   ✳
      ✳   ✳       ✳   ✳       ✳   ✳
      ✳ ✳ ✳       ✳ ✳ ✳       ✳ ✳ ✳
```

Diagram   1   2   3

a) How many asterisks will there be on the 4th rectangle? the 5th rectangle?
b) Write the general term for the sequence in the numbers of asterisks.
c) Predict the number of asterisks on the 25th rectangle.
d) Which rectangle is made from 92 asterisks?

Find the indicated sum for each arithmetic series.
29. S_{20} for $3 + 8 + 13 + ...$
30. S_{25} for $-20 - 18 - 16 - ...$

Find the sum of each arithmetic series.
31. $7 + 10 + 13 + ... + 70$
32. $65 + 59 + 53 + ... - 85$

Find S_n for each arithmetic series.
33. $a = 3, t_n = 147, n = 19$
34. $a = -1, t_n = -37, n = 10$

35. Find the sum of the 50 greatest negative integers.

36. Measurement The side lengths in a quadrilateral form an arithmetic sequence. The perimeter is 38 cm, and the shortest side measures 5 cm. What are the other side lengths?

37. Pattern What is the total number of asterisks in the first 16 diagrams of the pattern in question 28?

38. Piling logs In a triangular pile of logs, the top row contains one log. Each row below the top row contains one log more than the row above. If the bottom row contains 21 logs, how many logs are in the pile?

Exploring Math

Impossible Scores

1. Suppose that, in a flag football league, a team gets 3 points for a field goal and 5 points for a touchdown. No other ways of scoring are possible.
a) It is impossible for a team to score 1 or 2 points. What are the other impossible team scores?
b) What is the highest impossible score?
c) Explain why all the scores higher than your answer to part b) are possible.

2. Suppose that a team gets 3 points for a field goal and 7 points for a touchdown.
a) What team scores are impossible?
b) What is the highest impossible score?
c) Explain why all the scores higher than your answer to part b) are possible.

3. Repeat question 2, but with 4 points for a field goal and 7 points for a touchdown.

4. Suppose that a team gets 4 points for a field goal and 6 points for a touchdown.
a) List the first 10 impossible team scores.
b) Is there a highest impossible score? Explain.

5. Repeat question 4, but with 3 points for a field goal and 6 points for a touchdown.

6. a) Can you use the numbers of points for a field goal and a touchdown to predict whether there is a highest impossible team score? Explain.
b) Devise 3 new scoring systems that will give a highest impossible team score.
c) Without giving 1 point for either a touchdown or a field goal, devise 3 new scoring systems that will not give a highest impossible team score.
d) Test your scoring systems from parts b) and c).

Chapter Check

1. What is the amount of an investment of $8000 at the end of 3 years, if the annual interest rate is 8% compounded semi-annually?

2. Calculate the total price of a $695 printer, if the GST is 7%, the PST is 9%, and
a) the PST is calculated only on the price
b) the PST is calculated on the price plus the GST

Given the general term, state the first 5 terms of each sequence. Then, graph t_n versus n.
3. $t_n = 2n - 3$
4. $t_n = n^2 + 3$

Given the general term, find the twelfth term of each sequence.
5. $t_n = 4 + 3n$
6. $t_n = (n + 1)^2$

Write the first 4 terms determined by each recursion formula.
7. $t_1 = 7; t_n = t_{n-1} - 3$
8. $t_1 = -2; t_n = t_{n-1} + n$

9. Write the next 3 terms of the arithmetic sequence.
19, 14, 9, ...

10. Find t_{21} for the arithmetic sequence.
6, 10, 14, ...

11. Find t_n for the arithmetic sequence.
−5, −11, −17, ...

12. Find the number of terms in the arithmetic sequence.
11, 14, 17, ..., 146

13. Find 2 arithmetic means between −5 and −89.

14. Find S_{15} for the arithmetic series.
4 + 11 + 18 + ...

15. Find the sum of the arithmetic series.
−12 − 9 − 6 − ... + 39

16. Find S_{20} for an arithmetic series with $a = 2$ and $t_{20} = 78$.

17. Olympic medals The table shows the medals won by three countries at the Summer Olympics in Barcelona.

Country	Medals		
	Gold	Silver	Bronze
Britain	5	3	12
Canada	7	4	7
Italy	6	5	8

How could points be awarded for each type of medal so that
a) Britain seems to have the most successful team?
b) Britain seems to have the least successful team?
c) Italy seems to have the most successful team?
d) Canada seems to have the most successful team?

18. Space shuttle When Dr. Roberta Bondar flew on the space shuttle, it lifted off at about 10:00 and then orbited the Earth at regular intervals. It completed its third orbit at about 14:30 on the same day as it lifted off. At about what times did it complete its first and second orbits?

19. Repaying a loan The table shows the data for the repayment of a loan. The table shows the Year (Y), Opening Balance (OB), Annual Interest Rate (IR), Interest Charged (IC), Annual Payment (AP), and Closing Balance (CB).

Y	OB	IR (%)	IC	AP	CB
1	$25 000.00	6	$1500.00	$5934.91	$20 565.09
2	$20 565.09	6	$1233.91	$5934.91	$15 864.09
3	$15 864.09	6	$ 951.85	$5934.91	$10 881.03
4	$10 881.03	6	$ 652.86	$5934.91	$ 5 598.98
5	$ 5 598.98	6	$ 335.94	$5934.92	$ 0.00

a) What was the total interest paid in the first 3 years of the loan?
b) How much of the annual payment at the end of year 3 went toward the opening balance?
c) If an extra payment of $9684.06 were made at the end of year 1, when would the loan be paid off? Explain.

Using the Strategies

1. Suppose that you:
• placed 2 identical coins as shown in the diagram
• pressed firmly on the bottom coin
• rolled the top coin around the bottom one until the top coin returned to its original position

a) Predict how many times the rolling coin would turn around its centre as it rolled around the stationary coin.
b) Test your prediction.
c) Explain your observation.

2. Use an isometric grid with the dimensions shown. How many different-sized equilateral triangles can you make on the grid?

3. Five friends — three women and two men — met at a school reunion. The men were Smith and Wong. The women were Bevan, Lee, and Kostash. Their occupations were biologist, farmer, accountant, writer, and baker. Each one lived in a different city: London, England; Vancouver; Paris, France; Toronto; and Melbourne, Australia. Use the information to determine the occupation and city of residence for each person.
• Mr. Smith is not the baker.
• The five friends are Ms. Bevan, Mr. Wong, the woman who lives in Paris, the man who lives in Vancouver, and the biologist who lives in Melbourne.
• The accountant and Ms. Lee live outside Canada but not in Melbourne.
• Ms. Bevan does not live in London and is not the woman writer.

4. Estimate how many supermarkets there are in Canada.

5. Place the numbers 1, 2, 3, 4, 5, 6, 7, and 8 at the vertices of a cube, so that the sum of the numbers at the vertices of each face is the same.

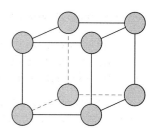

6. The first rectangle is 3 m long and 2 m wide. The second is 4 m long and 3 m wide. The third is 5 m long and 4 m wide. Find the area of the shaded region.

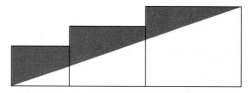

7. a) How many dominoes are there in a double-six set of dominoes?
b) How many spots are there on a double-six set of dominoes?

8. Place one of the digits 1, 2, 3, 4, 5, and 6 in each square to make the multiplication true.

9. The average age of the students in a group was 15. The average age of the teachers in the same group was 45. The average age of the whole group was 19. What was the smallest possible number of people in the group?

D A T A   B A N K

1. The coldest wind chill temperature recorded in Manitoba was −76°C. The air temperature was −41°C. What was the wind speed?

2. Suppose a road from Montreal to Denver is to be tiled with square tiles that have the same dimensions as processed cheese slices. About how many tiles will be needed, if the road is 120 tiles wide?

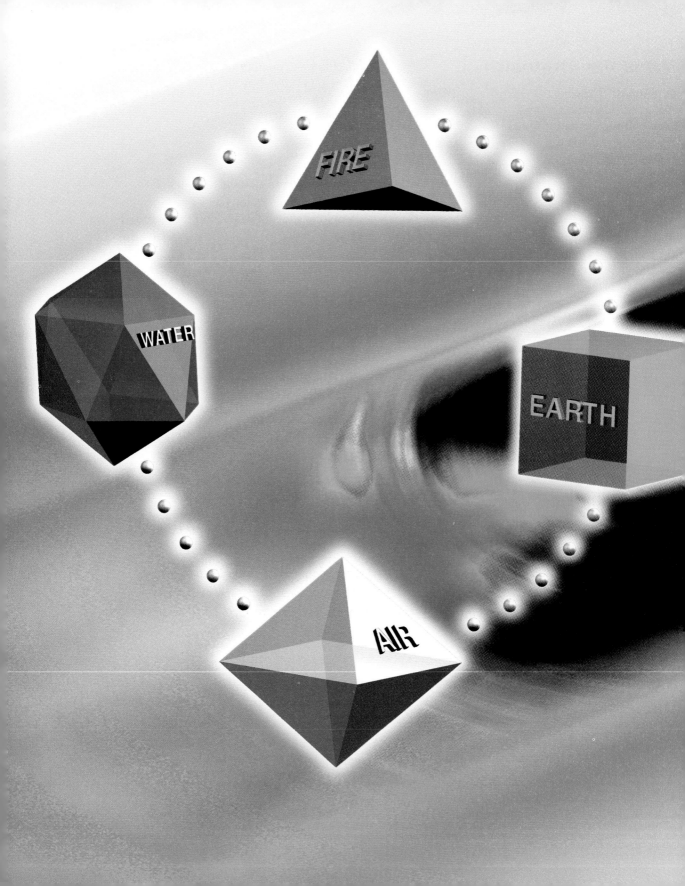

CHAPTER 3

Polynomials

The 5 regular polyhedra are the tetrahedron, cube, octahedron, dodecahedron, and icosahedron. Plato, a Greek philosopher who lived around 400 B.C., named 4 of the polyhedra for what were then believed to be the 4 elements of the universe.

Use the diagram to name the polyhedron that represents each element. Then, complete the following experiment.

Open this textbook at any place from page 1 to page 481. Write down both page numbers. If the page numbers are 3-digit numbers, drop the hundreds digits. For example, the numbers 276 and 277 would become 76 and 77. Find the sum of the two numbers. This sum is your first special number.

Now, open the textbook at a different place and write down both page numbers. Again, drop the hundreds digits. Find the sum of the two numbers. This sum is your second special number.

Subtract the smaller special number from the larger. The result is your magic number.

Start counting at Fire on the diagram and move clockwise. Fire is 1, Earth is 2, Air is 3, Water is 4, Fire is 5, and so on. Continue counting in this manner until you reach your magic number. Note the element on which you stopped.

Compare your result with your classmates' results. Explain why this trick always works.

🤝 Finding a Winning Strategy

Nim is a game for two players.

1 Nim With Two Piles

The game starts with 2 piles of counters.

The rules are as follows.
• Players take turns removing as many counters as they wish from one of the piles. At least 1 counter must be removed on a turn.
• The player who takes the last counter, or pile of counters, is the winner.

1. Start with 4 counters in one pile and 7 in the other. Work with a classmate to find a winning strategy for the player who goes first.

2. Derek and Nicole are playing Nim. One pile has 3 counters and one has 5. Derek plays first. Explain how Derek can always win.

3. Derek and Nicole are playing Nim. Both piles have 3 counters. Derek plays first. Explain how Nicole can always win.

4. Suppose the rule for winning is changed, and the player forced to pick up the last counter loses the game. If both piles have 3 counters and Derek plays first, explain how Nicole can always win.

2 Nim With Three Piles

In this game, there are 3 piles of counters.

The rules are the same as those for Nim with two piles. Players take turns removing counters and the player who takes the last counter, or pile of counters, is the winner.

1. Start with 3 counters in one pile, 2 in another, and 1 in the third. Work with a classmate to find a winning strategy for the player who plays second.

2. Suppose the rule for winning is changed, and the player forced to pick up the last counter loses the game. There are 3 counters in one pile, 2 in another, and 1 in the third. Work with a classmate to find a winning strategy for the player who plays second.

3 Straight-Line Nim

Use a row of counters that touch each other, like the ones shown. Any number of counters can be used.

The rules are as follows.
• Players take turns removing 1 or 2 counters. If 2 counters are removed, they must be counters that touch each other.
• The player who takes the last counter, or counters, is the winner.

1. Work with a classmate to find the winning strategy for the player who plays first, if there is an odd number of counters.

2. Work with a classmate to see if there is a winning strategy for the player who plays first, if there is an even number of counters.

Algebra Skills

Evaluate for $x = 2$, $y = 1$, and $z = 3$.

1. $5x - y + 2z$
2. $4xz - 3y + 9$
3. $2(4x + 7y)$
4. $yz - xz + xyz$
5. $5(z - x - y)$
6. $3y^2 + 4z^2 - 2x^2$
7. $3y(7 + 8z)$
8. $(xyz)^2$
9. $(x - y)(y - z)$
10. $(x - y - z)^3$

Evaluate for $a = -3$, $b = -1$, and $c = 2$.

11. $2ac - 5b$
12. $6a - 2b - 3c$
13. $4a(2b - c)$
14. $ab - bc - ac$
15. $2a^2 + 3b^2 - c^2$
16. $a^3 - b^3 + c^3$
17. $(2abc)^2$
18. $(a - b)(b - c)$
19. $(a + b - c)^2$
20. $4ab^2c - 15$

Simplify, using the exponent rules. Express each answer in exponential form.

21. $2^3 \times 2^4$
22. $3^2 \times 3^4 \times 3$
23. $a^2 \times a^3$
24. $y^4 \times y^5$
25. $m^2 \times m^4 \times m$
26. $2y^2 \times 4y^3$
27. $(-4a^3)(2a^4)$
28. $(n^4)^2$
29. $(2x^2)^3$
30. $(-6m^3)(-2m^4)$
31. $2^5 \div 2^3$
32. $3^2 \div 3^2$
33. $m^7 \div m^4$
34. $a^4 \div a$
35. $y^5 \times y^3 \div y^2$
36. $-20a^4 \div (-5a^2)$

Simplify, by collecting like terms.

37. $3x + 5x - 2x$
38. $3a + 4b - 6a + b$
39. $4a + 5b - 6 - 7b + 8 + 4a$
40. $2x^2 + 3x - 4 - 2x + 5x^2 - 7$
41. $4m - 3mn + 2n - n + 5m - 6mn$
42. $2x^2 - x^2 + 5y^2 - y^2 - 3x^2$

Expand.

43. $2(x - 5)$
44. $4(2a + 3b - 2c)$
45. $4(1 - 2x + 5y)$
46. $-2(3x - 7)$
47. $-3(x^2 - 2x - 1)$
48. $-(2a - 4b - c)$
49. $3x(x + 2)$
50. $4x(x^2 - 3x + 1)$
51. $-2y(y^2 + y - 3)$
52. $-a(2a^2 - 3a - 5)$

Expand and simplify.

53. $2(x + 1) + 3(x - 1)$
54. $5(2y + 1) - (5y - 2)$
55. $3(m^2 + m + 1) + 2(m^2 - 1)$
56. $4(1 + x^2) - 3(x^2 - 3x - 2)$
57. $2(n^2 + n - 2) + 5(n^2 - 3n - 1)$
58. $5x(x - 3) - 2x(x + 2)$
59. $t(t^2 + 3t - 2) - 3t(t^2 - t + 1)$
60. $3(3z^2 - 5z - 4) + 2z(4z - 1)$

Mental Math

Integers

Find two integers that have the given product and sum.

	Product	Sum		Product	Sum
1.	4	4	**2.**	6	5
3.	5	6	**4.**	8	6
5.	12	7	**6.**	12	8
7.	20	9	**8.**	24	11
9.	24	10	**10.**	30	17
11.	1	-2	**12.**	-2	-1
13.	-5	-4	**14.**	-9	0
15.	-6	1	**16.**	-10	3
17.	7	-8	**18.**	16	-8
19.	16	-10	**20.**	-15	-2

Multiplying Two Numbers That Differ by 2

To determine 16×14, first square their average, 15, then subtract 1.

$15^2 = 225$
$225 - 1 = 224$
So, $16 \times 14 = 224$.

Calculate.

1. 13×11
2. 14×12
3. 19×21
4. 29×31
5. 24×26
6. 41×39
7. 101×99
8. 59×61
9. 999×1001

To calculate 1.6×1.4 or 160×140, think 16×14, then place the decimal point.

$1.6 \times 1.4 = 2.24$
$160 \times 140 = 22\,400$

Calculate.

10. 1.2×1.4
11. 1.5×1.3
12. 260×240
13. 290×310
14. 10.1×9.9
15. 130×11
16. 210×1.9
17. 39×0.41
18. 510×49
19. 180×1.6
20. 89×910
21. 0.69×0.71

22. Describe a method for multiplying two numbers that differ by 4. Use your method to evaluate each of the following.

a) 11×15
b) 22×18
c) 17×13
d) 32×28
e) 27×23
f) 98×102
g) 1.4×18
h) 380×420
i) 520×4.8

INVESTIGATING MATH

Algebra Tiles

Each algebra tile is assigned dimensions that result in the following areas.

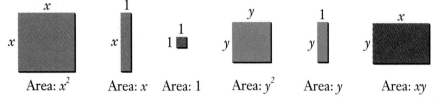

Area: x^2 Area: x Area: 1   Area: y^2 Area: y Area: xy

When the tiles are turned over to show the white side, the dimensions result in the following areas.

Area: $-x^2$ Area: $-x$ Area: -1 Area: $-y^2$ Area: $-y$ Area: $-xy$

1 Representing Expressions With Tiles

1. Write the expression represented by each group of tiles.

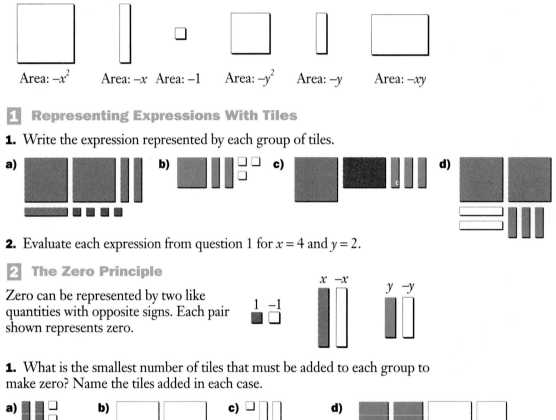

2. Evaluate each expression from question 1 for $x = 4$ and $y = 2$.

2 The Zero Principle

Zero can be represented by two like quantities with opposite signs. Each pair shown represents zero.

1. What is the smallest number of tiles that must be added to each group to make zero? Name the tiles added in each case.

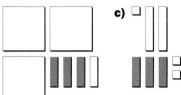

2. Model $2x^2 + 3x + 1$ in three different ways with eight tiles.

98

3 Adding Expressions

If two expressions are modelled by two sets of algebra tiles, you can combine the sets to model the sum of the expressions.

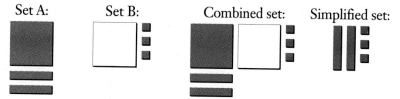

Set A: Set B: Combined set: Simplified set:

Adding the expressions $x^2 + 2x$ and $-x^2 + 3$ gives the simplified expression $2x + 3$.

Copy and complete the table. The first row has been completed.

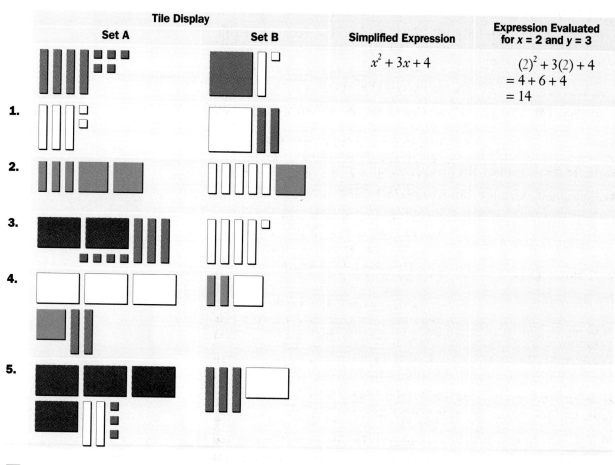

Tile Display		Simplified Expression	Expression Evaluated for x = 2 and y = 3
Set A	Set B		
		$x^2 + 3x + 4$	$(2)^2 + 3(2) + 4$ $= 4 + 6 + 4$ $= 14$
1.			
2.			
3.			
4.			
5.			

4 Designing Algebra Tiles

An x^2-tile represents the area of a square with dimensions x by x.
1. What shape of tile would you need to model x^3?
2. Sketch the shapes you would use to model each of the following expressions.
a) $x^3 + y^3$ **b)** $x^3 + x^2y + xy^2 + y^3$
3. Sketch algebra tiles that you could you use to model the expression
$x^2 + xy + y^2 + z^2 + xz + yz$.

3.1 Reviewing Polynomials

Explore: Evaluate the Expression

When a dolphin leaps vertically out of the water, the dolphin's speed just before it breaks the surface is about 7 m/s. The dolphin's height, in metres, above the water t seconds after breaking the surface can be found using the expression $7t - 5t^2$.

How high is the dolphin after 0.5 s? 1 s?

Inquire

1. After 1 s, is the dolphin still climbing or is it diving back toward the water? How do you know? *diving back toward the water.*

2. Calculate the time at which the dolphin reaches its maximum height. *0.5 s.*

3. What is the dolphin's maximum height? *2.25 m*

4. If a dolphin was swimming on the moon and left the water at 7 m/s, the dolphin's height after t seconds could be found using the expression $7t - 0.8t^2$. To the nearest second, find the time the dolphin would take to reach its maximum height. *0.9s*

A **monomial** is a number, a product of one or more variables, or the product of a number and one or more variables. A **numerical coefficient** is the number part of a monomial.

monomial

numerical coefficient →⟨$7t$⟩← variable

A **polynomial** is an algebraic expression formed by adding or subtracting monomials. Each monomial is called a **term** of the polynomial. Polynomials are classified in two ways. One way uses the number of terms they have.

Monomials have 1 term.	Binomials have 2 terms.	Trinomials have 3 terms.
7	$x + 17$	$x^2 + 4x - 3$
$5x$	$x - y$	$2a + 3b - 4c$
$-4m^2n$	$2m^2 - 4mn$	$2x^2 - 2xy + 3y^2$

Polynomials with more than three terms are simply called *polynomials*.

Polynomials are also classified using the exponents of their variables. The **degree** of a term is the sum of the exponents of its variables.

$3x^2$ is a term of degree 2, because the only exponent is 2.

$4m^2n^3$ is a term of degree 5, because the sum of the exponents 2 and 3 is 5.

The degree of a polynomial is the highest degree of any of its terms.

Polynomial	Number of Terms	Classified by Number of Terms	Degree	Classified by Degree
5	1	monomial	0	constant
3x	1	monomial	1	linear
4x + 2	2	binomial	1	linear
$x^2 + 4x + 1$	3	trinomial	2	quadratic
$3x^3 + x^2 - x - 7$	4	polynomial	3	cubic
$4x^4 - 5x - 4$	3	trinomial	4	quartic

The terms of a polynomial are usually written so that the exponents of the variable are in descending order or ascending order. Polynomials that include more than one variable are usually arranged in ascending order or descending order of one of the variables.

Descending order of x: $5x^3y + 4x^2y^2 + 4xy^3 - 3$

Descending order of y: $4xy^3 + 4x^2y^2 + 5x^3y - 3$

Terms that have the same variable factors, such as $7x$ and $5x$, are called **like terms**. Simplify an expression containing like terms by adding their coefficients. This process is known as **collecting like terms**.

$$7x + 3y + 5x - 2y = 12x + y \qquad 3x^2 + 4xy - 6xy + 8x^2 = 11x^2 - 2xy$$

Example 1   Adding Polynomials

Simplify $(x^2 + 4x - 2) + (2x^2 - 6x + 9)$.

Solution

To add polynomials, collect like terms.

Horizontal format: $(x^2 + 4x - 2) + (2x^2 - 6x + 9) = x^2 + 4x - 2 + 2x^2 - 6x + 9$

$$= x^2 + 2x^2 + 4x - 6x - 2 + 9$$
$$= 3x^2 - 2x + 7$$

Vertical format: $x^2 + 4x - 2$
$2x^2 - 6x + 9$
$\overline{3x^2 - 2x + 7}$

Example 2   Vertical Subtraction

Subtract $(4y^2 - 2y + 3) - (3y^2 + 5y - 2)$.

Solution

To subtract a polynomial, determine its opposite and add.

To determine the opposite, multiply each term by -1.

$4y^2 - 2y + 3 \qquad \rightarrow \qquad 4y^2 - 2y + 3$
$3y^2 + 5y - 2 \qquad \rightarrow \qquad -3y^2 - 5y + 2$
$\overline{y^2 - 7y + 5}$

Example 3   Horizontal Subtraction

Simplify $(6a^2 - ab + 4) - (7a^2 + 4ab - 2)$.

Solution

To subtract, add the opposite.

One way is to multiply each term to be subtracted by -1.

$(6a^2 - ab + 4) - (7a^2 + 4ab - 2) = (6a^2 - ab + 4) - 1(7a^2 + 4ab - 2)$
$$= 6a^2 - ab + 4 - 7a^2 - 4ab + 2$$
$$= -a^2 - 5ab + 6$$

Example 4   Multiplying Monomials

Multiply.

a) $(2x^2)(7x)$ **b)** $(-4a^2b)(3ab^3)$

Solution

Multiply the numerical coefficients. Then, multiply the variables using the exponent rules for multiplication.

a) $(2x^2)(7x) = (2)(7)(x^2)(x)$ 　　　　**b)** $(-4a^2b)(3ab^3) = (-4)(3)(a^2)(a)(b)(b^3)$
$= 14x^3$ 　　　　　　　　　　　　　　　　$= -12a^3b^4$

Example 5   Dividing Monomials

Simplify $\dfrac{20x^3 y^4}{-5x^2 y^2}$.

Solution

Divide the numerical coefficients. Then, divide the variables using the exponent rules for division.

$$\frac{20x^3 y^4}{-5x^2 y^2} = \left(\frac{20}{-5}\right)\left(\frac{x^3}{x^2}\right)\left(\frac{y^4}{y^2}\right)$$
$$= -4xy^2$$

Practice

Classify each polynomial by degree and by number of terms.

1. $4y + 7$ 　　　　　　　　**2.** $2x^2 - 8x$
3. $3a^3 - 3a - 1$ 　　　　**4.** $5w^3 - 3w^2 + w - 6$
5. $8x$ 　　　　　　　　　**6.** $9 - 6x^4 + 2x^2$
7. $6x^3 - 11x$ 　　　　　**8.** 78

Evaluate each expression for the given value(s) of the variable(s).

9. $x^2 - 7x + 12$ for $x = 3$
10. $4a^2 - 5a + 8$ for $a = 2$
11. $2x^2 - 3xy + 4y^2$ for $x = 2, y = -1$

Write each polynomial in descending order of x.

12. $6x^2 - 5x + 4x^3 - 7$
13. $1 - 8x^2 - 5x^5 + 7x^4 - 3x^3$
14. $4xy^4 + 9x^4y - 8x^3y^2 + 2x^5 - y^5$

Write each polynomial in ascending order of x.

here **15.** $2x - 5x^3 - 6x^2 + 11$
16. $x^5 - 4 + 3x^3 - 2x^2 - 5x^4$
17. $x^3y^3 - 2xy^5 + x^5y - x^6 + 2 + x^2y^4$

→ *Add.*

18. $4x + 8$　　　　　　　**19.** $3a + 7b$
　　　$2x + 9$　　　　　　　　　$9a - 3b$

20. $5x^2 - 4x - 2$　　　　**21.** $2x^2 - 6xy + 9$
　　　$8x^2 + 3x - 3$　　　　　　$8x^2 + 3xy - 4$

Subtract.

22. $7x + 9$　　　　　　　**23.** $4a - 2b$
　　　$3x + 5$　　　　　　　　　$3a + 2b$

24. $8y^2 + 5y - 7$　　　　**25.** $3m^2n + mn - 7n$
　　　$9y^2 + 3x - 3$　　　　　　$5m^2n + 3mn - 8n$

Simplify.

26. $(3x + 4) + (5x + 2)$　　**27.** $(7a - 6) + (2a + 9)$
28. $(2 - 3yz) + (7 + 6yz)$　**29.** $(m + n) + (5m - 2n)$
30. $(5x + 7) - (2x + 1)$　　**31.** $(6a - 2b) - (4a + b)$
32. $(c + d) - (c - d)$　　　**33.** $(2m - n) - (3m + 4n)$

Simplify.

34. $(3x^2 - 6x + 9) + (4x^2 + 7x + 8)$
35. $(7x^2y + 3xy - 2y^2) + (8x^2y - xy - y^2)$
36. $(2a^2 - 2a - 7) - (a^2 - 6a - 11)$
37. $(3t^2 - 12) - (4t^2 + 5t + 7)$

38. $(2x + 4) + (3x - 2) - (5x + 8)$
39. $(5x^2 - y^2) - (3x^2 + 2y^2) - (x^2 + 3y^2)$

Multiply.
40. $(4x)(7x^2)$
41. $(3ab)(-4ab^2)$
42. $(-6m^2n^3)(-7mn^2)$
43. $(-2xyz^3)(-4x^3y^2)$
44. $(8r^3s^2t)(4s^3t)$
45. $(2xy)(-3x^2y^3)(-3x^2)$

Simplify.
46. $\dfrac{24x^4}{6x}$
47. $\dfrac{18a^2b}{9ab}$
48. $\dfrac{-36xy^2}{4y^2}$

49. $\dfrac{-27a^2bc^4}{-3abc^2}$
50. $\dfrac{-40x^3y^4z^2}{8x^3y^3}$
51. $\dfrac{-75s^2t^5}{-25s^2t^2}$

Applications and Problem Solving

52. Write a polynomial with the given number of terms and degree.
a) 1 term, degree 3
b) 2 terms, degree 4
c) 3 terms, degree 2
d) 4 terms, degree 5

53. Free fall The distance, in metres, an object falls during free fall is given by the expression $4.9t^2$, where t seconds is the time during which the object falls. An object dropped from the top of the Canterra Tower building in Calgary would hit the ground after 6 s. Calculate the height of the building.

54. a) Subtract $2x^2 - 3x + 4$ from $3x^2 + 4x - 5$.
b) Subtract $3x^2 + 4x - 5$ from $2x^2 - 3x + 4$.
c) Compare your answers from parts a) and b). What can you conclude about reversing the order of subtraction?

55. Measurement Given the perimeter, P, and the lengths of two sides, find the length of the third side.
a)

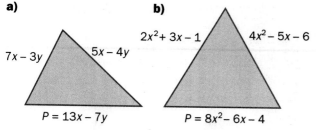

$7x - 3y$ $5x - 4y$ $2x^2 + 3x - 1$ $4x^2 - 5x - 6$
$P = 13x - 7y$ $P = 8x^2 - 6x - 4$

56. Measurement For the rectangular prism, write and simplify an expression that represents
a) the volume
b) the surface area

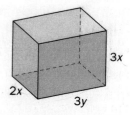

$3x$
$2x$
$3y$

57. A pizza of diameter d just fits inside a square box.
a) Write an expression for the area of the pizza in terms of d.
b) Write an expression for the area of the bottom of the box in terms of d.
c) Divide your expression from part a) by your expression from part b).
d) Use your result from part c) to find the percent of the area of the bottom of the box that the pizza covers. Round your answer to the nearest percent.

58. State two values of y for which the value of $3y^2 + 4y - 14$ is a perfect square.

59. Measurement The area of the shaded region is 400 cm^2. Find the dimensions of the large rectangle.

30 × 20

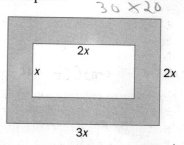

x $2x$ $2x$ $3x$

PROBLEM SOLVING

3.2 Look for a Pattern

Mathematics is a search for patterns. Many people look for patterns in their work. Detectives look for patterns at crime scenes. Wildlife biologists look for patterns in animal populations. Patterns can be used to make predictions.

Tutankhamun became king of ancient Egypt around 1334 B.C., when he was only 9 years old. The location of his tomb was unknown for over three thousand years, until a team of archeologists discovered it in 1922. Among the 5000 items found in the tomb were boards used for the ancient game of Senet. A Senet board consists of a 3 by 10 grid of squares. Find the total number of squares of all sizes on a Senet game board.

Understand the Problem

1. What information are you given?
2. What are you asked to find?
3. Do you need an exact or approximate answer?

Think of a Plan

Start with smaller grids and look for a pattern. Use the pattern to write a rule that relates the dimensions of the grid to the total number of squares.

Use diagrams to show the smaller grids. Then, set up a table.

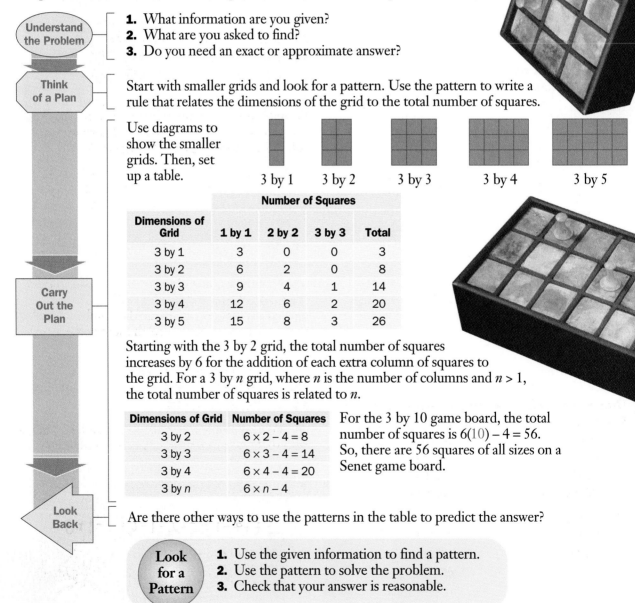

3 by 1   3 by 2   3 by 3   3 by 4   3 by 5

Dimensions of Grid	Number of Squares			
	1 by 1	2 by 2	3 by 3	Total
3 by 1	3	0	0	3
3 by 2	6	2	0	8
3 by 3	9	4	1	14
3 by 4	12	6	2	20
3 by 5	15	8	3	26

Carry Out the Plan

Starting with the 3 by 2 grid, the total number of squares increases by 6 for the addition of each extra column of squares to the grid. For a 3 by n grid, where n is the number of columns and $n > 1$, the total number of squares is related to n.

Dimensions of Grid	Number of Squares
3 by 2	$6 \times 2 - 4 = 8$
3 by 3	$6 \times 3 - 4 = 14$
3 by 4	$6 \times 4 - 4 = 20$
3 by n	$6 \times n - 4$

For the 3 by 10 game board, the total number of squares is $6(10) - 4 = 56$. So, there are 56 squares of all sizes on a Senet game board.

Look Back

Are there other ways to use the patterns in the table to predict the answer?

Look for a Pattern

1. Use the given information to find a pattern.
2. Use the pattern to solve the problem.
3. Check that your answer is reasonable.

Applications and Problem Solving

Find the total number of squares on each grid.
1. 4 squares by 10 squares
2. 5 squares by 12 squares
3. chessboard (8 by 8)
4. Snakes and Ladders board (10 by 10)

Find a rule that relates x and y. Then, copy and complete each table.

5.

x	y
0	4
1	5
2	6
3	7
9	
	20
21	

6.

x	y
0	1
1	3
2	5
3	7
8	
	25
50	

7.

x	y
1	8
2	7
5	4
6	
	1
9	

8.

x	y
19	12
16	9
14	7
12	
9	
	0

9.

x	y
1	2
2	5
3	10
4	17
	37
20	
100	

10.

x	y
1	3
2	7
3	11
4	15
10	
	63
30	

Determine the rule. Then, copy and complete each table.

11.

TREE	E
FOX	X
APPLE	P
LUNCH	
SCHOOL	

12.

CAR	2
GREEN	3
TWELVE	4
YELLOW	
CAMERA	

13.

HAZEL	E
TOWN	W
SALMON	O
STAIR	
PLACES	

14.

STREAM	2
PEOPLE	0
SEVEN	1
OCTOBER	
BANANA	

15. Predict the last digit if each number is expressed in standard form.
a) 4^{98} **b)** 7^{99}

16.

How many asterisks are there in
a) the 12th figure?   **b)** the 35th figure?

Write the next 3 terms in each sequence.
17. 6, 12, 11, 22, 21, 42, ▨, ▨, ▨
18. 2, 4, 6, 10, 16, 26, ▨, ▨, ▨
19. 1, 2, 5, 10, 17, 26, ▨, ▨, ▨

20. The numbers in each of the columns A, B, C, D, and E are found by applying different rules to the numbers in columns p and q. Determine each rule. Then, copy and complete the table.

p	q	A	B	C	D	E
−4	2	−6	6	−6	12	−7
2	−8	10	−10	−4	−60	−15
−5	6					
1	−4					
−3	−6					

21. a) How many odd numbers are in this series?
$1 + 3 + 5 + 7 + \ldots + 287 + 289 + 291$
b) Explain how you could use the number of odd numbers in the series to find the sum of these odd numbers.

Find the next term in each sequence.
22. 873, 168, 48, 32, ▨
23. 47, 29, 18, 11, 7, ▨

24. Write a problem that can be solved by finding a pattern. Have a classmate solve your problem.

PATTERN POWER

In an English class, the teacher wrote the whole numbers from 1 to 9 on the chalkboard. What was the rule for arranging the numbers?

3.3 Multiplying Polynomials

Canadian architect Arthur Erickson has designed
many buildings. He designed Simon Fraser
University around a walkway and meeting place
known as the Central Mall. The mall is covered
by a rectangular glazed roof supported on girders.
The dimensions of the roof can be represented
by the expressions $x + 1$ and $2x + 11$. The area
covered by the roof is the product of these
dimensions, $(x + 1)(2x + 11)$.

Explore: Use Algebra Tiles

To find the product $(x + 3)(x + 2)$,
make a rectangle with a length of $x + 3$
and a width of $x + 2$.

Use the sides of
the tiles to mark
off the dimensions
along two
perpendicular
lines.

Use the marks
as a guide and
make the
rectangle with
algebra tiles.

a) What is the area of the rectangle?
b) Copy and complete the statement $(x + 3)(x + 2) = $  .

Inquire

1. Use algebra tiles or draw diagrams to find each product.
a) $(x + 4)(x + 1)$ **b)** $(x + 2)(2x + 3)$ **c)** $(2x + 1)(3x + 2)$

2. Check your solutions to question 1 by substituting 1 for x in the
product of binomials and in the resulting trinomial.

3. Write a rule for multiplying two binomials.

4. a) Use your rule to multiply $(x + 1)(2x + 11)$ to find a trinomial that
represents the area of the roof of the Central Mall at Simon Fraser
University.
b) If x represents 40 m, what is the area of the roof, in square metres?

5. Describe how you could use algebra tiles or diagrams to multiply
each of the following. State the resulting trinomial in each case.
a) $(x + y)(2x + y)$ **b)** $(x + 2y)(x + y)$ **c)** $(3x + y)(x + 2y)$

106

Example 1   Multiplying Binomials

Find the product of the binomials.

a) $(x+6)(x+8)$ **b)** $(2x-y)(3x+y)$

Solution 1
Use the distributive property.

a) $(x+6)(x+8) = (x+6)(x+8)$
$= (x+6)x + (x+6)8$
$= x^2 + 6x + 8x + 48$
$= x^2 + 14x + 48$

b) $(2x-y)(3x+y) = (2x-y)(3x+y)$
$= (2x-y)3x + (2x-y)y$
$= 6x^2 - 3xy + 2xy - y^2$
$= 6x^2 - xy - y^2$

Solution 2
Multiply each term in the first binomial by each term in the second binomial.

a) $(x+6)(x+8) = (x+6)(x+8)$
$= x^2 + 8x + 6x + 48$
$= x^2 + 14x + 48$

b) $(2x-y)(3x+y) = (2x-y)(3x+y)$
$= 6x^2 + 2xy - 3xy - y^2$
$= 6x^2 - xy - y^2$

Recall that FOIL means First, Inside, Outside, Last.

Example 2   Multiplying a Binomial and a Trinomial
Find the product $(y-3)(y^2 - 4y + 7)$.

Solution
Multiply each term in the first polynomial $(y-3)(y^2 - 4y + 7) = (y-3)(y^2 - 4y + 7)$
by each term in the second polynomial.

$= y^3 - 4y^2 + 7y - 3y^2 + 12y - 21$
$= y^3 - 7y^2 + 19y - 21$

Example 3   Simplifying Expressions
Expand and simplify $4x(2x^2 + 5x - 3) - (2x^2 - 7)$.

Solution
$4x(2x^2 + 5x - 3) - (2x^2 - 7) = 4x(2x^2 + 5x - 3) - 1(2x^2 - 7)$
$= 8x^3 + 20x^2 - 12x - 2x^2 + 7$
$= 8x^3 + 18x^2 - 12x + 7$

Practice

Expand.

1. $2x(3x-4)$ **2.** $3a(4a-3b)$
3. $-4t(5s-t)$ **4.** $-x(2x^2 - 7x)$
5. $2y^2(3y-1)$ **6.** $-3m^2(m^2 - 6m)$
7. $(3x-1)2$ **8.** $(4a+3)(5a)$
9. $(1-6y)(-3)$ **10.** $(x^2 - 3x)(-4x)$

Expand and simplify.

11. $2(x-4) + 5(x+3)$ **12.** $3(m-3) - 6(m-7)$
13. $4(2x-7) - 5(4x+9)$ **14.** $5(3t-7) - (4t+1)$
15. $4x + 3(2x-5) + 6(1-5x)$
16. $8(1-3y) - 4 + 2(8y-7)$

Expand.

17. $6(3a - 4b - 9)$ **18.** $-4(3t^2 - t - 1)$
19. $-(4m^2 - m - 7)$ **20.** $7(9y^2 - 3y - 7)$
21. $-5(2x^2 + 3xy - y^2)$ **22.** $4y(2y^2 + 3y - 1)$
23. $-6x^2(3x^2 - 6x - 9)$ **24.** $2ab^2(4a^2 b - ab + 3ab^2)$
25. $abc(3a + 4b - 2c)$

Expand and simplify.

26. $3x(x-4) - x(x+5) - 2x(x-1)$
27. $4a(a+3b) + 2b(2a-b) - 6(a-b)$
28. $2x(x^2 - 3x - 4) - 3x(4x^2 - x + 5)$
29. $3y(y^2 - y - 1) - 2y(3y^2 - 6)$
30. $3s(2s^2 + st - t^2) - t(4s^2 - st + 3t^2)$

107

31. a) Explain how the diagram illustrates the product $(x + 4)(x + 5)$.
b) State the product in simplified form.

	x	$+$	5
x	x^2		$5x$
$+$			
4	$4x$		20

Find each product.
32. $(t + 5)(t + 4)$ **33.** $(m + 4)(m + 1)$
34. $(x - 3)(x - 4)$ **35.** $(w - 7)(w - 8)$
36. $(x - 4)(x - 4)$ **37.** $(y - 9)(y + 7)$
38. $(s + 1)(s - 4)$ **39.** $(a - 4)(a - 9)$
40. $(4 + x)(7 - x)$ **41.** $(2 - y)(3 - y)$
42. $(x + 7)(x + 7)$ **43.** $(b - 8)(b + 8)$

44. a) Explain how the diagram illustrates the product $(2x + y)(3x + 2y)$.
b) State the product in simplified form.

	$2x$	$+$	y
$3x$	$6x^2$		$3xy$
$+$			
$2y$	$4xy$		$2y^2$

Expand and simplify.
45. $(x + 3)(3x + 1)$ **46.** $(3a + 5)(a + 4)$
47. $(y - 3)(4y + 5)$ **48.** $(5m - 2)(m - 4)$
49. $(3x - 4)(3x - 4)$ **50.** $(1 - 6t)(4 + 5t)$
51. $(3a - 5)(3a + 5)$ **52.** $(3x + y)(x + 4y)$
53. $(4a - b)(2a - 5b)$ **54.** $(5m + 2n)(4m - 3n)$
55. $(4s - 3t)(5s - 6t)$ **56.** $(7a + 8b)(a - b)$
57. $(2x^2 - xy)(x^2 - 3xy)$ **58.** $(-3a + 4b)(2a + 3b)$

Expand and simplify.
59. $(x + 6)(x + 4) + (x + 2)(x + 3)$
60. $(y - 3)(y - 1) - (y + 2)(y - 6)$
61. $(2x - 3)(x + 5) + (3x + 4)(4x + 1)$
62. $3(b - 7)(b - 6)$
63. $2(m + 3)(m + 5) + 4(2m + 3)$
64. $3(x - 4)(x + 3) - 2(4x - 1)$
65. $5(3t - 4)(2t - 1) - (6t - 5)$
66. $2(3x + 2)(3x + 2) - 3(2x - 1)(2x - 1)$
67. $12 - 2(3y - 2)(3y + 2) - (2y + 5)(2y + 5)$

68. a) Explain how the diagram illustrates the product $(2x + 1)(x + 2y + 3)$.
b) State the product in simplified form.

	x	$+$	$2y$	$+$	3
$2x$	$2x^2$		$4xy$		$6x$
$+$					
1	x		$2y$		3

Expand and simplify.
69. $(x + 3)(x^2 + 2x + 4)$
70. $(y - 2)(y^2 - y - 5)$
71. $(3m + 2)(2m^2 + 3m - 4)$
72. $(t^2 - 5t - 7)(2t + 1)$
73. $(x^2 + 2x - 1)(x^2 - x - 4)$
74. $(y - 2)(y^3 - 2y^2 + 3y - 1)$
75. $(3a^2 - 4a + 2)(a^2 - a - 5)$
76. $(x^3 - 7)(3x^3 + 7)$
77. $(2x - 1)(x^3 - 2x^2 + 5x - 3)$

Applications and Problem Solving

Expand and simplify.
78. $3[5 + 4(x - 7)]$
79. $2[3(2t - 4) + 5(t + 3)]$
80. $4[1 - 2(3y - 1)] + 2[4(y - 6) - 1]$
81. $2x[x + 2(x - 3)] - x(3x - 4)$
82. $3y[1 - y(y - 3)] - [2 - y(y - 4)]$
83. $(3x - 5)[3 + (2x + 4)(x - 1)]$
84. $(2x + 1)[(3 - 2x)(1 - 3x) + 2(x - 3)]$

85. Diving Annie Pelletier won a bronze medal for Canada in women's springboard diving at the Summer Olympics in Atlanta. She dove from a springboard with dimensions that can be represented by the binomials $7x - 2$ and $x - 10$.
a) Multiply the binomials.
b) If x represents 70 cm, what was the area of the board, in square centimetres? in square metres?

86. Forts In the 1870s and 1880s, before Alberta and Saskatchewan became Canadian provinces, a number of forts were built on the Prairies. Most of the forts were rectangular in shape and were surrounded by a high fence, known as a palisade. The table includes expressions that can be used to represent the lengths and widths of the palisades of two forts.

Fort	Date Built	Length	Width
Walsh	1875	$x + 20$	$x - 10$
Macleod	1883	$2x + 8$	$x + 7$

a) Write a trinomial that represents the area within the palisade of each fort.
b) If x represents 70 m, calculate each area, in square metres.

87. Measurement Write and simplify an expression to represent the area of each figure.

a) $x^2 + x - 2$

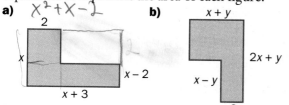

b)

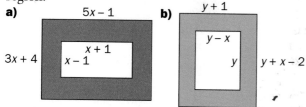

88. Measurement Write and simplify an expression to represent the area of each shaded region.

a)

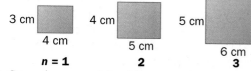

b)

$y + 1$

$y - x$

y $y + x - 2$

89. a) Multiply $(x + 1)(x + 2)$. Then, multiply the result by $x - 3$ and simplify.
b) Multiply $(x + 1)(x - 3)$. Then, multiply the result by $x + 2$ and simplify.
c) Multiply $(x - 3)(x + 2)$. Then, multiply the result by $x + 1$ and simplify.
d) Does the order in which you multiply three binomials affect the result?

90. Pattern The diagrams show the first 3 rectangles in a pattern.

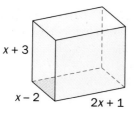

3 cm   4 cm   5 cm

4 cm   5 cm   6 cm

$n = 1$ 2   3

a) State the area of the 4th rectangle.
b) Write a product of two binomials to represent the area of the nth rectangle in terms of n.
c) Multiply the binomials from part b).
d) State the area of the 28th rectangle, in square centimetres.

91. Measurement The dimensions of a rectangular prism are represented by binomials, as shown.
a) Write, expand, and simplify an expression that represents the surface area of this prism.

$x + 3$

$x - 2$ $2x + 1$

b) Write, expand, and simplify an expression that represents the volume of the prism.
c) If x represents 5 cm, what is the surface area, in square centimetres?
d) If x represents 5 cm, what is the volume, in cubic centimetres?

92. Pattern a) Describe the pattern in words.
b) Copy and evaluate the numerical expressions.

$$1 \times 2 \times 3 + 2$$
$$2 \times 3 \times 4 + 3$$
$$3 \times 4 \times 5 + 4$$
$$4 \times 5 \times 6 + 5$$

c) How is each answer related to the middle number of the three that are multiplied together?
d) Generalize the pattern by letting the middle number in the three that are multiplied together be x. Write expressions for the other numbers in terms of x. Then, write an algebraic expression that matches the pattern in the numerical expressions. Expand and simplify the algebraic expression.
e) How does the simplified expression explain the answer to part c)?

93. Pattern a) Describe the pattern in words.
b) Copy and evaluate the numerical expressions.

$$(2 \times 3) - (1 \times 4)$$
$$(3 \times 4) - (2 \times 5)$$
$$(4 \times 5) - (3 \times 6)$$
$$(5 \times 6) - (4 \times 7)$$

c) Generalize the pattern by letting the first term in the second bracket be x. Write expressions for the other numbers in terms of x. Then, write an algebraic expression that matches the pattern in the numerical expressions. Expand and simplify the algebraic expression.
d) How does the simplified expression explain the answer to part b)?

94. Is the product of two binomials always a trinomial? Explain.

LOGIC POWER

What are the four moves that X should not play, if X wants to stop O from winning?

109

3.4 Special Products

The Haida have lived on the coast of the Queen Charlotte Islands for at least 6000 years. Haida houses, known as plank-houses, were built using large posts, beams, and planks cut from local cedars. The largest of the Haida houses was built around 1850 at Masset.

The base of the house at Masset was a square. The area of the base can be represented by the expression $(x + 2)^2$. Just as 4^2 means the square of 4, or 4×4, the expression $(x + 2)^2$ means the square of $x + 2$, or $(x + 2)(x + 2)$.

Explore: Look for Patterns

Copy and complete the table.

Binomial Squared	Simplified Trinomial
$(x + 3)^2$	$x^2 + 6x + 9$
$(x - 4)^2$	$x^2 - 8x + 16$
$(2x + 3)^2$	$4x^2 + 12x + 9$
$(3x - 2)^2$	$9x^2 - 6x + 4$
$(5x + 3y)^2$	$25x^2 + 30xy + 9y^2$
$(2x - 5y)^2$	$4x^2 - 60xy + 25y^2$
$(a + b)^2$	$a^2 + 2ab + b^2$
$(a - b)^2$	$a^2 - 2ab + b^2$

Inquire

1. Compare the first term in each trinomial to the first term in each binomial. What is the pattern?

2. Compare the last term in each trinomial to the second term in each binomial. What is the pattern?

3. Compare the middle term in each trinomial to the two terms in each binomial. What is the pattern?

4. Write a rule for squaring a binomial.

5. a) Use the rule to expand $(x + 2)^2$ to find an expression that represents the area of the base of the Haida house at Masset.
b) If x represents 15 m, what was the area of the base, in square metres?
c) How does the area you found in part b) compare with the area of your classroom floor?

Explore: Look for Patterns

Copy and complete the table.

Multiplication	Simplified Product
$(x + 4)(x - 4)$	$x^2 - 16$
$(y - 5)(y + 5)$	$y^2 - 25$
$(2x - 1)(2x + 1)$	$4x^2 - 1$
$(3x + 2y)(3x - 2y)$	$9x^2 - 4y^2$
$(x^2 - 1)(x^2 + 1)$	$x^4 - 1$
$(a + b)(a - b)$	$a^2 - b^2$

Inquire

1. How are the two binomials in each multiplication alike?

2. How are the two binomials in each multiplication different?

3. How do the terms in the product compare with the two binomials in each multiplication?

4. Explain why there are only two terms in the product.

5. Write a rule for finding the product of these types of binomials.

Example 1   Squaring Binomials

Expand. **a)** $(3x + 2y)^2$ **b)** $(3x - y)^2$

Solution

Use the patterns for squaring binomials.

a) $(a + b)^2 = a^2 + 2ab + b^2$
$(3x + 2y)^2 = (3x)^2 + 2(3x)(2y) + (2y)^2$
$\qquad\qquad = 9x^2 + 12xy + 4y^2$

b) $(a - b)^2 = a^2 - 2ab + b^2$
$(3x - y)^2 = (3x)^2 - 2(3x)(y) + (y)^2$
$\qquad\qquad = 9x^2 - 6xy + y^2$

A product of the type $(a + b)(a - b)$ is known as a product of the sum and difference of two terms.

Example 2   Product of a Sum and Difference

Expand $(4x + 3y)(4x - 3y)$.

Solution

Use the pattern for the product of a sum and difference.
$$(a + b)(a - b) = a^2 - b^2$$
$$(4x + 3y)(4x - 3y) = (4x)^2 - (3y)^2$$
$$= 16x^2 - 9y^2$$

Example 3   Simplifying Expressions

Expand and simplify $2(3x + 4)^2 - (4x + 5)(4x - 5)$.

Solution

$$2(3x + 4)^2 - (4x + 5)(4x - 5) = 2(9x^2 + 24x + 16) - (16x^2 - 25)$$
$$= 18x^2 + 48x + 32 - 16x^2 + 25$$
$$= 2x^2 + 48x + 57$$

Example 4   Cube of a Binomial

Expand and simplify.

a) $(x + 3)^3$ **b)** $(2x - 3y)^3$

Solution

a) $(x + 3)^3 = (x + 3)(x + 3)^2$
$\qquad\qquad = (x + 3)(x^2 + 6x + 9)$
$\qquad\qquad = x^3 + 6x^2 + 9x + 3x^2 + 18x + 27$
$\qquad\qquad = x^3 + 9x^2 + 27x + 27$

b) $(2x - 3y)^3 = (2x - 3y)(2x - 3y)^2$
$\qquad\qquad = (2x - 3y)(4x^2 - 12xy + 9y^2)$
$\qquad\qquad = 8x^3 - 24x^2y + 18xy^2 - 12x^2y + 36xy^2 - 27y^3$
$\qquad\qquad = 8x^3 - 36x^2y + 54xy^2 - 27y^3$

Practice

Expand.

1. $(x+5)^2$ **2.** $(y+1)^2$

3. $(x-6)^2$ **4.** $(m-3)^2$

5. $(x+3)(x-3)$ **6.** $(y+6)(y-6)$

7. $(m-7)(m+7)$ **8.** $(t-8)(t+8)$

Expand.

9. $(3x+2)^2$ **10.** $(5x-1)^2$

11. $(2x+3)(2x-3)$ **12.** $(2m+7)^2$

13. $(3y-2)(3y+2)$ **14.** $(4y-3)^2$

15. $(1+5m)(1-5m)$ **16.** $(2-3t)^2$

Expand.

17. $(2x+3y)(2x-3y)$ **18.** $(2x+3y)^2$

19. $(3a-b)(3a+b)$ **20.** $(4t-5s)^2$

21. $(4m-5n)(4m+5n)$ **22.** $(3c+7d)^2$

23. $(y+6x)(y-6x)$ **24.** $(a-8b)^2$

Expand and simplify.

25. $(x+4)^2+(x-2)^2$

26. $(y+6)(y-6)+(y+7)^2$

27. $(m-8)^2-(m-1)(m+1)$

28. $2(a+3)(a-3)+3(a+2)^2$

29. $4(2x+1)^2-2(3x-7)^2$

30. $5(3t-1)^2-4(4t-5)(4t+5)$

Expand and simplify.

31. $(x-7)(x+5)-2(x+6)^2$

32. $(2x-3)^2-(3x-1)(4x+5)$

33. $3(3m+1)(m-4)-4(2m+3)(2m-3)$

34. $3t^2-(3-2t)^2+5(2t-1)(2t+1)$

35. $12-2(3y+1)^2-(y-9)(3y+2)$

36. $2t(1-3t^2)-4(1-3t)(1+3t)$

37. $4(5s+t)(5s-t)-6(3t^2-t)$

38. $2(3m-n)^2-3(2m-5)(m+3)$

39. $(x+2y)(x-2y)+(2x+y)^2$

40. $3(a-2b)^2-4(2a+b)^2$

41. $(2x+5)(5x-3)(5x+3)$

42. $(2m-1)(3m+4)^2$

Expand and simplify.

43. $(x+2)^3$ **44.** $(y-1)^3$

45. $(4x+5)^3$ **46.** $(3-2t)^3$

47. $(2y-x)^3$ **48.** $(3a+2b)^3$

Applications and Problem Solving

49. Egyptian Pyramids The three giant pyramids near Giza in Egypt are included among the Seven Wonders of the Ancient World. Each of the three pyramids has a square base, whose side length can be represented by the expression shown in the table.

Pyramid	Side Length of Base
Menkaure	$x-1$
Khafre	$2x-4$
Khufu	$2x+10$

a) Write and expand an expression that represents the area of each base.

b) If x represents 110 m, what is the area of the base of each pyramid, in square metres?

50. Use $x=y-2$ to write each of the following in terms of y. Then, expand and simplify.

a) x^2-2x+3 **b)** $3x^2+5x-9$

51. Measurement

a) Write, expand, and simplify an expression that represents the volume of the rectangular prism.

b) If x represents 8 cm, what is the volume, in cubic centimetres?

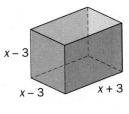

$x-3$

$x-3$ $x+3$

52. Mining Canada is one of the world's top producers of copper. If the copper produced in Canada in one year were made into a cube, the length of each edge could be represented by the expression $7x-2$.

a) Write, expand, and simplify an expression that represents the volume of the cube.

b) If x represents 6 m, calculate the volume of the cube, in cubic metres.

53. Measurement The length of an edge of a cube is represented by the expression $2x-y$.

a) Write, expand, and simplify an expression that represents the surface area of the cube.

b) If x represents 3 cm and y represents 2 cm, calculate the surface area, in square centimetres.

c) Write, expand, and simplify an expression that represents the volume of the cube.

d) If x represents 3 cm and y represents 2 cm, calculate the volume, in cubic centimetres.

54. Expand and simplify.
a) $(x^2 + 1)^2$ **b)** $(y^2 - 1)^2$
c) $(x^2 + y^2)^2$ **d)** $(x^2 - y^2)^2$
e) $(2x^2 + 3)^2$ **f)** $(3y^2 - 4)^2$
g) $(x^2 - 2y^2)^2$ **h)** $(4x^2 + 3y^2)^2$

55. Expand and simplify.
a) $(x^2 + 1)(x^2 - 1)$ **b)** $(y^2 - 2)(y^2 + 2)$
c) $(x^2 + y^2)(x^2 - y^2)$ **d)** $(8a^2 + 3)(8a^2 - 3)$
e) $(3x^2 + 2y^2)(3x^2 - 2y^2)$
f) $(4 - 3c^2)(4 + 3c^2)$

56. Expand and simplify.
a) $(6y - 1)(6y + 1)(y - 1)^2$
b) $(m - 2)^2(m + 3)^2$

57. Measurement The side length of a square is represented by x centimetres. The length of a rectangle is 3 cm greater than the side length of the square. The width of the rectangle is 3 cm less than the side length of the square. Which figure has the greater area and by how much?

58. Measurement The side length of the square is $a + b + c$.
a) Write, expand, and simplify an expression that represents the area of the square.
b) Use your result from part a) to expand $(2x + 3y + 1)^2$.

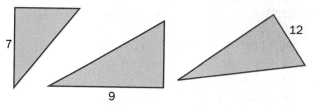

59. Pattern Here are the first 5 rows of Pascal's triangle.
a) Expand and simplify $(a + b)^2$. How do the coefficients of the terms in the result compare with the numbers in the third row of Pascal's triangle?
b) Expand and simplify $(a + b)^3$. How do the coefficients of the terms in the result compare with the numbers in the fourth row of Pascal's triangle?
c) Use your findings in parts a) and b) to expand and simplify $(a + b)^4$.
d) Use the pattern you found for $(a + b)^3$ in part b) to expand $(x + 3y)^3$.
e) Use the pattern you found for $(a + b)^4$ in part c) to expand $(x + 2)^4$.

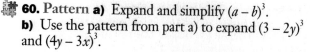

60. Pattern a) Expand and simplify $(a - b)^3$.
b) Use the pattern from part a) to expand $(3 - 2y)^3$ and $(4y - 3x)^3$.

61. Measurement If a right triangle has one leg of length 8 units, possible whole-number lengths of the other leg and the hypotenuse can be found as follows. List pairs of whole numbers whose product is 8^2 or 64.

$64 = 64 \times 1$ $64 = 16 \times 4$
$64 = 32 \times 2$ $64 = 8 \times 8$

Then, find whether each pair of whole numbers represents the product of the sum and difference of two whole numbers.

$64 = 32 \times 2$ $64 = 16 \times 4$
$= (17 + 15)(17 - 15)$ $= (10 + 6)(10 - 6)$
$= 17^2 - 15^2$ $= 10^2 - 6^2$

For the products 64×1 and 8×8, there are no pairs of whole numbers with either a sum of 64 and a difference of 1, or a sum of 8 and a difference of 8. Thus, if a right triangle has one leg of length 8 units, the other leg and the hypotenuse could measure 6 units and 10 units, or 15 units and 17 units.
a) Find the possible whole-number lengths of the second leg and the hypotenuse for each of the following right triangles.

b) Explain why the method works.

3.5 Use a Diagram

Diagrams can be useful for simplifying and solving problems of many types, including design, navigation, and scheduling problems.

For their final assignment, ten media students must write reviews of three television commercials. Each commercial is to run continuously for 10 min. To allow time for changeovers, 15-min time slots are needed. The commercials each student has chosen to review are shown in the table. Design a possible schedule for showing all the commercials in the minimum number of time slots, starting at 13:00. All students must be able to see their chosen commercials, so there must be no conflicts.

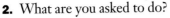

| | Student | | | | | | | | | |
Commercial	1	2	3	4	5	6	7	8	9	10
A	X			X				X	X	
B			X		X		X			X
C		X	X				X	X	X	
D				X		X	X			X
E		X		X		X		X		X
F			X		X		X		X	X
G				X		X		X		

Understand the Problem

1. What information are you given?
2. What are you asked to do?

Think of a Plan

Represent each commercial by a point. Since Student 1 wants to review commercials A, C, and E, join these points as shown in Figure 1 to indicate that these commercials cannot be shown at the same time. Complete the network by connecting the points that represent the commercials each of the other students wants to review, as shown in Figure 2.

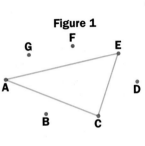

Figure 1

Carry Out the Plan

Now put a symbol at each point to represent a time slot. Because connected points represent commercials that cannot be shown at the same time, no two connected points can have the same symbol. Use the least possible number of symbols, as shown in Figure 3.

Since four different symbols are used, four time slots are needed. Commercials A and G will be shown at the same time, as will B and E, and C and D.

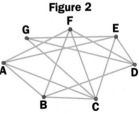

Figure 2

A possible schedule is as follows.
☐ Commercials A and G: 13:00 to 13:15
△ Commercials B and E: 13:15 to 13:30
○ Commercials C and D: 13:30 to 13:45
◊ Commercial F: 13:45 to 14:00

Look Back

At what times will Student 3 see the commercials? Can all students review the commercials they want? Is there another way to schedule the commercials into the four time slots, so that there are no conflicts? Is there another way to solve the problem?

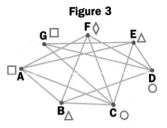

Figure 3

Use a Diagram

1. Draw a diagram to represent the situation.
2. Use the diagram to solve the problem.
3. Check that the answer is reasonable.

Applications and Problem Solving

1. Seven film students must write reviews of three short films, each lasting one-half hour. The films the seven students have chosen are shown in the table.

	Student						
Film	**1**	**2**	**3**	**4**	**5**	**6**	**7**
A	X		X	X		X	
B	X				X	X	X
C		X		X		X	X
D		X			X		
E	X	X	X	X			X
F		X			X		

Set up a schedule for showing the films in the minimum number of time periods, starting at 13:00. Allow 10 min for changeovers between time periods. All students must be able to see their three films, so there can be no conflicts.

2. There are 6 towns in Galleragas County — Alliston, Bevan, Dunstan, Clearwater, Flagstaff, and Gaston. Only the following pairs of towns are joined by roads.

Alliston is 25 km from Flagstaff.
Bevan is 32 km from Gaston.
Dunstan is 28 km from Flagstaff.
Clearwater is 19 km from Bevan.
Gaston is 33 km from Flagstaff.
Alliston is 24 km from Bevan.
Dunstan is 19 km from Alliston.
Clearwater is 27 km from Flagstaff.

Draw a map showing possible locations of the towns. Then, find the shortest route from
a) Alliston to Gaston
b) Clearwater to Dunstan
c) Gaston to Bevan if you must go through Flagstaff
d) Flagstaff to Alliston if you must go through Bevan

3. How many diagonals are there in
a) a regular octagon?
b) a regular decagon?

4. The diagram shows the outdoor pool at Lakeview Park. A cement deck 2 m wide surrounds the outside edge of the pool. Calculate the area of the deck.

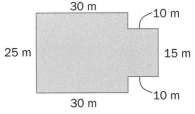

5. There are 6 teams in a summer basketball league. Each team plays 4 games against each of the other teams. How many games are played?

6. Sarah wants to cover the floor of a kitchen with 20-cm square floor tiles. The kitchen measures 3 m by 4 m. How many tiles does she need?

7. Ari has to paint the ceiling and walls of a room in his house. The room is 5 m long and 4 m wide. The ceiling height is 2.5 m. There are 2 doors, each measuring 1 m by 2 m. There are 3 windows, each measuring 1 m by 1.5 m. The windows and doors will not be painted. The room will get 2 coats of paint. One litre of paint covers 10 m^2. How many litres of paint should Ari buy?

8. Tara left her hotel to sightsee in Calgary. She walked 4 blocks north, 3 blocks east, 1 block south, 4 blocks east, 6 blocks north, 5 blocks west, 2 blocks south, and 6 blocks west. How many blocks and in which directions should Tara walk to get back to her hotel in the shortest distance?

9. Write a scheduling problem, like the one shown in the example. Have a classmate solve your problem.

LOGIC POWER

A box can hold 24 soup cans in 4 rows and 6 columns. Show how you can put 18 of the cans in the box so that the number of cans in each row and in each column is even.

TECHNOLOGY

Multiplying Polynomials With a Graphing Calculator

Complete each of the following with a graphing calculator that has the capability to expand and simplify the products of polynomials.

1 Multiplying Polynomials

Expand.
1. $(4x + 3)(2x + 9)$
2. $(2x - 5)(3x + 5)$
3. $(2y - 3)(10y - 11)$
4. $(4x + 3y)(8x - 7y)$
5. $(5x - 2y)(3y + 4x)$
6. $(3x^2 - 5)(4x^2 + 7)$
7. $(x^2 + 2y)(3x^2 - 5y)$
8. $(5x^3 - 4y)(2x^3 - y)$

Expand and simplify.
9. $(x + 2)(x^2 - 3x + 6)$
10. $(2x - 3)(2x^2 + 5x + 1)$
11. $(5m + 6)(m^2 - 8m + 3)$
12. $(x^2 + 3x + 1)(x^2 + 2x + 5)$
13. $(y^2 - y - 1)(3 - 4y + 2y^2)$
14. $2x^2 + (4x + 1)(5x^2 - 2x - 3)$
15. $(2x^2 + 4y)(3x^2 - y) - 3(x^2 - y^2)$
16. $4(3x^2 + 2) + (x + 1)(x^2 - 6x - 5)$

2 Special Products

Expand.
1. $(x + 15)^2$
2. $(t - 5)^2$
3. $(9 - y)^2$
4. $(3x - 8)^2$
5. $(4m^2 + 7)^2$
6. $(6 - 5r)^2$
7. $(8x + 3y)^2$
8. $(-5x + 2y)^2$
9. $(3x + 2)^3$
10. $(4x + 5y)^3$
11. $(2y - 7x)^3$
12. $(2x + 3y + z)^2$

Expand.
13. $(2x + 11)(2x - 11)$
14. $(4 + 5x)(4 - 5x)$
15. $(3y + 5x)(3y - 5x)$
16. $(6t^2 - 5s^2)(6t^2 + 5s^2)$

Expand and simplify.
17. $(x + 1)^2 - (x - 1)^2$
18. $(x + 3)(x - 3) - (x + 2)^2$
19. $3(4y + 3)^2 - (2y + 1)^2$
20. $2(3 + 2m)^2 - 3(4m - 1)^2$
21. $(2x + 5)(2x - 5)(x + 3)^2$
22. $3(y + 5)^2(y - 2)^2$

3 Multiplying and Evaluating

Use your graphing calculator to evaluate each of the following for the given value(s) of the variable(s).

1. $x^2 - 5; x = 2$

2. $y^2 + 10y - 6; y = 3$

3. $6x^2 + 18x - 31; x = 9$

4. $4y^2 - 15y - 2; y = -3$

5. $2x^2 - 5xy + 12y^2; x = -1, y = 4$

6. $20x^2 + 3xy + 9y^2; x = -10, y = 2$

7. $41 - 5x + 3x^2; x = 1.5$

8. $8u^2 - 9uv - 2v^2; u = 1.8, v = -0.5$

9. In archery competitions, two archers usually shoot side by side at two targets placed at the end of a rectangular area called a lane.

a) The dimensions of a lane in men's competition can be represented by the binomials $13x - 1$ and $x - 2$. Write and expand an expression that represents the area of the lane.

b) If x represents 7 m, what is the area of the lane, in square metres?

c) The dimensions of a lane in women's competition can be represented by the binomials $23y + 1$ and $2y - 1$. Write and expand an expression that represents the area of the lane.

d) If y represents 3 m, what is the area of the lane, in square metres?

10. Expo 86 in Vancouver featured the world's tallest freestanding flagpole, with a height of 86 m. On the Canadian flag flown from the flagpole, the side length of the white square could be represented by the expression $5x - 3$.

a) Write and expand an expression to represent the area within the white square, including the red maple leaf.

b) Half the area of a Canadian flag lies within the sides of the white square. If x represents 3 m, calculate the area of the whole flag, in square metres.

3.6 Reviewing Common Factors

The terms in the expression $2x + 2y$ have the common factor 2, because $2x = 2 \times x$ and $2y = 2 \times y$. Removing the common factor from the expression $2x + 2y$ gives the expression $2(x + y)$, which is written in factored form.

The minimum stopping distance of a car, in metres, on dry, level concrete, is modelled by the polynomial $0.2s + 0.007s^2$, where s is the car's speed in kilometres per hour. The terms in the polynomial $0.2s + 0.007s^2$ have the common factor s.

Explore: Use Algebra Tiles

To factor $2x^2 + 4x$, use two x^2-tiles and four x-tiles to make a rectangle. Use the rectangle to write the area as a product of the length and width. There are two possible rectangles.

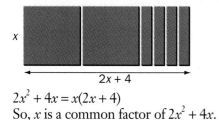

x

$2x + 4$

$2x^2 + 4x = x(2x + 4)$
So, x is a common factor of $2x^2 + 4x$.

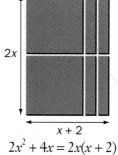

$2x$

$x + 2$

$2x^2 + 4x = 2x(x + 2)$
So, $2x$ is a common factor of $2x^2 + 4x$.

The greatest common factor is $2x$, so the second rectangle gives the answer. Thus, $2x^2 + 4x = 2x(x + 2)$.

Examine the following four rectangles made using algebra tiles. Then, copy and complete the table.

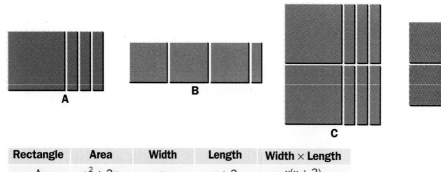

A

B

C

D

Rectangle	Area	Width	Length	Width × Length
A	$x^2 + 3x$	x	$x + 3$	$x(x + 3)$
B	$3x^2 + x$	x	$3x + 1$	$x(3x + 1)$
C	$2x^2 + 6x$	$2x$	$x + 3$	$2x(x + 3)$
D	$2x^3 + 2x^2$	$2x$	$x^2 + x$	$2x(x^2 + x)$

Inquire

1. For each rectangle, removing a common factor from the expression in the "Area" column gives the expression in the "Width × Length" column. Write a rule for removing a common factor from an expression.

2. Use your rule to factor each of the following polynomials.
   a) $5x^2 + 20x$ **b)** $6x^2 - 3x$ **c)** $12ab + 8ac - 4ad$

3. **a)** Factor the expression $0.2s + 0.007s^2$ for finding the minimum stopping distance of a car on dry, level concrete.
   b) Calculate the minimum stopping distance, in metres, for a speed of 50 km/h; of 90 km/h.

A polynomial is factored when it is written as the product of polynomials. It is always possible to factor a polynomial in some way. For example, here are 3 ways to factor $2x + 5$.

$$2x + 5 = 2(x + 2.5) \qquad 2x + 5 = 5\left(\frac{2}{5}x + 1\right) \qquad 2x + 5 = x\left(2 + \frac{5}{x}\right)$$

We will consider that a polynomial is completely factored when no more variable factors can be removed and no more integer factors, other than 1 or –1, can be removed. In factoring polynomials with integer coefficients, we will find factors with integer coefficients. This process is called **factoring over the integers**. The polynomial $2x + 5$ cannot be factored over the integers. Factoring $2x + 4$ over the integers gives $2(x + 2)$ as the factored form, not $4\left(\frac{x}{2} + 1\right)$ or $x\left(2 + \frac{4}{x}\right)$.

Example 1   Factoring a Polynomial
Factor $8x^3 - 6x^2y^2 + 4x^2y$.

Solution
The greatest common factor of the coefficients, 8, 6, and 4, is 2.
The greatest common factor of the variable parts, x^3, x^2y^2, and x^2y, is x^2.
Therefore, the greatest common factor of the polynomial is $2x^2$.
$$8x^3 - 6x^2y^2 + 4x^2y = 2x^2(4x) - 2x^2(3y^2) + 2x^2(2y)$$
$$= 2x^2(4x - 3y^2 + 2y)$$

Example 2   Binomial Common Factor
Factor $4x(w + 1) + 5y(w + 1)$.

Solution
Think of $(w + 1)$ as one number.
$$4x(w + 1) + 5y(w + 1) = (w + 1)(4x) + (w + 1)(5y)$$
$$= (w + 1)(4x + 5y)$$

Some polynomials do not have a common factor in all of their terms. These polynomials can sometimes be factored by grouping terms that do have a common factor.

Example 3   Factoring by Grouping
Factor $ac + bc + ad + bd$.

Solution
Group terms that have a common factor.
$$ac + bc + ad + bd = (ac + bc) + (ad + bd) \qquad \text{or} \qquad ac + bc + ad + bd = (ac + ad) + (bc + bd)$$
$$= c(a + b) + d(a + b) \qquad\qquad\qquad = a(c + d) + b(c + d)$$
$$= (c + d)(a + b) \qquad\qquad\qquad\qquad = (a + b)(c + d)$$

Practice

Factor, if possible.

1. $5x + 25$ **2.** $4x + 13$ **3.** $8x + 8$

4. $9y - 9$ **5.** $3x - 15y$ **6.** $25x^2 + 10x$

7. $4ax + 8ay - 6az$ **8.** $5pqr - pqs - 10pqt$

9. $2x^2 - 2x - 6$ **10.** $3y^2 - 9y - 20$

Factor completely, if possible.

11. $9a^3 + 27b^2$

12. $3x^5 - 6x^3 + 3x$

13. $12y - 8y^2 + 24y^3$

14. $24w^5 - 6w^3$

15. $6rst + 3rs - 7t$

16. $33ab + 22bc - 11b^2$

17. $24xy^2 + 16x^2y$

18. $35xy - 10y^2$

19. $5rst - 15ab + 7cd$

20. $24xy^2 - 12xy + 36x^2y$

21. $27a^2b^3 + 9a^2b^2 - 18a^3b^2$

22. $6m^3n^2 + 18m^2n^3 - 12mn^2 + 24mn^3$

Factor, if possible.

23. $5x(a + b) + 3(a + b)$

24. $3m(x - 1) + 5(x - 1)$

25. $7x(m + 4) - 3(m - 4)$

26. $4y(p + q) - x(p + q)$

27. $4t(m + 7) + (m + 7)$

28. $3t(x - y) - (x + y)$

29. $8x(m - n) + 6(m - n)$

Factor by grouping.

30. $wx + wy + xz + yz$ **31.** $xy + 12 + 4x + 3y$

32. $x^2 + x - xy - y$ **33.** $m^2 - 4n + 4m - mn$

34. $2x^2 + 6y + 4x + 3xy$ **35.** $5m^2t - 10m^2 + t^2 - 2t$

36. $3a^2 + 6b^2 - 9a - 2ab^2$

Applications and Problem Solving

37. Vertical motion If an object is thrown vertically upward at a speed of v metres per second, the height of the object, in metres, after t seconds is given by the expression $vt - 5t^2$.

a) A ball is thrown vertically upward at a speed of 20 m/s. Write the expression that gives the height of the ball, in metres, after t seconds.

b) Make a table to find the height of the ball after 0 s, 1 s, 2 s, 3 s, 4 s, and 5 s.

c) What is the maximum height of the ball?

d) Why does the height of the ball after 5 s have no meaning?

e) At what times is the height 0 m?

f) Factor the expression you wrote in part a).

g) How does the factored expression let you determine the times when the height of the ball is 0 m?

Measurement *Write an expression for the area of each shaded region*

a) *as a polynomial*   **b)** *in factored form*

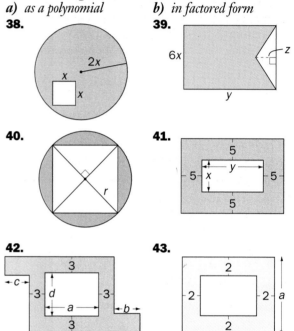

38. **39.** **40.** **41.** **42.** **43.**

44. If it is possible to remove a common factor from the expression $2x^2 + ky + 4$, where k is an integer, what can you state about the possible values of k? Explain.

45. a) Write a polynomial in which there are three terms, each numerical coefficient is 1, and the greatest common factor is st^2.

b) Write your polynomial in factored form.

46. a) Write a polynomial that has four terms with different numerical coefficients and a greatest common factor of $3xy$.

b) Write your polynomial in factored form.

47. Write a problem similar to questions 45 and 46. Have a classmate solve your problem.

Personal Fitness Training

Physical activity contributes to our overall health and well-being. However, according to the first National Health Survey from Statistics Canada, only 17% of Canadians over the age of 15 are physically active, with teenagers the most active.

Some people consult a personal fitness trainer at a recreation centre or a fitness club to increase their level of physical activity. A personal trainer assesses fitness levels, and designs and monitors programs to improve fitness levels. Fitness assessments include measures of cardiorespiratory fitness, muscle strength, flexibility, and endurance.

Most personal trainers study physical education and exercise science at university or college. Many are certified by the Canadian Association of Sports Science as fitness appraisers.

1 Cardiorespiratory Assessment

One index of cardiorespiratory fitness involves three 30-s heartbeat measurements taken at precise intervals after a timed step exercise. This index is found by substituting the duration of the exercise, d seconds, and the three heartbeat measurements, a, b, and c beats, into the expression $\dfrac{100d}{2(a+b+c)}$.

The index is interpreted using this scale.

Index	Fitness Level
< 55	poor
55 – 64	low average
65 – 79	average
80 – 89	good
> 90	excellent

1. a) Expand the denominator of the expression.
b) What do the terms in the expanded form of the denominator represent?

2. Write the expression $\dfrac{100d}{2(a+b+c)}$ in simplest form.

3. Find, and interpret, the cardiorespiratory fitness of each person.
a) 5 min of exercise and heartbeat measurements of 91, 88, and 83 beats
b) 4 min of exercise and heartbeat measurements of 52, 49, and 45 beats
c) 5 min of exercise and heartbeat measurements of 70, 66, and 59 beats

2 Burning Energy

The amount of energy you burn depends on how active you are. The energy, in kilojoules, a person burns during an activity can be found using the expression mdr. In this expression, m kilograms is the person's mass, d minutes is the duration of the activity, and r kilojoules per minute per kilogram is the rate.

This table shows the rates for some activities.

Activity	Rate (kJ/min/kg)
Aerobic Dancing	0.567
Free Weights	0.361
Running (7.2 min/km)	0.567
Running (5.0 min/km)	0.874
Running (3.7 min/km)	1.058
Swimming	0.655
Walking	0.336
Word Processing	0.113

1. Find the energy burned, to the nearest tenth of a kilojoule, in each situation.
a) a 53.5-kg person lifting free weights for 30 min
b) a 66-kg person word processing for 2 h
c) a 72.5-kg person running at 7.2 min/km for 45 min

2. Which burns more energy for a specific person, walking for 1 h or aerobic dancing for 30 min? How much more energy is burned?

121

PROBLEM SOLVING

3.7 Work Backward

You could take many routes to get to Mount Robson Provincial Park. Some routes would take longer than others. When you find the route that takes the least time, you are solving an *optimization problem*. Optimization problems are very common. Bus and trucking companies want to know the best routes to take. Cities want to make water and sewage systems as efficient as possible.

Many optimization problems can be expressed mathematically, so that every possible solution can be given a score. The object is to find the solution with the best score, which may indicate the shortest time, the largest profit, and so on.

Optimization problems can be solved by computer. In some difficult cases, even a powerful computer takes too long to work out every possible solution. The computer must then be programmed to use a more efficient method.

The map at the right is drawn on a 3 by 3 block. The map shows the possible routes between A and B. You can travel only north (up) or east (right). The time to travel each route is marked in minutes. The object is to find the route from A to B that takes the shortest time.

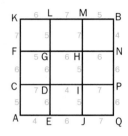

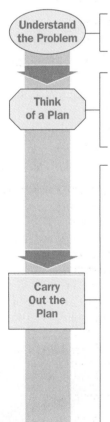

Understand the Problem

1. What information are you given?
2. What are you asked to find?

Think of a Plan

One way to solve the problem is to list the 20 routes from A to B and find the total time for each route. This method is called the "brute force" method. Another method, called **dynamic programming**, was developed by Richard Bellman. For this method, you start at B and work backward.

Carry Out the Plan

From M, there is 1 route to B. It takes 5 min. Write 5 at M and an arrow *right*, as shown in Figure 1.
From L, there is 1 route to B. It takes 7 + 5 or 12 min. Write 12 at L and an arrow *right*.
From N, there is 1 route to B. It takes 4 min. Write 4 at N and an arrow *up*.
From P, there is 1 route to B. It takes 6 + 4 or 10 min. Write 10 at P and an arrow *up*.
From H, there are 2 routes to B. Going up first takes 7 + 5 or 12 min. Going right first takes 6 + 4 or 10 min. Write 10 at H and an arrow *right*.

From K, there is 1 route. It takes 6 + 12 or 18 min, as shown in Figure 2.
Going up from G takes 5 + 12 or 17 min. Going right from G takes 6 + 10 or 16 min. Write 16 at G and an arrow *right*.
Going up from I takes 5 + 10 or 15 min. Going right from I

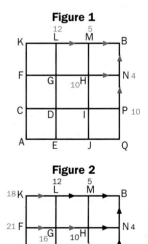

Figure 1

Figure 2

takes 7 + 10 or 17 min. Write 15 at I and an arrow *up*.
From Q, there is 1 route. It takes 6 + 10 or 16 min.
Going up from F takes 7 + 18 or 25 min. Going right from F
takes 5 + 16 or 21 min. Write 21 at F and an arrow *right*.
Going up from D takes 6 + 16 or 22 min. Going right from D
takes 4 + 15 or 19 min. Write 19 at D and an arrow *right*.
Going up from J takes 5 + 15 or 20 min. Going right from J
takes 7 + 16 or 23 min. Write 20 at J and an arrow *up*.

Repeating the process results in Figure 3.

The shortest route from A to B takes 29 min.
The route is shown by the arrows from A in Figure 4.
The route is A to E to D to I to H to N to B.

Check the time taken on the shortest route from A to B
against the data given in the problem.

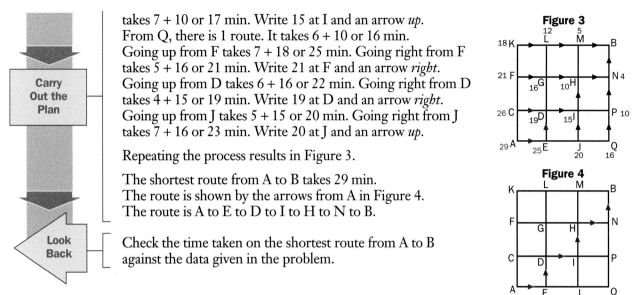

Figure 3

Figure 4

Using dynamic programming, 22 additions were needed to solve the problem. Using
the brute force method would have taken 100 additions. As the number of blocks
increases, the number of routes and additions increases dramatically. A
30 by 30 block problem would have about 1.2×10^{17} different routes and about
7.1×10^{18} additions. At 1 billion additions per second, a computer would take
about 225 years to solve this problem using the brute force method. Using dynamic
programming, a computer can solve a 30 by 30 block problem in less than a second.
Many real-world problems are larger than a 30 by 30 block problem.

Work Backward

1. Start with what you know.
2. Work backward to get an answer.
3. Check that the answer is reasonable.

Applications and Problem Solving

1. Copy each diagram. The time to travel each road is marked in
minutes. Use dynamic programming to find the route that takes the
least time to travel from A to B. Assume that north is up.

a)

Travel north or east.

b)

Travel south or west.

c)

Travel north or west.

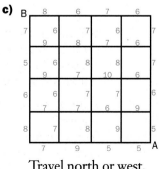

123

2. Terry put some baseballs into a pitching machine and tried to hit home runs. On the first round, he hit $\frac{1}{3}$ of the balls out of the park. Balls that were not hit for home runs were put back into the machine. On the second round, Terry hit $\frac{1}{4}$ of the remaining balls out of the park. On the third round, he hit $\frac{1}{6}$ of the remaining balls out of the park. After this round, 15 balls remained. How many balls were there at the start?

3. The bus taking the school band to the airport is leaving at 07:30. You have to allow 30 min to get your drums from the music room and put them on the bus. You take 25 min to walk to school. Before you leave home, you need 15 min to pack, and 25 min to shower and get dressed. Eating breakfast takes 15 min. Walking your dog takes 20 min. For what time should you set your alarm?

4. The map shows an area in Saskatchewan.

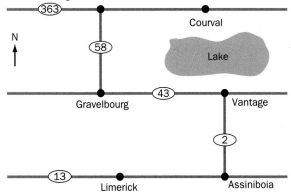

a) Crystal wants to drive from Courval to Limerick. Write the directions you would give her. Include highway numbers, towns, and the turns she should make.
b) Without looking at the map, use the directions you wrote in part a) to write the directions Crystal needs to drive from Limerick to Courval.

5. The population of the Yukon Territory was 2.25 times greater in 1951 than it was in 1931. The population doubled from 1951 to 1971 and increased by 10 000 from 1971 to 1991. In 1991, the population reached 28 000.
a) What was the population of the Yukon Territory in 1931?

b) In 1901, the population of the Yukon Territory was 27 000, but it dropped to 9000 ten years later. Explain why.
c) Use the information provided in parts a) and b) to sketch a graph of population versus year for the Yukon Territory from 1901 to 1991.

6. In a chess tournament, each player played a game in the first round. The losers dropped out. This process continued until a winner was declared. Jasmin won the tournament by winning 8 games. How many people were in the tournament?

7. Two numbers have a sum of 3 and a product of 4. Find the sum of the reciprocals of the two numbers.

8. Describe how an automobile accident investigator uses the work backward problem solving strategy to determine what happened.

9. Describe how the work backward problem solving strategy is used to solve equations.

10. There is a two-person game called "Say 21." The game starts at 1 and ends at 21. Players take turns increasing the total by 1 or 2. The loser is the player who is forced to say "21."
a) Play 4 games, alternating going first.
b) In pairs, work backward to find a winning strategy for the game. Write a description of your strategy, stating the reasons for your decisions.
c) Suppose the rules are changed so that the winner is the person who says "21." Play 4 games, alternating going first. Then, use the work backward strategy to find the winning strategy for the game.

11. Write a problem that can be solved by working backward. Have a classmate solve your problem.

LOGIC POWER

The figure is made from 9 toothpicks. The total number of triangles and parallelograms in the figure is 8. Make a new figure by moving 2 toothpicks so that the number of triangles changes, the number of parallelograms changes, but the total number of triangles and parallelograms is still 8.

3.8 Factoring $x^2 + bx + c$

The JumboTron in the Skydome is one of the largest video display boards in the world. The area of this rectangular board can be represented by the trinomial $x^2 + 26x + 25$. To find the length and width of the rectangle, the trinomial can be written as the product of two binomials. This process is known as factoring the trinomial.

Algebra tiles can be used as a model for factoring trinomials. If a rectangle can be formed to represent a trinomial, then the length and width of the rectangle represent the factors of the trinomial.

Explore: Use Algebra Tiles

To factor $x^2 + 4x + 3$, use one x^2-tile, four x-tiles, and three 1-tiles to make a rectangle.
a) What is the length of the rectangle?
b) What is the width of the rectangle?

Inquire

1. Copy and complete the statement $x^2 + 4x + 3 = ($ X+3 $)($ X+1 $)$.

2. Use algebra tiles or draw diagrams to find the factors of each of the following trinomials. Write each answer in the form $x^2 + $ ▢ $x + $ ▢ $= (x + $ ▢ $)(x + $ ▢ $)$.

a) $x^2 + 3x + 2$ **b)** $x^2 + 5x + 6$ **c)** $x^2 + 6x + 5$ **d)** $x^2 + 6x + 8$

3. From each of your answers to question 2, how are the following related?
a) the first term in the trinomial and the first terms in the binomials
b) the coefficient of the middle term in the trinomial and the second terms in the binomials
c) the last term in the trinomial and the second terms in the binomials

4. a) Factor the expression $x^2 + 26x + 25$ to find binomials that represent the length and width of the JumboTron.
b) If x represents 9 m, what are the length and the width of the JumboTron, in metres?

Another way to factor trinomials of the form $x^2 + bx + c$ is to study a general expansion.

$$(x + r)(x + s) = x^2 + sx + rx + rs = x^2 + \underbrace{(s + r)}_{\text{Sum of terms}}x + \underbrace{rs}_{\text{Product of terms}}$$

$$x^2 + \quad bx \quad + \quad c$$

So, you can factor a trinomial of the form $x^2 + bx + c$ by finding two terms whose sum is b and whose product is c.

Example 1   Factoring Trinomials

Factor $x^2 - 8x + 12$. Check by substitution.

Solution

For $x^2 - 8x + 12$, $b = -8$ and $c = 12$. Use a table to find two integers whose sum is -8 and whose product is 12. The only two integers with a sum of -8 and a product of 12 are -6 and -2. So, $x^2 - 8x + 12 = (x - 6)(x - 2)$.

From Example 1, note that factoring is the inverse operation of expanding.

$$\xleftarrow{}\text{Factoring}\xrightarrow{}$$
$$x^2 - 8x + 12 = (x - 6)(x - 2)$$
$$\xleftarrow{}\text{Expanding}\xrightarrow{}$$

Product of 12		Sum
12	1	13
-12	-1	-13
6	2	8
★ -6	-2	-8
4	3	7
-4	-3	-7

Check:
Let $x = 1$.
L.S. $x^2 - 8x + 12$
$= (1)^2 - 8(1) + 12$
$= 1 - 8 + 12$
$= 5$
R.S. $(x - 6)(x - 2)$
$= (1 - 6)(1 - 2)$
$= (-5)(-1)$
$= 5$

Example 2   Removing a Common Factor

Factor $3x^2 + 3x - 18$.

Solution

First, remove the common factor.
$3x^2 + 3x - 18 = 3(x^2 + x - 6)$
For $x^2 + x - 6$, $b = 1$ and $c = -6$.
The two integers whose sum is 1 and whose product is -6 are 3 and -2.
So, $x^2 + x - 6 = (x + 3)(x - 2)$
and $3x^2 + 3x - 18 = 3(x + 3)(x - 2)$.

Product of -6		Sum
6	-1	5
-6	1	-5
★ 3	-2	1
-3	2	-1

It may be quicker and easier to use mental math than to make a table.

Example 3   Trinomials With Two Variables

Factor $x^2 - 2xy - 15y^2$.

Solution

To factor $x^2 - 2xy - 15y^2$, find two integers with a product of -15 and a sum of -2.
The two integers are -5 and 3.
So, $x^2 - 2xy - 15y^2 = (x - 5y)(x + 3y)$.

Product of -15		Sum
15	-1	14
-15	1	-14
5	-3	2
★ -5	3	-2

Many trinomials, such as $x^2 + 3x + 5$, cannot be factored over the integers. No two integers have a sum of 3 and a product of 5.

Practice

If possible, find two integers with the given product and sum.

	Product	Sum			Product	Sum
1.	15	8		**2.**	15	8
3.	-30	7		**4.**	-30	7
5.	10	7		**6.**	10	7
7.	36	-13		**8.**	36	-13

Factor, if possible. Check by substituting $x = 1$ into the expanded form and the factored form.

9. $x^2 + 5x + 4$
10. $x^2 + 8x + 15$
11. $m^2 + 7m + 10$
12. $t^2 + 9t + 12$
13. $r^2 - 13r + 42$
14. $n^2 + 11n + 30$
15. $r^2 - 7r + 10$
16. $w^2 - 10w + 16$
17. $x^2 - 9x + 24$
18. $m^2 - 10m + 24$

Factor, if possible.

19. $y^2 - y - 20$

20. $x^2 + 7x - 18$

21. $t^2 + t - 18$

22. $x^2 + 5x - 14$

23. $n^2 - 10n - 24$

24. $m^2 - 4m - 21$

25. $r^2 + 7r - 20$

26. $x^2 - 8x - 20$

Factor, if possible.

27. $m^2 + 18m + 80$

28. $m^2 + m - 12$

29. $x^2 + 2x + 5$

30. $r^2 - 17r + 42$

31. $y^2 - 17y + 72$

32. $x^2 - 6x - 16$

33. $t^2 - 15t + 26$

34. $y^2 - 2y - 4$

35. $m^2 + 7m - 6$

36. $n^2 + 7n - 44$

37. $x^2 - 10x + 21$

38. $w^2 + 12w + 20$

39. $r^2 - r - 30$

40. $y^2 - 20y + 36$

41. $n^2 - 4n + 5$

42. $x^2 + 12xy + 35y^2$

43. $a^2 - 4ab - 77b^2$

44. $c^2 - cd - 2d^2$

45. $x^2 + 5xy - 36y^2$

46. $x^2 - 4xy + 6y^2$

47. $x^2y^2 + 3xy + 2$

48. $a^2b^2 - 5ab + 6$

49. $8 + 7y - y^2$

50. $16 - 6x - x^2$

Factor completely.

51. $2x^2 - 6x + 4$

52. $3x^2 + 12x + 9$

53. $5y^2 + 40y + 60$

54. $4t^2 - 8t - 60$

55. $6x^2 + 18x - 24$

56. $ax^2 + 10ax - 24a$

57. $x^3 + 18x^2 + 72x$

58. $2x^2 - 22x + 56$

59. $5w^2 + 20w - 60$

60. $3x - 2x^2 - x^3$

Applications and Problem Solving

61. Signboard On the outside of the Skydome, there is a large, rectangular electronic signboard. The approximate area of the signboard can be represented by the trinomial $x^2 - 3x - 4$.

a) Factor $x^2 - 3x - 4$ to find binomials that represent the length and width of the signboard.

b) If x represents 17 m, find the length and width of the signboard, in metres.

62. Visualization If you tried to use algebra tiles to make a rectangle that represents the polynomial $x^2 + 4x + 5$, what would you observe? Why?

63. Visualization Describe how you could use algebra tiles to factor

a) $x^2 + 4xy + 3y^2$

b) $x^2 + 5xy + 6y^2$

64. a) Movies Each letter shown in the table represents a different integer. The letter Y represents 5.

Letter	A	C	E	K	M	N	O	P	S	T	Y
Integer											5

Factor the following 5 trinomials. In each case, write the factored form with the larger binomial first. For example, the factored form of $x^2 + 2x - 8$ would be written as $(x + 4)(x - 2)$, because $x + 4 > x - 2$. Use the factored forms to find the integer represented by each capital letter. Then, copy and complete the table.

$x^2 + 20x - 96 = (x + M)(x + A)$
$x^2 - 27x + 72 = (x + N)(x + C)$
$x^2 - 16x - 80 = (x + E)(x + T)$
$x^2 - 25x - 84 = (x + S)(x + K)$
$x^2 + 9x - 90 = (x + P)(x + O)$

b) Replace each of the following integers with its corresponding letter from part a) to name a Canadian movie producer who was a silent-comedy pioneer, and find the films for which he was famous.

Name: 24   −4   −24   −28
     3   4   −3   −3   4   −20   −20
Films: −28   4   5   3   −20   −6   −3   4
      −28   −6   15   3

65. Find 3 values of k such that each trinomial can be factored over the integers.

a) $x^2 + 2x + k$

b) $x^2 - 5x + k$

66. List all values of k for which the trinomial can be factored over the integers.

a) $x^2 + kx + 6$

b) $x^2 + kx - 12$

c) $x^2 + kx + 24$

d) $x^2 + kx - 18$

67. Factor.

a) $x^4 + 2x^2 + 1$

b) $x^4 + x^2 - 6$

c) $x^4 - 3x^2 - 10$

d) $x^4 + 10x^2y + 9y^2$

68. Factor.

a) $(x + a)^2 + 3(x + a) + 2$ **b)** $(x - b)^2 + 4(x - b) - 5$

69. Equations a) Copy and complete each statement.

$x^2 - 2x - 35 = ($ $)($ $)$
$t^2 + 3t - 40 = ($ $)($ $)$

b) How do the factors help you determine the values of the variable that give the trinomial a value of zero?

127

3.9 Factoring $ax^2 + bx + c$, $a \neq 1$

Canadian swimmer Marianne Limpert won a silver medal in the 200-m individual medley at the Summer Olympics in Atlanta. She swam in a pool whose area can be represented by the trinomial $3x^2 + 17x + 10$.

In some trinomials, such as $3x^2 + 17x + 10$, the numerical coefficient of the x^2 term is not 1. Algebra tiles can be used as a model to factor trinomials of this type.

Explore: Use Algebra Tiles

To factor $2x^2 + 5x + 2$, use two x^2-tiles, five x-tiles, and two 1-tiles to make a rectangle.
a) What is the length of the rectangle?
b) What is the width of the rectangle?

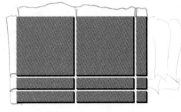

Inquire

1. Copy and complete the statement $2x^2 + 5x + 2 = ($ 2x+1 $)($ x+2 $)$.

2. Use algebra tiles or draw diagrams to find the factors of each of the following trinomials.
a) $2x^2 + 3x + 1$ **b)** $2x^2 + 5x + 3$ **c)** $3x^2 + 7x + 2$ **d)** $3x^2 + 8x + 4$

3. a) Factor the expression $3x^2 + 17x + 10$ to find binomials that represent the length and width of the Olympic swimming pool in Atlanta.
b) If x represents 16 m, what are the dimensions of the pool, in metres?

Another way to factor trinomials of the form $ax^2 + bx + c$, $a \neq 1$, is to use the guess and check strategy.

Example 1   Factoring by Guess and Check
Factor $3x^2 - 5x - 2$.

Solution
For the trinomial $3x^2 - 5x - 2$, the possible first terms of the binomial factors are $3x$ and x.

$3x^2 - 5x - 2 = (3x \qquad)(x \qquad)$

The product of the second terms of the binomials must be -2. The terms could be 2 and -1 or -2 and 1.

List all the possible pairs of factors and expand to see which pair gives the correct middle term of the trinomial.

	GUESS		CHECK
Possible Factors	Expansion	Trinomial	Is the middle term correct?
$(3x + 2)(x - 1)$	$3x^2 - 3x + 2x - 2$	$3x^2 - x - 2$	No
$(3x - 1)(x + 2)$	$3x^2 + 6x - x - 2$	$3x^2 + 5x - 2$	No
$(3x - 2)(x + 1)$	$3x^2 + 3x - 2x - 2$	$3x^2 + x - 2$	No
$(3x + 1)(x - 2)$	$3x^2 - 6x + x - 2$	$3x^2 - 5x - 2$	Yes

Therefore, $3x^2 - 5x - 2 = (3x + 1)(x - 2)$.

A third way to factor trinomials of the form $ax^2 + bx + c$, $a \neq 1$, is to break up the middle term into two parts, and then factor by grouping.

$$2x^2 + 11x + 12$$

Break up the middle term:	$= 2x^2 + 8x + 3x + 12$	$11x = 8x + 3x$
Group terms:	$= (2x^2 + 8x) + (3x + 12)$	
Remove common factors:	$= 2x(x + 4) + 3(x + 4)$	
Use the distributive property:	$= (2x + 3)(x + 4)$	

So, $2x^2 + 11x + 12 = (2x + 3)(x + 4)$.

Note that, when the middle term is broken up, the resulting two terms can be written in either order before terms are grouped. Reversing the above order gives

$$2x^2 + 11x + 12 = 2x^2 + 3x + 8x + 12$$
$$= (2x^2 + 3x) + (8x + 12)$$
$$= x(2x + 3) + 4(2x + 3)$$
$$= (x + 4)(2x + 3)$$

So, $2x^2 + 11x + 12 = (x + 4)(2x + 3)$, which is the same result as before.

The two terms to substitute for the middle term of the trinomial are not obvious. There are many pairs of terms that add to give $11x$, yet only one pair allows you to factor by grouping.

To find the correct pair of terms, look at the expansion of $(2x + 3)(x + 4)$. Compare the result to the trinomial $2x^2 + 11x + 12$. Note that this trinomial is of the form $ax^2 + bx + c$, where $a = 2$, $b = 11$, and $c = 12$.

$$(2x + 3)(x + 4) = (2x + 3)(x + 4)$$

$= 2x^2 + 8x + 3x + 12$	$8 + 3 = 11$, or b
$= 2x^2 + 11x + 12$	$8 \times 3 = 2 \times 12$, or $a \times c$

$$8 + 3 = 11$$
$$2x^2 + 11x + 12 = 2x^2 + 8x + 3x + 12$$
$$8 \times 3 = 24$$
$$2 \times 12 = 24$$

So, to factor $2x^2 + 11x + 12$, the middle term was replaced by two terms whose coefficients have a sum of 11, or b, and a product of 24, or $a \times c$.

Example 2 Breaking up the Middle Term → OMIT

Factor $6m^2 + 13m - 5$.

Solution

For $6m^2 + 13m - 5$, $a = 6$, $b = 13$, and $c = -5$.
Find two integers whose product is $a \times c$, or -30, and whose sum is b, or 13.
The only two integers that have a product of -30 and a sum of 13 are 15 and -2.

Product of –30		Sum
30	–1	29
–30	1	–29
15	–2	13
–15	2	–13
10	–3	7
–10	3	–7
6	–5	1
–6	5	–1

$$6m^2 + 13m - 5$$

Break up the middle term:	$= 6m^2 - 2m + 15m - 5$
Group terms:	$= (6m^2 - 2m) + (15m - 5)$
Remove common factors:	$= 2m(3m - 1) + 5(3m - 1)$
Use the distributive property:	$= (2m + 5)(3m - 1)$

So, $6m^2 + 13m - 5 = (2m + 5)(3m - 1)$.

129

Example 3 Removing a Common Factor

Factor $10x^2 - 22x + 4$.

Solution

Remove the common factor, then proceed as before.

$10x^2 - 22x + 4 = 2[5x^2 - 11x + 2]$

To factor $5x^2 - 11x + 2$, find two integers whose product is 10 and whose sum is -11. The integers are -10 and -1.

$$\begin{aligned}
2[5x^2 - 11x + 2] &= 2[5x^2 - 10x - x + 2] \\
&= 2[(5x^2 - 10x) + (-x + 2)] \\
&= 2[5x(x - 2) - 1(x - 2)] \\
&= 2[(5x - 1)(x - 2)] \\
&= 2(5x - 1)(x - 2)
\end{aligned}$$

Product of 10		Sum
10	1	11
* −10	−1	−11
5	2	7
−5	−2	−7

So, $10x^2 - 22x + 4 = 2(5x - 1)(x - 2)$.

Example 4 Trinomials With Two Variables

Factor $4x^2 - 5xy - 6y^2$.

Solution

To factor $4x^2 - 5xy - 6y^2$, find two integers with a product of -24 and a sum of -5. The integers are -8 and 3.

$$\begin{aligned}
4x^2 - 5xy - 6y^2 &= 4x^2 - 8xy + 3xy - 6y^2 \\
&= (4x^2 - 8xy) + (3xy - 6y^2) \\
&= 4x(x - 2y) + 3y(x - 2y) \\
&= (4x + 3y)(x - 2y)
\end{aligned}$$

Product of −24		Sum
24	−1	23
−24	1	−23
12	−2	10
−12	2	−10
8	−3	5
* −8	3	−5
6	−4	2
−6	4	−2

So, $4x^2 - 5xy - 6y^2 = (4x + 3y)(x - 2y)$.

Many trinomials of the form $ax^2 + bx + c$, $a \neq 1$, such as $2x^2 + 3x + 6$, cannot be factored over the integers. No two integers have a product of 12 and a sum of 3.

Practice

Factor, if possible. Check each factored form by substituting $x = 1$ into the expanded form and the factored form.

1. $2y^2 + 9y + 9$ **2.** $3m^2 + 10m + 3$

3. $5t^2 + 7t + 2$ **4.** $4r^2 + 12r + 3$

5. $2x^2 + 11x + 14$ **6.** $6x^2 + 11x + 3$

Factor, if possible.

7. $2x^2 - 5x + 3$ **8.** $3x^2 - 5x + 2$

9. $3t^2 - 10t + 8$ **10.** $5m^2 - 11m + 2$

11. $6m^2 - 13m + 6$ **12.** $4y^2 - 11x + 9$

Factor, if possible.

13. $2x^2 - x - 6$ **14.** $6x^2 - 5x - 4$

15. $2t^2 + 9t - 5$ **16.** $15n^2 - n - 2$

17. $3x^2 + x - 4$ **18.** $4x^2 - x - 3$

19. $5y^2 - 14y - 3$ **20.** $8x^2 - 10x - 3$

21. $9x^2 - 15x - 4$ **22.** $10t^2 + 11t - 6$

Factor, if possible.

23. $4t^2 + 8t + 3$ **24.** $10x^2 - 17x + 3$

25. $5t^2 + 2t - 2$ **26.** $2y^2 + 11y + 15$

27. $8y^2 - 22y + 12$ **28.** $6x^2 + x - 4$

29. $6r^2 + 15r + 9$ **30.** $12y^2 - 11y + 2$

31. $4x^2 - 18x - 10$ **32.** $2m^3 + 7m^2 - 30m$
33. $2t^3 + 9t^2 + 4t$ **34.** $18s^2 - 7s - 1$
35. $12r^2 + 27r + 15$ **36.** $5r^2s - 7rs + 2s$
37. $6 + 5y - 4y^2$ **38.** $2 - 7m + 3m^2$
39. $12 + 18t + 8t^2$ **40.** $6 + 5y - 6y^2$

Factor.

41. $6m^2 + mn - 2n^2$ **42.** $3x^2 + 7xy + 2y^2$
43. $10a^2 - 3ab - b^2$ **44.** $2x^2 - 11xy + 5y^2$
45. $6c^2 + 13cd + 2d^2$ **46.** $4m^2 - 5mn + n^2$
47. $6x^2 - 9xy + 3y^2$ **48.** $2m^2 - 4mn - 6n^2$
49. $4y^2 + 4xy - 8x^2$ **50.** $6a^2 + 14ab - 12b^2$

Applications and Problem Solving

51. Transportation Sydney Harbour Bridge in Australia is unusually wide for a long-span bridge. It carries two rail lines, eight road lanes, a cycle lane, and a walkway.
a) Factor the expression $10x^2 - 7x - 3$ to find binomials that represent the length and width of the bridge.
b) If x represents 50 m, what are the length and width of the bridge, in metres?

52. Visualization Describe how you would use algebra tiles to factor
a) $3x^2 + 4xy + y^2$ **b)** $2x^2 + 5xy + 2y^2$

53. Famous Canadians The letters shown in the table each represent a different integer. The letter Y represents –7, as shown.

Letter	A	C	D	E	H	I	J	L	M
Integer									
Letter	N	O	R	S	T	U	W	Y	
Integer								–7	

a) Factor the following 5 trinomials. In each case, write the factored form with the smaller coefficient of x first. For example, the factored form of $6x^2 + 11x + 4$ would be written as $(2x + 1)(3x + 4)$, because $2 < 3$. Use the factored forms to find the integer represented by each capital letter. Then, copy and complete the table.

$10x^2 - 29x + 10 = (Ax + U)(Cx + D)$
$4x^2 - 27x + 18 = (Wx + L)(Ex + H)$
$18x^2 - 27x + 4 = (Jx + I)(Mx + N)$
$56x^2 + 15x + 1 = (Ox + W)(Rx + W)$
$10x^2 - 91x + 9 = (Wx + S)(Tx + N)$

b) Replace each of the following integers with its corresponding letter from part a) to find the names of four famous Canadians. State what each of them became famous for.

Name 1:   3   7   –1   –4
           6   –4   10   5   –3   4   –6   –6

Name 2:   –1   4   –2
           –3   2   –1   –6   2   –1

Name 3:   6   2   8   –9   –3   2   –6   –6
           6   5   –6   –5   –3   2   –1

Name 4:   4   6   –4   –6   –7
           –9   10   7   1   4

54. List all values of k such that each trinomial can be factored over the integers.
a) $3x^2 + kx + 5$ **b)** $5x^2 + kx + 8$
c) $3x^2 + kx - 2$ **d)** $4x^2 + kx - 9$

55. Find 3 values of k such that each of the following trinomials can be factored over the integers.
a) $2x^2 + 3x + k$ **b)** $3x^2 - 8x + k$

56. Factor.
a) $2x^4 + 3x^2 + 1$ **b)** $2x^4 + 5x^2 - 3$
c) $3x^4 - x^2 - 4$ **d)** $6x^4 - 13x^2 + 6$
e) $2x^4 + 5x^2y + 2y^2$ **f)** $3x^4 + 11x^2y - 4y^2$

LOGIC POWER

Assume that no cubes are missing from the back or the base of the stack.
1. How many cubes are in the stack?
2. If the outside of the stack was painted green, how many cubes would have
a) 4 green faces?   **b)** 1 green face?

3.10 Factoring Special Quadratics

A space shuttle returning to Earth must withstand very high temperatures when it enters the atmosphere. About 70% of the shuttle is covered with two sizes of square insulating tiles. There are about 20 000 of the smaller tiles on its lower surface and about 7000 of the larger tiles on the tops of the wings and the sides of the fuselage.

If the area of one of the larger square tiles is represented by a^2 and the area of one of the smaller square tiles is represented by b^2, the difference in their areas is $a^2 - b^2$. A polynomial written in the form $a^2 - b^2$ is called a **difference of squares**.

Explore: Study the Diagrams

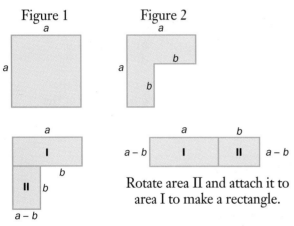

Rotate area II and attach it to
area I to make a rectangle.

Divide Figure 2
into areas I and II.

What is
a) the area of the square in Figure 1? a^2
b) the area of Figure 2, with the square area b^2 removed?
c) the length of the rectangle formed from areas I and II?
d) the width of the rectangle formed from areas I and II?

Inquire

1. Write an expression for the area of the rectangle in terms of the length and width. Do not expand the expression.

2. How is your expression from question 1 related to the expression you wrote, above, for the area of Figure 2?

3. Write a rule for factoring the difference of squares.

4. The two sizes of square tiles on the space shuttle have side lengths of 20 cm and 15 cm. Use your rule from question 3 to calculate the difference in their areas without squaring any numbers.

Example 1   Factoring a Difference of Squares
Factor $9x^2 - 16$.

Solution
Use the pattern for the difference of squares.

$$a^2 - b^2 = (a + b)(a - b)$$
$$9x^2 - 16 = (3x)^2 - 4^2$$
$$= (3x + 4)(3x - 4)$$

Example 2   Removing a Common Factor
Factor $8x^2 - 18y^2$.

Solution
Remove the common factor. Then, factor the difference of squares.
$$8x^2 - 18y^2 = 2(4x^2 - 9y^2)$$
$$= 2(2x + 3y)(2x - 3y)$$

The trinomial that results from squaring a binomial is called a **perfect square trinomial**. Perfect square trinomials can be factored using the patterns from squaring binomials.

$$(a + b)^2 = a^2 + 2ab + b^2$$
$$(a - b)^2 = a^2 - 2ab + b^2$$

In a perfect square trinomial, the following conditions are met:
• The first and last terms are perfect squares.
• The middle term is twice the product of the square roots of the first and last terms.

Example 3   Perfect Square Trinomials
a) Is $4x^2 + 20x + 25$ a perfect square trinomial?   **b)** If it is, factor it.

Solution
a) $4x^2 = (2x)^2$ and $25 = 5^2$, so the first and last terms are perfect squares. $20x = 2(2x)(5)$, so the middle term is twice the product of the square roots of the first and last terms. So, $4x^2 + 20x + 25$ is a perfect square trinomial.
b) $4x^2 + 20x + 25 = (2x)^2 + 2(2x)(5) + 5^2$
$$= (2x + 5)^2$$

Practice

Factor, if possible. Check each factored form by substituting $x = 1$ into the expanded form and the factored form.

1. $x^2 - 9$ **2.** $y^2 - 16$
3. $z^2 + 81$ **4.** $25a^2 - 36$
5. $1 - 64t^2$ **6.** $36 - 49a^2$
7. $49 + x^2$ **8.** $25x^2 - 64y^2$
9. $4t^2 - 9s^2$ **10.** $100p^2 - 121q^2$
11. $16^2 - 81y^2$ **12.** $225b^2 - a^2$

State whether each trinomial is a perfect square trinomial. If it is, factor it.

13. $x^2 + 6x + 9$ **14.** $y^2 - 10y + 25$
15. $x^2 - 8x + 4$ **16.** $4t^2 + 4t + 1$ – Yes
17. $m^2 - 20m + 100$ **18.** $16t^2 + 24t + 9$

19. $49 + 14x + x^2$ **20.** $1 - 16t + 64t^2$
21. $9x^2 - 24x + 16$ **22.** $4 + 28r + 49r^2$
23. $81x^2 - 72xy + 64y^2$ **24.** $36m^2 + 60mn + 25n^2$
25. $121m^2 - 22m + 1$ **26.** $9a^2 + 12ab + 4b^2$

Factor fully, if possible.
27. $y^2 - 144$ **28.** $25x^2 + 5y + 1$
29. $9a^2 - 24a + 16$ **30.** $2x^2 - 32$
31. $y^2 + 36$ **32.** $3x^2 + 6x + 3$
33. $m^2 - 14m + 49$ **34.** $4p^2 + 20pq + 25q^2$
35. $49x^2 - 121y^2$ **36.** $80a^2 - 45b^2$
37. $100x^2 + 10x + 1$ **38.** $y^3 - 36y$
39. $y^3 - 18y^2 + 81y$ **40.** $36x^2 + 100y^2$
41. $3x^3 - 48x$ **42.** $5m^3 - 40m^2 + 80m$
43. $81x^2 - 144$ **44.** $3b^2 - 300$

Applications and Problem Solving

45. Visualization Describe how you would use algebra tiles to factor each of the following.
a) $x^2 + 2xy + y^2$ **b)** $4x^2 + 4xy + y^2$

Factor.
46. $(x + 2)^2 - 9$ **47.** $16 - (y - 3)^2$
48. $(m + 1)^2 - (m + 2)^2$ **49.** $x^4 + 22x^2 + 121$

50. $t^6 - 18t^3 + 81$ **51.** $\dfrac{x^2}{4} - \dfrac{1}{9}$

52. $25x^4 - 81$ **53.** $(2x + y)^2 - (2x - y)^2$

54. Numbers Evaluate by factoring the difference of squares.
a) $53^2 - 47^2$ **b)** $45^2 - 35^2$ **c)** $820^2 - 180^2$

55. Determine the value(s) of k such that each trinomial is a perfect square.
a) $x^2 + kx + 16$ **b)** $9x^2 + kx + 49$
c) $x^2 + 4x + k$ **d)** $4x^2 - 12x + k$
e) $kx^2 + 40x + 16$ **f)** $kx^2 - 24xy + 9y^2$

56. Volleyball The area of a volleyball court, excluding the service areas, can be represented by the trinomial $2x^2 - 4x + 2$.
a) Factor the trinomial completely.
b) If the length of the court is twice the width, use the factors from part a) to write expressions that represent the length and the width.
c) If x represents 10 m, what are the length and the width of the court, in metres?

57. Measurement The volume of a rectangular prism is represented by the polynomial $2x^3 - 24x^2 + 72x$.
a) Factor the polynomial completely.
b) If the expression for each dimension of the prism includes x, what are the expressions that represent the possible sets of dimensions?
c) If x represents 8 cm, what are the possible dimensions of the prism?
d) Could x represent 5 cm? Explain.

58. Integers If a and b are integers, find values of a and b such that $a^2 - b^2$ is 21.

59. Measurement The volume of a cylinder is given by the expression $\pi r^2 h$, where r is the radius and h is the height. Two cylinders have equal heights of 6 cm. Their radii are 5.6 cm and 4.4 cm.

a) Without squaring any numbers, write an expression to represent the difference between the volumes of the cylinders. Express your answer in the form of a monomial multiplied by a difference of squares.
b) Use the difference of squares to write the difference in the volumes in the form ▓▓ π, where ▓▓ represents a whole number.
c) Evaluate the difference in the volumes to the nearest cubic centimetre.

60. Pattern

The four grids have side lengths of 1, 2, 3, and 4 units, respectively.
a) Describe the pattern in the number of shaded grid squares.
b) If s represents the side length of the grid, write an expression in terms of s for finding the number of shaded grid squares. Write your expression in factored form.
c) Use your expression to find the number of shaded grid squares on a 12 by 12 grid; a 91 by 91 grid.
d) If the number of shaded grid squares is 529, what is the side length of the grid?
e) The total number of grid squares on a grid of side length s is s^2. Subtract the expression you found in part b) from s^2 to find an expression that represents the number of unshaded squares in terms of s.
f) Use your expression from part e) to find the number of unshaded squares on a 53 by 53 grid.

61. Measurement The area of a square is represented by the expression $49 - 28x + 4x^2$, where x represents a positive integer. What are the possible values for the perimeter of the square?

62. Factor.
a) $(x + 3)^2 - y^2$ **b)** $x^2 - 4x + 4 - 9y^2$
c) $4x^2 + 12xy + 9y^2 - 4z^2$ **d)** $x^4 - 2x^2y + y^2 - z^2$

63. Measurement A circle has an area of $(9x^2 + 30x + 25)\pi$ square centimetres, where x is a positive integer. Determine the smallest diameter that the circle can have.

Changing Arrangements of Algebra Tiles

In each of the following activities, you are to change a rectangular arrangement of algebra tiles. You must rearrange the tiles, rather than rotating the existing arrangement. Compare each of your new arrangements with a classmate's.

1 Factoring $x^2 + bx + c$

1. For the rectangle shown, state
a) the trinomial that represents the area
b) the binomial factors of the trinomial

2. a) Use algebra tiles or draw diagrams
to make two other rectangular arrangements of the tiles.
b) Do the factors of the trinomial change with different arrangements?
c) Check your findings from part b) by making all the rectangular
arrangements that represent $x^2 + 7x + 12$.

2 Factoring $ax^2 + bx + c, a \neq 1$

1. For the rectangle shown, state
a) the trinomial that represents the
area
b) the binomial factors of the
trinomial

2. a) Use algebra tiles or draw
diagrams to make three other
rectangular arrangements of the tiles.
b) Do the factors of the trinomial change with different
arrangements?
c) Check your findings from part b) by making all the
rectangular arrangements that represent $3x^2 + 5x + 2$.

3 Factoring Special Trinomials

1. For the rectangle shown, state
a) the trinomial that represents the area
b) the binomial factors of the trinomial

2. a) Use algebra tiles or draw diagrams to
make two other rectangular arrangements
of the tiles.
b) Do the factors of the trinomial change with different arrangements?
c) If a trinomial is represented by a rectangle that is a square, what
can you state about the trinomial?
d) Check your findings from part c) by making all the rectangles
that represent $x^2 + 6x + 9$ and $x^2 + 8x + 16$.

TECHNOLOGY

Factoring Polynomials With a Graphing Calculator

Complete each of the following with a graphing calculator that has the capability to factor polynomials.

1 Factoring Polynomials

Remove the common factor.
1. $6x^2 + 15x - 12$
2. $14y^2 - 42y + 21$
3. $20 + 8m^2 + 12m$
4. $20x - 15x^2 + 10$
5. $4x^2y + 6xy - 8xy^2$
6. $3p^3q + 18p^2q^2 + 6pq^3$
7. $12a^3b^2 + 4a^2b^3 + 8ab^4 - 6b^5$
8. $10x^5y^2 + 5x^4y^3 - 15x^3y^4 + 25x^2y^5$

Factor, if possible.
9. $x^2 + 19x + 34$
10. $x^2 - 6x - 72$
11. $x^2 - 24x + 40$
12. $15 - 8t + t^2$
13. $x^2 + 7xy + 10y^2$
14. $28y^2 - 16xy + x^2$
15. $4n^2 + 13n + 9$
16. $2m^2 - 5m + 6$
17. $5x^2 - 17x - 12$
18. $15y^2 + 11y - 14$
19. $20 + 36z + 9z^2$
20. $x^4 + 2x^2 - 15$
21. $12x^4 - 5x^2 - 25$
22. $3x^2 - 14xy + 8y^2$
23. $15a^2 - ab - 6b^2$
24. $14x^2 + 55xy - 36y^2$
25. $(x + a)^2 + 6(x + a) + 8$
26. $(x - y)^2 - 5(x - y) + 6$

Factor completely.
27. $3x^2 - 30x + 27$
28. $4x^2 + 10x - 24$
29. $21x^2 + 119x - 42$
30. $75y^2 + 215y + 40$
31. $2u^2 - 6uv + 4v^2$
32. $36x^2 + 42xy - 18y^2$
33. $x^3 + 3x^2 + 2x$
34. $4t^3 - 26t^2 - 14t$
35. $30x^4 + 87x^2 + 30$
36. $24x^4 - 16x^2 - 8$
37. $8x^4 - 44x^2y^2 + 20y^4$
38. $99a^4 + 275a^2b^2 - 66b^4$

2 Factoring Special Products

Factor.
1. $25x^2 + 60x + 36$
2. $9y^2 - 30y + 25$
3. $4m^2 + 36mn + 81n^2$
4. $49x^2 - 56xy + 16y^2$
5. $4a^4 + 20a^2b^2 + 25b^4$
6. $9x^2 - 64$
7. $25 - 169x^2$
8. $4x^4 - 9y^2$
9. $(x^2 + 2)^2 - 4y^4$
10. $36x^2 - (y^2 - 5)^2$

Factor completely.
11. $16m^2 - 64$
12. $36 - 16x^2$
13. $125x^4 - 80$
14. $72x^2 - 98y^4$
15. $2x^2 - 28x + 98$
16. $12x^2 + 60x + 75$
17. $32w^3 - 160w^2 + 200w$
18. $300 - 48x^4$
19. $4x^5 - 12x^4y + 9x^3y^2$
20. $36y^4 + 120x^2y^2 + 100x^4$

21. Write two special products that can be factored. Have a classmate factor them.

136

3 Egyptian Pyramids

1. The first of the giant pyramids built in Egypt was the Step Pyramid, built by the architect Imhotep for the pharaoh Djoser. The pyramid, which was built from over a million tonnes of limestone, is 62 m high. The area of its rectangular base can be represented by the expression $20x^2 + 23x - 21$.
a) Factor the expression $20x^2 + 23x - 21$ to write expressions that represent the length and width of the base.
b) If x represents 25 m, what are the length and width of the base, in metres?

2. The Bent Pyramid, which is about 98 m high, has faces that are steeper at the bottom than at the top. The area of the square base of the pyramid can be represented by the expression $36x^2 - 48x + 16$.
a) Factor $36x^2 - 48x + 16$ to write an expression that represents the side length of the base.
b) If x represents 32 m, what is the side length of the base, in metres?

Economics

Economics is the study of how societies use their resources to meet the needs of people.

Canada has a long economic history. Before European settlement, aboriginal peoples were active traders. Hunters might trade with crop growers, for example. In today's market economy, we decide which products and services to buy from businesses. Governments are also part of the economy. They supply many services, such as education and health care, which are financed with tax dollars.

Many organizations, including government departments, banks, industrial companies, and universities, employ economists. Their studies include debt management, trade, employment trends, wages, prices, and spending patterns. Economists have university degrees in economics or business.

1 Inflation

The Consumer Price Index (CPI) is compiled by Statistics Canada to show inflationary trends. The CPI is a measure of the average prices of household goods and services.

Year	CPI	Year	CPI
1976	47.5	1987	104.4
1977	51.3	1988	108.6
1978	55.9	1989	114.0
1979	61.0	1990	119.5
1980	67.2	1991	126.2
1981	75.5	1992	128.1
1982	83.7	1993	130.4
1983	88.5	1994	130.7
1984	92.4	1995	133.5
1985	96.0	1996	135.6
1986	100.0		

The table shows that goods and services costing $100 in 1986 would have cost $47.50 in 1976 and $135.60 in 1996.

1. Suppose a selection of goods and services cost $350.00 in 1985. Explain how to use the CPI to show that the equivalent cost in 1995 was $486.72.

2. For every $500.00 spent on household goods and services in the year you were born, how much was spent in
a) 1978?   **b)** 1994?

3. a) The CPI is related to Canada's annual inflation rate. The inflation rate in 1991 was 5.6%. Describe how this rate can be calculated from the CPI for 1990 and 1991.
b) Calculate the inflation rate for 1992, to the nearest tenth of a percent.
c) Which of the years shown in the table had the greatest inflation rate? What was the rate?

4. The expression $c(1 + r)^n$ can be used to predict costs in n years from now, using the present cost, c dollars, and assuming a constant inflation rate, r. Predict the cost of $1750.00 worth of goods and services 5 years from now, if the inflation rate throughout this time is
a) 2.5%   **b)** 4.5%   **c)** 9.2%

2 Locating Information

1. a) What is the meaning of the term *national debt*?
b) What is the size of Canada's national debt?
c) If the national debt was shared equally among all Canadians, how much would each person owe?

2. Find information about a part of Canada's economic history, such as aboriginal trade, farming, settlement of the West, or growth of cities.

3. John Kenneth Galbraith is a world-famous Canadian economist. Find information about his life and his contributions to economics.

CONNECTING MATH AND ARCHITECTURE

Gwennap Pit

The parish of Gwennap is located in Cornwall in south-western England. In past centuries, Gwennap was an important tin-mining centre.

In the sixteenth century, part of one of the mines collapsed, forming a large hole in the ground. In 1806, terraces were cut to create an outdoor amphitheatre, with a circular stage surrounded by rings of seats. Each ring of seats is about 1 m wide. The amphitheatre is known as Gwennap Pit.

1 Writing Expressions for Areas

Each ring of seats has an inner radius and an outer radius. The area of each ring is the difference between the area of a circle with the outer radius and the area of a circle with the inner radius.

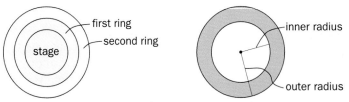

1. If the circular stage has a radius of r metres, write an expression for its area in terms of r.

2. For the first ring of seats around the stage, what is the inner radius in terms of r? the outer radius in terms of r?

3. a) Write an expression in terms of r for the area of the first ring of seats.
b) Simplify the expression from part a). Write the result in factored form.

4. Repeat question 3 for
a) the second ring   **b)** the third ring
c) the fourth ring   **d)** the fifth ring

5. a) Describe the pattern in the factored expressions for the areas of the first five rings.
b) Use the pattern to write an expression for the area of the nth ring in terms of n and r.
c) Use the expression from part b) to write an expression for the area of the eighth ring in terms of r.

6. If the radius of the stage is about 2.5 m, find the area of each of the following, to the nearest square metre.
a) the fifth ring   **b)** the seventh ring
c) the tenth ring   **d)** the twelfth ring

7. a) Write and simplify an expression in terms of r for the total area of all 13 rings of seats. Express the answer in factored form.
b) Find the total area of all 13 rings of seats, to the nearest 10 m^2.

2 Writing Expressions for Circumferences

Each ring of seats has an inner circumference and an outer circumference.

1. If the circular stage has a radius of r metres, write an expression in terms of r for the inner circumference of the first ring of seats.

2. Write an expression in terms of r for the inner circumference of each of the following rings of seats. Leave each expression in factored form.
a) the second ring   **b)** the third ring
c) the sixth ring   **d)** the thirteenth ring

3. a) Write and simplify an expression in terms of r to represent the total of the inner circumferences of all 13 rings of seats. Express the answer in factored form.
b) Find the total of the inner circumferences of all 13 rings of seats, to the nearest 10 m.

3 Estimating Seating Capacities

1. Estimate the number of people who can comfortably sit in each of the following. Explain your reasoning.
a) the first ring   **b)** the fifth ring
c) the eighth ring   **d)** the whole amphitheatre

2. The world's largest known crater is near Sudbury and has a radius of 125 km. About how many people could comfortably sit on a 1-m wide ring around the top of the crater?

Review

3.1 Simplify.

1. $(2x + y) + (3x - 4y)$
2. $(5x^2 - 3x + 4) + (3x^2 - x - 1)$
3. $(3a^2 - a - 2) - (4a^2 + 5a + 6)$
4. $(2m^2 + 2mn - n^2) - (m^2 - mn - 2n^2)$

Multiply.

5. $(3xy^2)(-4x^3y^2)$
6. $(-4rs^3t^2)(-6rst^4)$

Simplify.

7. $\dfrac{20a^2b^3c}{-5ab^3c}$
8. $\dfrac{-36m^3n^4p^2}{-9m^3np}$

3.3 Expand and simplify.

9. $3(x - 4) + 5(x + 6)$
10. $6(a + 3) - 2(a - 5)$
11. $2t(3t - 4) + t(2t + 5)$
12. $3y(y^2 - y - 1) - y(2y^2 - 3y + 4)$

13. Number a) Write a 2-digit number in which the units digit is larger than the tens digit. Subtract the tens digit from the units digit and multiply the result by 9. Add the product to the original number. How does the resulting number compare to the original number?
b) Let the original number be $10t + u$ and repeat the process. How does the result explain what happened in part a)?

Find the product.

14. $(x - 2)(x + 4)$
15. $(a + 5)(a - 6)$
16. $(2y - 3)(3y + 4)$
17. $(3x + y)(x - 4y)$

Expand and simplify.

18. $(y + 4)(y - 3) + (y - 2)(y - 3)$
19. $(2x - 1)(x - 4) - (3x + 2)(3x - 1)$
20. $3(2a + 3)(2a - 1) - 4(a^2 - 7)$
21. $(x - 3)(x^2 - 3x - 2)$
22. $(2t + 1)(3t^2 - t - 1)$

3.4 Expand.

23. $(x + 4)^2$
24. $(y - 4)(y + 4)$
25. $(a - 5)^2$
26. $(3t + 1)(3t - 1)$
27. $(2x - 3y)^2$
28. $(5a + 3b)(5a - 3b)$
29. $2(3m + 1)^2$
30. $(1 - 2x)^3$

Expand and simplify.

31. $(m - 3)(m + 3) + (m - 4)^2$
32. $(x - 6)^2 - (x + 5)(x - 5)$
33. $3(2t + 1)^2 + 2(3t - 1)(3t + 1)$
34. $2(3x + 2y)(3x - 2y) - 3(3x - y)^2$

35. Number a) In the Mental Math column on page 97, two numbers that differ by 2 were multiplied by squaring their average, then subtracting 1. How does the product of the sum and difference $(x + 1)(x - 1)$ explain the method?
b) Develop a similar method for multiplying two numbers that differ by 6.
c) Show how the product of a sum and difference explains your method from part b).

3.6 Factor, if possible.

36. $5t - 35$
37. $5y^2 - 20y$
38. $2xy - 8xy^2$
39. $7st - 22mn$

Factor, if possible.

40. $m(x + 4) - 3(x + 4)$
41. $5a(y - 2) + 7(y + 2)$
42. $2x(m + n) - (m + n)$

Factor by grouping.

43. $mx + my + 2x + 2y$
44. $x^2 - xy - x + y$
45. $2m^2 - 3t - 6m + mt$

46. Measurement The diagram shows two concentric circles of radii r and $r + 3$.
a) Write an expression involving r for the area of the smaller circle.
b) Write an expression involving r for the area of the larger circle. Expand and simplify the expression.
c) Subtract the expression you wrote in part a) from the simplified expression in part b). Factor the result.
d) If r represents 5 cm, calculate the area of the shaded part of the diagram, to the nearest tenth of a square centimetre.

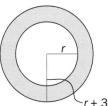

3.8 Factor, if possible.

47. $x^2 - x - 12$
48. $y^2 + 3y - 18$
49. $m^2 + 11m + 24$
50. $t^2 - 8t + 15$
51. $x^2 + 3x + 4$
52. $n^2 - 13n + 40$
53. $w^2 - w - 30$
54. $14 + 5m - m^2$
55. $x^2 + 9xy + 8y^2$
56. $c^2 - 10cd + 16d^2$

Factor fully.

57. $2x^2 - 2x - 40$ **58.** $ay^2 + 12ay - 28a$

59. $3x^2 + 12x - 36$ **60.** $5x^2 - 15x + 10$

61. Write two different trinomials that have $x + 3$ as a factor.

62. Write two different trinomials that have $x - 2$ as a factor.

.9 *Factor, if possible.*

63. $5m^2 + 17m + 6$ **64.** $6x^2 + 7x + 2$

65. $2x^2 - 7x + 5$ **66.** $3t^2 + 4t - 20$

67. $2m^2 + 2m - 3$ **68.** $6y^2 + y - 1$

69. $6x^2 - x - 1$ **70.** $9z^2 - 9z + 2$

71. $2x^2 + 11xy + 5y^2$ **72.** $4p^2 - 3pq - 7q^2$

Factor fully.

73. $4x^2 + 6x + 2$ **74.** $9t^2 + 3t - 6$

75. $20m^2 - 8m - 12$ **76.** $3y^3 - 7y^2 + 2y$

77. Basketball The backboard behind a basketball net is a rectangle whose area can be represented by the trinomial $77x^2 - 38x - 7$.

a) Factor $77x^2 - 38x - 7$ to find binomials that represent the length and width of the backboard.

b) If x represents 17 cm, find the dimensions of the backboard, in centimetres.

0 *Factor, if possible.*

78. $x^2 - 25$ **79.** $1 - 49m^2$

80. $m^2 + 16$ **81.** $49t^2 - 81s^2$

82. $4a^2 - 16b^2$ **83.** $144p^2 - q^2$

Determine whether the trinomial is a perfect square trinomial. If it is, factor it.

84. $x^2 + 10x + 25$ **85.** $y^2 - 12y + 36$

86. $9t^2 - 6t + 1$ **87.** $m^2 + 6m + 16$

88. $4x^2 + 12x + 9$ **89.** $25r^2 - 20rs + 4s^2$

Factor fully.

90. $5x^2 - 5$ **91.** $16m^2 - 36n^2$

92. $18y^2 + 60y + 50$ **93.** $5x^2 - 20xy + 20y^2$

94. Egyptian pyramid The North Stone Pyramid at Dahshur in Egypt has a square base with an area that can be represented by the trinomial $9x^2 - 12x + 4$.

a) Factor the trinomial to find a binomial that represents the side length of the base.

b) If x represents 74 m, what is the side length of the base, in metres?

Exploring Math

Designing With Panels

A rectangular wall is to be built from panels of wood that are each 1 m wide and 2 m long. The height of the wall must be 2 m. A 2 m high wall built from 1 panel has only 1 possible design.

A 2 m high wall built from 2 panels has 2 possible designs.

If there are 5 panels, the length of the wall is 5 m. One way to assemble the 5 panels is as shown.

Designs that are reflections of each other are considered different, so this design with 5 panels is different from the one above.

1. Copy and complete the table by finding the number of designs for 2 m high walls with each number of panels.

Number of Panels	Number of Designs
1	1
2	2
3	
4	
5	
6	
7	

2. Describe the relationship between the number of panels and the number of designs.

3. The pattern in the number of designs is called the Fibonacci sequence. Use the pattern to predict the number of designs with

a) 10 panels   **b)** 15 panels

Chapter Check

Simplify.
1. $(3y^2 - 2y + 1) + (5y^2 - y - 4)$
2. $(2x^2 - 3x - 5) - (2x^2 + 4x - 7)$

Simplify.
3. $(-2ab^2)(-5a^3b^3)$
4. $\dfrac{-45x^3yz^2}{-9x^2y}$

Expand and simplify.
5. $2(m - 5) - 4(2m + 1)$ **6.** $4x(2x - 3) - x(3x - 1)$
7. $2a(a^2 - 2a - 1) - 3a(2a^2 + a - 2)$

Find the product.
8. $(y - 3)(y + 5)$ **9.** $(2a + b)(a - 3b)$

Expand and simplify.
10. $(3m + 2)(m - 3) - (2m - 1)(m + 3)$
11. $2(3x + 1)(x + 2) - 3(4x - 5)$
12. $(3x + 2)(x^2 - 4x - 3)$

Expand and simplify.
13. $(x - 1)^2$ **14.** $(2y - 3)(2y + 3)$ **15.** $(x - 2y)^3$
16. $2(3x + 1)^2 + 3(2x - 1)(2x + 1)$

Factor.
17. $4m^2 - 28m$
18. $3ab - 9ab^2 + 6a^2b$
19. $x(m - 2) - 4(m - 2)$
20. $y^2 + 2x + 2y + xy$

Factor.
21. $x^2 - 7x + 12$
22. $a^2 + 4a - 21$
23. $y^2 + 9y + 20$
24. $t^2 - 6t - 27$
25. $x^2 + 6xy + 5y^2$
26. $m^2 - 9mn + 14n^2$

Factor fully.
27. $3x^2 - 3x - 6$
28. $4t^2 - 28t + 40$
29. $ay^2 + ay - 12a$

Factor.
30. $3t^2 + 8t + 5$
31. $2m^2 - 9m + 4$
32. $6x^2 - x - 2$
33. $4y^2 + 4y - 3$
34. $5r^2 - 11rs + 2s^2$
35. $4x^2 - 8xy - 5y^2$

Factor.
36. $x^2 - 4$ **37.** $1 - 36m^2$
38. $36t^2 - 49s^2$ **39.** $121a^2 - b^2$
40. $x^2 + 8x + 16$ **41.** $y^2 - 6y + 9$
42. $4t^2 - 4t + 1$ **43.** $4m^2 + 20m + 25$

Factor fully.
44. $3y^2 - 27$ **45.** $9t - 4t^3$
46. $8x^2 + 24x + 18$ **47.** $x^3 - 4x^2 + 4x$

48. Baseball In baseball, the first base bag is a square. Its side length can be represented by the expression $5x + 3$.
a) Write and expand an expression to represent the area of the top of the bag.
b) If x represents 7 cm, what is the area of the top of the bag, in square centimetres?

49. Bank note The face of a Canadian $20 bill has an area that can be represented by the expression $10x^2 + 9x - 40$.
a) Factor $10x^2 + 9x - 40$ to find expressions that represent the dimensions of the bill.
b) If x represents 32 mm, what are the dimensions of the bill, in millimetres?

PEANUTS reprinted by permission of United Features Syndicate, Inc.

Using the Strategies

1. Fred and Fran are in a rubber raft 1 km from shore. Fred punctures the raft, and water starts coming in at a rate of 10 L/min. The raft will sink if it has just over 30 L of water in it. Fran paddles toward the shore at a speed of 5 km/h. Find the slowest rate, in litres per minute, at which Fred must bail water so that they reach the shore without having to swim.

2. St. John's High School entered a basketball tournament. In the first round, each team played one game, and the losers dropped out. Each winner from the first round played one game in the second round, and the losers dropped out. This process continued until a winner was declared. St. John's won the tournament by winning 5 games. How many teams were in the tournament?

3. Adding the ages of Pedro and Lucia gives a total of 80 years. Pedro is five times as old as Lucia was when Pedro was as old as Lucia is now. How old are they?

4. If the pattern continues, what will be the number directly below 100?

```
            1
         2  3  4
      5  6  7  8  9
  10 11 12 13 14 15 16
```

5. A sector with an angle of 80° rotates clockwise about the centre of a circle. The sector moves 80° on each move. How many 80° moves will it take for the sector to arrive back at its original position?

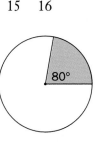

6. Copy the diagram. Fill in the boxes with the digits 2, 3, 4, and 5 to make the largest possible sum.

7. If you list the digits used to write the numbers from 1 to 100,
a) what is the median of the digits?
b) what is the mode of the digits?

8. George, Rena, Patricia, and Julian line up for a picture. If Patricia and George do not stand beside each other, in how many different ways can the group line up?

9. The number 12 is the first "abundant number," because it is the smallest whole number for which the sum of its factors, not including itself, is greater than itself.
$$1 + 2 + 3 + 4 + 6 = 16$$
What are the next three abundant numbers?

10. The length, width, and height of a rectangular prism are whole numbers of centimetres. Three faces of the prism have areas of 144 cm², 72 cm², and 32 cm². What are the dimensions of the prism?

11. Irina's shop prints and sells calendar pages. The pages have the days of the week and the dates on them, but not the names of the months. The person using the pages writes the name of the month on each page. There is a different page for each possible arrangement of dates in a month. How many different calendar pages must Irina print?

12. For all the books in your school library, estimate the total number of words.

D A T A   B A N K

1. You are leaving Calgary at 13:00 on a Friday to fly to Zurich via Toronto. If your plane averages 850 km/h, and you stop for 1 h in Toronto, at about what time on what day will you land in Zurich?

2. a) Name 3 Canadian cities that have a ratio of annual hours of sunshine to annual hours of blizzards of at least 100:1.
b) Name 2 Canadian cities that have a ratio of annual hours of sunshine to annual hours of blizzards of less than 50:1.

CHAPTER 4

Rational Expressions and Equations

Each coded expression includes a number and the first letters of some words. Decode each expression so that the number agrees with the words. You will find clues to some of the answers in the pictures.

a) 366 = D. in a L. Y.

b) 60 = M. in an H.

c) 3 = A. in a T.

d) 0 = F. P. of W. in D. C.

e) 6 = H. a D.

f) 4 = S. on a Q.

g) 3 = E. S. on an E. T.

h) 1 = T. on an E.

i) 3 = W. on a T.

j) 8 = L. on a S.

k) 18 = H. on a G. C.

l) 9 = L. of a C.

m) 88 = K. on a P.

n) 10 = P. in C.

o) 4 = T. on a C.

p) 5 = O. R.

q) 6 = F. on a C.

r) 4 = S. on a V.

s) 12 = T. at N.

t) 12 = T. on a T-T. S.

u) 8 = S. on a S. S.

v) 1 = S. in our S. S.

w) 50 = S. on the S. and S.

GETTING STARTED

Developing Formulas From Patterns

1. The three figures show cubes made from toothpicks and marshmallows.

a) Copy the table. Complete it by finding the numbers of marshmallows in figures with the given numbers of cubes.
b) Write a formula in the form $M =$ ▮▮▮ to find the number of marshmallows, M, needed to make a number of cubes, c.
c) Use your formula to calculate the number of marshmallows needed to make 10 cubes.
d) Copy the table. Complete it by finding the numbers of toothpicks in figures with the given numbers of cubes.
e) Write a formula in the form $T =$ ▮▮▮ to find the number of toothpicks, T, needed to make a number of cubes, c.
f) Use your formula to calculate the number of toothpicks needed to make 10 cubes.

Number of Cubes	Number of Marshmallows
1	
⋮	
5	

Number of Cubes	Number of Toothpicks
1	
⋮	
5	

2. The figures are made by connecting pairs of regular hexagons with side lengths of 1 unit.

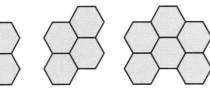

a) Copy the table. Complete it by finding the perimeters of figures with the given numbers of hexagons.
b) Let P represent the perimeter and let n represent the number of hexagons. Write a formula in the form $P =$ ▮▮▮ to find the perimeter from the number of hexagons.
c) Use your formula to calculate the perimeter of the figure that contains 22 hexagons.

Number of Hexagons	Perimeter
2	
4	
⋮	
10	

3. The three figures are made from squares and diagonals.

a) Copy the table. Complete it by finding the total numbers of triangles in figures with the given numbers of squares.
b) Write a formula in the form $T =$ ▮▮▮ to find the total number of triangles, T, formed from a number of squares, s.
c) Use your formula to calculate the total number of triangles formed from 15 squares.

Number of Squares	Number of Triangles
1	
⋮	
5	

146

Using Unit Rates

Some problems, such as the following example, can be solved using unit rates.

The average person speaks about 220 500 words in a week. About how many words does the average person speak in a year?

• Statement of fact:
Number of words spoken in 7 days is 220 500.

• Words in 1 day:
Number of words spoken in 1 day is $\dfrac{220\,500}{7}$.

• Words in 365 days:
Number of words spoken in 365 days is

$$365 \times \dfrac{220\,500}{7} = 11\,497\,500.$$

The average person speaks about 11 500 000 words in a year.

Solve the following problems.

1. A car uses gasoline at a rate of 15 L/100 km. About how many litres does the car use to travel 560 km?

2. An expert guitar maker takes about 1000 h to make 3 guitars. If she works 2400 h in a year, how many guitars can she make?

3. If 220 L of maple sap give 5.5 L of maple syrup, how many litres of maple syrup do 4000 L of maple sap give?

4. A spine-tailed swift can fly about 240 m in 5 s. About how long does it take to fly 600 m?

5. A dripping tap can waste up to 90 L of water a day. At this rate, how many hours does it take to waste 27 L?

6. A 16-g portion (one tablespoon) of peanut butter gives about 400 kJ of food energy. How much food energy does a 500-g jar of peanut butter give?

7. In Canada, there are about 9 km^2 of forested land for every 10 km^2 of land that is not forested. If about 4 850 000 km^2 of land is not forested, what area of land is forested?

8. Light takes about 8.5 min to travel 150 000 000 km from the sun to the Earth. About how many seconds does light take to travel 400 000 km from the moon to the Earth?

Mental Math

Solving Equations

Solve by inspection.

1. $x - 1 = 0$		**2.** $y - 2 = 3$
3. $a - 3 = 1$		**4.** $t - 5 = 2$
5. $m + 1 = 3$		**6.** $n + 2 = 4$
7. $p + 3 = 5$		**8.** $r + 5 = 8$
9. $2q = 6$		**10.** $4w = 8$
11. $5d = 15$		**12.** $3z = 12$
13. $\dfrac{k}{2} = 1$		**14.** $\dfrac{c}{3} = 2$
15. $\dfrac{e}{5} = 3$		**16.** $\dfrac{v}{4} = 4$

Adding Using Compatible Numbers

Compatible numbers are numbers that you can compute mentally. Suppose that you are adding two numbers, and one of them is just below a multiple of 10. The mental addition is easier if you round this number up to a multiple of 10, then add, then subtract the number you added when rounding.

For 79 + 64, think $79 + 1 + 64 = 80 + 64$
$$= 144$$
and $144 - 1 = 143$
So, $79 + 64 = 143$

Calculate.

1. $47 + 34$	**2.** $69 + 53$	**3.** $34 + 78$
4. $127 + 69$	**5.** $114 + 167$	**6.** $235 + 148$

Adapt the method to calculate the following. Describe how you adapted the method.

7. $1.8 + 2.3$	**8.** $3.4 + 1.9$	**9.** $7.8 + 4.5$
10. $8.4 + 3.9$	**11.** $12.7 + 6.5$	**12.** $120 + 190$
13. $240 + 180$	**14.** $150 + 270$	**15.** $660 + 450$

16. a) Complete the following additions, where x, y, n, m, and b represent whole numbers.
$(10x + y) + (10n + m)$
$(10x + y + b) + (10n + m) - b$

b) Explain how these additions are related to the method shown above.
c) For the method to work as shown, what is the value of $y + b$?

4.1 Dividing Polynomials

Canada officially has two national games, lacrosse and hockey. Lacrosse is thought to have originated with the Algonquin tribes in the St. Lawrence Valley. The game was very popular in the late nineteenth century and was at one time an Olympic sport. Canadian lacrosse teams won gold medals at the Summer Olympics in 1904 and 1908.

There are two forms of lacrosse — box lacrosse, which is played indoors, and field lacrosse.

Explore: Use Algebra Tiles

When field lacrosse is played under international rules, the rectangular field can be represented by the algebra tiles shown. If one dimension of the field is represented by the expression $2x$, write an expression to represent
a) the other dimension   **b)** the area

$x + 5$

$2x^2 + 10x$

Inquire

1. Write an expression that shows the area divided by $2x$.

2. How could you simplify your expression from question 1 to give the expression for the other dimension?

$x + 5$

3. Write a rule for dividing a polynomial by a monomial.

4. Divide.

a) $\dfrac{4t^2 + 8t}{4t}$ **b)** $\dfrac{10m^3 + 5m^2 + 15m}{5m}$ **c)** $\dfrac{6r^4 - 3r^3 + 9r^2}{3r^2}$

5. The expressions for the dimensions of a lacrosse field apply to both men's and women's games.
a) For the men's game, played 10-a-side, x represents 50 m. What are the dimensions of the field, in metres?
b) For the women's game, played 12-a-side, x represents 55 m. What are the dimensions of the field, in metres?

The distributive property applies to division.

$$\frac{3}{7} = \frac{2+1}{7}$$
$$= \frac{2}{7} + \frac{1}{7}$$

In general, $\dfrac{a+b}{c} = \dfrac{a}{c} + \dfrac{b}{c}$

Note that the expressions $\dfrac{a+b}{c}$, $\dfrac{a}{c}$, and $\dfrac{b}{c}$ have a variable in the denominator. Since division by zero is not defined, replacements that make the denominator zero are not allowed. Therefore, c cannot equal zero, that is, $c \neq 0$.

Example 1   Dividing by a Monomial

Simplify $\dfrac{24x^3 - 12x^2 + 18x}{6x}$. State the restriction on the variable.

Solution
Divide each term in the polynomial by the monomial.

$$\frac{24x^3 - 12x^2 + 18x}{6x} = \frac{24x^3}{6x} - \frac{12x^2}{6x} + \frac{18x}{6x}$$
$$= 4x^2 - 2x + 3$$

Because division by zero is not defined, $6x \neq 0$, so $x \neq 0$.

So, $\dfrac{24x^3 - 12x^2 + 18x}{6x} = 4x^2 - 2x + 3$, $x \neq 0$.

To divide a polynomial by a binomial, use a long division process like the one used with numbers, such as $176 \div 11$.

$$\text{Divisor} \overline{)\,\text{Dividend}}^{\,\text{Quotient}}_{\,\text{Remainder}}$$

In standard notation

```
        Think 17 ÷ 11 gives 1
          ↓
       16  ← Think 66 ÷ 11 = 6
   11)176
       11  ← Multiply 1 × 11.
       66  ← Subtract. Bring down the 6.
       66  ← Multiply 6 × 11.
        0  ← Subtract to get the remainder.
```

In expanded notation

```
          Think 100 ÷ 10 = 10
             ↓
          10 + 6  ← Think 60 ÷ 10 = 6
   10 + 1)100 + 70 + 6
          100 + 10
               60 + 6
               60 + 6
                    0
```

If 10 is represented by x, the division becomes $(x^2 + 7x + 6) \div (x + 1)$.

```
        Think x² ÷ x = x
           ↓
         x + 6   ← Think 6x ÷ x = 6
   x+1)x² + 7x + 6
       x² +  x        ← Multiply x(x + 1).
            6x + 6    ← Subtract. Bring down the 6.
            6x + 6    ← Multiply 6(x + 1).
                 0    ← Subtract to get the remainder.
```

Example 2 Dividing by a Binomial

Divide $x^3 + 5x - 18$ by $x - 2$.
State the restriction on the variable.

Solution

When there are missing terms in the dividend, include them with zero as the coefficient.

$$
\require{enclose}
\begin{array}{r}
x^2 + 2x + 9 \\[-2pt]
x - 2 \enclose{longdiv}{x^3 + 0x^2 + 5x - 18} \\
\underline{x^3 - 2x^2} \\
2x^2 + 5x \\
\underline{2x^2 - 4x} \\
9x - 18 \\
\underline{9x - 18} \\
0
\end{array}
$$

Because division by zero is not defined, $x - 2 \neq 0$, so $x \neq 2$.

So, $(x^3 + 5x - 18) \div (x - 2) = x^2 + 2x + 9$, $x \neq 2$.

In Example 2, $x - 2$ divides into $x^3 + 5x - 18$ evenly, or with a remainder of zero. In other words, $x - 2$ is a factor of $x^3 + 5x - 18$. Another way of writing the result is $x^3 + 5x - 18 = (x - 2)(x^2 + 2x + 9)$. When the divisor is not a factor of the dividend, the remainder is not zero. For the division $32 \div 5$, the quotient is 6, and the remainder is 2.

$$\frac{32}{5} = \frac{30}{5} + \frac{2}{5} = 6 + \frac{2}{5}$$

Another way to write the result is $32 = (5)(6) + 2$.

In general, for division,

$$\frac{\text{Dividend}}{\text{Divisor}} = \text{Quotient} + \frac{\text{Remainder}}{\text{Divisor}} \quad \text{or} \quad \text{Dividend} = (\text{Divisor})(\text{Quotient}) + \text{Remainder}$$

The situation is similar for the division of polynomials.

$$
\require{enclose}
\begin{array}{r}
x + 2 \\[-2pt]
x + 1 \enclose{longdiv}{x^2 + 3x + 5} \\
\underline{x^2 + x} \\
2x + 5 \\
\underline{2x + 2} \\
3
\end{array}
$$

So, $\dfrac{x^2 + 3x + 5}{x + 1} = x + 2 + \dfrac{3}{x + 1}$

The result can also be written as $x^2 + 3x + 5 = (x + 1)(x + 2) + 3$.

In general, for the division of polynomials,

$$\frac{\text{Polynomial}}{\text{Divisor}} = \text{Quotient} + \frac{\text{Remainder}}{\text{Divisor}} \quad \text{or} \quad \text{Polynomial} = (\text{Divisor})(\text{Quotient}) + \text{Remainder}$$

$$\frac{P}{D} = Q + \frac{R}{D} \qquad\qquad\qquad P = DQ + R$$

Example 3 **Division With a Non-zero Remainder**

a) Divide $6y^3 - 4y^2 - 9y - 3$ by $2y^2 - 3$. Express the answer in two forms.

b) State the restrictions on the variable.

Solution

a)

$$
\begin{array}{r}
3y - 2 \\
2y^2 - 3 \overline{)6y^3 - 4y^2 - 9y - 3} \\
\underline{6y^3 + 0y^2 - 9y} \\
-4y^2 + 0y - 3 \\
\underline{-4y^2 + 0y + 6} \\
-9
\end{array}
$$

So, $\dfrac{6y^3 - 4y^2 - 9y - 3}{2y^2 - 3} = 3y - 2 - \dfrac{9}{2y^2 - 3}$.

Also, $6y^3 - 4y^2 - 9y - 3 = (2y^2 - 3)(3y - 2) - 9$.

b) Since $2y^2 - 3 \neq 0$

$$2y^2 \neq 3$$

$$y^2 \neq \frac{3}{2}$$

$$y \neq \pm\sqrt{\frac{3}{2}}$$

The value of a polynomial containing the variable x depends on the value of x. The polynomial can be named using **function notation** as $P(x)$.

Read $P(x)$ as "the value of P at x" or "P of x."

For the polynomial $P(x) = x^2 - 5x + 3$,

When $x = 1$, $P(1) = 1^2 - 5(1) + 3$
$ = -1$

When $x = -1$, $P(-1) = (-1)^2 - 5(-1) + 3$
$ = 9$

If the divisor and the quotient also contain x, and the remainder is a number, the result of the division of a polynomial can be represented as

$$\frac{P(x)}{D(x)} = Q(x) + \frac{R}{D(x)} \qquad \text{or} \qquad P(x) = D(x)Q(x) + R$$

Example 4 **Using Function Notation**

a) Divide $2x^3 + x^2 - 13x + 9$ by $2x - 1$. Express the answer in the forms $\dfrac{P(x)}{D(x)} = Q(x) + \dfrac{R}{D(x)}$ and $P(x) = D(x)Q(x) + R$.

b) State the restriction on the variable.

Solution

a)

$$
\begin{array}{r}
x^2 + x - 6 \\
2x - 1 \overline{)2x^3 + x^2 - 13x + 9} \\
\underline{2x^3 - x^2} \\
2x^2 - 13x \\
\underline{2x^2 - x} \\
-12x + 9 \\
\underline{-12x + 6} \\
3
\end{array}
$$

b) $\quad 2x - 1 \neq 0$

$$2x \neq 1$$

$$x \neq \frac{1}{2}$$

So, in the form $\dfrac{P(x)}{D(x)} = Q(x) + \dfrac{R}{D(x)}$, $\dfrac{2x^3 + x^2 - 13x + 9}{2x - 1} = x^2 + x - 6 + \dfrac{3}{2x - 1}$.

Also, in the form $P(x) = D(x)Q(x) + R$, $2x^3 + x^2 - 13x + 9 = (2x - 1)(x^2 + x - 6) + 3$.

Practice

In each of the following, state any restrictions on the variables.

Divide.

1. $\dfrac{20xy}{4x}$ **2.** $\dfrac{30m^2n^2}{2mn^2}$ **3.** $\dfrac{-36x^4y^3}{9x^2y}$

4. $\dfrac{24abc^3}{-3bc}$ **5.** $\dfrac{-75a^3b^4c}{-15a^3bc}$ **6.** $\dfrac{-12x^4y}{12x^4y}$

Divide.

7. $\dfrac{(4x^2)(5x^3)}{10x^4}$ **8.** $\dfrac{(2a^3)(-4a^5)}{8a^7}$

9. $\dfrac{(-8m^2n^3)(-2mn)}{4mn^3}$ **10.** $\dfrac{(4ab)(2a^3)(3b^3)}{-6a^2b^2}$

Divide.

11. $\dfrac{2x+18}{2}$ **12.** $\dfrac{10a-15b}{5}$

13. $\dfrac{6x^2-18x-24}{-6}$ **14.** $\dfrac{14a+7b-7}{7}$

15. $\dfrac{4y^2-16y}{4y}$ **16.** $\dfrac{3a^2b-12ab^2}{-3ab}$

17. $\dfrac{12x^2y^3-15x^4y^2-18x^5y^4}{-3x^2y^2}$

Divide.

18. $(x^2+8x+15)\div(x+3)$
19. $(a^2-7a+10)\div(a-5)$
20. $(y^2-y-12)\div(y-4)$
21. $(t^2-4)\div(t+2)$

Divide.

22. $(x^3+2x^2+3x+2)\div(x+1)$
23. $(t^3+3t^2-5t-4)\div(t+4)$
24. $(a^3-3a^2-a+3)\div(a-3)$
25. $(x^3+4x^2-13x+8)\div(x-1)$
26. $(m^3+3m^2-4)\div(m+2)$
27. $(y^3-4y^2-2y+8)\div(y-4)$

Divide.

28. $(n^3+2n^2-n-2)\div(n^2-1)$
29. $(x^3-3x^2+4x-12)\div(x^2+4)$
30. $(y^3+5y^2+3y+15)\div(y^2+3)$
31. $(a^3-a^2-4a+4)\div(a^2-4)$

Divide.

32. $(2x^2+11x+15)\div(2x+5)$
33. $(3y^2+8y-3)\div(3y-1)$
34. $(5r^2-31r+6)\div(5r-1)$
35. $(4t^2-4t-3)\div(2t+1)$
36. $(6r^2-25r+14)\div(3r-2)$
37. $(10s^2+21s+9)\div(5s+3)$
38. $(8x^2+14x-15)\div(4x-3)$

Divide.

39. $(2x^3-2x^2+3x-3)\div(x-1)$
40. $(3z^3+6z^2+5z+10)\div(z+2)$
41. $(4m^3+2m^2-6m-3)\div(2m+1)$
42. $(6n^3-9n^2-8n+12)\div(2n-3)$
43. $(10d^3+15d^2+4d+6)\div(5d^2+2)$
44. $(8g^3-6g^2+4g-3)\div(2g^2+1)$
45. $(12s^3+3s^2-20s-5)\div(3s^2-5)$
46. $(21t^3-28t^2-6t+8)\div(7t^2-2)$

Divide. Write each answer in the forms $\dfrac{P}{D}=Q+\dfrac{R}{D}$ *and* $P=DQ+R$.

47. $(t^2+4t+2)\div(t+4)$
48. $(p^2-3p-8)\div(p-3)$
49. $(y^2-7y+10)\div(y-4)$
50. $(2w^2+w-3)\div(w+2)$
51. $(4a^2-7a-7)\div(a-2)$
52. $(x^2-8)\div(x-3)$
53. $(5c^2+14c+11)\div(5c+4)$
54. $(6m^2-5m-5)\div(3m-4)$
55. $(8n^2-18n+13)\div(2n-3)$
56. $(4y^2-29)\div(2y-5)$

Divide. Write each answer in the forms

$\dfrac{P(x)}{D(x)}=Q(x)+\dfrac{R}{D(x)}$ *and* $P(x)=D(x)Q(x)+R$.

57. $(6x^3+4x^2+9x+8)\div(3x+2)$
58. $(8y^3-4y^2+6y-9)\div(2y-1)$
59. $(12f^3-10f^2-18f+25)\div(6f-5)$
60. $(6m^3+2m^2-5)\div(3m+1)$
61. $(6n^3+14n^2+9n+12)\div(2n^2+3)$
62. $(10x^3+2x^2-5x-2)\div(2x^2-1)$
63. $(8m^3-4m^2+2m)\div(4m^2+1)$
64. $(9z^3+24z^2-12z-10)\div(3z^2-4)$

152

Applications and Problem Solving

65. Identify the dividend, divisor, quotient, and remainder.

a) $\dfrac{156}{12} = 13$

b) $\dfrac{350}{9} = 38 + \dfrac{8}{9}$

c) $\dfrac{a^2 - 5a + 6}{a - 3} = a - 2$

d) $\dfrac{2t^2 + 5t - 2}{2t - 1} = t + 3 + \dfrac{1}{2t - 1}$

e) $\dfrac{x^3 + x^2 + x - 3}{x^2 + 1} = x + 1 + \dfrac{4}{x^2 + 1}$

f) $\dfrac{6x^3 + 3x^2 - x}{3x^2 - 2} = 2x + 1 + \dfrac{3x + 2}{3x^2 - 2}$

66. Identify the dividend, divisor, quotient, and remainder. Can you be sure about the quotient and the divisor? Explain.

a) $255 = (11)(23) + 2$

b) $8y^3 + 6y^2 - 4y - 5 = (4y + 3)(2y^2 - 1) - 2$

c) $(x)(x + 1) + 3 = x^2 + x + 3$

67. Visualization Use algebra tiles or a sketch to show each of the following. State the result of each division.

a) $(x^2 + 5x + 4)$ is divisible by $(x + 1)$

b) $(3x^2 + 7x + 2)$ is divisible by $(x + 2)$

c) $(2x^2 + 7x + 3)$ is divisible by $(2x + 1)$

68. Table tennis The area of a table tennis table can be represented by the expression $2x^2 + 18x - 72$. The length of the table can be represented by $2x - 6$.

a) Write an expression that represents the width of the table.

b) Divide the expression from part a).

c) If x represents 140 cm, what are the dimensions of the table, in centimetres?

69. Pattern a) Divide each of the following.

$$\dfrac{x^3 - 1}{x - 1} \qquad \dfrac{x^3 - 8}{x - 2} \qquad \dfrac{x^3 - 27}{x - 3}$$

b) Describe the pattern in the coefficients of the quotients.

c) Predict the quotient for $\dfrac{x^3 - 64}{x - 4}$ and $\dfrac{x^3 - 125}{x - 5}$. Then, divide to check your predictions.

70. Divide.

a) $(x^5 - 1) \div (x - 1)$ **b)** $(x^5 - 32) \div (x - 2)$

71. Divide.

a) $(x^4 + 3x^3 + 2x^2 + 3x + 1) \div (x^2 + 1)$

b) $(y^4 + 2y^2 - 3) \div (y^2 - 1)$

c) $(2z^4 - 4z^2 - 16) \div (z^2 - 4)$

d) $(6a^4 - 17a^2 + 7) \div (3a^2 - 7)$

e) $(15t^4 + 25t^3 - 9t^2 + 10t - 6) \div (5t^2 + 2)$

72. Divide. State whether the value of the remainder depends on the value of the variable.

a) $(6x^3 + 4x^2 + 4x + 3) \div (2x^2 + 1)$

b) $(3y^3 - 9y^2 - 3) \div (3y^2 + 1)$

c) $(2t^3 + 4t^2 - 4t - 19) \div (2t^2 - 4)$

d) $(6d^4 - 13d^2 + d + 4) \div (2d^2 - 3)$

73. Find the whole-number value of k such that

a) $x + 3$ divides $x^2 - x - k$ evenly

b) $2t - 1$ is a factor of $6t^2 + t - k$

c) the remainder is 3 when $4x^2 - 9x + k$ is divided by $x - 1$

d) the remainder is -5 when $2x^3 + 7x^2 + 5x - k$ is divided by $2x + 1$

74. Measurement
The area of the triangle is represented by the expression $6m^2 - 5m - 4$. If the height is $3m - 4$, what is the base?

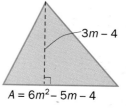

$A = 6m^2 - 5m - 4$

75. Measurement
The area of the trapezoid is represented by the expression $12n^2 - 11n + 2$. The bases are $3n + 1$ and $5n - 3$. What is the height?

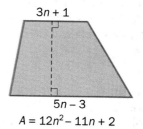

$A = 12n^2 - 11n + 2$

76. Dividing $5x^2 + 14x - 3$ by $x + 3$ gives a quotient of $5x - 1$. Explain why $\dfrac{5x^2 + 14x - 3}{x + 3}$ is not the same as $5x - 1$.

INVESTIGATING MATH

Using Synthetic Division

1 Dividing Polynomials by $x - b$

Synthetic division is a shortcut for dividing polynomials by divisors of the form $x - b$, where x is a variable and b is a constant.

Using long division to divide $4x^3 - 11x^2 + 12$ by $x - 2$ is as shown.

$$\begin{array}{r} 4x^2 - 3x - 6 \\ x-2 \overline{)\,4x^3 - 11x^2 + 0x + 12} \\ \underline{4x^3 - 8x^2} \\ -3x^2 + 0x \\ \underline{-3x^2 + 6x} \\ -6x + 12 \\ \underline{-6x + 12} \\ 0 \end{array}$$

For synthetic division, use the coefficients of the polynomial. Include the coefficient zero for a missing term. Divide by the constant, b, from the divisor.
$x - b = x - 2$, so $b = 2$

Bring down the first coefficient.

$$\begin{array}{c|cccc} 2 & 4 & -11 & 0 & 12 \\ & \downarrow & & & \\ \hline & 4 & & & \end{array}$$

Multiply the first coefficient by b. Write the product under the second coefficient and find the sum.

$$\begin{array}{c|cccc} 2 & 4 & -11 & 0 & 12 \\ & & {}^{\times 2}8 & & \\ \hline & 4 & -3 & & \end{array}$$

Multiply the sum by b. Write the product under the third coefficient and find the sum.

$$\begin{array}{c|cccc} 2 & 4 & -11 & 0 & 12 \\ & & 8 & {}^{\times 2}{-6} & \\ \hline & 4 & -3 & -6 & \end{array}$$

Multiply the latest sum by b. Write the product under the fourth coefficient and find the sum.

$$\begin{array}{c|cccc} 2 & 4 & -11 & 0 & 12 \\ & & 8 & -6 & {}^{\times 2}{-12} \\ \hline & 4 & -3 & -6 & 0 \end{array}$$

The final number in the bottom row is the remainder. The other numbers in the bottom row are the coefficients of the quotient. So, the quotient is $4x^2 - 3x - 6$, and the remainder is 0.

$$\frac{4x^3 - 11x^2 + 12}{x - 2} = 4x^2 - 3x - 6 \text{ or } 4x^3 - 11x^2 + 12 = (x - 2)(4x^2 - 3x - 6)$$

Note that, for a divisor such as $x + 3$, b is negative. $x - b = x + 3$, so $b = -3$

Divide using synthetic division.

1. $(x^2 + 4x - 5) \div (x - 1)$
2. $(y^2 + 8y + 12) \div (y + 2)$
3. $(n^2 - 9) \div (n + 3)$
4. $(2t^2 - 2t - 12) \div (t - 3)$
5. $(5n^2 + 7n - 6) \div (n + 2)$
6. $(4r^3 - 10r^2 + 3r + 2) \div (r - 2)$
7. $(6a^3 + 24a^2 + a + 4) \div (a + 4)$
8. $(3c^3 - 16c^2 + 25) \div (c - 5)$
9. $(x^4 - x^3 + x - 1) \div (x - 1)$

Divide using synthetic division.

10. $(w^3 + 3w - 8) \div (w - 2)$
11. $(3x^2 + 16x + 18) \div (x + 3)$
12. $(4y^3 - 4y^2 + y) \div (y - 1)$
13. $(2z^3 + 19z^2 - 11z - 15) \div (z + 10)$
14. $(8g^3 - 23g^2 - 19) \div (g - 3)$
15. $(q^4 - 2q^3 - 8q^2 + 2q - 1) \div (q - 4)$
16. $(10x^4 + 80x^3 - x) \div (x + 8)$
17. $(9y^4 + 18y^3 - y^2 + 7) \div (y + 2)$

2 Dividing Polynomials by $ax - b$

To divide by $ax - b$, rewrite the divisor in the form $a\left(x - \dfrac{b}{a}\right)$.

To divide $6x^3 - x^2 + 10x - 9$ by $3x - 2$, rewrite $3x - 2$ as $3\left(x - \dfrac{2}{3}\right)$.

Use synthetic division to divide by $\left(x - \dfrac{2}{3}\right)$:

$$
\begin{array}{r|rrrr}
\tfrac{2}{3} & 6 & -1 & 10 & -9 \\
 & & 4 & 2 & 8 \\
\hline
 & 6 & 3 & 12 & -1
\end{array}
$$

$$\frac{6x^3 - x^2 + 10x - 9}{x - \dfrac{2}{3}} = (6x^2 + 3x + 12) - \frac{1}{x - \dfrac{2}{3}}$$

Now divide by 3:

$$\frac{6x^3 - x^2 + 10x - 9}{3\left(x - \dfrac{2}{3}\right)} = \frac{6x^2 + 3x + 12}{3} - \frac{1}{3\left(x - \dfrac{2}{3}\right)}$$

$$\frac{6x^3 - x^2 + 10x - 9}{3x - 2} = 2x^2 + x + 4 - \frac{1}{3x - 2}$$

or $6x^3 - x^2 + 10x - 9 = (3x - 2)(2x^2 + x + 4) - 1$

Divide using synthetic division.

1. $(2x^2 + x - 1) \div (2x - 1)$
2. $(9t^2 - 6t - 8) \div (3t + 2)$
3. $(6a^2 - 5a - 4) \div (3a - 4)$
4. $(8y^3 + 8y^2 + 8y + 3) \div (2y + 1)$
5. $(15c^3 + 9c^2 - 5c - 3) \div (5c + 3)$
6. $(10m^3 + 7m^2 + 18) \div (2m + 3)$
7. $(16u^4 - 8u^3 + 2u - 1) \div (2u - 1)$
8. $(18t^4 - 26t^2 - 7t - 20) \div (3t + 4)$

Divide using synthetic division.

9. $(12x^2 + 25x - 4) \div (4x - 1)$
10. $(20e^3 + 15e^2 + 24e + 10) \div (4e + 3)$
11. $(9f^3 - 31f + 15) \div (3f - 5)$
12. $(30n^3 + 12n^2 - 25n - 20) \div (5n + 2)$
13. $(10r^4 + 12r^3 + 15r^2 + 43r + 31) \div (5r + 6)$

3 Finding a Pattern

1. a) If $P(x) = x^3 - 4x^2 + 3x - 10$, divide $P(x)$ by $x - 4$ and state the remainder.
b) Evaluate $P(4)$ and compare with the remainder from part a).

2. a) If $P(y) = 2y^3 + 3y^2 - 5y + 7$, divide $P(y)$ by $y + 3$ and state the remainder.
b) Evaluate $P(-3)$ and compare with the remainder from part a).

3. a) If $P(z) = 6z^3 - 3z^2 - 8z$, divide $P(z)$ by $2z - 1$ and state the remainder.

b) Evaluate $P\left(\dfrac{1}{2}\right)$ and compare with the remainder from part a).

4. Write a rule for using substitution to find the remainder when a polynomial is divided by a divisor of the form
a) $x - b$ **b)** $ax - b$

5. Use substitution to find the remainder from each division.
a) $(2x^3 - 15x + 15) \div (x - 2)$
b) $(4n^4 - 4n^3 - 5n + 3) \div (n - 1)$
c) $(y^3 + 6y^2 + 9y + 4) \div (y + 4)$
d) $(6m^3 - 3m^2 - 7m - 2) \div (2m + 1)$

155

4.2 Simplifying Rational Expressions

A rectangular flag of a country, state, or province can be made in many sizes, but the length and width of any given flag are always in the same ratio.

The area of a British Columbia flag can be represented by the polynomial $2x^2 + 3x + 1$, and the length can be represented by $2x + 1$. The width of the flag is represented by the expression $\dfrac{2x^2 + 3x + 1}{2x + 1}$. This is an example of a **rational expression**, which is an algebraic fraction whose numerator and denominator are polynomials. The following are also rational expressions.

$$\frac{3}{x+2} \qquad \frac{x+1}{x+3} \qquad \frac{5y}{y^2-1} \qquad \frac{a^2+b^2}{a^2-b^2}$$

Explore: Complete the Table

Copy the table. Complete it by evaluating each rational expression for the different values of x.

Rational Expression	x-value					
	−1	0	1	2	3	4
$\dfrac{2x+4}{x^2+5x+6}$	.2	0	.5	.4	.3	.28
$\dfrac{2}{x+3}$	1	.6	.5	.4	.3	.25

Inquire

1. What does the completed table tell you about the two rational expressions?

2. Factor the numerator and the denominator of the first rational expression.

3. How do the factors you wrote in question 2 explain your findings in question 1?

4. a) The expressions $\dfrac{2x+4}{x^2+5x+6}$ and $\dfrac{2}{x+3}$ are known as **equivalent forms** of a rational expression. Explain why.

b) Which of the two forms in part a) is simpler? Explain.

5. a) Factor the numerator in the expression $\dfrac{2x^2+3x+1}{2x+1}$, which represents $\dfrac{\text{area}}{\text{length}}$ or the width of the British Columbia flag.

b) Which simpler expression is an equivalent form of the expression $\dfrac{2x^2+3x+1}{2x+1}$?

c) If x represents 2 units of length, what is the ratio length:width for the British Columbia flag?

156

A rational number is said to be **reduced** or **simplified** if the numerator and denominator have no common factors other than 1.

$$\frac{8}{10} = \frac{\overset{1}{\cancel{2}} \times 4}{\underset{1}{\cancel{2}} \times 5}$$
$$= \frac{4}{5}$$

A rational expression is in **simplest form** or **lowest terms** when the numerator and denominator have no common factors other than 1.

Because division by 0 is not defined, the values of variables in the denominators of rational expressions may be restricted. In the expression $\dfrac{x+3}{x(x-2)}$, x cannot equal 0 or 2, because either value makes the denominator 0. We state the restrictions as $x \neq 0, 2$.

Example 1 Simplifying Rational Expressions
Express in simplest form. State the restrictions on the variable.

a) $\dfrac{2x}{x^2 + x}$ **b)** $\dfrac{x^2 + 3x - 10}{x^2 + 8x + 15}$

Solution

a) Factor the denominator. Divide by the common factor.

$$\frac{2x}{x^2+x} = \frac{\overset{1}{2\cancel{x}}}{\underset{1}{\cancel{x}}(x+1)}$$
$$= \frac{2}{x+1}, \ x \neq 0, -1$$

b) Factor the numerator and the denominator. Divide by the common factor.

$$\frac{x^2+3x-10}{x^2+8x+15} = \frac{\overset{1}{(\cancel{x+5})}(x-2)}{\underset{1}{(\cancel{x+5})}(x+3)}$$
$$= \frac{x-2}{x+3}, \ x \neq -3, -5$$

Example 2 Simplifying Rational Expressions
Reduce to lowest terms. State the restrictions on the variable.

a) $\dfrac{2y^2 - y - 15}{4y^2 - 13y + 3}$ **b)** $\dfrac{4t^2 - 12t + 9}{4t^2 - 9}$

Solution

a) $\dfrac{2y^2-y-15}{4y^2-13y+3} = \dfrac{\overset{1}{(\cancel{y-3})}(2y+5)}{(4y-1)\underset{1}{(\cancel{y-3})}}$
$$= \frac{(2y+5)}{(4y-1)}, \ y \neq \frac{1}{4}, 3$$

b) $\dfrac{4t^2-12t+9}{4t^2-9} = \dfrac{(2t-3)\overset{1}{(\cancel{2t-3})}}{(2t+3)\underset{1}{(\cancel{2t-3})}}$
$$= \frac{(2t-3)}{(2t+3)}, \ t \neq \frac{3}{2}, -\frac{3}{2}$$

Practice

In each of the following, state any restrictions on the variables.

Simplify.

1. $\dfrac{-6x^2y^3}{-18x^3y}$

2. $\dfrac{16a^2bc}{4a^2b^2c^2}$

3. $\dfrac{-4x^4y^2z}{20x^3y^3z}$

4. $\dfrac{5x}{5(x+4)}$

5. $\dfrac{21m(m-4)}{7m^2}$

6. $\dfrac{8t^2(t+5)}{4t(t-5)}$

7. $\dfrac{7x(x-3)}{14x^2(x-3)}$

8. $\dfrac{(m-1)(m+2)}{(m+4)(m-1)}$

Write in simplest form.

9. $\dfrac{2x}{2x+8}$

10. $\dfrac{y^2}{y^2+2y}$

11. $\dfrac{10x}{5x^2-15x}$

12. $\dfrac{4m^2-8mn}{4mn}$

13. $\dfrac{6a^2+9a}{12a^2}$

14. $\dfrac{3xy}{6x^2y-12xy^2}$

15. $\dfrac{3t^3+6t^2-15t}{3t}$

16. $\dfrac{4x}{16x^3-12x}$

17. $\dfrac{14n^4-4n^3+6n^2+8n}{2n^2}$

18. $\dfrac{10y^4+5y^3-15y^2}{5y}$

Reduce to lowest terms.

19. $\dfrac{4x+4y}{5x+5y}$

20. $\dfrac{a^2+2a}{a^2-3a}$

21. $\dfrac{6t-36}{t-6}$

22. $\dfrac{4m+24}{8m-24}$

23. $\dfrac{5x-10}{3x-6}$

24. $\dfrac{8x^2+4x}{6x^2+3x}$

25. $\dfrac{2x^2-2x}{2x^2+2x}$

26. $\dfrac{4a^2b+8ab}{6a^2-6a}$

27. $\dfrac{5xy+10x}{2y^2+4y}$

Express in simplest equivalent form.

28. $\dfrac{m-2}{m^2-5m+6}$

29. $\dfrac{y^2+10y+25}{y+5}$

30. $\dfrac{2x+6}{x^2-6x-27}$

31. $\dfrac{r^2-4}{5r+10}$

32. $\dfrac{a^2+a}{a^2+2a+1}$

33. $\dfrac{x^2-9}{2x^2y-6xy}$

34. $\dfrac{2w+2}{2w^2+3w+1}$

35. $\dfrac{3t^2-8t+4}{6t^2-4t}$

36. $\dfrac{8z+6z^2}{9z^2-16}$

37. $\dfrac{5x^2+3xy-2y^2}{3x^2+3xy}$

Simplify.

38. $\dfrac{x^2+4x+4}{x^2+5x+6}$

39. $\dfrac{a^2-a-12}{a^2-9a+20}$

40. $\dfrac{m^2-5m+6}{m^2+2m-15}$

41. $\dfrac{y^2-8y+15}{y^2-25}$

42. $\dfrac{x^2-10x+24}{x^2-12x+36}$

43. $\dfrac{n^2-n-2}{n^2+n-6}$

44. $\dfrac{p^2+8p+16}{p^2-16}$

45. $\dfrac{2t^2-t-1}{t^2-3t+2}$

46. $\dfrac{6v^2+11v+3}{4v^2+8v+3}$

47. $\dfrac{6x^2-13x+6}{8x^2-6x-9}$

48. $\dfrac{3z^2-7z+2}{9z^2-6z+1}$

49. $\dfrac{2m^2-mn-n^2}{4m^2-4mn-3n^2}$

Applications and Problem Solving

50. Simplify, if possible.

a) $\dfrac{1-x}{x-1}$

b) $\dfrac{x-1}{x+1}$

c) $\dfrac{y^2+1}{y^2-1}$

d) $\dfrac{3t-7}{3t-7}$

e) $\dfrac{t^2-s^2}{(s+t)^2}$

f) $\dfrac{x^3-2x^2+3x}{2x^2-4x+6}$

51. For which values of x are the following rational expressions not defined?

a) $\dfrac{2x-y}{x-y}$

b) $\dfrac{4x}{3x+y}$

c) $\dfrac{3}{x^3}$

d) $\dfrac{x^2}{x^3-8}$

e) $\dfrac{x^2+3x-11}{x^4-1}$

f) $\dfrac{3x^2+5xy+2y^2}{4x^2-9y^2}$

52. State whether each of the following rational expressions is equivalent to the expression $\dfrac{x+1}{x-1}$.

a) $\dfrac{x+2}{x-2}$ **b)** $\dfrac{x^2+x}{x^2-x}$ **c)** $\dfrac{4+4x}{4x-4}$

d) $\dfrac{3x+1}{3x-1}$ **e)** $\dfrac{(x+1)^2}{(x-1)^2}$ **f)** $\dfrac{1+x}{1-x}$

53. Saskatchewan flag The area of a Saskatchewan flag can be represented by the polynomial $x^2 + 3x + 2$ and its width by $x + 1$.
a) Write a rational expression that represents the length.
b) Write the expression in simplest form.
c) If x represents 1 unit of length, what is the ratio length:width for a Saskatchewan flag?

54. Measurement The length of an edge of a cube is $x + 1$. Write and simplify a rational expression that represents the ratio of the volume to the surface area.

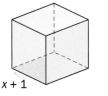

$x + 1$

55. Express in simplest equivalent form.

a) $\dfrac{x^4+2x^2+1}{x^4-1}$ **b)** $\dfrac{9a^4-16}{3a^4+a^2-4}$

c) $\dfrac{4y^4-9z^4}{(2y^2+3z^2)^2}$ **d)** $\dfrac{(2m-3n)^2(2m+3n)}{16m^4-81n^4}$

e) $\dfrac{x^4-y^4}{(x^4+2x^2y^2+y^4)(x^2-2xy+y^2)}$

56. Pattern The diagrams show the numbers of asterisks in the first 4 diagrams of two patterns.

Pattern 1

Pattern 2

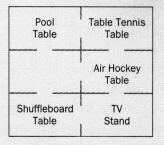

a) For pattern 1, express the number of asterisks in the nth diagram in terms of n.
b) For pattern 2, the number of asterisks in the nth diagram is given by the binomial product $(n + \blacktriangle)(n + \blacksquare)$, where $\blacktriangle$ and $\blacksquare$ represent whole numbers. Replace $\blacktriangle$ and $\blacksquare$ in the binomial product by their correct values.
c) Divide your polynomial from part b) by your expression from part a).
d) Use your result from part c) to calculate how many times more asterisks there are in the 10th diagram of Pattern 2 than there are in the 10th diagram of Pattern 1.
e) If a diagram in Pattern 1 has 20 asterisks, how many asterisks are in the corresponding diagram of Pattern 2?
f) If a diagram in Pattern 2 has 1295 asterisks, how many asterisks are in the corresponding diagram in Pattern 1?

57. Write rational expressions in one variable so that the restrictions on the variables are as follows.
a) $x \neq 1$ **b)** $y \neq 0, -3$
c) $a \neq \dfrac{1}{2}, -\dfrac{3}{4}$ **d)** $t \neq -1, \pm\sqrt{3}$

LOGIC POWER

The diagram shows the locations of 5 pieces of furniture in rooms in a recreation centre. You need to switch the locations of the TV stand and the shuffleboard table. Because of lack of space, you can only move each piece of furniture into an adjacent empty room. What is the smallest number of moves you need to switch the TV stand and the shuffleboard table?

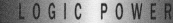

4.3 Multiplying and Dividing Rational Expressions

The game of badminton originated in England around 1870. Badminton is named after the Duke of Beaufort's home, Badminton House, where the game was first played. The International Badminton Federation now has over 50 member countries, including Canada.

Explore: Use the Diagram

In a doubles game of badminton, there are four service courts. The width of each service court is half the width of the whole court. The length of each service court is one third the distance between the long service lines. The width of the whole court and the distance between the long service lines can be modelled by the polynomials shown.

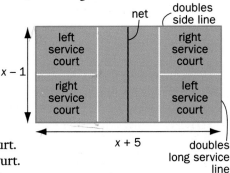

a) Write an expression that represents the width of each service court.
b) Write an expression that represents the length of each service court.
c) Write an expression that represents the area of each service court.

Leave your answer in the form $\dfrac{x-1}{\bullet} \times \dfrac{x+5}{\blacktriangle}$, where each numerator is a binomial, and $\bullet$ and $\blacktriangle$ represent whole numbers.

Inquire

1. a) Use the expressions $x - 1$ and $x + 5$ to write an expression that represents the whole area shown.
b) Expand and simplify the expression.

2. a) What fraction of the area does each service court cover?
b) Write this fraction of the expression you wrote in question 1b).

3. How is the expression you wrote in question 2b) related to your expression for the area of each service court? Explain.

4. Write a rule for multiplying rational expressions.

5. Multiply. Simplify the product, if possible.

a) $\dfrac{x}{3} \times \dfrac{y}{4}$

b) $\dfrac{3}{a} \times \dfrac{a^2}{b}$

c) $\dfrac{x^2}{2} \times \dfrac{4}{xy}$

d) $\dfrac{x+1}{3} \times \dfrac{x-1}{4}$

e) $\dfrac{3t}{2t+1} \times \dfrac{4}{t+2}$

f) $\dfrac{4x^2}{3y^3} \times \dfrac{9y^4}{8x^4}$

6. In the expressions that model the badminton court, x represents about 7 m. Find the area of each service court, in square metres.

Rational expressions can be multiplied in the same way that fractions are multiplied.	$\dfrac{3}{4} \times \dfrac{5}{6} = \dfrac{3 \times 5}{4 \times 6}$ ← Multiply the numerators. ← Multiply the denominators.

$$= \dfrac{15}{24}$$

$$= \dfrac{{}^1\!3 \times 5}{{}_1\!3 \times 8} \quad \text{Factor. Then, divide by the common factor.}$$

$$= \dfrac{5}{8}$$

Example 1 Multiplying Rational Expressions

Simplify $\dfrac{3a^3}{2b^2} \times \dfrac{10b^3}{9a^2}$. State the restrictions on the variables.

Solution

$$\dfrac{3a^3}{2b^2} \times \dfrac{10b^3}{9a^2} = \dfrac{30a^3b^3}{18a^2b^2} \quad \leftarrow \text{Multiply the numerators.}$$
$$\quad\quad\quad\quad\quad\quad\quad \leftarrow \text{Multiply the denominators.}$$

$$= \dfrac{5 \times \overset{1}{6} \times \overset{1}{a^2} \times a \times \overset{1}{b^2} \times b}{3 \times 6 \times a^2 \times b^2} \quad \begin{array}{l}\text{Factor. Then divide by}\\ \text{the common factors.}\end{array}$$

$$= \dfrac{5ab}{3}, a, b \neq 0$$

When multiplying some rational expressions, you may find it simpler to factor the numerators and denominators first.

Example 2 Multiplying Rational Expressions Involving Polynomials

Simplify $\dfrac{x^2+x-6}{x^2+2x-15} \times \dfrac{x-3}{x-2}$. State the restrictions on the variable.

Solution

$$\dfrac{x^2+x-6}{x^2+2x-15} \times \dfrac{x-3}{x-2} = \dfrac{(x+3)(x-2)}{(x+5)(x-3)} \times \dfrac{x-3}{x-2} \quad \text{Factor.}$$

$$= \dfrac{(x+3)(x-2)(x-3)}{(x+5)(x-3)(x-2)} \quad \text{Multiply the numerators and denominators.}$$

$$= \dfrac{(x+3)\overset{1}{(x-2)}\overset{1}{(x-3)}}{(x+5)(x-3)(x-2)} \quad \text{Divide by the common factors.}$$

$$= \dfrac{x+3}{x+5}, x \neq 2, 3, -5$$

Rational expressions can be divided in the same way that fractions are divided.

$$\dfrac{2}{3} \div \dfrac{5}{7} = \dfrac{2}{3} \times \dfrac{7}{5} \quad \text{Multiply by the reciprocal.}$$
$$= \dfrac{14}{15}$$

Example 3 Dividing Rational Expressions

Simplify $\dfrac{x^2-x-20}{x^2-6x} \div \dfrac{x^2+9x+20}{x^2-12x+36}$. State the restrictions on the variable.

Solution

$$\dfrac{x^2-x-20}{x^2-6x} \div \dfrac{x^2+9x+20}{x^2-12x+36} = \dfrac{x^2-x-20}{x^2-6x} \times \dfrac{x^2-12x+36}{x^2+9x+20} \quad \text{Multiply by the reciprocal.}$$

$$= \dfrac{(x-5)(x+4)}{x(x-6)} \times \dfrac{(x-6)(x-6)}{(x+4)(x+5)} \quad \text{Factor.}$$

$$= \dfrac{(x-5)(x+4)(x-6)(x-6)}{x(x-6)(x+4)(x+5)} \quad \text{Multiply the numerators and denominators.}$$

$$= \dfrac{(x-5)\overset{1}{(x+4)}\overset{1}{(x-6)}(x-6)}{x\underset{1}{(x-6)}\underset{1}{(x+4)}(x+5)} \quad \text{Divide by the common factors.}$$

$$= \dfrac{(x-5)(x-6)}{x(x+5)}, x \neq 0, 6, -4, -5$$

161

Note that, in a division of the form $\dfrac{a}{b} \div \dfrac{c}{d}$, variables may need to be restricted in expressions b, c, and d. In the division in Example 3, the expressions $x^2 - 6x$, $x^2 + 9x + 20$, and $x^2 - 12x + 36$ cannot equal 0, because division by 0 is not defined.

A **complex fraction** is a rational expression that contains a fraction in its numerator or denominator. Examples include

$$\dfrac{\dfrac{1}{3}}{2 + \dfrac{3}{5}} \qquad \dfrac{3xy}{\dfrac{x}{2}} \qquad \dfrac{\dfrac{x^2 - 1}{2x}}{\dfrac{x + 1}{4}}$$

To simplify a complex fraction, rewrite it as a division problem.

$$\dfrac{\dfrac{4}{3}}{\dfrac{2}{9}} = \dfrac{4}{3} \div \dfrac{2}{9}$$

$$= \dfrac{4}{3} \times \dfrac{9}{2} \qquad \text{Multiply by the reciprocal.}$$

$$= 6$$

Example 4   Simplifying Complex Fractions

Simplify $\dfrac{\dfrac{a^2 - 4}{a + 3}}{\dfrac{2a - 4}{a^2 + 2a - 3}}$. State the restrictions on the variable.

Solution

$$\dfrac{\dfrac{a^2 - 4}{a + 3}}{\dfrac{2a - 4}{a^2 + 2a - 3}} = \dfrac{a^2 - 4}{a + 3} \div \dfrac{2a - 4}{a^2 + 2a - 3}$$

$$= \dfrac{a^2 - 4}{a + 3} \times \dfrac{a^2 + 2a - 3}{2a - 4} \qquad \text{Multiply by the reciprocal.}$$

$$= \dfrac{(a + 2)(a - 2)}{a + 3} \times \dfrac{(a + 3)(a - 1)}{2(a - 2)} \qquad \text{Factor.}$$

$$= \dfrac{(a + 2)\overset{1}{\cancel{(a - 2)}}\overset{1}{\cancel{(a + 3)}}(a - 1)}{2\underset{1}{\cancel{(a + 3)}}\underset{1}{\cancel{(a - 2)}}} \qquad \text{Divide by the common factors.}$$

$$= \dfrac{(a + 2)(a - 1)}{2}, \; a \neq 1, 2, -3$$

Note the inclusion of $a \neq 1$ in the restrictions on the variable.

Because division by 0 is not defined, $a + 3 \neq 0$, so $a \neq -3$
$$2a - 4 \neq 0, \text{ so } a \neq 2$$
$$a^2 + 2a - 3 \text{ or } (a + 3)(a - 1) \neq 0, \text{ so } a \neq -3, 1$$

Practice

In each of the following, state any restrictions on the variables.

Simplify.

1. $\dfrac{y^2}{3} \times \dfrac{8}{y}$

2. $\dfrac{7}{2x^3} \times \dfrac{-x^4}{14}$

3. $\dfrac{-5n^2}{12} \times \dfrac{4}{-15n^5}$

4. $\dfrac{-4m}{9} \times 6$

Simplify.

5. $\dfrac{3}{x} \div \dfrac{12}{x^2}$

6. $\dfrac{y^3}{6} \div \dfrac{y^2}{-3}$

7. $\dfrac{-15}{2m^2} \div \dfrac{10}{3m^4}$

8. $\dfrac{-8t^4}{3} \div \dfrac{-6t^2}{5}$

9. $\dfrac{20}{3x^5} \div \dfrac{-15}{8x^2}$

10. $\dfrac{4r^3}{-3} \div 2r^4$

Simplify.

11. $\dfrac{3x^3}{2y} \times \dfrac{8y^2}{9x}$

12. $\dfrac{8m^3}{3n^2} \div \dfrac{5m^2}{6n}$

13. $\dfrac{21xy}{4t^2} \times \dfrac{12}{7x^2y}$

14. $\dfrac{-4a}{7b^3} \div \dfrac{-8a^4}{7}$

15. $\dfrac{12m}{-5t} \div \dfrac{8m^2}{-15}$

16. $\dfrac{15a^2b}{4c} \div \dfrac{8abc}{-3}$

Simplify.

17. $\dfrac{16ab}{9x^4y^2} \times \dfrac{3x^5y^4}{8a^2b^2}$

18. $\dfrac{6x^2y}{5mn^3} \div \dfrac{9xy}{10mn^4}$

19. $\dfrac{5xy}{6x^2y} \div \dfrac{10xy^2}{9x^3y^2}$

20. $-12a^2b \times \dfrac{4ab^2}{-3ab^3}$

21. $6x^3y^4 \div \dfrac{2xy}{-3}$

22. $\dfrac{4a^2b^2c}{-3ab} \div 6c^2$

Simplify.

23. $\dfrac{3}{x-4} \times \dfrac{x-4}{6}$

24. $\dfrac{m+2}{5} \div \dfrac{y+1}{10}$

25. $\dfrac{5(y-2)}{y+1} \times \dfrac{y+1}{10}$

26. $\dfrac{2(x+1)}{x-2} \div \dfrac{x+1}{x-2}$

27. $\dfrac{4a^2b}{3(a+b)} \div \dfrac{-8ab^2}{a+b}$

28. $\dfrac{3(m+4)}{5m} \times \dfrac{6m^3}{2(m+4)}$

Simplify.

29. $\dfrac{4x+4}{3x-3} \times \dfrac{6x-6}{5x+5}$

30. $\dfrac{6m^3}{m+3} \times \dfrac{5m+15}{8m^3}$

31. $\dfrac{3a+6}{9a^2} \div \dfrac{a+2}{-3a}$

32. $\dfrac{x^2-4}{x+3} \div \dfrac{4x-8}{3x+9}$

33. $\dfrac{7y^2}{y^2-9} \times \dfrac{4y+12}{14y^3}$

34. $\dfrac{m^2-25}{m^2-16} \div \dfrac{2m-10}{4m+16}$

35. $\dfrac{4x-6}{8x^2y} \times \dfrac{4xy}{6x-9}$

36. $\dfrac{2x^2-8}{6x+3} \div \dfrac{6x-12}{18x+9}$

Simplify.

37. $\dfrac{x^2+5x+6}{x^2-6x+5} \times \dfrac{x^2+x-30}{x^2+9x+18}$

38. $\dfrac{a^2+7a+12}{a^2+4a+4} \times \dfrac{a^2-a-6}{a^2-9}$

39. $\dfrac{m^2-3m-4}{m^2+5m} \div \dfrac{m^2-7m+12}{m^2+2m-15}$

40. $\dfrac{x^2-xy-20y^2}{x^2-8xy+15y^2} \div \dfrac{x^2+2xy-8y^2}{x^2-xy-6y^2}$

41. $\dfrac{x^2+3xy}{x^2-xy-42y^2} \times \dfrac{x^2-10xy+21y^2}{x^2-9y^2}$

42. $\dfrac{a^2+15ab+56b^2}{a^2-3ab-54b^2} \div \dfrac{a^2+6ab-16b^2}{a^2+4ab-12b^2}$

Simplify.

43. $\dfrac{12a^2-19a+5}{4a^2-9} \times \dfrac{2a-3}{3a-1}$

44. $\dfrac{2x^2-5x-3}{2x^2-11x+15} \times \dfrac{4x^2-8x-5}{4x^2+4x+1}$

45. $\dfrac{12w^2-5w-2}{8w^2+2w-21} \div \dfrac{12w^2+w-6}{8w^2-2w-15}$

46. $\dfrac{9s^2+30st+25t^2}{25s^2-25st-6t^2} \times \dfrac{20s^2-49st+30t^2}{12s^2+5st-25t^2}$

47. $\dfrac{3x^3+14x^2-5x}{4x^2+7x+3} \div \dfrac{6x^2-2x}{8x^2+2x-3}$

48. $\dfrac{4x^2-9y^2}{4x^2-12xy+9y^2} \div \dfrac{4x^2+12xy+9y^2}{9y^2-4x^2}$

163

Simplify.

49. $\dfrac{\dfrac{5x}{x}}{2}$

50. $\dfrac{\dfrac{3x^2y^3}{-6xy^2}}{\dfrac{4x^3y}{2x^2y^2}}$

51. $\dfrac{\dfrac{2t-4}{3t+5}}{\dfrac{4t-8}{2t-1}}$

52. $\dfrac{\dfrac{(s-1)(s-2)}{(s+3)(s+1)}}{\dfrac{(s-2)^2}{s^2+3s}}$

Simplify.

53. $\dfrac{\dfrac{3n+6}{2n+2}}{\dfrac{n+2}{n^2-1}}$

54. $\dfrac{\dfrac{6x^2-37x+6}{x^2-9x+18}}{\dfrac{12x^2+16x-3}{5x^2-12x-9}}$

Applications and Problem Solving

55. Soccer On a soccer pitch, the goal area or goal box is inside the penalty area or penalty box and forms part of it. The dimensions of the goal box and the penalty box can be represented as shown.

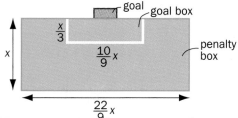

a) Write an expression that represents the area of the goal box.

b) Write an expression that represents the area of the penalty box.

c) Determine how many times greater the area of the penalty box is than the area of the goal box.

d) Does the fact that x represents 16.5 m affect your answer to part c)? Explain.

56. Measurement Write the area of the rectangle in simplest form.

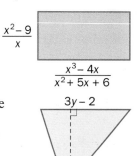

57. Measurement The area of the trapezoid is $6y^2-5y-6$. What is its height?

164

58. Measurement

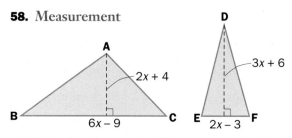

a) Write, but do not simplify, an expression for the area of $\triangle ABC$.

b) Write, but do not simplify, an expression for the area of $\triangle DEF$.

c) Write and simplify an expression that represents the ratio of the area of $\triangle ABC$ to the area of $\triangle DEF$.

59. Measurement Write and simplify an expression that represents the fraction of the area of the large rectangle covered by the shaded rectangle.

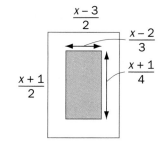

60. In divisions of the form $\dfrac{a}{b} \div \dfrac{c}{d}$, the expressions b, c, and d must all be examined for possible restrictions on the variables. Explain why.

61. Simplify. $\dfrac{(x^2-4)^{-1}}{(x-3)^{-1}} \times \dfrac{(x^2-x-6)^{-1}}{(x^2+4x+4)^{-1}}$

62. As the value of y increases, what happens to the value of each of the following expressions? Explain.

a) $\dfrac{15y^2-2y-1}{6y^2+7y-3} \times \dfrac{2y-1}{10y^2-3y-1}$

b) $\dfrac{8y^2+2y-1}{6y^2-y-2} \div \dfrac{8y-2}{9y-6}$

63. Write 2 different pairs of rational expressions with a product of $\dfrac{3x^2+7xy+2y^2}{x^2-y^2}$.

64. Write 4 different pairs of rational expressions with a product of $\dfrac{4x^2-8x+4}{2x^2+5x-3}$. Compare your expressions with a classmate's.

4.4 Adding and Subtracting Rational Expressions, I

The blimp that provided overhead television coverage
of the first World Series played in Canada was based
in Miami. The blimp flew 1610 km from Miami to
Washington, D.C., then 634 km to Toronto.

The time taken to fly from Miami to Washington was $\dfrac{1610}{s}$ hours,

where s was the average speed in kilometres per hour. The time

taken to fly from Washington to Toronto was $\dfrac{634}{s}$ hours.

The total flying time from Miami to Toronto was

$\dfrac{1610}{s} + \dfrac{634}{s}$ hours. The expression $\dfrac{1610}{s} + \dfrac{634}{s}$ is the sum
of two rational expressions with the same denominator.

Explore: Use the Diagrams

Rectangle A and Rectangle B have different areas but the same width.
Rectangle C is formed by placing Rectangles A and B end to end.

Rectangle A

$x + 2$ | area = $x^2 + 3x + 1$

Rectangle B

$x + 2$ | area = $x^2 + 4x + 2$

Rectangle C

a) Using the areas of rectangles A and B, write and simplify an expression
that represents the area of rectangle C.
b) What is the width of rectangle C? $x + 2$

Inquire

1. Write, but do not divide, a rational expression that represents the length of
a) rectangle A   **b)** rectangle B $(x^2 + 3x + 1)(x + 2)$ $(x^2 + 4x + 2) - (x + 2)$

2. Using the area and the width of rectangle C, write, but do not divide, a
rational expression that represents the length of rectangle C.

3. How does the expression you wrote in question 2 compare with the two
expressions you wrote in question 1? Explain.

4. Write a rule for adding two rational expressions with the same denominator.

5. Add.

a) $\dfrac{x}{3} + \dfrac{4x}{3}$ **b)** $\dfrac{5}{3t} + \dfrac{2}{3t}$ **c)** $\dfrac{n+1}{n+3} + \dfrac{n-1}{n+3}$ **d)** $\dfrac{x^2+1}{x^2} + \dfrac{2x^2+1}{x^2}$

6. The flying time of the blimp from Miami to Toronto was $\dfrac{1610}{s} + \dfrac{634}{s}$ hours.
a) Add the rational expressions.
b) If the average speed, s, of the blimp was 85 km/h, what was the total flying time,
in hours?

165

To add or subtract fractions with a common denominator, add or subtract the numerators and place the result over the common denominator. The same method can be used to add or subtract rational expressions with common denominators.

$$\frac{5}{7}+\frac{1}{7}-\frac{2}{7}=\frac{5+1-2}{7}$$
$$=\frac{4}{7}$$

Example 1 Adding and Subtracting With Common Denominators

Simplify.

a) $\dfrac{3}{x^2}+\dfrac{5}{x^2}-\dfrac{2}{x^2}$

b) $\dfrac{4x-1}{x+2}-\dfrac{x+3}{x+2}$

Solution

a) $\dfrac{3}{x^2}+\dfrac{5}{x^2}-\dfrac{2}{x^2}=\dfrac{3+5-2}{x^2}$

$$=\dfrac{6}{x^2},\ x\neq0$$

b) $\dfrac{4x-1}{x+2}-\dfrac{x+3}{x+2}=\dfrac{(4x-1)-(x+3)}{x+2}$

$$=\dfrac{4x-1-x-3}{x+2}$$

$$=\dfrac{3x-4}{x+2},\ x\neq-2$$

To add or subtract fractions with different denominators, rewrite the fractions as equivalent fractions with a common denominator.

$$\frac{1}{6}+\frac{3}{4}=\frac{2\times1}{2\times6}+\frac{3\times3}{3\times4}$$
$$=\frac{2}{12}+\frac{9}{12}$$
$$=\frac{11}{12}$$

$$\frac{3}{5}-\frac{1}{2}=\frac{2\times3}{2\times5}-\frac{5\times1}{5\times2}$$
$$=\frac{6}{10}-\frac{5}{10}$$
$$=\frac{1}{10}$$

The lowest common denominator (LCD) is normally used but is not necessary. If a greater common denominator is used, the result will reduce.

$$\frac{1}{6}+\frac{3}{4}=\frac{4}{24}+\frac{18}{24}$$
$$=\frac{22}{24}$$
$$=\frac{11}{12}$$

Example 2 Adding and Subtracting With Whole-Number Denominators

Simplify $\dfrac{3x+2}{4}+\dfrac{x-4}{8}-\dfrac{2x-1}{6}$.

Solution
To find the LCD, find the least common multiple (LCM) of the denominators 4, 8, and 6. The LCM can be found by factoring. It must contain all the separate factors of 4, 8, and 6.

$4=2\times2$
$8=2\times2\times2$ The LCM is $2\times2\times2\times3=24$
$6=2\times3$

So, the LCD is 24.

$$\frac{3x+2}{4}+\frac{x-4}{8}-\frac{2x-1}{6}=\frac{6(3x+2)}{6(4)}+\frac{3(x-4)}{3(8)}-\frac{4(2x-1)}{4(6)}$$

Rewrite with a common denominator.

$$=\frac{6(3x+2)}{24}+\frac{3(x-4)}{24}-\frac{4(2x-1)}{24}$$

$$=\frac{6(3x+2)+3(x-4)-4(2x-1)}{24}$$

$$=\frac{18x+12+3x-12-8x+4}{24}$$

Expand the numerator.

$$=\frac{13x+4}{24}$$

Simplify.

Practice

In each of the following, state any restrictions on the variables.

Simplify.

1. $\dfrac{2}{y}+\dfrac{4}{y}-\dfrac{5}{y}$

2. $\dfrac{5}{x^2}-\dfrac{3}{x^2}+\dfrac{6}{x^2}$

3. $\dfrac{4}{x+3}+\dfrac{5}{x+3}$

4. $\dfrac{x}{x-2}-\dfrac{y}{x-2}$

Simplify.

5. $\dfrac{x+7}{2}+\dfrac{x+4}{2}$

6. $\dfrac{2y-1}{3}+\dfrac{3y-6}{3}$

7. $\dfrac{3a-1}{a}-\dfrac{4a+2}{a}$

8. $\dfrac{5x-y}{3x}-\dfrac{4x+y}{3x}$

9. $\dfrac{x^2+4}{x+1}+\dfrac{2x^2}{x+1}$

10. $\dfrac{6t-8}{7}-\dfrac{3-5t}{7}$

Find the LCM.

11. 4, 5, 6

12. 4, 9, 12

13. 8, 10, 12

14. 20, 15, 10

Simplify.

15. $\dfrac{2x}{2}+\dfrac{x}{3}$

16. $\dfrac{3a}{4}+\dfrac{a}{2}-\dfrac{2a}{6}$

17. $\dfrac{x}{5}-\dfrac{y}{2}+\dfrac{7}{10}$

18. $\dfrac{3m}{8}-\dfrac{m}{6}-\dfrac{2m}{3}$

Simplify.

19. $\dfrac{2m+3}{2}+\dfrac{3m+4}{7}$

20. $\dfrac{4x-3}{4}+\dfrac{x+2}{3}$

21. $\dfrac{y-5}{6}-\dfrac{2y-3}{4}$

22. $\dfrac{2x+3y}{5}-\dfrac{4x-y}{2}$

23. $\dfrac{4t-1}{6}+\dfrac{3t+2}{2}-\dfrac{2t+1}{3}$

24. $\dfrac{3a-b}{9}-\dfrac{a-2b}{3}-\dfrac{4a-3b}{6}$

25. $\dfrac{5x-1}{5}+1-\dfrac{4x-3}{6}$

Applications and Problem Solving

26. Backgammon A backgammon game board consists of two rectangles of the same size, known as tables, separated by a divider, called the bar.

a) The area of each table on a backgammon board can be modelled by the expression $6x^2-5x+1$, and the width of the board by $3x-1$. Write and simplify an expression that represents the width, w, of each table in terms of x.

b) If the width of the bar is $\dfrac{x+1}{3}$, write and simplify an expression that represents the length of the whole board in terms of x.

c) If x represents 8 cm, what are the dimensions of each table? of the whole board?

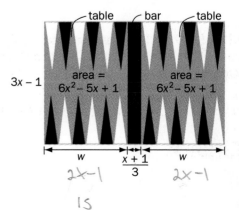

27. Flying times a) Write an expression that represents the time, in hours, it takes a plane to fly 1191 km from Winnipeg to Calgary at an average speed of s kilometres per hour.
b) Write an expression that represents the time, in hours, it takes a plane to fly 685 km from Calgary to Vancouver at the same speed as in part a).
c) Write and simplify an expression that represents the total flying time for a trip from Winnipeg to Vancouver via Calgary.
d) If the average speed of the plane is 700 km/h, what is the total flying time, in hours, for the trip in part c)?

28. Measurement The diameter of the smaller circle is d. The diameter of the larger circle is $d + 1$.
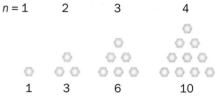
a) Write an expression that represents the area of the smaller circle in terms of d.
b) Write an expression that represents the area of the larger circle in terms of d.
c) Write and simplify an expression that represents the area of the shaded part of the diagram in terms of d.
d) If d represents 10 cm, find the area of the shaded part of the diagram, to the nearest tenth of a square centimetre.

29. Measurement The length of a rectangle is represented by $\dfrac{3x+1}{2}$ and its width by $\dfrac{2x-1}{3}$.
a) Write and simplify an expression that represents the perimeter of the rectangle in terms of x.
b) Find the three smallest values of x that give whole-number values for the perimeter.

30. Measurement The diagram shows trapezoid ABCD divided into rhombus ABCE and isosceles triangle ADE.
a) Write an expression that represents the area of the triangle in terms of x.
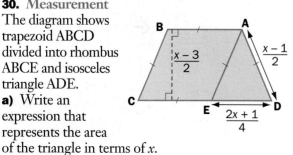
b) Write an expression that represents the area of the rhombus in terms of x.
c) Add and simplify the expressions you wrote in parts a) and b).
d) Write and simplify an expression that represents the longer base of the trapezoid in terms of x.
e) Use the formula for the area of a trapezoid to write and simplify an expression that represents the area of the trapezoid in terms of x.
f) Compare your expressions from parts c) and e).

31. Pattern Triangular numbers of objects can be arranged to form triangles. The first four triangular numbers are as follows.
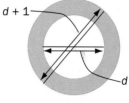

n = 1   2   3   4

1   3   6   10

a) An expression for finding the nth triangular number can be written in the form $\dfrac{n(n+\blacktriangle)}{\blacksquare}$, where $\blacktriangle$ and $\blacksquare$ represent whole numbers. Copy and complete the expression by finding the numbers represented by $\blacktriangle$ and $\blacksquare$.
b) Write the 5th, 6th, 7th, 8th, and 9th triangular numbers.
c) Add any two consecutive triangular numbers. What kind of number results?
d) Write an expression that represents the $(n + 1)$th triangular number.
e) Add your expressions from parts a) and d). Simplify the result and express it in factored form.
f) How does your result from part e) explain your result from part c)?

WORD POWER

Copy the following sentence.

I do not have quite enough ▩▩▩ factored.

Then, use the pattern in the sentence to choose one of the following words to complete the sentence. Explain your reasoning.

integers   numbers   terms   expressions

4.5 Adding and Subtracting Rational Expressions, II

The three-toed sloth of South America moves very slowly. It can travel twice as fast in a tree as it can on the ground. If its speed on the ground is s metres per minute, its speed in a tree is $2s$ metres per minute. The sloth can travel 15 m on the ground in $\dfrac{15}{s}$ minutes and 15 m in a tree in $\dfrac{15}{2s}$ minutes. The total time it takes to travel 15 m on the ground and 15 m in a tree is $\dfrac{15}{s} + \dfrac{15}{2s}$ minutes. The expression $\dfrac{15}{s} + \dfrac{15}{2s}$ is the sum of two rational expressions with different denominators. Adding these rational expressions involves finding the LCM of two monomials that include variables.

Explore: Complete the Table

To find the LCM of each group of monomials, copy and complete the table.

	Monomials	Factored Form	LCM
1.	$2a^2b$ $6b^2$	$2 \times a \times a \times b$ $2 \times 3 \times b \times b$	$6b$
2.	$10x^3$ $15x^2y^2$		
3.	$3xy$ $6yz$ $9xz$		
4.	$5x^3$ $8x^2y$ $10xy^2$		

Inquire

1. Describe a method for mentally finding the LCM of monomials that include variables.

2. a) Factor the binomials $2x + 4$ and $3x + 6$.
b) Write the LCM of the binomials in factored form.

3. Write the LCM of each pair of expressions in factored form.
a) $3a - 9, 4a - 12$ **b)** $x^2 + x, 2x^2 + 2x$
c) $2y - 6, y^2 - 9$ **d)** $x^2 + 5x + 6, x^2 + x - 2$

4. The total time a three-toed sloth takes to travel 15 m on the ground and 15 m in a tree is $\dfrac{15}{s} + \dfrac{15}{2s}$ minutes.

a) State the common denominator of the two rational expressions.
b) Add the expressions.
c) If s represents 2.5 m/min, what is the total time, in minutes, the sloth takes to travel 15 m on the ground and 15 m in a tree?

Example 1   Adding and Subtracting With Monomial Denominators

Simplify $\dfrac{4}{5a} - \dfrac{3}{2a^2} + \dfrac{1}{a^3}$.

Solution

Find the LCD.

$5a = 5 \times a$
$2a^2 = 2 \times a \times a$
$a^3 = a \times a \times a$

The LCD is $5 \times 2 \times a \times a \times a$ or $10a^3$.

Rename each expression using the common denominator.

$$\frac{4}{5a} - \frac{3}{2a^2} + \frac{1}{a^3} = \frac{2a^2(4)}{2a^2(5a)} - \frac{5a(3)}{5a(2a^2)} + \frac{10(1)}{10(a^3)}$$

$$= \frac{8a^2}{10a^3} - \frac{15a}{10a^3} + \frac{10}{10a^3}$$

$$= \frac{8a^2 - 15a + 10}{10a^3}, \ a \neq 0$$

Example 2   Denominators With a Common Binomial Factor

Simplify $\dfrac{m}{2m-4} - \dfrac{3}{3m-6} + 1$.

Solution

$2m - 4 = 2(m - 2)$
$3m - 6 = 3(m - 2)$

The LCD is $2 \times 3 \times (m - 2)$ or $6(m - 2)$.

$$\frac{m}{2m-4} - \frac{3}{3m-6} + 1 = \frac{m}{2(m-2)} - \frac{3}{3(m-2)} + \frac{1}{1}$$

$$= \frac{3(m)}{3 \times 2(m-2)} - \frac{2(3)}{2 \times 3(m-2)} + \frac{6(m-2)(1)}{6(m-2)(1)}$$

$$= \frac{3m}{6(m-2)} - \frac{6}{6(m-2)} + \frac{6(m-2)}{6(m-2)}$$

$$= \frac{3m - 6 + 6(m-2)}{6(m-2)}$$

$$= \frac{3m - 6 + 6m - 12}{6(m-2)} \qquad \text{Expand the numerator.}$$

$$= \frac{9m - 18}{6(m-2)} \qquad \text{Simplify.}$$

$$= \frac{3 \times \overset{1}{\cancel{3}} \times \overset{1}{(\cancel{m-2})}}{2 \times \cancel{3} \times (\cancel{m-2})} \qquad \begin{array}{l}\text{The numerator and denominator have} \\ \text{common factors, so simplify further.}\end{array}$$

$$= \frac{3}{2}, \ m \neq 2$$

Example 3   Denominators With Different Binomial Factors

Simplify $\dfrac{x}{6x+6}+\dfrac{5}{4x-12}$.

Solution

$6x + 6 = 6(x + 1)$

$4x - 12 = 4(x - 3)$

The LCD is $12(x + 1)(x - 3)$.

$$\dfrac{x}{6x+6}+\dfrac{5}{4x-12}=\dfrac{x}{6(x+1)}+\dfrac{5}{4(x-3)}$$

$$=\dfrac{2(x-3)}{2(x-3)}\times\dfrac{x}{6(x+1)}+\dfrac{3(x+1)}{3(x+1)}\times\dfrac{5}{4(x-3)}\qquad\text{Rewrite with a common denominator.}$$

$$=\dfrac{2x(x-3)}{12(x+1)(x-3)}+\dfrac{15(x+1)}{12(x+1)(x-3)}$$

$$=\dfrac{2x(x-3)+15(x+1)}{12(x+1)(x-3)}$$

$$=\dfrac{2x^2-6x+15x+15}{12(x+1)(x-3)}\qquad\text{Expand the numerator.}$$

$$=\dfrac{2x^2+9x+15}{12(x+1)(x-3)},\,x\neq-1,3\qquad\text{Simplify.}$$

Example 4   Trinomial Denominators

Simplify $\dfrac{4}{y^2+5y+6}-\dfrac{5}{y^2-y-12}$.

Solution

$y^2 + 5y + 6 = (y + 2)(y + 3)$

$y^2 - y - 12 = (y + 3)(y - 4)$

The LCD is $(y + 2)(y + 3)(y - 4)$.

$$\dfrac{4}{y^2+5y+6}-\dfrac{5}{y^2-y-12}=\dfrac{4}{(y+2)(y+3)}-\dfrac{5}{(y+3)(y-4)}$$

$$=\dfrac{y-4}{y-4}\times\dfrac{4}{(y+2)(y+3)}-\dfrac{y+2}{y+2}\times\dfrac{5}{(y+3)(y-4)}\qquad\text{Rewrite with a common denominator.}$$

$$=\dfrac{4(y-4)-5(y+2)}{(y+2)(y+3)(y-4)}$$

$$=\dfrac{4y-16-5y-10}{(y+2)(y+3)(y-4)}\qquad\text{Expand the numerator.}$$

$$=\dfrac{-y-26}{(y+2)(y+3)(y-4)},\,y\neq-2,-3,4\qquad\text{Simplify.}$$

You can check your solution by substituting 1 for the variable in the original expression and in the simplified expression. The results will be the same if the solution is correct.

Practice

In each of the following, state any restrictions on the variables.

Write an equivalent expression with a denominator of $12x^2y^2$.

1. $\dfrac{2}{xy}$ **2.** $\dfrac{x}{y}$ **3.** $\dfrac{5}{3xy^2}$ **4.** $\dfrac{-y}{6x^2}$

Find the LCM.

5. $10a^2b,\ 4ab^3$ **6.** $3m^2n,\ 2mn^2,\ 6mn$

7. $2x^3,\ 6xy^2,\ 4y$ **8.** $10s^2t^2,\ 20s^2t,\ 15st^2$

Simplify.   ~~monomials denominator~~

9. $\dfrac{3}{2x}+\dfrac{4}{5x}$ **10.** $\dfrac{2}{4y}+\dfrac{3}{3y}-\dfrac{1}{2y}$

11. $\dfrac{1}{2x^2}+\dfrac{3}{3x}-\dfrac{2}{x^3}$ **12.** $\dfrac{3}{2m^2n}-\dfrac{1}{m^2n^3}+\dfrac{4}{5mn}$

13. $x-\dfrac{2}{x}+5$ **14.** $\dfrac{3m+4}{mn}-\dfrac{1}{m}-2$

15. $\dfrac{4x-1}{3x^2}-\dfrac{2x+3}{x}+\dfrac{5x+2}{5x^2}$

16. $\dfrac{x-2y}{x}-\dfrac{4x+y}{xy}-\dfrac{3x-4y}{y}$

Find the LCM of each of the following. Leave answers in factored form.

17. $3m+6,\ 2m+4$ **18.** $3y-3,\ 5y+10$

19. $4m-8,\ 6m-18$ **20.** $8x-12,\ 10x-15$

Simplify.   ~~common~~ ~~binomials denominator~~

21. $\dfrac{4}{x+3}+\dfrac{5}{4x+12}$ **22.** $\dfrac{1}{3y-15}-\dfrac{2}{y-5}$

23. $\dfrac{t}{t-4}-\dfrac{2t}{3t-12}$ **24.** $\dfrac{2}{2m+2}+\dfrac{5}{3m+3}$

25. $\dfrac{3}{4y-8}-\dfrac{2}{3y-6}$ **26.** $\dfrac{1}{4a+2}+\dfrac{4}{6a+3}$

Simplify.

27. $\dfrac{2}{x+1}+\dfrac{3}{x+2}$ **28.** $\dfrac{m}{m-3}-\dfrac{5}{m+2}$

29. $\dfrac{3}{x}+\dfrac{5}{x-1}$ **30.** $\dfrac{2}{t-1}+\dfrac{1}{5}+2$

31. $\dfrac{2x}{x-2}-\dfrac{3x}{x+2}$ **32.** $\dfrac{4}{3n-1}-\dfrac{3}{2n+3}$

Simplify.

33. $\dfrac{1}{2x-2}+\dfrac{3}{4x-8}$ **34.** $\dfrac{t}{3t+15}-\dfrac{1}{6t-24}$

35. $\dfrac{4}{2s-12}-\dfrac{s}{5s-5}$ **36.** $\dfrac{2m}{3m-15}+\dfrac{m}{4m-8}$

State the LCM in factored form.

37. $x+2,\ x^2+4x+4$ **38.** $y^2+6y+8,\ y^2-4$

39. $t^2-t-12,\ t^2-3t-4$ **40.** $2x-4,\ x^2-3x-4$

41. $m^2+6m+9,\ m^2-2m-15$

Simplify.

42. $\dfrac{2}{x+3}+\dfrac{3}{x^2+5x+6}$ **43.** $\dfrac{y}{y^2-16}-\dfrac{4}{y+4}$

44. $\dfrac{3x}{x-5}+\dfrac{2x}{x^2-4x-5}$ **45.** $\dfrac{a}{a^2-7a+12}-\dfrac{2a}{a-3}$

46. $\dfrac{4}{2x^2+3x+1}+\dfrac{2}{2x+1}$ **47.** $\dfrac{6}{2n-1}-\dfrac{3}{6n^2-5n+1}$

Simplify.   ~~trinomials~~

48. $\dfrac{2}{m^2+4m+3}+\dfrac{1}{m^2+7m+12}$

49. $\dfrac{1}{x^2+4x+4}-\dfrac{3}{x^2-4}$

50. $\dfrac{a}{a^2-25}-\dfrac{2}{a^2-9a+20}$

51. $\dfrac{4m}{m^2-9m+18}+\dfrac{2m}{m^2-11m+30}$

52. $\dfrac{5}{3x^2+4x+1}+\dfrac{2}{3x^2-2x-1}$

53. $\dfrac{3y}{4y^2-9}-\dfrac{2y}{4y^2-12y+9}$

Simplify.

54. $\dfrac{t+1}{t-1}+\dfrac{2}{t^2-5t+4}$ **55.** $\dfrac{y+1}{y-1}+\dfrac{y-1}{y^2+y-2}$

56. $\dfrac{x-2}{x^2+4x+3}-\dfrac{2x+1}{x+3}$ **57.** $\dfrac{n^2+4n-3}{n^2-16}+\dfrac{4-3n}{3n-12}$

58. $\dfrac{m+4}{m^2-m-12} - \dfrac{m}{m^2-5m+4}$

59. $\dfrac{a+2}{a^2-1} - \dfrac{a-1}{a^2+2a+1}$

60. $\dfrac{3w-4}{w^2+5w+4} + \dfrac{2w-3}{w^2+2w-8}$

61. $\dfrac{2x-1}{2x^2+3x+1} + \dfrac{2x+1}{3x^2+4x+1}$

62. $\dfrac{2z-1}{4z^2-25} - \dfrac{2z+5}{4z^2-8z-5}$

Applications and Problem Solving

63. a) Copy and complete the table. The first line has been completed.

Expressions	Product	LCM	GCF	LCM × GCF
3x, 5x	15x²	15x	x	15x²
12, 8				
15y², 9y				
a + 1, a − 1				
2t − 2, 3t − 3				

b) How is the product LCM × GCF related to the product of each pair of expressions?

c) Explain why the relationship you found in part b) exists.

64. Patrol boat a) An RCMP patrol boat left Tofino on Vancouver Island and travelled for 45 km along the coast at a speed of s kilometres per hour. Write an expression that represents the time taken, in hours.

b) The boat returned to Tofino at a speed of $2s$ kilometres per hour. Write an expression that represents the time taken, in hours.

c) Write and simplify an expression that represents the total time, in hours, the boat was at sea.

d) If s represents 10 km/h, for how many hours was the boat at sea?

65. Animal speeds A zebra can run 100 m at a speed of x metres per second. A lion can run 100 m at a speed that is 4 m/s faster than the zebra's speed.

a) Write an expression in terms of x to represent the number of seconds the zebra takes to run 100 m.

b) Write an expression in terms of x to represent the number of seconds the lion takes to run 100 m.

c) Write and simplify an expression in terms of x to represent how many seconds longer the zebra takes than the lion to run 100 m.

d) If x represents 18 m/s, how many seconds longer does the zebra take than the lion to run 100 m, to the nearest second?

66. Simplify.

a) $\dfrac{2+\dfrac{1}{x}}{2-\dfrac{1}{x}}$

b) $\dfrac{y+\dfrac{1}{4}}{y-\dfrac{1}{2}}$

c) $\dfrac{1-\dfrac{1}{m^2}}{2-\dfrac{2}{m}}$

d) $\dfrac{\dfrac{t}{2}+1}{\dfrac{1}{4}+\dfrac{t+1}{t^2}}$

e) $\dfrac{\dfrac{n^2+9}{2}+3n}{n^2-9}$

f) $\dfrac{x^2-\dfrac{(5x+3)}{2}}{x^2-\dfrac{(8x+3)}{3}}$

67. Write two rational expressions with binomial denominators and with each of the following sums. Compare your answers with a classmate's.

a) $\dfrac{5x+8}{(x+1)(x+2)}$

b) $\dfrac{5x-5}{6x^2-13x+6}$

c) $\dfrac{x^2-3}{(x-1)(x-3)}$

d) $\dfrac{4x^2}{4x^2-9}$

LOGIC POWER

Assume that no cubes are missing from the back or base of the stack.

1. How many cubes are in the stack?

2. If the outside of the stack was painted yellow, how many cubes would have
a) 1 yellow face?
b) 2 yellow faces?

Rational Expressions and the Graphing Calculator

Complete the following with a graphing calculator that has the capability
to simplify and perform operations on rational expressions.

When you enter a rational expression into the calculator, be careful with
your use of brackets. Check that the display shows the intended expression.

1 Simplifying Rational Expressions

Simplify.

1. $\dfrac{3x^2 y}{6x^5 y^3}$ **2.** $\dfrac{-42a^9 b^2 c}{14a^4 b^2 c^3}$ **3.** $\dfrac{15p^6 q^3 rs^2}{-6pq^4 r^2 s^6}$

Simplify.

4. $\dfrac{4x^3 - 6x^2 + 8x}{2x^2}$ **5.** $\dfrac{3t^3}{12t^4 + 6t^3 - 3t^2}$

6. $\dfrac{5m^2 + 25m}{5m(3-m)}$ **7.** $\dfrac{8x^3 + 6x^2 - 4x}{10x^3 + 2x^2 + 4x}$

Simplify.

8. $\dfrac{x^2 + 7x + 10}{x^2 - 3x - 10}$ **9.** $\dfrac{x^2 + 2xy + y^2}{x^2 - y^2}$ **10.** $\dfrac{12n^2 - 13n + 3}{8n^2 + 14n - 15}$

11. $\dfrac{10m^2 - 17m + 6}{8m^2 - 14m + 5}$ **12.** $\dfrac{x^4 - 4}{x^4 - 4x^2 + 4}$ **13.** $\dfrac{(9a^3 + 4ab^2)(3a^2 + ab - 2b^2)}{81a^4 - 16b^4}$

2 Dividing Polynomials

When there is no remainder, polynomials can be divided on the calculator in the same
way that rational expressions are simplified. The proper fraction function also can be
used to divide polynomials, whether or not there is a remainder.

Divide.
1. $(y^2 - 11y + 30) \div (y - 6)$ **2.** $(x^3 + 4x^2 + 7x + 6) \div (x + 2)$
3. $(8t^2 + 10t - 3) \div (2t + 3)$ **4.** $(3v^3 + 12v^2 - 2v - 8) \div (3v^2 - 2)$

Divide.
5. $(x^2 + 15x + 26) \div (x + 3)$ **6.** $(n^3 - 5n^2 + 14) \div (n - 2)$
7. $(18d^2 + 15d - 29) \div (6d - 5)$ **8.** $(14m^3 - 12m^2 + 49m - 35) \div (2m^2 + 7)$

3 Operations on Rational Expressions

Multiply.

1. $\dfrac{-2m^2 n^5}{15} \times \dfrac{3}{m^3 n^4}$ **2.** $\dfrac{12x^2 y^3}{25ab} \times \dfrac{5ab^2}{6x^3 y^3}$ **3.** $\dfrac{3a}{4a - 12} \times \dfrac{2a - 6}{9a^2}$

4. $\dfrac{x^2 + 2x + 1}{x^2 - 5x + 6} \times \dfrac{x - 3}{x + 1}$ **5.** $\dfrac{9x^2 - 4}{8x^2 - 6x - 9} \times \dfrac{4x^2 - 9}{6x^2 + 13x + 6}$ **6.** $\dfrac{8p^2 - 22p + 5}{15p^2 + 14p + 3} \times \dfrac{10p^2 + p - 3}{6p^2 - 11p - 10}$

Divide.

7. $\dfrac{3s^3t^4}{4} \div \dfrac{9s^2t^5}{2}$

8. $\dfrac{36y^3z^2}{-7mn^2} \div \dfrac{24yz}{21m^2n}$

9. $\dfrac{2x+4}{3x-9} \div \dfrac{6x+12}{4x-12}$

10. $\dfrac{y^2-3y-18}{y^2-9y+14} \div \dfrac{y^2+7y+12}{y^2-2y-35}$

11. $\dfrac{6m^2+13m-5}{8m^2+16m-10} \div \dfrac{9m^2-4}{6m^2+m-2}$

12. $\dfrac{\dfrac{20x^2+17x+3}{18x^2-3x-10}}{\dfrac{25x^2+30x+9}{18x^2-9x-5}}$

13. a) Add $\dfrac{5q-1}{6} + \dfrac{2q+1}{4}$.

b) If your calculator gives the answer in the form of two rational expressions with different denominators, use the common denominator function to write the answer as a single rational expression.

Add. Use the common denominator function as necessary.

14. $\dfrac{2x+1}{5} + \dfrac{4x-3}{2}$

15. $\dfrac{2}{y} + \dfrac{5}{y^2} + \dfrac{3}{y^3}$

16. $\dfrac{4}{3n-9} + \dfrac{3}{4n-12}$

17. $\dfrac{t}{4t+4} + \dfrac{3t}{4t-4}$

18. $\dfrac{n+1}{n^2+4n+4} + \dfrac{4n}{n^2-4}$

19. $\dfrac{c+1}{4c^2-2c-2} + \dfrac{c-1}{2c^2+3c+1}$

20. a) Subtract $\dfrac{5q-1}{6} - \dfrac{2q+1}{4}$.

b) If necessary, use the common denominator function to write the answer as a single rational expression.

Subtract. Use the common denominator function as necessary.

21. $\dfrac{3z+1}{10} - \dfrac{4z-5}{15}$

22. $\dfrac{3}{2x} - \dfrac{1}{3x^2}$

23. $\dfrac{5}{6r+9} - \dfrac{2}{12r+18}$

24. $\dfrac{5x}{3x+4} - \dfrac{3x}{2x-1}$

25. $\dfrac{2y}{y^2-3y+2} - \dfrac{3y}{y^2-4y+3}$

26. $\dfrac{2t-3}{9t^2+6t+1} - \dfrac{t}{6t^2-19t-7}$

4 Problem Solving

1. The side lengths in a rectangle are $\dfrac{x+3}{4}$ and $\dfrac{x-1}{3}$.

a) Write an expression that represents the area.
b) Write and simplify an expression that represents the perimeter.
c) Write and simplify an expression that represents the ratio of the perimeter to the area.
d) Can $x = 1$ in your expression from part c)? Explain.

2. The surface area of a cube is represented by $\dfrac{3y^2-12y+12}{2}$.
a) Find the length of each edge of the cube.
b) Find the ratio of the volume to the surface area.

INVESTIGATING MATH

Using a Model to Review Linear Equations

Recall the meanings of these algebra tiles.

1   −1 x −x

1 Representing Equations

1. Write the equation represented by each balance scale.

a)

b)

2. Use tiles or a sketch to model each of the following equations.
a) $x + 1 = 2$ **b)** $3x - 2 = 4$ **c)** $2x - 1 = -3$

2 Solving Equations by Addition and Subtraction

A pair of tiles that includes one small red tile and one small white tile represents zero. The balance scale represents the equation $x + 2 = 3$, which can be solved as follows.

zero

Isolate the variable by adding two −1-tiles to each side.

$x + 2 + (-2) = 3 + (-2)$
or $x + 2 - 2 = 3 - 2$

Remove zero pairs and show the result.

$x = 1$

The equation $x + 2 = 3$ has the solution $x = 1$.

1. a) Write the equation represented by the balance scale.
b) How many red 1-tiles must you add to each side to isolate the variable?
c) Show the tiles that remain when you remove any zero pairs from the model.
d) What is the solution to the equation?

2. Use tiles or a sketch to solve each of the following.
a) $x + 1 = 4$ **b)** $5 = x + 3$ **c)** $x + 2 = 2$
d) $x + 2 = -3$ **e)** $x + 2 = -2$ **f)** $-4 = x + 1$

3. Use tiles or a sketch to solve each of the following.
a) $x - 1 = 1$ **b)** $2 = x - 3$ **c)** $x - 2 = -1$
d) $x - 4 = -2$ **e)** $-2 = x - 2$ **f)** $x - 1 = 3$

3 Solving Equations by Division

The balance scale represents the equation $2x = 4$, which can be solved as follows.

Group the tiles to show the value of one variable.

 $\dfrac{2x}{2} = \dfrac{4}{2}$

Show the result.

 $x = 2$

The equation $2x = 4$ has the solution $x = 2$.

1. Use tiles or a sketch to solve each of the following.
a) $3x = 6$ **b)** $2x = 6$ **c)** $8 = 4x$

2. Use tiles or a sketch to solve each of the following.
a) $3x = -3$ **b)** $-4 = 2x$ **c)** $4x = -8$

4 Solving Equations in More Than One Step

1. a) Write the equation represented by the balance scale.
b) How many red 1-tiles must you add to each side to isolate the variable?
c) Show the tiles that remain when you remove any zero pairs from the model.
d) Group the remaining tiles to show the value of one variable.
e) What is the solution to the equation?

2. Use tiles or a sketch to solve each of the following.
a) $2x - 1 = 5$ **b)** $2x - 3 = 1$ **c)** $2x + 1 = 3$
d) $3x - 2 = 1$ **e)** $6 = 2x + 2$ **f)** $3x - 1 = 5$

5 Solving Equations With Variables on Both Sides

A pair of tiles that includes one long green tile and one long white tile represents zero.

zero

1. a) Use tiles or a sketch to model the equation $3x = 2x + 1$.
b) How many $-x$-tiles must you add to each side to isolate the variable on one side?
c) Show the tiles that remain when you remove any zero pairs from the model.
d) What is the solution to the equation?

2. Use tiles or a sketch to solve each of the following.
a) $4x = 3x + 2$ **b)** $2x = x - 2$ **c)** $3x = 2x - 3$
d) $5x = 3x + 4$ **e)** $2x + 2 = x + 3$ **f)** $2x - 1 = 3x + 2$
g) $3x + 1 = x + 5$ **h)** $4x - 1 = x + 2$ **i)** $-2 + 3x = -4 + x$

4.6 A Review of Solving Equations

Explore: Use the Information

The first modern Summer Olympic Games were held in Athens, Greece, in 1896. When the Summer Olympics were held in Montreal, 92 nations were represented. This number is 1 more than 7 times the number of nations represented at the games in Athens. You can use the equation $7x + 1 = 92$ to find the number of nations represented at the first Summer Olympics.

Inquire

1. Solve the equation $7x + 1 = 92$ to find the number of nations represented at the games in Athens.

2. Describe the method you used to solve the equation.

3. Solve.

a) $x + 9 = 16$ **b)** $y - 4 = -12$ **c)** $4b = 12$
d) $3m + 1 = 13$ **e)** $2t + 10 = 14$ **f)** $3z - 2 = -8$

An **equation** is a statement of equality between two expressions. The equation $2x + 3 = 5$ is called an **open sentence**, because it may be true or false. A value of a variable that makes an equation true is known as a **solution** or a **root** of the equation. For example, the open sentence $2x + 3 = 5$ is true only for $x = 1$, so $x = 1$ is the solution of the equation. This value of the variable is said to *satisfy* the equation.

Performing the same operation on both sides of an equation gives an **equivalent equation** with the same solution. This property of equations is shown in the following table.

Operation	Example	Symbolic Form
Addition	$6 = 6$, so $6 + 3 = 6 + 3$	If $a = b$, then $a + c = b + c$
Subtraction	$6 = 6$, so $6 - 3 = 6 - 3$	If $a = b$, then $a - c = b - c$
Multiplication	$6 = 6$, so $6 \times 3 = 6 \times 3$	If $a = b$, then $a \times c = b \times c$
Division	$6 = 6$, so $\dfrac{6}{3} = \dfrac{6}{3}$	If $a = b$, then $\dfrac{a}{c} = \dfrac{b}{c}$, $c \neq 0$

To solve an equation, isolate the variable on one side of the equation.

Example 1 Variable on Both Sides
Solve and check $3x - 6 = x + 4$.

Solution

$$3x - 6 = x + 4$$

Add 6 to both sides:
$$3x - 6 + 6 = x + 4 + 6$$
$$3x = x + 10$$

Subtract x from both sides:
$$3x - x = x + 10 - x$$
$$2x = 10$$

Divide both sides by 2:
$$\frac{2x}{2} = \frac{10}{2}$$
$$x = 5$$

Check: L.S. $= 3x - 6$ R.S. $= x + 4$
$= 3(5) - 6$ $= 5 + 4$
$= 15 - 6$ $= 9$
$= 9$
The solution is $x = 5$.

Example 2 Equation With Brackets
Solve and check $4(3a + 6) + 2 = 3a - 10$.

Solution

$$4(3a + 6) + 2 = 3a - 10$$

Expand to remove brackets:
$$12a + 24 + 2 = 3a - 10$$
$$12a + 26 = 3a - 10$$

Subtract 26 from both sides:
$$12a + 26 - 26 = 3a - 10 - 26$$
$$12a = 3a - 36$$

Subtract $3a$ from both sides:
$$12a - 3a = 3a - 36 - 3a$$
$$9a = -36$$

Divide both sides by 9:
$$\frac{9a}{9} = \frac{-36}{9}$$
$$a = -4$$

Check: L.S. $= 4(3a + 6) + 2$ R.S. $= 3a - 10$
$= 4[3(-4) + 6] + 2$ $= 3(-4) - 10$
$= 4(-12 + 6) + 2$ $= -12 - 10$
$= 4(-6) + 2$ $= -22$
$= -24 + 2$
$= -22$
The solution is $a = -4$.

Practice

Solve.

1. $x + 5 = 7$ **2.** $y - 4 = 6$
3. $3a = 15$ **4.** $4k = -8$
5. $2 = t + 5$ **6.** $1 = -6 + z$
7. $-27 = 3b$ **8.** $16 = -4c$

Solve and check.

9. $7x + 5 = 19$ **10.** $9s - 4 = 5$
11. $2n + 5 = -3$ **12.** $3r - 15 = -3$
13. $-4 = 8 + 6p$ **14.** $5 = 10b - 15$
15. $-10 = 3k - 1$ **16.** $-6 = 8 + 7d$

Solve and check.

17. $2x + 5 = x + 9$ **18.** $5y - 2 = 3y + 4$
19. $5k + 4 = 2k - 5$ **20.** $x + 8 = 10 + 3x$
21. $z - 2 = -6 + 3z$ **22.** $3a + 6 - 4 = 10 + a$
23. $4t + 7 + 2t = 6t + 4t - 9$
24. $17 + 6n = -2n + 5n + 5 + 9$

Solve and check.

25. $2(y + 2) = 6$ **26.** $3(p + 1) = 0$
27. $-8 = 4(m - 3)$ **28.** $3(x - 1) = 2x + 5$
29. $2r - 8 + 4 = 5(r + 1)$ **30.** $3(2 - z) = 7z + 12 - 4z$
31. $10 + 6v - 12 = 4(2v - 1) + 2$
32. $2(3w + 1) = 2w + 1 + 7 + w$

Solve.

33. $3(a-1) = 2(a+1)$ **34.** $2(3-b) = 3(b-3)$

35. $3(x+1) + 2 = 2(x-1)$

36. $2(u-1) - (u-1) = 1 - u$

37. $4(y-5) + 2(y-2) = 3(y-3)$

38. $2(2s+1) + 6(2-s) + 3(3s-1) = -3$

Applications and Problem Solving

39. Nobel Prizes Marie Curie was one of the first winners of the Nobel Prize for Physics. Ninety-one years later, in 1994, Canada's Bertram Brockhouse won the Nobel Prize for Physics. Solve the equation $c + 91 = 1994$ to find the year in which Marie Curie won the prize.

40. Animal lengths The average length of an Arctic wolf is 150 cm. This is 10 cm longer than twice the average length of an otter. Solve the equation $2l + 10 = 150$ to find the average length of an otter, l centimetres.

41. Planets Uranus rotates about its axis once every 17 h. This is 3 h less than twice the length of time Jupiter takes to rotate once about its axis. Solve the equation $2t - 3 = 17$ to find the number of hours, t, that Jupiter takes to rotate once about its axis.

42. Arctic weather Alert is a community north of the Arctic Circle on Canada's Ellesmere Island. Adding seven to the number of frost-free days per year in Alert gives the same result as multiplying the number of frost-free days by three, then subtracting one. Solve the equation $f + 7 = 3f - 1$ to find the number of frost-free days per year in Alert.

43. Subway stations The number of stations in the subway networks of Toronto and Montreal is the same. Forty-seven less than 5 times the number of stations in Toronto is the same as 83 more than 3 times the number of stations in Montreal. Solve the equation $5s - 47 = 3s + 83$ to find the number of stations, s, in each subway network.

44. Highest points Subtracting 98 m from the height of the highest point in Manitoba, and multiplying the result by two, gives the height of the highest point in Saskatchewan. The highest point in Saskatchewan is 1468 m above sea level.

Solve the equation $2(h - 98) = 1468$ to find the height, h metres, of the highest point in Manitoba.

45. Measurement
The perimeter of the rectangle is 14 units. Solve the equation $2(6 + 2x) + 2(2 - x) = 14$ to find

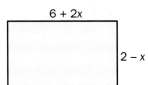

6 + 2x

2 − x

a) the value of x **b)** each side length

46. Basketball court
The length of a basketball court is 2 m less than twice the width. That is, if the width is x metres, the length is $2x - 2$ metres. The perimeter of the court is 80 m. Solve the equation $2x + 2(2x - 2) = 80$ to find the dimensions of a basketball court.

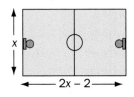

x

2x − 2

47. Measurement The areas of the rectangles are the same.

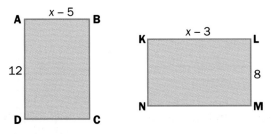

A   x − 5   B

12

D   C

K   x − 3   L

8

N   M

Solve the equation $12(x - 5) = 8(x - 3)$ to find
a) the dimensions of each rectangle
b) the area of each rectangle

48. Quadratic equations Find two solutions for each of the following equations.
a) $x^2 + 1 = 17$ **b)** $2x^2 - 1 = 17$

49. In Example 1, the variable appears on both sides of the equation. Write an equation similar to the one shown in Example 1, but with the solution $x = -7$.

50. Write an equation with brackets on one side of the equal sign and with the solution $x = -4$.

51. Write an equation with brackets on both sides of the equal sign and with the solution $x = 5$.

Photography

Are you aware of the special lighting that nature creates at different times of the day? Do you have an eye for detail? Are you a visual person? If so, you might be interested in a career in photography. Many colleges offer programs in photography.

Many photographers work freelance. Others work for a commercial studio, for a large company with its own photography department, or for a newspaper or magazine. Most photographers specialize in fields such as photojournalism, portrait photography, or aerial photography.

One of the world's most famous portrait photographers is Yousuf Karsh of Ottawa. Karsh is known for his dramatic lighting and his ability to show the personality of a subject. His photograph of former British Prime Minister Winston Churchill is widely regarded as one of the finest portrait photographs ever taken.

Lighting is one of the conditions a photographer must consider when shooting. Two of the variables affecting how much light strikes the film in a camera are the **shutter speed** and the size of the **aperture**. The aperture is the circular opening in the lens that light passes through on its way to the film. The shutter speed is the length of time the aperture is open and the film exposed.

Winston Churchill by Karsh

1 Shutter Speed

A shutter speed is expressed as a fraction of a second. The shutter speed selector shows the denominators of fractions. Selecting 1000 means the aperture will be open for $\frac{1}{1000}$ s.

1. Would a low number, such as 2, or a high number, such as 500, give the faster shutter speed? Explain.

2. Would you use a low number or a high number to shoot each of the following? Explain.
a) an unblurred "stop action" sports shot
b) a still life or stationary object

2 F-stops

The aperture is opened or closed to a chosen diameter by turning a ring. The size of the opening is indicated by an f-stop.

1. The f-stops, 1.4, 2, 2.8, 4, 5.6, ... , are terms of a sequence. The decimal terms have been truncated. The exact values are $\sqrt{2}$, 2, $2\sqrt{2}$, 4, $4\sqrt{2}$, ...

a) What type of sequence do the f-stops form?
b) Describe the sequence in words.
c) What are the next five terms expressed as exact values?
d) Write all ten f-stops in decimal form. Truncate decimals at the tenths place.

2. a) Write a formula for the area of a circle in terms of its diameter, d.
b) The diameter of the aperture, d, is related to the f-stop, f, and the focal length of the lens, F, by the formula $d = \frac{F}{f}$. Express the area of the aperture in terms of F and f.
c) Find the area of the aperture, to the nearest square millimetre, for a lens of focal length 140 mm and f-stops of 2.8, 4, and 5.6.
d) How does the area change as the f-stops increase?

3. Would you use a low f-stop or a high f-stop when shooting in each situation? Explain.
a) bright sunlight   **b)** dim light

4.7 Solving Rational Equations

As a result of the moon's gravitational pull, the tides on the Earth's surface rise and fall about every 12 h. The average height of tides around the world is 0.8 m. The world's highest tides occur in Canada's Bay of Fundy.

Explore: Solve an Equation

Multiplying the maximum height of tides in the Bay of Fundy by 0.05 gives the average height of tides around the world. Therefore, the equation $0.05x = 0.8$ can be used to find the maximum height, x metres, of tides in the Bay of Fundy.

Inquire

1. Solve the equation $0.05x = 0.8$ by dividing both sides by 0.05.

2. Solve the equation $0.05x = 0.8$ by first multiplying both sides by 100.

3. Another way to write the equation $0.05x = 0.8$ is $\dfrac{x}{20} = \dfrac{4}{5}$. Solve this equation by first multiplying both sides by 20 to clear the fractions.

4. What is the maximum height of tides in the Bay of Fundy?

5. Solve.

a) $0.3y + 0.6 = 1.2$ **b)** $0.02z - 0.11 = 0.01$ **c)** $\dfrac{x}{8} = \dfrac{1}{2}$

d) $\dfrac{2t}{3} + 4 = 2$ **e)** $\dfrac{c}{2} - \dfrac{4}{5} = \dfrac{1}{5}$ **f)** $\dfrac{a}{4} - \dfrac{2}{3} = \dfrac{5}{6}$

Equations that involve rational expressions are known as **rational equations**. They can be solved by performing the same operations on both sides.

Example 1 Equations With Decimals
Solve $3.5y - 4.2 = 1.4 - 16.8$.

Solution 1
Solve the equation by clearing the decimal fractions.

$$3.5y - 4.2 = 1.4 - 16.8$$
Simplify:
$$3.5y - 4.2 = -15.4$$
Multiply both sides by 10: $10 \times (3.5y - 4.2) = 10 \times (-15.4)$
$$35y - 42 = -154$$
Add 42 to both sides:
$$35y - 42 + 42 = -154 + 42$$
$$35y = -112$$
Divide both sides by 35:
$$\frac{35y}{35} = \frac{-112}{35}$$

Estimate
$-120 \div 40 = -3$

$$y = -3.2$$

The solution is $y = -3.2$.

Solution 2

Keep the decimal fractions.

$$3.5y - 4.2 = 1.4 - 16.8$$

Simplify: $\qquad\qquad\qquad\qquad 3.5y - 4.2 = -15.4$

Add 4.2 to both sides: $\qquad\ \ 3.5y - 4.2 + 4.2 = -15.4 + 4.2$

$$3.5y = -11.2$$

Divide both sides by 3.5: $\qquad\quad \dfrac{3.5y}{3.5} = \dfrac{-11.2}{3.5}$

$$y = -3.2$$

Estimate

$$-12 \div 4 = -3$$

The solution is $y = -3.2$.

Equations involving more than one rational expression can be solved by multiplying both sides by the lowest common denominator (LCD).

Example 2   Using the LCD

Solve and check $\dfrac{x}{2} + 3 = 2 + \dfrac{3x}{4}$.

Solution

The LCD of $\dfrac{1}{2}$ and $\dfrac{3}{4}$ is 4.

$$\dfrac{x}{2} + 3 = 2 + \dfrac{3x}{4}$$

Multiply both sides by 4: $\qquad 4 \times \left(\dfrac{x}{2} + 3\right) = 4 \times \left(2 + \dfrac{3x}{4}\right)$

$$2x + 12 = 8 + 3x$$

Subtract 8 from both sides: $\qquad 2x + 12 - 8 = 8 + 3x - 8$

$$2x + 4 = 3x$$

Subtract $2x$ from both sides: $\qquad 2x + 4 - 2x = 3x - 2x$

$$4 = x$$

Check: L.S. $= \dfrac{x}{2} + 3$

$= \dfrac{4}{2} + 3$

$= 2 + 3$

$= 5$

R.S. $= 2 + \dfrac{3x}{4}$

$= 2 + \dfrac{3(4)}{4}$

$= 2 + \dfrac{12}{4}$

$= 2 + 3$

$= 5$

The solution is $x = 4$.

Some equations, such as $\frac{3}{x}=1$ and $\frac{4}{x-1}=\frac{2}{x+1}$, have the variable in the denominator of one or more rational expressions. The value of the variable must be restricted in these cases, because division by zero is not defined. In $\frac{3}{x}=1$, x cannot equal zero. In $\frac{4}{x-1}=\frac{2}{x+1}$, x cannot equal 1 or –1.

Example 3   Variable in the Denominator

Solve $\frac{4}{x}+6=2$. State the restriction on the variable.

Solution

$$\frac{4}{x}+6=2$$

Multiply both sides by x:

$$x\times\left(\frac{4}{x}+6\right)=x\times 2$$

$$4+6x=2x$$

Subtract $6x$ from both sides:

$$4+6x-6x=2x-6x$$

$$4=-4x$$

Divide both sides by –4:

$$\frac{4}{-4}=\frac{-4x}{-4}$$

$$-1=x$$

The solution is $x=-1$, $x\neq 0$.

Example 4   Binomial Denominators

Solve $\frac{1}{m-2}=\frac{5}{m+4}$. State the restrictions on the variable.

Solution
The LCD is $(m-2)(m+4)$.

$$\frac{1}{m-2}=\frac{5}{m+4}$$

Multiply both sides by $(m-2)(m+4)$:

$$(m-2)(m+4)\left[\frac{1}{(m-2)}\right]=(m-2)(m+4)\left[\frac{5}{(m+4)}\right]$$

$$m+4=5(m-2)$$

$$m+4=5m-10$$

Subtract m from both sides:

$$m+4-m=5m-10-m$$

$$4=4m-10$$

Add 10 to both sides:

$$4+10=4m-10+10$$

$$14=4m$$

Divide both sides by 4:

$$\frac{14}{4}=\frac{4m}{4}$$

$$\frac{7}{2}=m$$

The solution is $m=\frac{7}{2}$, $m\neq 2, -4$.

Practice

In each of the following, state any restrictions on the variable.

Solve and check.

1. $\dfrac{x}{3} = 2$ **2.** $\dfrac{x}{2} = -4$ **3.** $\dfrac{2}{5}a = -1$ **4.** $-2 = -\dfrac{x}{3}$

Solve and check.

5. $\dfrac{x}{4} + 6 = 8$ **6.** $\dfrac{y}{2} - 5 = 1$ **7.** $-2 = \dfrac{1}{5}x - 1$

8. $2m + \dfrac{1}{2} = 1$ **9.** $2 = 3 + \dfrac{2x}{5}$ **10.** $\dfrac{3t}{2} - 1 = 1$

Solve and check.

11. $m + 4.6 = 5.9$ **12.** $3.2 = y + 7.5$
13. $0.5x = 1.8$ **14.** $-5.4 = 3.6z$
15. $2e + 1.2 = 2.8$ **16.** $-10.3 = 2.8x + 6.5$

Solve.

17. $2(a + 1.5) = 4.2$ **18.** $3.6 = 4(x + 1.2)$
19. $12.5 = 6.5 + 2.5(a - 1)$ **20.** $2.1x - 3(x + 1.5) = 0$

Solve. Round each answer to the nearest hundredth.
21. $1.7(2w + 1.1) = 3.1$ **22.** $4.5(1.6c + 1.2) = 4.1$
23. $1.4(z - 1.6) + 2.3(z + 1.2) = 4.4$

Solve and check.

24. $\dfrac{n}{2} + \dfrac{1}{3} = \dfrac{1}{6}$ **25.** $\dfrac{2y}{5} - \dfrac{1}{2} = \dfrac{1}{10}$

26. $\dfrac{d-1}{4} = -1\dfrac{1}{2}$ **27.** $\dfrac{3r+2}{2} = \dfrac{1}{6}$

28. $\dfrac{3-2c}{2} = \dfrac{4-2c}{3}$ **29.** $\dfrac{4k+1}{5} + \dfrac{1}{5} = -\dfrac{2}{3}$

Solve.

30. $\dfrac{1}{3}x + \dfrac{1}{3} = x - 1$ **31.** $\dfrac{3}{4}m - 2 = \dfrac{m}{4} - \dfrac{1}{4}$

32. $3 - \dfrac{1}{2}x = \dfrac{x+2}{2}$ **33.** $\dfrac{3y}{5} + 2 = 4 - \dfrac{3y}{5}$

34. $\dfrac{t+5}{8} = t + \dfrac{3}{2}$ **35.** $\dfrac{x+1}{4} + \dfrac{x+2}{2} = 1$

36. $\dfrac{3}{4}(q - 2) = 3 - 6q$ **37.** $x - \dfrac{x+2}{2} = \dfrac{x}{3} + 1$

Solve. Variable denominator

38. $\dfrac{3}{x} = 2$ **39.** $-\dfrac{4}{n} = 2$ **40.** $\dfrac{2}{a} = 4$

41. $\dfrac{1}{5x} = \dfrac{1}{10}$ **42.** $\dfrac{2}{3m} = -4$ **43.** $\dfrac{-5}{2x} = -\dfrac{1}{4}$

Solve.

44. $\dfrac{3}{y} + 1 = 4$ **45.** $4 + \dfrac{1}{2x} = 5$ **46.** $15 - \dfrac{3}{x} = 6$

47. $\dfrac{1}{d} - 2 = \dfrac{3}{d}$ **48.** $\dfrac{4}{s} + \dfrac{1}{2} = 1$ **49.** $0 = \dfrac{2}{3} + \dfrac{3}{x}$

Solve and check.

50. $\dfrac{x+1}{x} = \dfrac{1}{2}$ **51.** $\dfrac{3(t-1)}{t} = \dfrac{3}{2}$

52. $\dfrac{1}{4} = \dfrac{1-y}{2y}$ **53.** $\dfrac{4}{5} = \dfrac{2(x+1)}{x}$

54. $\dfrac{2(z+2)}{4z} + 1 = \dfrac{7}{4}$ **55.** $\dfrac{x-1}{2x} + \dfrac{1}{x} = 2$

56. $\dfrac{2}{x} - \dfrac{3}{2x} = \dfrac{1}{3}$ **57.** $\dfrac{5}{2y} + \dfrac{11}{12} = \dfrac{2}{3y}$

Solve. Binomial denominators

58. $\dfrac{1}{y} = \dfrac{2}{y+1}$ **59.** $\dfrac{3}{k-3} = \dfrac{2}{k}$

60. $\dfrac{3}{x} = \dfrac{5}{x-1}$ **61.** $\dfrac{6}{y} + \dfrac{2}{y-2} = 0$

62. $\dfrac{4}{x+1} = \dfrac{2}{x-1}$ **63.** $\dfrac{-2}{x-1} = \dfrac{1}{x+5}$

Solve.

64. $\dfrac{3}{x+1} = \dfrac{5}{3x-1}$ **65.** $\dfrac{4}{2x-1} = \dfrac{1}{x-2}$

66. $\dfrac{3w+5}{3w+1} = \dfrac{1}{2}$ **67.** $\dfrac{1}{2w+3} + \dfrac{3}{2w+3} = 2$

68. $\dfrac{2}{t-1} + 3 = \dfrac{4}{t-1}$ **69.** $\dfrac{-1}{2m-1} = 2 + \dfrac{5}{2m-1}$

Applications and Problem Solving

70. Human heart The average mass of the heart of a 14-year-old human is 156 g. This mass is 6 g more than half the average mass of an adult's heart. Solve the equation $\dfrac{m}{2} + 6 = 156$ to find the average mass, m grams, of an adult's heart.

185

71. Planets Pluto is the coldest planet in our solar system. Its lowest temperature is –230°C. Dividing the lowest temperature on Venus by 2 and then subtracting the result from 2°C gives the lowest temperature on Pluto. Solve the equation

$2 - \dfrac{v}{2} = -230$ to find the lowest temperature on Venus, v degrees Celsius.

72. Sleeping koalas Of all the world's mammals, the koala sleeps the greatest number of hours each day. Increasing the number of hours it sleeps by 3 h and multiplying the result by 0.4 gives the same result as multiplying the number of hours it sleeps by 0.5 and then subtracting 1 h. Solve the equation $0.4(t + 3) = 0.5t - 1$ to find the number of hours, t, a koala sleeps in a day.

73. Lowest temperatures Increasing the lowest January temperature recorded in Edmonton by 8°C and dividing the result by 2 gives the lowest January temperature recorded in Vancouver. The reading for Vancouver was –18°C. Solve the equation $\dfrac{x+8}{2} = -18$ to find the lowest January temperature in Edmonton, x degrees Celsius.

74. Human bones The average length of a human femur bone is 50.5 cm. Half the sum of the average length of the femur bone and the average length of the fibula bone is 45.5 cm. Use the equation $\dfrac{b + 50.5}{2} = 45.5$ to find the average length of the fibula bone, b centimetres.

75. Ostrich's speed Dividing the top speed of an ostrich by 5 and adding 10 km/h gives the same result as subtracting 2 km/h from the top speed, then dividing by 3. Solve the equation $\dfrac{s}{5} + 10 = \dfrac{s-2}{3}$ to find the top speed of an ostrich, s kilometres per hour.

76. Number Dividing 45 by a number, then subtracting 3, gives a result of –8. Solve the equation $\dfrac{45}{x} - 3 = -8$ to find the number.

77. Number Dividing 20 by a number gives the same result as dividing 12 by 2 less than the number. Solve the equation $\dfrac{20}{x} = \dfrac{12}{x-2}$ to find the number.

78. Measurement Two rectangles have the same width. The smaller rectangle has an area of 20 cm^2 and a length l centimetres. The larger rectangle has an area of 30 cm^2 and a length $(l + 5)$ centimetres.

a) Explain why the equation $\dfrac{30}{l+5} = \dfrac{20}{l}$ can be used to represent the information.

b) Solve the equation $\dfrac{30}{l+5} = \dfrac{20}{l}$ to find the length of each rectangle.

c) What is the width of each rectangle?

79. Solve each equation. State the restrictions on the variable.

a) $\dfrac{x^2 + 2}{x} = \dfrac{2x+1}{2}$

b) $\dfrac{x+1}{2x} = \dfrac{x+1}{2x-1}$

c) $\dfrac{x+1}{x^2 - 4} = \dfrac{1}{x}$

d) $\dfrac{2x-1}{x^2 - 1} = \dfrac{2}{x+2}$

e) $\dfrac{2}{x^2 - 16} + \dfrac{1}{x+4} = \dfrac{3}{x-4}$

f) $\dfrac{x}{x-1} = \dfrac{x^2}{x^2 - 2x + 1} + \dfrac{1}{x-1}$

80. a) Solve the equation $\dfrac{2}{x-1} = \dfrac{1}{2x-2}$ and state the restrictions on the variable.

b) Does the equation $\dfrac{2}{x-1} = \dfrac{1}{2x-2}$ have a solution? Explain.

81. Write each of the following equations. Have a classmate check that each equation gives the correct solution.

a) an equation involving at least two decimals and with the solution $x = 4$

b) an equation with a rational expression on each side and the solution $x = -1$

c) an equation with the variable in the denominator and the solution $x = \dfrac{1}{2}$

4.8 Problem Solving With Equations

Explore: Use an Equation

The bobcat and the cougar are two wild cats found in Canada. The average mass of a cougar is 2 kg less than nine times the average mass of a bobcat. The average mass of a cougar is 70 kg.

a) Let the mass of a bobcat be *m* kilograms. Write an expression involving *m* to represent nine times the mass of a bobcat.

b) Write an expression involving *m* to represent 2 kg less than nine times the mass of a bobcat.

c) Write an equation that relates the mass of a cougar to the expression you wrote in part b).

Inquire

1. Solve your equation from c), above, to find the average mass, in kilograms, of a bobcat.

2. The lynx is another wild cat found in Canada. The average mass of a cougar is 6 kg more than four times the average mass of a lynx. Write and solve an equation to find the average mass of a lynx.

3. How do the masses of a lynx and a bobcat compare?

Example 1 Lake Elevations

The elevation of Lake Erie is 173 m above sea level. This is 25 m higher than twice the elevation of Lake Ontario. What is the elevation of Lake Ontario?

Solution

Let *h* represent the elevation of Lake Ontario, in metres.

Then, the elevation of Lake Erie is represented by $2h + 25$.

Since the elevation of Lake Erie is 173 m,

$$2h + 25 = 173$$

Subtract 25 from both sides: $\quad 2h + 25 - 25 = 173 - 25$

$$2h = 148$$

Divide both sides by 2: $\quad \dfrac{2h}{2} = \dfrac{148}{2}$

$$h = 74$$

The elevation of Lake Ontario is 74 m above sea level.

Check: Elevation of Lake Erie is $2(74) + 25$ m or 173 m.

Many problems can be solved using rational equations.

Example 2 Areas of Islands

The Queen Charlotte Islands form part of British Columbia. The two largest islands in the group are Graham Island and Moresby Island. Their combined area is about 9000 km^2. The area of Moresby Island is 600 km^2 less than half the area of Graham Island. What is the area of each island?

Solution

Let a represent the area, in square kilometres, of Graham Island. Then, the area of Moresby Island is $\frac{a}{2} - 600$.

Since the combined area is 9000 km^2,

$$a + \frac{a}{2} - 600 = 9000$$

Multiply both sides by 2:

$$2 \times \left(a + \frac{a}{2} - 600 \right) = 2 \times 9000$$
$$2a + a - 1200 = 18\,000$$
$$3a - 1200 = 18\,000$$

Add 1200 to both sides:

$$3a - 1200 + 1200 = 18\,000 + 1200$$
$$3a = 19\,200$$

Divide both sides by 3:

$$\frac{3a}{3} = \frac{19\,200}{3}$$
$$a = 6400$$

For Moresby Island,
$$\frac{a}{2} - 600 = \frac{6400}{2} - 600$$
$$= 3200 - 600$$
$$= 2600$$

The area of Graham Island is 6400 km^2.
The area of Moresby Island is 2600 km^2.

Check: The combined area is $6400 + 2600$ or 9000 km^2.

Example 3 Comparing Speeds

The speed of a plane is seven times as great as the speed of a car. The car takes 3 h longer than the plane to travel 315 km. Find the speed of the car and the speed of the plane, in kilometres per hour.

Solution

Let the speed of the car be s kilometres per hour.

Then, the speed of the plane is $7s$ kilometres per hour.

$$\text{distance} = \text{speed} \times \text{time, so time} = \frac{\text{distance}}{\text{speed}}$$

The car travels 315 km in $\frac{315}{s}$ hours.

The plane travels 315 km in $\frac{315}{7s}$ hours.

The car takes 3 h longer than the plane, so $\frac{315}{s} = \frac{315}{7s} + 3, s \neq 0$

The LCD is $7s$.

Multiply both sides by $7s$:

$$7s \times \frac{315}{s} = 7s \times \left(\frac{315}{7s} + 3 \right)$$
$$2205 = 315 + 21s$$

Subtract 315 from both sides:	$2205 - 315 = 315 + 21s - 315$
	$1890 = 21s$
Divide both sides by 21:	$\dfrac{1890}{21} = \dfrac{21s}{21}$
	$90 = s$
	$630 = 7s$

The speed of the car is 90 km/h. The speed of the plane is 630 km/h.

Check: The car takes $\dfrac{315}{90}$ hours or 3.5 h to travel 315 km.

The plane takes $\dfrac{315}{630}$ hours or 0.5 h to travel 315 km.

So, the car takes 3 h longer than the plane.

Example 4   Driving Speeds

Melanie drove 404 km from Edmonton to Banff in the same length of time as Heidi took to drive 364 km from Edmonton to Jasper. Melanie drove 10 km/h faster than Heidi. At what speed did Heidi drive, in kilometres per hour?

Solution

Let Heidi's speed be s kilometres per hour.
Then, Melanie's speed was $s + 10$ kilometres per hour.

$$\text{distance} = \text{speed} \times \text{time, so time} = \dfrac{\text{distance}}{\text{speed}}$$

Heidi drove 364 km in $\dfrac{364}{s}$ hours.

Melanie drove 404 km in $\dfrac{404}{s+10}$ hours.

They drove for the same length of time, so	$\dfrac{364}{s} = \dfrac{404}{s+10}$
The LCD is $s(s + 10)$.	
Multiply both sides by $s(s + 10)$:	$s(s+10) \times \dfrac{364}{s} = s(s+10) \times \dfrac{404}{s+10}$
	$364(s+10) = 404s$
	$364s + 3640 = 404s$
Subtract $364s$ from both sides:	$364s + 3640 - 364s = 404s - 364s$
	$3640 = 40s$
Divide both sides by 40:	$\dfrac{3640}{40} = \dfrac{40s}{40}$
	$91 = s$

So, Heidi drove at a speed of 91 km/h.

Check: Heidi drove for $\dfrac{364}{91}$ hours or 4 h. Melanie drove for $\dfrac{404}{101}$ hours or 4 h.

The driving times were equal.

Practice

Write an algebraic expression for each statement.

1. 2 kg less than the mass of a lion

2. one metre more than six times the height of a door

3. two dollars less than three fifths the price of a CD

4. the number of cubes plus twice the number of cubes

5. the sum of two consecutive whole numbers

6. the dollar value of a number of nickels

Write an equation to find each unknown value. Do not solve the equation.

7. The length of a rectangle is 10 cm. The length is 1 cm more than twice the width. What is the width of the rectangle?

8. Erin bought a pack of pens for $4.50. The price of each pen is 50¢. How many pens are in the pack?

9. Luc's age is one third of his father's age. Luc is 16. How old is Luc's father?

10. Farida had $40. After buying 4 tickets for a baseball game, she had $16 left over. What was the price of each ticket?

11. The average life span of a woodland caribou is 5 years longer than half the average life span of a moose. The sum of their life spans is 35 years. What is the life span of a moose?

Applications and Problem Solving

12. Primates The average length of a chimpanzee is 92 cm. This is 8 cm less than half the average length of a gorilla. Use the following steps to determine the average length of a gorilla.
a) Let *g* represent the average length of a gorilla, in centimetres. Write an expression involving *g* to represent half the average length of a gorilla.
b) Write an expression involving *g* to represent 8 cm less than half the average length of a gorilla.
c) Write an equation that can be solved to find the average length of a gorilla.
d) Solve the equation and state the average length of a gorilla.

13. Skating Roma's skates cost $175. Every two weeks that she skates, she spends $4.50 on sharpening the blades. One year, she spent a total of $256 on the cost of skates and sharpening. Find the number of weeks she skated that year.

14. Driving distances The driving distance from Regina to Moose Jaw is 4 km less than one fifth the driving distance from Moose Jaw to Medicine Hat. The total driving distance from Regina to Medicine Hat through Moose Jaw is 470 km. What is the driving distance from Moose Jaw to Medicine Hat?

15. Numbers Determine the following numbers.
a) three consecutive whole numbers with a sum of 39
b) two consecutive integers with half of the smaller equal to 4 more than one third of the larger

16. Planets Venus orbits the sun in 4 days less than one third of the time Mars takes to orbit the sun. Venus orbits the sun in 225 days. How many days does Mars take to orbit the sun?

17. Coins Maylin put $1.75 in nickels, dimes, and quarters into a vending machine. The number of quarters she used was one more than the number of nickels. The number of dimes was one less than the number of nickels. How many of each type of coin did she use?

18. Areas of lakes The two biggest lakes in Manitoba are Lake Winnipeg and Lake Winnipegosis. Their combined area is about 29 800 km². The area of Lake Winnipegosis is 700 km² less than one quarter of the area of Lake Winnipeg. What is the area of each lake?

19. Mixed nuts What mass of peanuts should be added to 1 kg of cashews so that the mixture is 75% peanuts, by mass?

20. Driving speeds Nadia and Kyle shared the driving on a 1250-km trip from Edmonton to Vancouver. Nadia drove for 6 h and Kyle drove for 8 h. Nadia drove 10 km/h faster than Kyle. How fast did Kyle drive?

21. Jogging Raoul jogged from home to his sister's house at 12 km/h. For the return journey over the same route, he walked at 8 km/h. The two journeys took an hour altogether.
a) How far is Raoul's home from his sister's?
b) For how many minutes did Raoul jog?

22. Speeds on water A hydrofoil travels 120 km in the same time it takes a boat to travel 80 km. The hydrofoil is 10 km/h faster than the boat. Find the speed of the hydrofoil and the speed of the boat, in kilometres per hour.

23. Driving distance Indira and Ben drove from Winnipeg to Dauphin Lake and back. Indira drove from Winnipeg to Dauphin Lake at 100 km/h. Ben drove on the return trip at 80 km/h. Their total driving time was 6.75 h. What is the distance, in kilometres, from Winnipeg to Dauphin Lake?

24. Coins A box contains a total of 14 quarters and dimes. The total value of the coins is $2.15. How many dimes and how many quarters are in the box?

25. Cars overtaking Sara left Kootenay National Park at 08:00 and drove toward Vancouver at 85 km/h. Allana left the park an hour later and drove along the same route at 100 km/h. At what time did Allana overtake Sara?

26. Friends meeting Alain left Lethbridge at 09:00 and drove toward Whitecourt at 90 km/h. His friend Joshua left Whitecourt at the same time and drove toward Lethbridge at 80 km/h. The distance from Whitecourt to Lethbridge is 680 km.
a) At what time did Alain and Joshua meet?
b) Alain and Joshua met in Red Deer. How far is Red Deer from Lethbridge?

27. Flying distance A plane flew from Calgary to Montreal at 750 km/h. If it had flown at 600 km/h, the trip would have taken an hour longer. What is the flying distance from Calgary to Montreal?

28. Flying speed A plane takes the same amount of time to fly 1500 km at 100 km/h below its top speed as it takes to fly 1250 km at 200 km/h below its top speed. What is the plane's top speed, in kilometres per hour?

29. Comparing speeds The average speed of a plane is eight times as great as the average speed of a train. The plane takes $8\frac{3}{4}$ hours less than the train to travel 1050 km. Find the average speed of the plane and the average speed of the train.

30. Number Dividing 108 by one more than a number gives the same result as dividing 72 by three less than the number. What is the number?

31. Flying speeds David flew 300 km on a commuter plane, then 2000 km on a passenger jet. The passenger jet flew twice as fast as the commuter plane. The total flying time for the journey was $3\frac{1}{4}$ hours. What was the speed of each plane, in kilometres per hour?

32. Write a word problem that could be solved using each of the following equations. Compare your problems with a classmate's.
a) $0.25x = 6$ **b)** $x + (x + 2) = 18$
c) $\frac{x}{80} = 3$ **d)** $20 - 3x = 8$
e) $\frac{5}{x-1} = \frac{8}{x+2}$ **f)** $\frac{x}{90} - \frac{x}{100} = 1$

33. Write a word problem that can be solved with a rational equation. Have a classmate solve your problem.

PATTERN POWER

The diagrams show coloured 3 by 3, 4 by 4, and 5 by 5 grids.

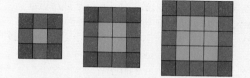

If the pattern continues, how many grid squares of each colour are there on grids with the following dimensions?
a) 6 by 6   **b)** 7 by 7
c) n by n **d)** 100 by 100

4.9 Equations With Literal Coefficients

In some equations, numerical coefficients are replaced by letters. The equation $3x + 4 = 1$ can be written as $ax + b = c$. A **formula** is an equation in which one variable is written in terms of other variables. For example, the formula for the area of a triangle is $A = \frac{1}{2}bh$, where b is the base and h is the height.

When you rewrite a formula to isolate a variable, you are *solving* for that variable.

For the formula $A = \frac{1}{2}bh$, multiplying both sides by 2 gives $2A = bh$. Then, dividing both sides by b gives $\frac{2A}{b} = h$. Thus, solving $A = \frac{1}{2}bh$ for h gives $h = \frac{2A}{b}$.

Explore: Solve the Formula

Though we stop growing taller in our teens, an adult's ears keep growing. The length, l millimetres, of a person's ears is given by the formula $l = e + \frac{g}{5}$, where e millimetres is the length of the ears at age 15, and g is the number of years of growth since age 15.
a) Solve the formula for g.   **b)** Solve the formula for e.

Inquire

1. Melina's ears were 47.8 mm long when she was 15. Use the formula $l = e + \frac{g}{5}$ to predict the length of her ears when she reaches 35.

2. Rory is 15, and his ears are 55 mm long. Use the formula solved for g to determine how old he will be when his ears are 64 mm long.

3. Kathy is 40 years old, and her ears are 54 mm long. Use the formula solved for e to calculate the length of her ears when she was 15.

4. How old will you be when your ears are 1 cm longer than they are now?

Example 1   Solving Equations With Literal Coefficients
Solve each equation for the variable indicated. State the restrictions in each case.
a) $v = u + at$, for t **b)** $ax + ay = cz$, for a

Solution
a)
$$v = u + at$$
Subtract u from both sides:
$$v - u = u + at - u$$
$$v - u = at$$
Divide both sides by a:
$$\frac{v - u}{a} = \frac{at}{a}$$
$$\frac{v - u}{a} = t$$
Since the denominator cannot equal zero, $a \neq 0$.

b)
$$ax + ay = cz$$
Remove the common factor a from the two terms in which it occurs:
$$a(x + y) = cz$$
Divide both sides by $x + y$:
$$\frac{a(x + y)}{x + y} = \frac{cz}{x + y}$$
$$a = \frac{cz}{x + y}$$
Since the denominator cannot equal zero, $x + y \neq 0$.

Example 2 Printing T-Shirts

The total cost of printing custom-designed T-shirts is given by the formula $T = s + cn$, where s is the cost to set up the printer, c is the cost of an unprinted T-shirt, and n is the number of T-shirts printed. It costs $60 to set up the printer, and an unprinted T-shirt costs $6.50. How many T-shirts can be printed for $840?

Solution 1

Solve the formula for n.

$$T = s + cn$$

Subtract s from both sides:
$$T - s = s + cn - s$$
$$T - s = cn$$

Divide both sides by c:
$$\frac{T - s}{c} = \frac{cn}{c}$$
$$\frac{T - s}{c} = n$$

Substitute the values of T, s, and c: $n = \dfrac{840 - 60}{6.50}$
$$= 120$$

So, 120 T-shirts can be printed for $840.

Solution 2

Substitute the values of T, s, and c into the formula $T = s + cn$.

$$840 = 60 + 6.50n$$

Subtract 60 from both sides:
$$840 - 60 = 60 + 6.50n - 60$$
$$780 = 6.50n$$

Divide both sides by 6.50:
$$\frac{780}{6.50} = \frac{6.50n}{6.50}$$
$$120 = n$$

Practice

Solve each equation for x. State any restrictions.

1. $2ax = b$

2. $ax + by = 3$

3. $ax - cx = 4 + b$

4. $\dfrac{x}{2} + 5y = c$

5. $2(x + 3) = x - b$

6. $\dfrac{2}{x - 1} = 3y$

Solve each formula for the variable indicated.

7. $d = st$; s

8. $C = 2\pi r$; r

9. $Ax + By + C = 0$; y

10. $y = mx + b$; x

11. $A = \dfrac{h}{2}(a + b)$; a

12. $\dfrac{1}{R} = \dfrac{1}{R_1} + \dfrac{1}{R_2}$; R_1

Applications and Problem Solving

13. Density The formula for the density of a substance is $D = \dfrac{m}{V}$, where m is its mass and V is its volume.
a) Solve for V.
b) Find the volume in cubic centimetres, of a 63-g lump of silver, which has a density of 10.5 g/cm^3.

14. Measurement The formula for the perimeter, P, of a rectangle is $P = 2(l + w)$, where l is the length and w is the width.
a) Solve for l.
b) Find the length for a width of 15.5 cm and a perimeter of 51 cm.

15. Heart rate The maximum safe heart rate for a fit and healthy person while exercising is given approximately by the formula $r = 176 - \dfrac{4}{5}a$, where r is the rate in beats per minute, and a is the age in years. Vlad has a maximum safe heart rate of 140 beats/min. How old is he?

16. Food energy The amount of energy, E kilojoules, in a portion of food is given by $E = 16.8(P + C + 2.25F)$, where P is the number of grams of protein, C is the number of grams of carbohydrate, and F is the number of grams of fat in the portion.
a) A bagel provides 840 kJ of energy. It contains 7 g of protein and 2 g of fat. How many grams of carbohydrate does it contain?
b) A serving of potato salad provides 1512 kJ of energy. Like the bagel, it contains 7 g of protein. The mass of carbohydrate is 29 g. How many grams of fat does it contain?

17. Example 2 shows two methods for solving a numerical problem that involves a formula. When you answered questions 15 and 16, which method(s) did you choose? Explain why.

TECHNOLOGY

Solving Linear Equations on a Graphing Calculator

Complete each of the following with a graphing calculator that has the capability to solve linear equations algebraically.

1 Solving Equations

1. Solve.

a) $3x + 2 = 11$ **b)** $4a - 8 = 15$

c) $6 + 12b = 3$ **d)** $8 = 2 - 12t$

e) $4n - 10 = -22$ **f)** $-3p + 10 = -2$

g) $6 = 5z + 14$ **h)** $6 + 5m = 10$

i) $\dfrac{y}{2} + 5 = 8$ **j)** $4 - \dfrac{c}{3} = -1$

2. Solve.

a) $1.2r + 5.3 = 7.1$ **b)** $-1.4 = 3.5k + 4.2$

c) $4(q - 1) = 2$ **d)** $3w + 2 = 2(w - 1)$

e) $\dfrac{y+1}{2} - 3 = 4$ **f)** $\dfrac{4}{e} - 1 = 7$

g) $\dfrac{2}{x+1} = 4$ **h)** $\dfrac{3t}{2t-1} = \dfrac{1}{2}$

i) $\dfrac{5}{1-y} = \dfrac{3}{y-2}$ **j)** $\dfrac{2}{4x+3} - \dfrac{1}{4x+3} = -3$

3. Solve. Round each answer to the nearest hundredth.

a) $1.2u + 5 = -2$ **b)** $2.4(f - 3) = 1.6$

c) $5g - 2(1 - g) = 3.5$ **d)** $6.5 - \dfrac{2.2}{d} = -3.6$

4. a) Solve the equation $\dfrac{4}{4x^2 - 1} + \dfrac{1}{2x+1} = \dfrac{2}{2x-1}$ with paper and pencil, then with a graphing calculator.
b) Explain the calculator display.

2 Solving Formulas

1. Solve each formula for the variable indicated.

a) $A = lw$; w **b)** $V = lwh$; l

c) $y = mx + b$; x **d)** $P = 2l + 2w$; w

e) $\dfrac{1}{p} + \dfrac{1}{q} = \dfrac{1}{r}$; r **f)** $\dfrac{1}{p} + \dfrac{1}{q} = \dfrac{1}{r}$; q

2. State any restrictions on the variables in each part of question 1.

1. The flying distance from Winnipeg to Vancouver is 360 km more than the flying distance from Toronto to Winnipeg. The total flying distance from Toronto to Vancouver with a stop in Winnipeg is 3364 km. Write and solve an equation to find the flying distance from Toronto to Winnipeg.

2. When a groundhog hibernates in winter, its heart rate drops to 15 beats/min. This rate is 7 beats/min less than one quarter of its normal heart rate. Write and solve an equation to find a groundhog's normal heart rate.

3. On a 100-question multiple-choice test, each question has 5 possible answers. The test is scored with the formula $S = r - 0.25(100 - r)$, where S is the score and r is the number of right answers.
a) Solve the formula for r.
b) How many right answers would give a score of 100? 80? 50? 0?
c) Is this scoring system fair? Explain.

4. The rate at which a cricket chirps depends on the air temperature. A formula that shows the approximate relationship between the temperature in degrees Celsius, T, and the average number of chirps per second, n, is as follows.

$$n = \frac{7T}{60} - 0.5$$

a) Solve the formula for T.
b) To the nearest degree, at what temperature does a cricket chirp once a second? twice a second?

5. Robyn and Karl shared the driving on a trip from Cranbrook to Revelstoke. They drove for equal lengths of time, but Karl drove 20 km/h faster than Robyn. Karl drove for 225 km, and Robyn drove for 175 km.
a) At what speed did each of them drive, in kilometres per hour?
b) How long did the trip take?

CONNECTING MATH AND TRANSPORTATION

Planning Driving Routes

On the road map, primary highways are named with 2 digits and secondary highways with 3 digits. Local roads are marked with an L. Driving distances are shown in kilometres. Assume that you can average 80 km/h on primary highways, 60 km/h on secondary highways, and 40 km/h on local roads.

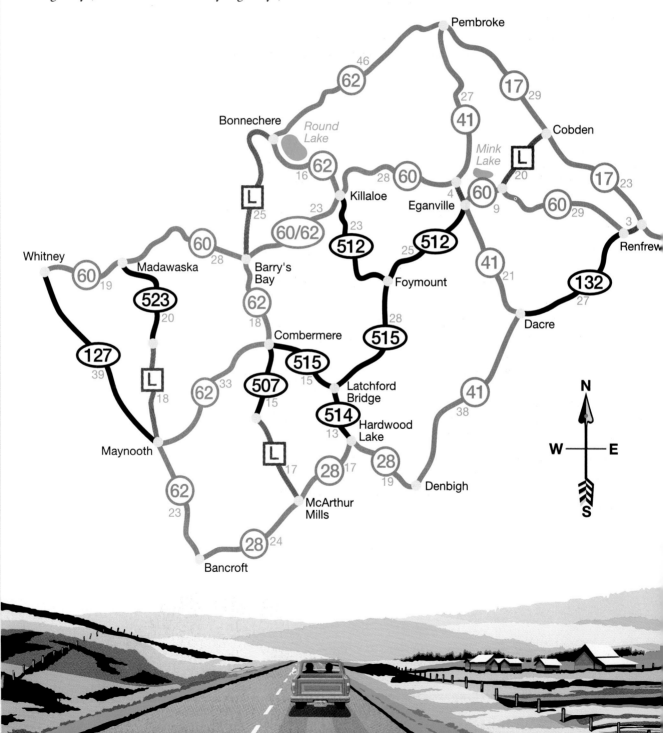

1 Writing Directions

1. The compass directions north, south, east, and west are marked on the map. Using these compass directions, and the highways and towns on the map, write a set of directions for someone who wants to drive from Barry's Bay to Latchford Bridge, passing through Pembroke and not travelling on the same road twice.

2. Write a set of directions for someone who wants to drive from Whitney to Combermere to McArthur Mills and then to Dacre, without travelling on any local roads.

2 Comparing Routes

1. a) Find the shortest driving distance from Cobden to Maynooth using any of the three types of roads.
b) Find the shortest driving distance from Cobden to Maynooth using at least one primary highway, at least one secondary highway, and no local roads.
c) Find the shortest driving distance from Cobden to Maynooth using only primary highways.

2. a) Which of the routes you found in question 1 takes the least time? How long does it take, to the nearest minute?
b) Which route takes the longest time? How long does it take, to the nearest minute?

3. a) Find the shortest driving distance from Whitney to Denbigh, if you can use any of the three types of roads and you must pick up a parcel in Bonnechere on your way.
b) Find the shortest driving distance from Whitney to Denbigh through Bonnechere using at least one primary highway, at least one secondary highway, and no local roads.
c) Find the shortest driving distance from Whitney to Denbigh through Bonnechere using only primary highways.

4. a) Which of the routes you found in question 3 takes the least time? How long does it take, to the nearest minute?
b) Which route takes the longest time? How long does it take, to the nearest minute?

COMPUTER DATA BANK

Healthy Eating

Health Canada recommends that no more than 30% of our daily food energy should come from fat, and we should eat at least 25 g to 30 g of fibre a day.

To achieve a balanced diet, *Canada's Food Guide to Healthy Eating* recommends different numbers of servings per day from the four food groups.

Enjoy a variety of foods from each group every day.

Choose lower-fat foods more often.

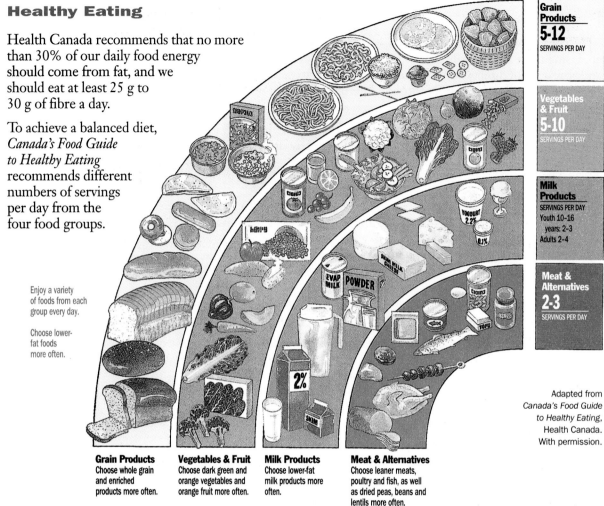

Grain Products
5-12
SERVINGS PER DAY

Vegetables & Fruit
5-10
SERVINGS PER DAY

Milk Products
SERVINGS PER DAY
Youth 10-16 years: 2-3
Adults 2-4

Meat & Alternatives
2-3
SERVINGS PER DAY

Adapted from
Canada's Food Guide to Healthy Eating,
Health Canada.
With permission.

Grain Products
Choose whole grain and enriched products more often.

Vegetables & Fruit
Choose dark green and orange vegetables and orange fruit more often.

Milk Products
Choose lower-fat milk products more often.

Meat & Alternatives
Choose leaner meats, poultry and fish, as well as dried peas, beans and lentils more often.

Use the *Nutrition* database, from the Computer Databank for ClarisWorks, to complete the following.

1 Fat and Fibre

1. Create a table to display only these fields for all records.

Food   Type   Serving   Energy, kJ   Fat, g   Fibre, g

2. Find the records for *ice cream, vanilla*. What is the mass of fat in a serving of each vanilla ice cream?

3. a) Find all the records for which one serving contains more fat than one serving of the higher-fat vanilla ice cream. How many records are displayed?
b) Are you surprised by any of the foods you found in part a)? Explain.
c) What do you notice about the fibre content of the foods you found in part a)?

4. Sort all the records from greatest to least fibre content. What do you notice about the fat content of foods with high fibre content?

198

5. a) Find the record for *potato chips*. What are the masses of fat and fibre in a serving?
b) Compare the serving size with the amount that you think of as a "normal portion." What are the masses of fat and fibre in your idea of a "normal portion"?

6. Repeat question 5 for these foods.
a) eclair **b)** strawberries, raw **c)** pizza, cheese **d)** cookies, peanut butter

2 Energy From Fat

One gram of fat provides about 38 kJ of energy. For a food, the percent of energy from fat, *p*, is given by the formula $p = \dfrac{38f}{e} \times 100$, where *f* is the mass of fat, in grams, and *e* is the energy from the food, in kilojoules.

1. Add a calculation field called *Energy from Fat, %*, rounding the percents to 1 decimal place.

2. Find all the records where over 60% of the energy is from fat. Explain why some percents are greater than 100%.

3 Planning Meals

1. Sort all the records alphabetically by type, and, within each type, alphabetically by food.

2. Plan two different breakfast meals by selecting foods from the database. Remember to include such items as added butter and sugar. Also, increase serving sizes to your idea of a "normal portion" by copying records if appropriate.

3. For each meal, create summary fields called *Total Energy, kJ, Total Fat, g,* and *Total Fibre, g.*

4. For each meal, create another summary field called *Total Energy from Fat, %,* rounded to 1 decimal place, using the total fat and total energy values.

5. What might you add to each meal to increase the mass of fibre without increasing the mass of fat?

6. Repeat questions 2 to 5 for two different lunch meals.

4 Planning a Day's Menu

1. Select foods that you enjoy from the database and use them to plan a whole day's menu, including snacks.

2. Create summary fields called *Total Energy, kJ, Total Fat, g,* and *Total Fibre, g.*

3. Create another summary field called *Total Energy from Fat, %,* rounded to 1 decimal place, using the total fat and total energy values.

4. The typical Canadian gets 40% or more of energy from fat and eats 15 g or less of fibre in a day. Is your menu typical? Explain.

5. If your menu does not have 30% or less of energy from fat and 25 g to 30 g of fibre, adjust it to meet these recommendations.

6. Which is easier for you — eating enough fibre or not eating too much fat? Explain.

Review

In each of the following, state any restrictions on the variables.

4.1 *Divide.*

1. $\dfrac{6x^2 - 9x}{-3x}$

2. $\dfrac{8a^3 - 4a^2 + 10a}{2a}$

3. $(n^2 + 10n + 16) \div (n + 2)$
4. $(y^3 + y^2 - 2) \div (y - 1)$
5. $(4w^2 + 4w - 1) \div (2w + 3)$
6. $(t^3 + t^2 - 3t - 9) \div (t + 3)$
7. $(3x^3 - 2x^2 + 12x - 9) \div (x^2 + 4)$

8. Business The area of a business card can be represented by the expression $28y^2 - 15y + 2$, and the length by the expression $7y - 2$.
a) Write an expression that represents the width.
b) Divide the expression from part a).
c) If y represents 13 mm, what are the dimensions of the business card, in millimetres?

4.2 *Simplify.*

9. $\dfrac{3x}{3x + 9}$

10. $\dfrac{8y^2 - 10xy}{4y}$

11. $\dfrac{5x - 5y}{7x - 7y}$

12. $\dfrac{3m^2 - 3m}{4m^2 - 4m}$

13. $\dfrac{t - 2}{t^2 - 3t + 2}$

14. $\dfrac{2a^2 - 7a - 15}{a - 5}$

15. $\dfrac{y^2 - 9}{y^2 + y - 12}$

16. $\dfrac{6n^2 - 7n - 3}{12n^2 + 7n + 1}$

17. Alberta flag The area of an Alberta flag can be represented by the expression $2x^2 + 4x + 2$, and its width by $x + 1$.
a) Write and simplify an expression for the length.
b) Write and simplify an expression that represents the ratio length:width for an Alberta flag.

4.3 *Simplify.*

18. $\dfrac{5x^3}{2y} \times \dfrac{8y}{15x^2}$

19. $\dfrac{-4a^3}{3b} \div \dfrac{2a}{3b^2}$

20. $\dfrac{3a^2b}{-4xy} \times \dfrac{-5x^2y}{6ab^2}$

21. $\dfrac{b^2}{8x^3y} \div \dfrac{3b}{4xy}$

22. $\dfrac{3x - 3}{2x + 2} \times \dfrac{5x + 5}{6x - 6}$

23. $\dfrac{4m + 8}{3n - 3} \div \dfrac{2m + 6}{7n - 7}$

24. $\dfrac{t^2 + 4t + 4}{t - 2} \div \dfrac{3t + 6}{t^2 - 5t + 6}$

25. $\dfrac{2x^2 - 5x - 3}{2x^2 - 5x + 2} \times \dfrac{2x^2 + 3x - 2}{x^2 - 4x + 3}$

26. $\dfrac{\dfrac{6y^2 - 5y + 1}{12y^2 + 5y - 2}}{\dfrac{3y^2 - 4y + 1}{4y^2 + 3y - 1}}$

27. Measurement
Rectangles B and C are attached to rectangle A, as shown.

The area of rectangle B is $2t^2 - 3t + 1$.
The area of rectangle C is $3t^2 - 2t - 1$.
The lengths of these two rectangles are as shown in the diagram.

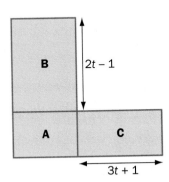

a) Write a rational expression that represents the width of rectangle B.
b) Write a rational expression that represents the width of rectangle C.
c) Write and simplify the product of the expressions you wrote in parts a) and b).
d) What type of rectangle is A? Explain.

4.4 *Simplify.*

28. $\dfrac{5}{x} + \dfrac{1}{x} - \dfrac{8}{x}$

29. $\dfrac{2m + 1}{m - 2} + \dfrac{3m - 5}{m - 2}$

30. $\dfrac{4z - 3}{z^2} - \dfrac{3z - 1}{z^2}$

31. $\dfrac{2t}{3} - \dfrac{3t}{4} + \dfrac{t}{6}$

32. $\dfrac{4x + 1}{5} + \dfrac{2x - 1}{4}$

33. $\dfrac{2a - 3b}{6} - \dfrac{3a - 2b}{4}$

34. Measurement The perimeter of a triangle is $\dfrac{9x + 1}{4}$. If two of the side lengths are $\dfrac{x + 1}{2}$ and $\dfrac{2x - 1}{2}$, what is the third side length?

5 *Simplify.*

35. $\dfrac{2}{y} + \dfrac{4}{y^2} - \dfrac{1}{y}$

36. $\dfrac{4}{x^2} - \dfrac{5}{xy} + \dfrac{2}{y^2}$

37. $\dfrac{a}{2a-2} + \dfrac{2}{3a-3}$

38. $\dfrac{2}{x+3} - \dfrac{4}{x+1}$

39. $\dfrac{2}{t^2+3t+2} - \dfrac{1}{t^2+t-2}$

40. $\dfrac{x+1}{3x^2+4x+1} + \dfrac{2x-1}{3x^2-5x-2}$

6 *Solve.*

41. $3a - 2 = -8$

42. $4z + 3 = 6z - 5$

43. $2(m + 4) = m - 1$

44. $6(2 - w) = 2(1 + 2w)$

45. Rivers The Columbia River has a length of 2000 km, which is 140 km less than twice the length of the Assiniboine River. Solve the equation $2x - 140 = 2000$ to find the length of the Assiniboine River, x kilometres.

7 *Solve.*

46. $\dfrac{x}{3} + 1 = -2$

47. $-1 = 3 + \dfrac{2}{x}$

48. $2.9 + 5z = -4.6$

49. $2.1 = 1.4(t + 1)$

50. $\dfrac{a-1}{4} = \dfrac{2a+1}{5}$

51. $\dfrac{x+1}{2} - \dfrac{x-1}{3} = \dfrac{3}{2}$

52. $\dfrac{2}{2m-1} + \dfrac{1}{2m-1} = -3$

8 53. National parks The combined area of Jasper National Park and Banff National Park is about 17 500 km². The area of Banff National Park is 900 km² greater than half the area of Jasper National Park. Find the area of each park.

54. Driving distance Cory drove from Saskatoon to Regina at an average speed of 80 km/h. Because he stopped at a restaurant, his average speed on the return trip was only 65 km/h. Cory's total driving time from Saskatoon to Regina and back was 7.25 h. What is the driving distance from Saskatoon to Regina?

9 *Solve each formula for the variable indicated.*

55. $P = \dfrac{F}{A}; A$

56. $V = \dfrac{1}{3}bh; h$

57. $P = 2a + b; a$

58. $a = \dfrac{b+c}{2}; b$

Exploring Math

Expressions for Numbers

One way to write an expression for 12 using five 6s is:

$$\dfrac{66}{6} + \dfrac{6}{6} = 12$$

Another way is $6 + 6 + \dfrac{6-6}{6} = 12$

1. Use five 6s and any of the arithmetic operations +, −, ×, and ÷ to write an expression for each whole number from 1 to 9. Compare your solutions with a classmate's.

2. Repeat question 1 using four 4s instead of five 6s.

3. Repeat question 1 using five 5s instead of five 6s.

4. One way to write an expression for 1 using seven ns is:

$$\dfrac{n}{n} - \dfrac{n-n+n-n}{n} = 1$$

Use seven ns and any of the arithmetic operations +, −, ×, and ÷ to write expressions for the numbers from 2 to 6. Compare your solutions with a classmate's.

5. Write a problem like problems 1 to 3, above. Your problem should include the use of the same one-digit number several times in each expression. Your problem should also require expressions for at least four consecutive whole numbers. Check that your problem can be solved. Then, have a classmate solve your problem. Compare your classmate's solutions with the solutions you expected.

Chapter Check

In each of the following, state any restrictions on the variables.

Divide.

1. $\dfrac{9y^3 + 6y^2 - 12y}{3y}$

2. $(12t^2 - 5t - 3) \div (3t + 1)$
3. $(x^3 + x^2 - 5x + 2) \div (x - 2)$
4. $(2n^3 - 2n^2 + n) \div (2n^2 + 1)$

Simplify.

5. $\dfrac{3x - 3y}{5x - 5y}$

6. $\dfrac{2y^2 + 4y}{3y^2 + 6y}$

7. $\dfrac{t^2 - 16}{t^2 - t - 12}$

8. $\dfrac{2m^2 + m - 3}{3m^2 + 2m - 5}$

Simplify.

9. $\dfrac{4x^3}{3y} \times \dfrac{9y}{10x^2}$

10. $\dfrac{-8y^3}{3x} \div \dfrac{4y}{3x^2}$

11. $\dfrac{y^2 - 9}{y - 4} \div \dfrac{y^2 + 6y + 9}{3y - 12}$

12. $\dfrac{x^2 + 2x - 3}{x^2 + 6x + 8} \times \dfrac{x^2 + 2x - 8}{x^2 + x - 6}$

13. $\dfrac{2a^2 - a - 1}{3a^2 + a - 2} \div \dfrac{2a^2 - 3a - 2}{3a^2 - 11a + 6}$

Simplify.

14. $\dfrac{n + 2}{3} + \dfrac{2n - 1}{4}$

15. $\dfrac{3}{x} - \dfrac{2}{x^2} + \dfrac{4}{x^3}$

16. $\dfrac{1}{2z - 3} + \dfrac{2}{z + 1}$

17. $\dfrac{2}{x^2 + 5x + 4} - \dfrac{3}{x^2 - 3x - 4}$

Solve and check.

18. $4w + 3 = -13$
19. $3g - 6 = -5 + 2g$
20. $2(n - 1) - 5n = 4$
21. $4(2m - 1) = 5(m + 1)$

Solve.

22. $-1 = \dfrac{3y}{2} - 2$

23. $1.5(x - 2) + 2.3 = -3.1$

24. $\dfrac{2m + 3}{4} = -1$

25. $\dfrac{3a - 2}{2} = \dfrac{4a + 3}{3}$

26. $11 - \dfrac{6}{x} = 13$

27. $\dfrac{2}{z + 2} = \dfrac{5}{z + 3}$

Solve for the variable indicated.

28. $F = ma;\ m$

29. $A = P + Prt;\ P$

30. Running times A turkey can run 50 m at a speed of s metres per second. A chicken's speed over the same distance is 3 m/s less than the turkey's speed.
a) Write an expression in terms of s to represent the time, in seconds, the turkey takes to run 50 m.
b) Write an expression in terms of s to represent the time, in seconds, the chicken takes to run 50 m.
c) Write and simplify an expression that represents how many seconds longer the chicken takes than the turkey to run 50 m.
d) If s represents 7 m/s, how much longer does the chicken take than the turkey to run 50 m, to the nearest tenth of a second?

31. Earth science The average mass of aluminum per tonne of the Earth's crust is 80 kg. This mass is 10 kg more than one quarter of the mass of silicon per tonne of the Earth's crust. Find the mass of silicon per tonne of the Earth's crust.

32. Comparing speeds An express train travels 440 km in the same time that a car takes to travel 380 km. The train travels 15 km/h faster than the car. Find the speed of the car and the speed of the train, in kilometres per hour.

Using the Strategies

1. If each of the following numbers is written in standard form, what is the final digit?
a) 9^{32} **b)** 12^{33}

2. The difference between the squares of two consecutive whole numbers is 63. What are the numbers?

3.

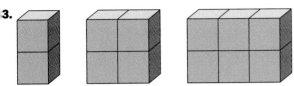

The first diagram shows two 1-cm cubes, with one stacked on top of the other. If the lower cube sits on a table, the area of the exposed faces is 9 cm². The second diagram shows four cubes, and the area of the exposed faces is 14 cm². The third diagram shows six cubes, and so on.
a) What is the total area of the exposed faces for a figure made from 36 cubes?
b) If the total area of the exposed faces of a figure is 129 cm², how many cubes make up the figure?

4. A dart that lands on a regulation dart board can score a whole number of points from 1 to 20, or double or triple each number. If the dart hits the bulls-eye, it can score 25 or 50 points. Ray threw 3 darts at the board. One of them missed the board, but Ray scored a total of 35 points with the other 2 darts. In how many different ways could he have scored 35 points with 2 darts?

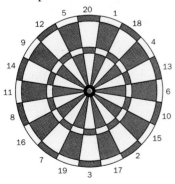

5. Find the two smallest whole numbers that give
a) a remainder of 1 when divided by 3, 4, 5, or 8
b) a remainder of 2 when divided by 5, 7, or 9

6. In Edmonton, avenues run east-west, and streets run north-south. How many ways are there to walk west and south from the intersection of 123rd Avenue and 122nd Street to the intersection of 118th Avenue and 126th Street?

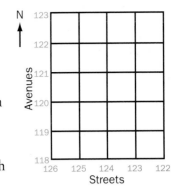

7. A kite is a quadrilateral with two pairs of adjacent sides equal. If a kite has diagonals that are 10 cm and 12 cm long, what is the area of the kite?

8. The population of Kamloops, British Columbia, was 67 000 in 1991. The population increased by 5000 from 1986 to 1991, decreased by 2000 from 1981 to 1986, and increased by 6000 from 1976 to 1981. What was the population of Kamloops in 1976?

9. About how many piano tuners are there in Canada?

D A T A   B A N K

1. a) If you were driving from Edmonton to Portland, Oregon, would the shorter route be through Vancouver and Seattle or through Calgary and Spokane?
b) If you drove at an average speed of 90 km/h, how much driving time would you save by taking the shorter route rather than the longer route?

2. On a typical summer afternoon on Canada's West Coast, the relative humidity is about 75%. Around the Lower Great Lakes, the typical summer afternoon humidity is about 60%. What air temperature would give a humidex reading of 38°C
a) on the West Coast?
b) around the Lower Great Lakes?

(handwritten notes in margin: $2^3 = 8$, $3^3 = 27$, $4^3 = 64$)

Chapter 1

Given that x is a real number, graph each of the following on a number line.

1. $x > 1$ **2.** $x \le |-3|$ **3.** $-1 < x < 4$

Simplify.

4. $\sqrt{28}$ **5.** $\sqrt{150}$ **6.** $\sqrt{5} \times \sqrt{35}$

7. $\sqrt{6} \times 4\sqrt{3}$ **8.** $2\sqrt{14} \times 3\sqrt{2}$ **9.** $\dfrac{\sqrt{33}}{\sqrt{11}}$

Estimate. Then, find an approximate value, to the nearest hundredth.

10. $\sqrt{55}$ **11.** $\sqrt{8550}$ **12.** $\sqrt{0.65}$

13. $\sqrt{0.078}$ **14.** $\sqrt{40} - 2\sqrt{3}$ **15.** $\dfrac{\sqrt{91}}{\sqrt{15}}$

Simplify.

16. $\sqrt{54} + \sqrt{24} - \sqrt{96}$

17. $2\sqrt{20} - \sqrt{8} + \sqrt{18} - 3\sqrt{45}$

18. $\sqrt[3]{16} + \sqrt[3]{54}$ **19.** $4\sqrt[3]{32} - \sqrt[3]{108}$

Expand and simplify.

20. $\sqrt{5}(\sqrt{10} + \sqrt{15})$ **21.** $(2\sqrt{6} - 5\sqrt{2})^2$

22. $(3\sqrt{2} + 2\sqrt{3})(\sqrt{2} - \sqrt{3})$

23. $(\sqrt{7} - 4\sqrt{5})(2\sqrt{7} + \sqrt{5})$

Simplify.

24. $\dfrac{5}{\sqrt{3}}$ **25.** $\dfrac{2\sqrt{7}}{1 + \sqrt{5}}$ **26.** $\dfrac{3\sqrt{10} + 7\sqrt{2}}{2\sqrt{10} - \sqrt{2}}$

Evaluate.

27. $(-2)^0$ **28.** 4^{-2} **29.** $(2^{-3})^{-1}$ **30.** $\dfrac{1}{-(3^{-2})}$

Simplify.

31. $x^5 \times x^{-3}$ **32.** $y^6 \div y^4$ **33.** $(x^{-1}y^2)^3$

34. $\dfrac{(a^4)^{-2}}{b^{-2}}$ **35.** $(3m^2n^{-4})(-2m^{-3}n^5)$

36. $(-9p^{-3}q^6) \div (-3pq^{-1})$ **37.** $\dfrac{-30s^2t^{-3}}{2s^4t^{-2} \times 5s^{-3}t^5}$

Evaluate.

38. $16^{-\frac{1}{4}}$ **39.** $\left(\dfrac{8}{27}\right)^{-\frac{1}{3}}$ **40.** $-25^{-\frac{3}{2}}$ **41.** $\sqrt{\sqrt[3]{64}}$

42. Measurement The area of a rectangle is 8 square units. The width is $\sqrt{5} - 1$ units. Write and simplify an expression for the length.

204

Chapter 2

1. Find the amount of an investment of $10 000 at the end of 4 years, if the interest rate is 8% compounded semi-annually.

2. Calculate the total price of a $1295 TV set, if the GST is 7%, the PST is 8%, and the PST is calculated

a) only on the price

b) on the price plus the GST

Given the general term, state the first 6 terms of each sequence. Then, graph t_n versus n.

3. $t_n = 2(n - 1)$ **4.** $t_n = n^2 + 5$

5. Write the first 5 terms determined by the following recursion formula.
$t_1 = -6; \; t_n = t_{n-1} + 5$

6. For the arithmetic sequence 9, 15, 21, ..., find

a) t_n **b)** t_{25}

7. Find the number of terms in the arithmetic sequence.
$-1, -8, -15, ..., -190$

8. Find 3 arithmetic means between 14 and -6.

9. Find S_{12} for the arithmetic series.
$-35 - 25 - 15 - ...$

10. Find the sum of the arithmetic series.
$21 + 23 + 25 + ... + 43$

11. Repaying a loan The table shows the data for the repayment of a loan. The columns show the year (Y), Opening Balance (OB), Interest Rate (IR), Interest Charged (IC), Annual Payment (AP), and Closing Balance (CB).

Y	OB	IR (%)	IC	AP	CB
1	$9000.00	7	$630.00	$2195.02	$ 7434.98
2	$7434.98	7	$520.45	$2195.02	$ 5760.41
3	$5760.41	7	$403.23	$2195.02	$ 3968.62
4	$3968.62	7	$277.80	$2195.02	$ 2051.40
5	$2051.40	7	$143.60	$2195.00	$ 0.00

a) How much of the fourth annual payment goes toward the opening balance?

b) What is the total interest paid on the loan?

c) If an extra payment of $3709.01 were made at the end of year 2, when would the loan be paid off?

Chapter 3

Simplify.
1. $(3x^2 + 4x - 2) + (x^2 - 5x + 4)$
2. $(2y^2 - 8y + 7) - (5y^2 - 6y - 2)$
3. $(-5a^3b^4)(2a^2b)$
4. $\dfrac{-12x^3 y^5 z^{-4}}{-4xy^3 z^2}$

Expand and simplify.
5. $3(2m - n) - 2(m + 3n)$
6. $4p(p + 2) + 3p(2p - 5)$
7. $2w(w^2 + w - 3) - 3w(2w^2 + 1)$
8. $(2x + 1)(3x - 1)$
9. $(5z + 1)(z - 2) + (4z - 3)(2z + 3)$
10. $(y + 3)(2y^2 - 5y - 4)$

Expand and simplify.
11. $(4x + 1)(4x - 1)$
12. $(2m + n)^2$
13. $(y - 4)^2 - (y + 2)(y - 2)$
14. $4(t + 1)^2 + 2(2t + 1)(2t - 1)$

Factor.
15. $5yz^2 - 35yz + 10y^2z$
16. $2mn + 3mp - 4n - 6p$
17. $x^2 - 12x + 20$
18. $6y^2 + 11y + 4$
19. $25a^2 - b^2$
20. $49b^2 - 28b + 4$

Factor fully.
21. $2s^2 + 14s - 16$
22. $9x^2 + 24x + 12$
23. $ax^2 + 6ax + 9a$
24. $40c^2 - 10d^2$

25. Parthenon The Parthenon in Athens has a rectangular base whose dimensions can be represented by the expressions $5x + 2$ and $3x - 8$.
a) Write and simplify an expression that represents the area of the base of the Parthenon.
b) If x represents 14 m, what is the area of the base of the Parthenon, in square metres?

26. Calculator screen On one model of graphing calculator, the area of the screen can be represented by the expression $8x^2 + 10x - 25$.
a) Factor the expression $8x^2 + 10x - 25$ to find binomials that represent the dimensions of the screen.
b) If x represents 17 mm, what are the dimensions of the screen, in millimetres?

Chapter 4

State any restrictions on the variable.

Divide.
1. $\dfrac{6x^3 - 8x^2 + 12x}{-2x}$
2. $(m^2 + 5m - 6) \div (m + 6)$
3. $(10y^2 + 17y - 20) \div (5y - 4)$
4. $(x^3 + 4x^2 + 3x + 12) \div (x^2 + 3)$

Simplify.
5. $\dfrac{5n^2 + 5n}{10n^2 + 5n}$
6. $\dfrac{3b^2 - 4b - 4}{b - 2}$
7. $\dfrac{t^2 - 9}{t^2 + 7t + 12}$
8. $\dfrac{6d^2 + d + 1}{8d^2 - 2d - 3}$

Simplify.
9. $\dfrac{4x^2 y}{9mn^2} \times \dfrac{-3mn}{-2xy}$
10. $\dfrac{a^2b^2}{10p^2q} \div \dfrac{2ab}{-5pq^2}$
11. $\dfrac{t^2 + 6t + 9}{t^2 - 6t + 9} \times \dfrac{3t - 9}{2t + 6}$
12. $\dfrac{2x^2 + 5x + 2}{2x^2 - 3x - 9} \div \dfrac{2x^2 + 3x - 2}{2x^2 + x - 3}$

Simplify.
13. $\dfrac{3y + 1}{3} + \dfrac{2y - 1}{2}$
14. $\dfrac{x - 1}{5} - \dfrac{3x - 2}{10}$
15. $\dfrac{3}{2n + 1} + \dfrac{4}{3n - 2}$
16. $\dfrac{4}{2m^2 - m - 1} - \dfrac{2}{m^2 + 2m - 3}$

Solve.
17. $4a + 1 = 2a - 5$
18. $3(2z + 1) = -2(z + 1)$
19. $\dfrac{x}{2} + \dfrac{1}{3} = 1$
20. $3.1(k + 1) = k + 7.3$
21. $\dfrac{3}{x + 1} = \dfrac{-4}{2x - 3}$

Solve for the variable indicated.
22. $y = mx + b; m$
23. $A = P + Prt; r$

24. Lengths of rivers The length of the Fraser River is 159 km less than half the length of the St. Lawrence River. The total length of the two rivers is 4428 km. What is the length of each river?

CHAPTER 5

Relations and Functions

There is a relationship between how exciting an amusement park ride is and the number of people who ride it. The more exciting the ride, the more riders there are.

Some people enjoy the excitement of a roller coaster ride. Canada has many roller coasters, including the world's largest triple-loop indoor roller coaster. This coaster, called the *Mindbender*, is in the West Edmonton Mall. The *Mindbender* turns riders upside down three times and reaches a top speed of over 80 km/h.

For rides with loops, there is a relationship between the minimum speed of the car through a loop and the radius of the loop. The relationship is given by the equation

$$s = \sqrt{9.8r}$$

where r is the radius of the loop in metres, and s is the speed of the car in metres per second.

1. Why is there a minimum speed?

2. What is the minimum speed for a loop with a radius of 15 m? 20 m?

3. To double the minimum speed, by what factor must the radius of a loop be increased? Explain.

4. What do you think is the smallest radius a loop could have? Explain.

5. What other rides at an amusement park have a minimum speed to work properly?

Using Graphs

1 Writing Coordinates

The Cartesian system for graphing ordered pairs on a coordinate grid was developed by René Descartes in the seventeenth century. Points are plotted on a grid defined by two perpendicular number lines. The horizontal line is called the *x*-axis, and the vertical line, the *y*-axis.

1. Copy the following letters into your notebook in the order shown.

> B C D E F G I K L N O P R S U

2. Above each letter, write the coordinates of the corresponding point shown on the grid.

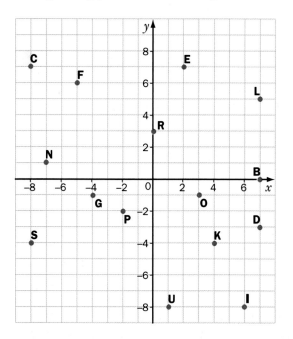

3. For each letter, add the two integers in the ordered pair and write the total below the letter.

4. Arrange the totals in order from smallest to largest. Write the largest total twice.

5. Write the letter that corresponds to each total above the total. You will get the last names of three Canadians known all over the world. What kind of work made each of them famous?

2 Writing Statements

Write as many statements as you need to describe fully the information shown on each graph.

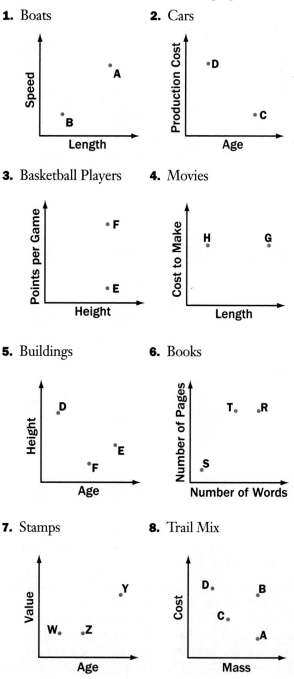

1. Boats

2. Cars

3. Basketball Players

4. Movies

5. Buildings

6. Books

7. Stamps

8. Trail Mix

3 Sketching and Interpreting Graphs

1. Sketch a graph of the amount of daylight versus the time of the year, starting on March 21 and ending on March 20.

2. The graph shows the speed of a dirt bike versus distance travelled from the start line during one lap of a race. If the rider slows down at corners, draw a possible diagram of the track.

3. Water is poured into the bottle at a constant rate. Draw a graph of the height of the water in the bottle versus time.

4. Sketch a graph of a batter's distance from home plate versus time, as she runs around the bases after hitting a home run.

5. During a baseball game, Marco hit a ground ball to the shortstop. The ball bounced 3 times after Marco hit it. The shortstop fielded it and threw Marco out at first base. Sketch a graph of the height of the ball versus time, starting when the ball left the pitcher's hand and ending when the ball was caught at first base.

Mental Math

Expressions and Equations

Evaluate each expression for the given values of x.
1. $2x + 1$; $x = 1, 3, 5, 9, 12$
2. $3x - 1$; $x = 3, 4, 7, 11, 20$
3. $x + 4$; $x = 3, 2, 1, -1, -4$
4. $5x - 2$; $x = -1, -2, -3, -5, -10$
5. $6 - 2x$; $x = 5, 4, -2, -3, -4$
6. $x^2 + 3$; $x = 4, 2, 0, -2, -4$
7. $x^2 - 6$; $x = 5, 3, 1, -1, -3$

Find 4 ordered pairs that satisfy each equation.
8. $x + y = 8$ 9. $x - y = 2$
10. $x + y = -1$ 11. $x - y = -2$

Subtracting Using Compatible Numbers

Suppose that you are subtracting one number from another and that the number you are subtracting is close to a multiple of 10. The mental subtraction is easier if you adjust both numbers so that you subtract a multiple of 10.

For $83 - 49$, think
$$(83 + 1) - (49 + 1) = 84 - 50$$
$$= 34$$
So, $83 - 49 = 34$

For $91 - 42 = 49$, think
$$(91 - 2) - (42 - 2) = 89 - 40$$
$$= 49$$
So, $91 - 42 = 49$

Calculate.
1. $55 - 19$ 2. $73 - 28$ 3. $80 - 39$
4. $81 - 22$ 5. $70 - 31$ 6. $61 - 23$

Adapt the method to calculate the following. Describe how you adapted the method.
7. $6.4 - 2.9$ 8. $10 - 3.8$ 9. $7.2 - 4.3$
10. $650 - 280$ 11. $1800 - 990$ 12. $710 - 320$

13. a) Complete the following subtractions, where x, y, n, m, b, and c represent whole numbers.
$(10x + y) - (10n + m)$
$(10x + y + b) - (10n + m + b)$
$(10x + y - c) - (10n + m - c)$
b) Explain how the subtractions in part a) are related to the method shown above.
c) For the method to work as shown, what is the value of $m + b$? the value of $m - c$?

INVESTIGATING MATH

Relationships in Polygons

In this section, you will make polygons from squares. The squares must be joined along whole edges.

These polygons are allowed. These figures are not allowed.

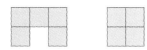 

1 Polygons From 8 Squares

1. Polygon 1 is made up of 8 squares. The area, A, of the polygon is 8 square units. A common vertex within a polygon is called an **interior point**. Polygon 1 has 2 interior points, marked 1 and 2. The perimeter, P, of the polygon is 14 units. Half the perimeter, H, is 7 units.

Copy and complete the table for the other polygons shown.

Polygon	Area (A)	Interior Points (I)	Perimeter (P)	Half the Perimeter (H)
1	8	2	14	7
2				
3				
4				
5				

2. Describe the relationship between half the perimeter, H, the area, A, and the number of interior points, I.

3. Write an expression for H in terms of A and I.

$H =$

4. Write an expression for the perimeter, P, in terms of A and I.

$P =$

5. Draw 4 more polygons made up of 8 squares. Use the polygons to test your expressions.

2 Other Polygons

1. On grid paper, draw 7 different polygons made up of 6, 7, 9, 10, 11, 12, and 13 squares. Draw at least one polygon with 4 interior points, one with 3, one with 2, one with 1, and one with no interior points.

2. Number the polygons from 1 to 7. Then, copy and complete the table.

Polygon	Area (A)	Interior Points (I)	Perimeter (P)
1			
2			
3			
4			
5			
6			
7			

3. Use your earlier expression for P in terms of A and I to calculate the perimeter of each polygon. Does the expression work for all the polygons?

3 Using an Expression

1. Use your expression for P in terms of A and I to calculate the perimeter of a polygon with an area of 16 square units and with 5 interior points. Check your answer by drawing the polygon and finding the perimeter.

2. Calculate the perimeter of a polygon made up of 50 squares and with 10 interior points.

3. Calculate the perimeter of a polygon made up of 200 squares and with 1 interior point.

5.1 Binary Relations

Halley's comet enters our solar system about every 75 years. It will return next in the year 2061.

A comet's tail always points away from the sun. There is a relationship between the length of the tail and the comet's distance from the sun. The closer the comet is to the sun, the longer the tail is.

Much of mathematics is a study of how things relate to each other.

Explore: Look for a Pattern

Each large square is made with red border squares and white interior squares.

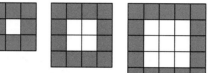

Copy and complete the table. Assume that each small square has a side length of 1 unit.

Side Length of Large Square (s)	Number of Red Squares (n)	Ordered Pair (s, n)
3	8	(3, 8)
4	12	(4, 12)
5	16	(5, 16)
6		
7		

Inquire

1. What is the greatest common factor of the numbers of red squares in the second column of the table? 4

2. Use the number you found in question 1 to write an expression for the number of red squares in terms of the side length, s.

$n = 4(n)$

3. Describe the relation in words.

4. How many red squares are there if the large square has a side length of 14? 40?

5. What does (8, 28) mean?

6. Does (36, 11) have any meaning for this pattern? Explain.

A **relation** is a set of ordered pairs. They are called ordered pairs because the order of the two values or **elements** is important. The set of the first elements in the ordered pairs is called the **domain** of the relation. The set of the second elements is called the **range** of the relation.

For example, for the relation (1, 2), (2, 4), (3, 6), the domain is {1, 2, 3}, and the range is {2, 4, 6}.

Relations can be described in other ways. Four of the ways of describing the relation (−2, 5), (1, 2), (4, −1) are as follows.

Table of Values

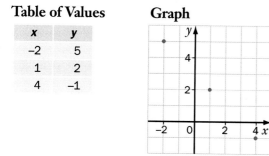

x	y
−2	5
1	2
4	−1

Graph

Words
The sum of two numbers is three.

Equation
$x + y = 3$
or $y = 3 - x$
or $x = 3 - y$

Example   The Gemini Project

The Gemini project involves Canada, Argentina, Brazil, the United Kingdom, Chile, and the United States. These countries cooperated to build two 8-m telescopes. One is on Mauna Kea in Hawaii, the other on Cerro Pachon in Chile. There is a relationship between the amount of money a country invested and the number of nights in a year that the country's astronomers get to use the two telescopes.

Country	Money Invested (millions of U.S. dollars)	Total Nights on the Two Telescopes
Brazil	4.5	15
Argentina	4.5	15
Chile	9	30
Canada	27	90
United Kingdom	45	150
United States	90	300

a) Write the relation as a set of ordered pairs.
b) What is the domain? the range?
c) How many nights on the telescopes does each 1 million dollars provide?
d) Write an equation in the form $N = \rule{1cm}{0.3cm}\, m$ to describe the relation, where N is the number of nights on the telescopes, and m is the number of millions of dollars invested.
e) Describe the relation in words.

Solution
a) (4.5, 15), (9, 30), (27, 90), (45, 150), (90, 300)
b) The domain is {4.5, 9, 27, 45, 90}.
The range is {15, 30, 90, 150, 300}.

c) As 4.5 or $\dfrac{9}{2}$ million dollars provides 15 nights,

1 million dollars provides $\dfrac{15}{\frac{9}{2}}$ or $3\dfrac{1}{3}$ nights.

d) $N = 3\dfrac{1}{3}m$ or $N = \dfrac{10}{3}m$

e) For every million dollars invested, a country's astronomers get three and one third nights on the telescopes.

Another way to describe a relation is with an arrow diagram, which shows how each element in the domain is paired with an element in the range. The arrow diagram for the previous example is as shown.

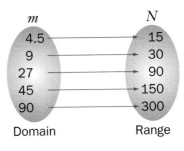

Domain   Range

Practice

For each relation,
a) *write the domain and range*
b) *draw the relation as an arrow diagram*
c) *describe the relation in words*
1. (2, 6), (3, 9), (5, 15), (8, 24)
2. (9, 7), (12, 10), (18, 16), (32, 30)
3. (4, 9), (5, 9), (7, 9), (21, 9)
4. (3, 4), (3, 7), (3, 11), (3, 30)

For each arrow diagram,
a) *write the relation as a set of ordered pairs*
b) *write the domain and range*

5. **6.**
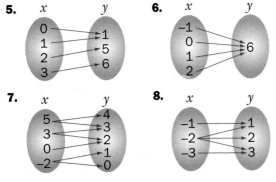

7. **8.**

Draw an arrow diagram for the relation shown in each table of values.

9.

x	y
1	2
2	4
3	6
4	8
5	10

10.

x	y
-2	1
-1	1
0	3
1	3
2	1

Write 5 ordered pairs for each relation.
11. The second number is 4 times the first number.
12. The second number is 2 less than the first number.

For each graph,
a) *express the relation as a set of ordered pairs*
b) *write the domain and range*

13.
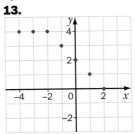

14.
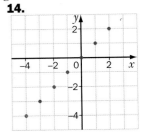

15.
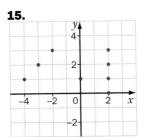

16.
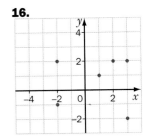

Applications and Problem Solving

17. Pattern Each large square is made with red border squares and blue interior squares.

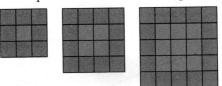

Copy and complete the table.

Side Length of Large Square (s)	Number of Blue Squares (b)
3	1
4	4
5	9
6	16
7	25

a) What is the pattern in the numbers of blue squares?
b) Write the results as a set of ordered pairs of the form (s, b).
c) Write the domain and range of the relation.
d) Show the relation as an arrow diagram.
e) Describe the relation in words.
f) Write an expression for the number of blue squares, b, in terms of the side length, s.
$$b = (s-2)^2$$
g) How many blue squares are there if the side length of the large square is 12? 50?
h) What is the side length of the large square if there are 529 blue squares? 1936 blue squares?
i) Is it possible to have a large square with 408 blue squares? Explain.

18. Mirrors To see a complete reflection of yourself in a mirror fixed to a flat wall, you must use a mirror that is long enough. The table shows the minimum mirror lengths, l, for people of different heights, h.

Height of Person, h (cm)	Length of Mirror, l (cm)
140	70
150	75
160	80
170	85
180	90

a) Write the relation as a set of ordered pairs of the form (h, l).
b) Write the domain and range of the relation.
c) Show the relation as an arrow diagram.
d) Describe the relation in words.
e) Write an equation to describe the relation.

19. Months a) Write a set of ordered pairs for the relation (number of letters in the name of the month, number of days in the month in a leap year).
b) Write the domain and range of the relation.
c) Draw an arrow diagram for the relation.
d) Can this relation be described by an equation? Explain.

20. Baseball The table gives the win-loss records for the American League baseball champions for 6 consecutive years.

Team	Won	Lost
Minnesota	85	77
Oakland	104	58
Oakland	99	63
Oakland	103	59
Minnesota	95	67
Toronto	96	66

a) Write the relation as a set of ordered pairs, (won, lost).
b) Write the domain and range of the relation.
c) Show the relation as an arrow diagram.
d) Describe the relation in words.
e) Let w be the number of wins and l the number of losses. Write an equation to describe the relation.

21. Social Insurance Numbers a) A Social Insurance Number is the first element of a set of ordered pairs that defines a relation. What is the second element of the set of ordered pairs?
b) What is the range of this relation?

22. Geography a) Use an atlas to determine the longitude and latitude of each of Canada's provincial capitals, to the nearest degree.
b) Write your findings as 10 ordered pairs of the form (longitude, latitude).
c) Plot the 10 points on a graph of latitude versus longitude.
d) In what ways is your graph similar to a map of Canada? different from a map of Canada? Explain.

LOGIC POWER

A container made of interlocking cubes has outside dimensions $5 \times 5 \times 5$. The walls and the base of the container are all 1 cube thick. How many cubes were used to make the container?

5.2 Linear Relations and the Line of Best Fit

The *Pathfinder* is an experimental plane. Solar energy powers the propellers during daylight. At night, solar energy stored in the plane's batteries is used. This "eternal plane" can stay up almost forever. It is guided from the ground and will be used for search and rescue and for tracking storms. With the motors off, the plane can glide 22 km for every 1 km of altitude.

Explore: Draw a Graph

Copy and complete the table.

Altitude (km)	Glide Distance (km)
5	110
10	220
15	330
20	440
25	550

Plot the ordered pairs for this relation on a grid like the one shown. Join the points.

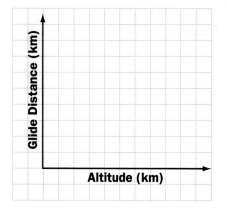

Inquire

1. Why is this relation called a linear relation?
2. Use the graph to find how far the plane glides from an altitude of
a) 18 km   b) 30 km

Using a graph to find values between the ordered pairs is known as **interpolation**.

Using a graph to find values outside the ordered pairs is known as **extrapolation**.

In many cases, data do not produce a straight line when the points are plotted. However, there are times when the points are close to a straight line.

Explore: Draw a Graph

The table gives the number of storeys and heights of 7 buildings in Canadian cities.

Building	Number of Storeys	Height (m)
First Canadian Place, Toronto	72	290
Manulife Place, Edmonton	36	146
Petro-Canada Centre, W Tower, Calgary	52	210
Place de Ville, Tower C, Ottawa	29	112
Royal Centre Tower, Vancouver	36	140
1100 Rue de la Gauchetière, Montreal	45	204
Toronto Dominion Centre, Winnipeg	33	126

Plot the points on a grid like the one shown. The result is called a **scatter plot**. Then, draw a straight line as close as possible to the most points. This line is called the **line of best fit.** There should be about as many points above the line as there are below the line.

Inquire

1. Use the graph to find the approximate number of storeys in the Commerce Court West building, if it is 239 m tall.

2. Scotia Plaza has 68 storeys. About how tall is it?

3. Why is a break mark used on the vertical axis between 0 and 100 m?

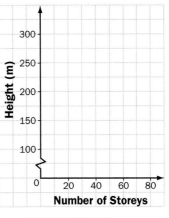

Example 1 Graphing a Linear Relation
Graph $y = 2x - 3$ for each of the following domains.
a) $\{3, 2, 1, 0, -1\}$ **b)** the real numbers, R

Solution 1
a) Complete a table of values.

x	2x – 3	y	(x, y)
3	2(3) – 3	3	(3, 3)
2	2(2) – 3	1	(2, 1)
1	2(1) – 3	–1	(1, –1)
0	2(0) – 3	–3	(0, –3)
–1	2(–1) – 3	–5	(–1, –5)

Plot the points on a grid. Do not join them, because the domain does not include values of x between the given values.

b) Plot the same 5 points determined in part a). Draw a line through the points, because the domain includes all real values of x.

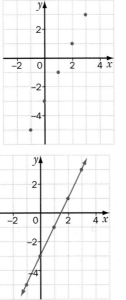

Solution 2

Use a graphing calculator.

a) Enter the data to be plotted. Depending on your model of calculator, you may need to enter the data as ordered pairs or lists. Graph the scatter plot as a set of unconnected points.

b) Without entering the data again, generate a line through the points.

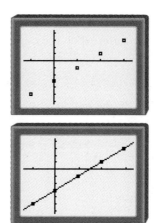

In part a) of Example 1, the graph is a series of separate points. This type of graph is said to be **discrete**. In part b) of Example 1, the graph is an unbroken line. This type of graph is said to be **continuous**.

Example 2 Graphing a Linear Relation

a) Graph $2x + 3y = 12$. The domain is R.

b) Use the graph to find the value of y, to the nearest tenth, when $x = 2.3$.

Solution 1

a) Solve the equation for y.

$$2x + 3y = 12$$
$$3y = 12 - 2x$$
$$y = \frac{12 - 2x}{3}$$

Choose values of x that give convenient values for y. Complete a table of values.

x	y
0	4
3	2
6	0

Plot the points on a grid. Because the domain is R, the graph is continuous. Join the points with a line.

b) When $x = 2.3$, $y = 2.5$, to the nearest tenth.

Solution 2

Use a graphing calculator.

a) Set the viewing window to suitable values. If your calculator has the capability to solve equations, use it to solve $2x + 3y = 12$ for y. Alternatively, solve for y using pencil and paper, as shown in Solution 1, part a). Use your calculator to graph the equation $y = \dfrac{12 - 2x}{3}$, without calculating a table of values.

b) Use the trace feature of your calculator to find the value of y when $x = 2.3$. The value of y is 2.5, to the nearest tenth.

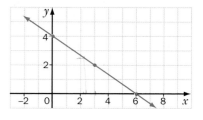

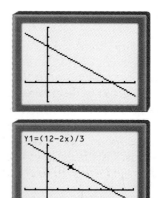

Y1=(12-2x)/3

x=2.3   y=2.4666667

Practice

The domain of each of the following relations is {2, 1, 0, −1, −2}. Complete a table of values. Then, graph each relation.

1. $y = x + 1$ **2.** $y = x - 2$
3. $y = 2x + 1$ **4.** $y = 2x - 1$
5. $y = 3x + 4$ **6.** $y = 3x - 2$

Graph each relation. The domain is R.

7. $y = x + 4$ **8.** $y = 3x + 2$
9. $y = 2x - 4$ **10.** $y = 4 - x$
11. $y = 5 - 3x$ **12.** $y = -2x + 7$

Graph each relation. The domain is R.

13. $x + y = 6$ **14.** $x - y = 1$
15. $x + y = -2$ **16.** $3x + y = 7$
17. $x + 2y = 4$ **18.** $3x + 2y = 6$
19. $x + 3y - 12 = 0$ **20.** $2x - 3y - 12 = 0$

Graph each of the following. The domain is R. For each graph, find the value of x when y = 1.3 and y = −2.1. Round each value to the nearest tenth.

21. $y = 4x - 3$ **22.** $2y = 3x + 1$
23. $2x + 3y = 1$ **24.** $5x - y - 2 = 0$

Applications and Problem Solving

25. Space probe The *Pioneer 10* space probe was launched in 1972 and recorded the first close-up images of Jupiter. *Pioneer 10* was the first space probe to leave our solar system. It now travels through space at 60 000 km/h.

a) Copy and complete the table to show the distance travelled by the space probe in different lengths of time.

Time (min)	Distance (km)
15	
30	
45	
60	

b) Plot distance versus time for this relation. Join the points.
c) The Earth has a circumference of about 40 000 km. Use the graph to determine how long it would take *Pioneer 10* to fly this distance.
d) The greatest east-west distance in Canada is 5514 km from Cape Spear, Newfoundland, to the Yukon-Alaska border. Use the graph to determine how long it would take *Pioneer 10* to cover this distance.

26. Baseball and softball The table gives the horizontal distance a baseball and softball travel through the air when launched at different speeds at an angle of about 40°.

Speed (km/h)	Distance (m) Baseball	Distance (m) Softball
130	88	75
145	104	87
160	119	98
175	133	109

a) Plot distance versus speed for these data.
b) Draw the line of best fit for each type of ball.
c) The average major league pitcher throws a baseball at 135 km/h. Use the graph to estimate how far a pitcher could throw a baseball.
d) A major league baseball player can hit a ball at a speed of 190 km/h. When hit at this speed, what is the distance travelled by a baseball? a softball?
e) Why does a baseball travel farther than a softball when they are both hit or thrown at the same speed?

27. Experimental data Vinegar is a dilute solution of acetic acid in water. Pure acetic acid is a liquid. The table shows experimental data for the mass, to the nearest gram, and volume, to the nearest millilitre, of several samples of pure acetic acid.

Volume of Acid (mL)	Mass of Beaker + Acid (g)	Mass of Acid (g)
0	80	
20	101	
40	122	
60	143	
80	164	
100	185	

a) What is the mass of the beaker?
b) Copy the table. Complete it by finding the mass of each sample of acid.
c) Plot mass of acid versus volume of acid for these data.
d) Use the graph to find the mass of 32 mL of acid.
e) If the pattern continues, what is the mass of the beaker and acid when the volume of acid is 120 mL?
f) Finding the mass, in grams, of 1 mL of acid gives the density of the acid, in grams per millilitre. What is the density of the acid?

28. Olympic high jump Ethel Catherwood of Canada won a gold medal at the 1928 Olympic Games in the high jump. She jumped 1.59 m.

The table gives the winning height of the women's high jump for the first 15 Olympic Games in which the high jump was held.

Year	Height (m)	Year	Height (m)
1928	1.59	1964	1.90
1932	1.66	1968	1.82
1936	1.60	1972	1.92
1940		1976	1.93
1944		1980	1.97
1948	1.68	1984	2.02
1952	1.67	1988	2.03
1956	1.76	1992	2.02
1960	1.85	1996	2.05

a) Plot height versus year for these data.
b) Draw the line of best fit.
c) Why are there no heights for 1940 and 1944? Use your graph to find what the winning heights might have been in 1940 and 1944.
d) Use your graph to find what the winning height might be in the year 2012.
e) Do you think that your answer from part d) is reasonable? Explain.
f) The winning height has generally increased with time. What do you think are the reasons?
g) Do you think that the winning height will ever reach a maximum value? Explain.

29. Running elk An elk can run at 20 m/s. If you drew a graph of distance travelled versus time for an elk running at this speed, would the graph be discrete or continuous? Explain.

30. Hopping kangaroo A kangaroo can cover a horizontal distance of 13 m in a single hop. If you drew a graph of distance travelled versus number of hops, would the graph be discrete or continuous? Explain.

31. Determine the domain and range of each of the following linear relations.

a)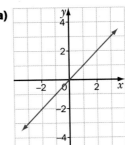

b)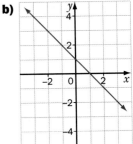

c)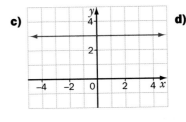

d)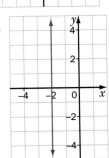

a) Copy and complete the pattern.

$$11^2 = \rule{2cm}{0.4cm}$$
$$101^2 = \rule{2cm}{0.4cm}$$
$$1001^2 = \rule{2cm}{0.4cm}$$

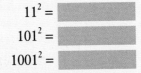

b) Describe the pattern in words.
c) Explain why the pattern works.
d) Write the next 2 lines of the pattern.
e) Use the pattern to find
$\sqrt{100\,000\,020\,000\,001}$.

The Environment

Many people have careers that are related to the environment. Some are scientists studying subjects such as rain forest destruction, acid rain, or the extinction of plant and animal species. Others work in jobs that have a direct environmental impact, including garbage disposal, sewage treatment, or recycling programs.

A team of Canadian scientists, including professors Donna Mergler and Marc Lucotte of the University of Quebec, has done environmental research in Brazil. The team has studied mercury pollution along the Tapajos River, a tributary of the Amazon.

There are two sources of mercury pollution in the Tapajos. The first is gold mining, in which liquid mercury is used to extract gold from sediments dredged from the river. The resulting amalgam is caught in wooden sluices and heated to evaporate the mercury, leaving the gold behind. Half of the mercury ends up in the river. The second source of mercury in the river is deforestation, which causes the release of naturally occurring mercury in the soil.

The Canadian team has found that villagers who eat the fish from the river are beginning to show signs of mercury poisoning.

1 Gold Mining

1. The graph shows the relationship between the mass of gold extracted and the mass of mercury used in Brazil's gold mining industry.

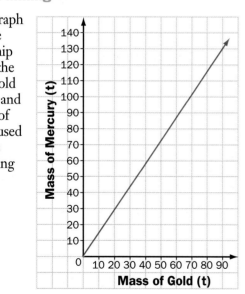

Mass of Gold (t)

a) Brazil produces about 90 t of gold per year. Use the graph to estimate the mass of mercury used per year.
b) What mass of mercury, to the nearest tenth of a tonne, is used in extracting each tonne of gold?

2. By redesigning the sluices, gold miners could boost gold production by 50% and reclaim 95% of the mercury. If the sluices were redesigned,
a) what mass of gold, in tonnes, would Brazil produce per year?
b) what mass of mercury, to the nearest tenth of a tonne, would be used per year?
c) what mass of mercury, to the nearest hundredth of a tonne, would be needed to extract each tonne of gold?

3. Use your results from question 2, parts a) and b), to plot a graph of mass of mercury versus mass of gold for the redesigned sluices.

2 Locating Information

1. Where would you look to find information about an environmental career or an environmental issue that interests you?

2. Find information about mercury contamination of fish in Canada's North. What causes the mercury pollution? Which people are most affected by it?

3. Mercury poisoning is sometimes called Minamata disease. Find out why, and describe the symptoms of mercury poisoning.

TECHNOLOGY

Scatter Plots and Computer Spreadsheets

Computer spreadsheets can be used to draw scatter plots.

The process involves the following general steps.
• Enter the data.
• Select the cells that contain data to be displayed on the scatter plot.
• Create the scatter plot.
• If necessary, draw the line of best fit. If the spreadsheet program does not have the capability to draw the line, print the scatter plot and draw the line with a pencil.

1 Members of Parliament

The table shows the approximate populations of 6 provinces at the time of a federal election and the number of Members of Parliament (MPs) elected from each province.

Province	Approximate Population (millions)	Number of MPs
Alberta	2.7	26
British Columbia	3.6	34
New Brunswick	0.7	10
Nova Scotia	0.9	11
Ontario	10.6	103
Quebec	7.1	75

1. Did each province elect its fair share of MPs? Explain.

2. a) Use a computer spreadsheet to draw a scatter plot of the number of MPs versus population. Draw the line of best fit.

b) Write a set of instructions another student could follow to complete part a).

3. Use the graph to copy and complete the following table.

Province	Approximate Population (millions)	Number of MPs
Manitoba	1.1	
Newfoundland	0.6	
PEI		4
Saskatchewan		14

4. The population of Saskatchewan at the time of the election was 1 million. The number of MPs elected from Manitoba was 14. Explain why your answers from question 3 did not agree exactly with these values.

5. a) Repeat questions 2 and 3 using a graphing calculator instead of a computer spreadsheet.
b) When creating scatter plots, what are the advantages and disadvantages of using a computer spreadsheet? a graphing calculator?

2 Olympic Winning Times

The table gives the winning times in the women's 100-m final at 7 Summer Olympics.

Year	Winning Time (s)
1928	12.2
1936	11.5
1952	11.5
1960	11.0
1968	11.0
1976	11.08
1984	10.97

1. Use a computer spreadsheet to draw a scatter plot of winning time versus year. Draw the line of best fit.

2. Use the graph to predict the winning time in 1980. Use your research skills to check the accuracy of your prediction.

3. Use the graph to predict the winning time in the year 2012.

4. Use the graph to predict what the winning time would have been in 1904. Why is it impossible to check the accuracy of this prediction?

3 Selecting Data

1. Locate data on the winning distances for different years in an Olympic throwing event, such as the javelin or discus.

2. Use the data to write questions that require predictions from a scatter plot.

3. Have a classmate answer your questions using a computer spreadsheet.

5.3 Non-Linear Relations

Non-linear relations are relations whose graphs are not straight lines.

Explore: Study the Graph

The Leaning Tower of Pisa was built in the twelfth century. The graph shows the relationship between the angle of tilt of the tower and the year.

Inquire

1. What was the tilt angle in 1700?

2. In what year was the tilt angle about 3°?

3. During what time period did the tilt angle increase at the fastest rate?

 4. Does the tower appear to have stopped leaning further? Explain.

5. Write a reasonable domain and range for this relation.

6. Do you think Pisans want the tower pushed upright?

7. If the tower suddenly fell over tomorrow, how would you show this change on the graph?

When drawing the graphs of some relations you will need to draw a curve through the points. This curve is known as the **curve of best fit.**

Example 1   Speeds of Runners

The table shows the average speeds of runners for races of different lengths.

Length of Race (m)	Speed (km/h)
100	36
200	37
400	33
800	28
1500	25

a) Plot speed versus length of race on a grid, and draw a smooth curve through the points.

b) Use the graph to find the speed for a 1000-m race.

c) Use the graph to find the speed for a 2000-m race.

d) Why is the shortest race not run at the fastest speed?

e) Write a reasonable domain and range for this relation.

Solution

a)

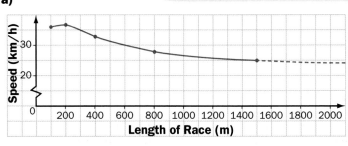

b) About 27 km/h.

c) Extending the graph to the right gives a speed of about 24 km/h for a 2000-m race.

d) In a 200-m race, the runners are at top speed for a greater distance than the runners in a 100-m race.

e) The domain is $100 \le d \le 1500$, where d is the length of the race in metres.

The range is $25 \le s \le 37$, where s is the speed in kilometres per hour.

Example 2   Graphing an Equation

a) Graph the relation $y = x^2 + 3$. The domain is R.

b) Find the range.

Solution 1

a) Complete a table of values.

Plot the points and join them with a smooth curve.

x	$x^2 + 3$	y	(x, y)
0	$0^2 + 3$	3	(0, 3)
1	$1^2 + 3$	4	(1, 4)
−1	$(−1)^2 + 3$	4	(−1, 4)
2	$2^2 + 3$	7	(2, 7)
−2	$(−2)^2 + 3$	7	(−2, 7)
3	$3^2 + 3$	12	(3, 12)
−3	$(−3)^2 + 3$	12	(−3, 12)

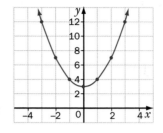

b) The variable y can have all real values greater than or equal to 3. So, the range is $y \geq 3$.

Solution 2

Use a graphing calculator.

a) Set the viewing window to suitable values. Then, use your calculator to graph the equation $y = x^2 + 3$, without calculating a table of values.

b) The range is $y \geq 3$.

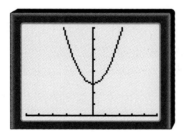

Practice

Graph each relation using a graphing calculator, or complete the table of values for each relation and draw its graph on a grid. The domain is R. Find the range.

1. $y = x^2 + 1$

x	y
0	1
1	2
−1	2
2	5
−2	5
3	10
−3	10

2. $y = x^2 − 4$

x	y
0	−4
1	−3
−1	−3
2	0
−2	0
3	5
−3	5

3. $y = −x^2 + 1$

x	y
0	1
1	0
−1	0
2	−3
−2	−3
3	−8
−3	−8

4. $y = −x^2 − 4$

x	y
0	−4
1	−5
−1	−5
2	−8
−2	−8
3	−13
−3	−13

225

Applications and Problem Solving

5. a) Measurement Copy and complete the table for squares of the given side lengths.

Side Length	Area	Perimeter
0	0	0
1	1	4
2	4	8
3	9	12
4	16	16
5	25	20

b) Plot the points on a grid and draw a smooth curve through the points.

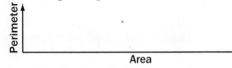

c) Use the graph to find the area of a square with a perimeter of 14 units.
d) Use the graph to find the perimeter of a square with an area of 20 square units.
e) Write the domain and range of this relation.
f) Write an expression for the area, A, in terms of the perimeter in the form $A = \dfrac{P^2}{6}$.

6. Stopping distance The stopping distance of a car is directly related to the car's speed. The table gives the minimum stopping distance on dry, level concrete. This distance includes the reaction time the driver takes to apply the brakes.

Speed (km/h)	Stopping Distance (m)
0	0
20	10
40	18
60	32
80	55
100	85

a) Plot the points on a grid and draw a smooth curve through the points.

Speed (km/h)

b) From the graph, what is the stopping distance at a speed of 50 km/h? 70 km/h?
c) A soccer field is 100 m long. How fast would a car be travelling if it took the length of the soccer field to stop?

7. a) On the same set of axes, sketch the graphs of $y = x^2$, $y = x^2 + 1$, $y = x^2 + 4$, $y = x^2 - 1$, and $y = x^2 - 4$ for x-values of 0, 1, −1, 2, −2, 3, −3. Alternatively, graph these relations in the same viewing window of a graphing calculator.
b) Describe how the graphs are the same and how they are different.

8. Pizza prices Rosina's Pizza Place uses the following price structure for a basic cheese and sauce pizza.

Size	Diameter (cm)	Cost ($)
Personal	20	5.00
Small	25	7.80
Medium	30	11.25
Large	35	15.30
Party	40	20.00

a) Plot a graph of cost versus diameter.
b) Calculate the area of each size of pizza, to the nearest square centimetre.
c) Plot a graph of cost versus area.
d) Compare the graphs from parts a) and c).
e) Is the pricing structure sensible? Explain.

9. Falling objects
By dropping objects from the Leaning Tower of Pisa, Galileo showed that different objects fall at the same rate. The table gives the total distance travelled by a baseball dropped from a tower after different lengths of time.

Time (s)	Distance (m)
0	0
1	4.9
2	19.6
3	44.1
4	78.4

a) Plot the points on a grid and draw a smooth curve through the points.

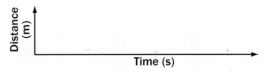

Time (s)

b) Acapulco cliff divers dive from a height of 36 m. How long does it take them to hit the water?
c) Using the data in the table, write an equation of the form $d = \boxed{4.9}\, t^2$, where ▓ represents a number.

d) In the equation from part c), replace d with y and replace t with x.

e) Graph the equation from part d) using positive and negative values of x.

f) Compare the graphs from parts a) and e). How are they the same? How are they different?

g) Explain the differences in the two graphs.

10. Graph each relation using a graphing calculator, or complete the table of values for each relation and draw its graph. If the domain is not stated, assume that it is R. Find the range.

a) $y = \sqrt{x}$

$x \geq 0$

x	y
0	0
1	1
4	2
9	3
16	4

b) $y = \dfrac{4}{x^2 + 1}$

x	y
0	4
1	2
-1	2
2	0.8
-2	0.8
3	0.4
-3	0.4

c) $y = 2^x$

x	y
0	1
1	2
2	4
3	8
4	16

11. One ordered pair that satisfies the relation $x^2 + y^2 = 25$ is $(3, 4)$. Substituting 3 for x and 4 for y gives $(3)^2 + (4)^2 = 9 + 16$

$$= 25$$

Another ordered pair is $(3, -4)$ because

$(3)^2 + (-4)^2 = 9 + 16$

$$= 25$$

a) Copy and complete the table of values for the relation $x^2 + y^2 = 25$.

x	y		x	y
3	4		-4	
3	-4		-4	
4			5	
4			-5	
-3			0	
-3			0	

There are two → values for y. →

b) Plot the points on a grid and join them with a smooth curve.

c) Use the same reasoning to draw a graph of $x^2 + y^2 = 100$.

d) Use your research skills to find out how to graph the relation $x^2 + y^2 = 25$ using a graphing calculator. Describe the steps required.

e) Explain why each of the steps you described in part d) is necessary.

12. Graph each relation using a graphing calculator, or complete the table of values and draw the graph. The domain is R, except where stated otherwise.

a) $y = |x|$

x	y
0	0
1	1
-1	1
2	2
-2	2
3	3
-3	3

b) $y = |x| + 4$

x	y
0	4
1	5
-1	5
2	6
-2	6
3	7
-3	7

c) $y = \dfrac{6}{x}$

$x \neq 0$

x	y
1	6
-1	-6
2	3
-2	-3
3	2
-3	-2
6	1
-6	-1

13. State the domain and range of the relation shown in each graph.

a)

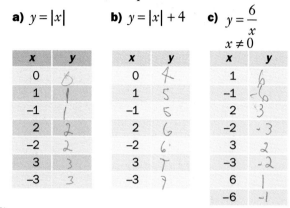

b)

c)

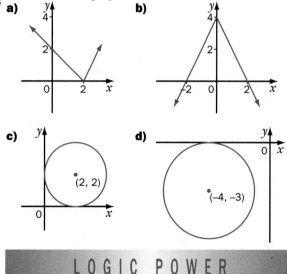

(2, 2)

d)

(-4, -3)

LOGIC POWER

Put a number in each blank so that the statement is true.

"In this sentence, the number of occurrences

of 1 is ___ , of 2 is ___ , of 3 is ___ ,

of 4 is ___ , and of 5 is ___ ."

5.4 General Relations

The 110-m hurdles race is one of the most exciting events in track. Mark McKoy was the second Canadian to win an Olympic gold medal in this event. The first was Earl Thomson who won gold in 1920.

Explore: Study the Graph

The graph shows an imaginary race between Mark McKoy and Earl Thomson, using their winning times at the Olympics.

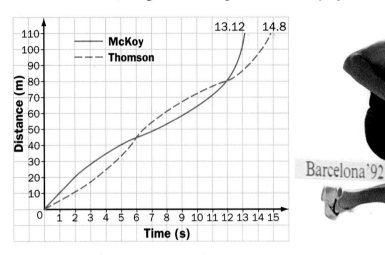

Inquire

1. Who took the lead at the start of the race?

2. After how many seconds was the leader passed?

3. At what distance from the finish did the winner take the lead?

4. Who won the race? What was his time?

5. What was the loser's time?

6. Draw a graph of an imaginary 110-m hurdles race between 3 runners. Have the lead change 3 times. Have one of the runners fall, then get up and complete the race. Compare your graph with a classmate's.

Example   Height of Water in a Spa

The diagram shows the side view of a personal spa. Your job as the spa tester is to run water into the spa up to the red mark, get into the spa, test the jets, get out of the spa, and then drain the spa. Sketch a graph of the height of the water in the deep end of the spa versus time from when you start to fill the spa to when it is empty.

Solution

From A to B, the height increases quickly as water fills only the deep end.

From B to C, the height increases more slowly as water fills the deep and shallow ends.

From C to D, the height increases rapidly when you get in.

From D to E, the height stays constant while you stay in.

From E to F, the height decreases rapidly when you get out.

From F to G, the height decreases slowly as you drain the top part of the spa.

From G to H, the height decreases rapidly as you drain the deep end.

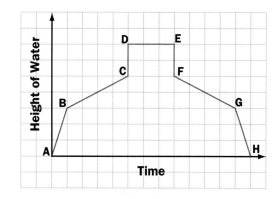

Applications and Problem Solving

1. Water-skiing The graph shows the distance a water-skier is from a dock versus time.

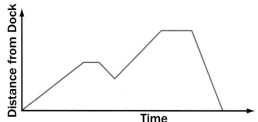

a) Describe what the water-skier did.
b) Draw a diagram to show the skier's possible path on the lake.

2. Bicycle trips State which of the following graphs can represent a trip taken by a student on a bicycle and which ones cannot. Give reasons for your decisions.

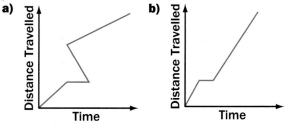

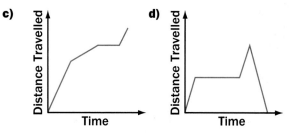

3. Visualization Each graph shows distance from school versus time. Describe a situation that could be represented by each graph.

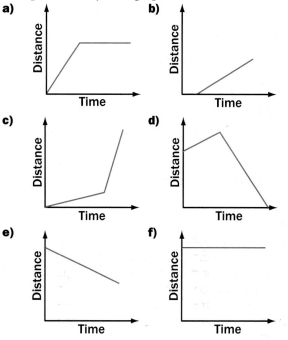

4. Whale watching The graph shows the distance from port versus time for three boats taking people to watch whales. The boats stop when whales are sighted.

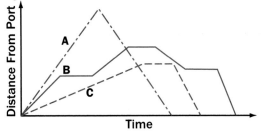

a) Describe what each boat did.
b) Which boat spent the most time stopped?
c) From which boat were no whales sighted?

5. Cross-country skiing The picture shows the elevation of a cross-country ski trail.

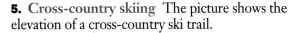

Sketch a graph of the possible speed of a skier versus the distance travelled along the trail.

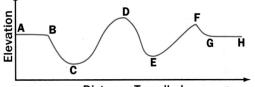

6. Montreal metro In the Montreal metro, you can ride a subway train anywhere on the system for the cost of one ticket. All tickets are the same price.
a) Sketch a graph of the cost versus the distance travelled for one trip on the metro.
b) Sketch a graph of the cost versus the number of metro trips. Is the graph discrete or continuous?

7. Measurement Sketch each of the following graphs.
a) perimeter of a square versus its side length
b) area of a circle versus its radius
c) volume of a cube versus the length of an edge

8. Shopping trip Josette leaves school and rides her bike 2 km to a pet shop to buy food for her dog. She passes her house on her way to the store. After buying the food, she rides 1 km to her house.

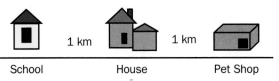

a) Sketch a graph of the distance Josette rides versus time, starting when she leaves school.

b) Sketch a graph of the distance Josette is from her house versus time, starting when she leaves school.
c) Sketch a graph of the distance Josette is from school versus time, starting when she leaves school.
d) Sketch a graph of the distance Josette is from the pet shop versus time, starting when she leaves school.

9. Go-kart track A test track for new go-karts is shown. Drivers drive at top speed on the straight parts and slow down for the corners.

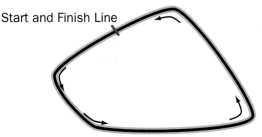

a) Sketch a graph of speed versus time from the start to the end of the first lap.

b) Sketch a graph of speed versus time from the start of the second lap to the end of the third lap.
c) For the second lap, sketch a graph of distance from the start line versus the time.

230

10. Tourist site An observatory overlooks a waterfall. A path leads from the observatory to a restaurant 500 m from the observatory. The path passes the waterfall 200 m from the observatory.

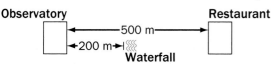

The graph shows the actions of 4 tourists during a 12-min period, starting at 09:00.

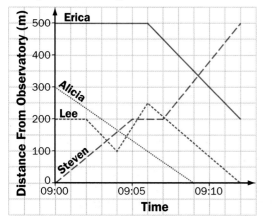

Describe what the 4 tourists did between 09:00 and 09:12. Include:
• where they were at 09:00
• the directions and distances they walked between 09:00 and 09:12
• the people each of them met or passed on the path during the 12-min period

11. Visualization Sketch each of the following graphs.
a) the temperature of boiling water versus the time for which it boils
b) the average daily temperature in your town versus the date for a year beginning on April 1
c) the height of the average person versus the person's age

12. Visualization Describe a practical situation that could be represented by each graph.

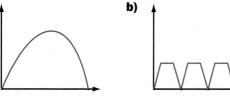

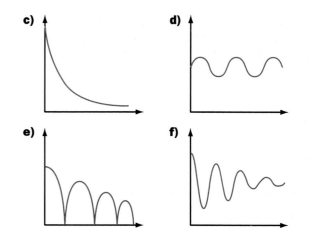

13. Visualization Sketch a graph similar to the one in question 4. Have a classmate describe the scene.

14. Swimming race Canadian swimmer Anne Ottenbrite won an Olympic gold medal in the 200-m breaststroke in a time of 2 min 30.38 s. Another Canadian, Lisa Flood, won a Commonwealth Games bronze medal in the 200-m breaststroke in 2 min 31.85 s.
a) Draw a graph of distance swum versus time for an imaginary race between these two swimmers. Use the times given and remember that, in a 200-m race, the swimmers turn after 50 m, 100 m, and 150 m. Have a classmate interpret your graph to describe the race.
b) For this imaginary race, how would the graph be different if you plotted distance from the start versus time, instead of distance swum versus time?

NUMBER POWER

It has been estimated that the material used to build the three large pyramids at Giza in Egypt could be used to construct a wall 3 m high and 0.3 m thick around the whole of France. Use estimation to suggest possible dimensions of a wall that could be built around your province with this amount of material.

5.5 Functions

A relation is a set of ordered pairs, (x, y).
The following are two relations.
Relation A: (2, 3), (4, 5), (6, 7), (8, 9)
Relation B: (6, 2), (6, 4), (8, 6), (10, 8)

A function is a special relation. A **function** is a set of
ordered pairs in which, for every x, there is only one y.

Relation A, above, is a function because, for each x,
there is only one y. Relation B is not a function because,
when x is 6, there are two values of y, namely 2 and 4.

Another way to define a function is that, for each element in the domain
(first elements), there is exactly one element in the range (second elements).

Explore: Complete the Table

The blue whale is the largest animal that ever roamed our planet.
A mature blue whale has a mass of about 150 t and is about 30 m long.
The blue whale has a cruising
speed of 20 km/h. Copy and
complete the table for the
distance travelled in different
lengths of time by a blue
whale at cruising speed.

Time (h)	Distance (km)
2	40
3	60
4	80
7	140
8	200

Inquire

1. Write the relation as a set of ordered pairs,
(time, distance).

2. For a blue whale at cruising speed, can there be
two different distances travelled in the same length of time? Explain.

3. Is the relation shown in the table a function? Explain.

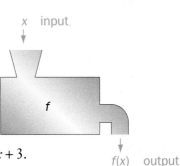

The ordered pairs (x, y) that satisfy the equation $y = 2x + 3$ form a function.
An equation that is a function can be named using function notation.

x-y notation	function notation
$y = 2x + 3$	$f(x) = 2x + 3$

In function notation, f names a function. Observe that the
symbol $f(x)$ is another name for y. The symbol $f(x)$ is the value
of the function f at x. Read $f(x)$ as "the value of f at x" or "f of x."

A function is like a machine. When an x value in the domain
of the function f enters, the machine produces the output $f(x)$.
The output $f(x)$ is determined by the rule of the function.

To find $f(5)$ for the function $f(x) = 2x + 3$, substitute 5 for x in $f(x) = 2x + 3$.
When $x = 5$, the value of y, or $f(5)$, is 13, because $2 \times 5 + 3 = 13$.

Example 1   Evaluating a Function

If $f(x) = 3x + 2$, find

a) $f(5)$ **b)** $f(-1)$ **c)** $f(0)$

Solution

a) $f(x) = 3x + 2$
$\quad f(5) = 3(5) + 2$
$\qquad\; = 15 + 2$
$\qquad\; = 17$

b) $f(x) = 3x + 2$
$\quad f(-1) = 3(-1) + 2$
$\qquad\;\; = -3 + 2$
$\qquad\;\; = -1$

c) $f(x) = 3x + 2$
$\quad f(0) = 3(0) + 2$
$\qquad\; = 0 + 2$
$\qquad\; = 2$

Example 2   Speed of a Blue Whale

A blue whale has a fleeing speed of about 9 m/s. The equation for
the distance travelled, d metres, in a time of t seconds is $d = 9t$, or $f(t) = 9t$.

a) Find $f(11)$.

b) Can any human run this distance in 11 s?

Solution

a) $f(t) = 9t$
$\quad f(11) = 9(11)$
$\qquad\;\; = 99$

b) Yes. The fastest humans can run 100 m in under 10 s.

In Example 2, since the distance, d, depends on the time, t,
we call d the **dependent variable** and t the **independent variable.**

When graphing a function, it is customary to plot the dependent variable
versus the independent variable. For the function in Example 2, the graph is
as shown. The distance travelled is plotted as a function of the time.

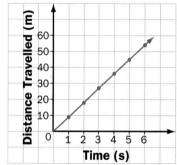

It is sometimes helpful to show a function as an arrow diagram.

Example 3   A Function as an Arrow Diagram

The function $f(x) = x^2$ has a domain $\{-1, 0, 1, 2\}$.

a) Show the function as an arrow diagram.   **b)** Find the range of f.

Solution

a) Complete a table of values for $f(x) = x^2$. Then, use the values to
draw the arrow diagram.

x	x^2	f(x)
-1	$(-1)^2$	1
0	$(0)^2$	0
1	$(1)^2$	1
2	$(2)^2$	4

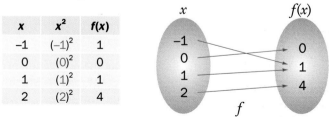

b) The range is $\{0, 1, 4\}$.

Practice

State whether each set of ordered pairs represents a function.

1. (2, 3), (3, 4), (4, 5), (5, 6), (6, 7) yes
2. (5, −3), (6, −4), (7, −5), (8, −6) yes
3. (4, 2), (5, 3), (4, 6), (6, 7) no
4. (6, 3), (7, 3), (8, 3) yes
5. (7, −1), (8, 0), (9, 1) yes
6. (5, 4), (5, 5), (5, 6), (5, 7) no

7. If $f(x) = 4x - 5$, find
a) $f(2)$ **b)** $f(7)$ **c)** $f(0)$
d) $f(-2)$ **e)** $f(-4)$ **f)** $f(100)$
g) $f(0.5)$ **h)** $f(-0.5)$ **i)** $f(1000)$

8. If $f(x) = 8 - 2x$, find
a) $f(1)$ **b)** $f(3)$ **c)** $f(5)$
d) $f(-3)$ **e)** $f(0)$ **f)** $f(10)$
g) $f(0.5)$ **h)** $f(-0.1)$ **i)** $f(4)$

9. If $g(x) = x^2 + 5$, find
a) $g(2)$ **b)** $g(0)$ **c)** $g(-2)$
d) $g(10)$ **e)** $g(-10)$ **f)** $g(0.5)$
g) $g(-0.1)$ **h)** $g(\sqrt{2})$ **i)** $g(\sqrt{6})$

10. If $f(n) = 2n^2 - 4n + 3$, find
a) $f(0)$ **b)** $f(1)$ **c)** $f(-2)$
d) $f(10)$ **e)** $f(-10)$ **f)** $f(0.5)$
g) $f(1.5)$ **h)** $f(\sqrt{5})$ **i)** $f(\sqrt{2})$

State the domain and range of each relation. State whether each is a function.

11.

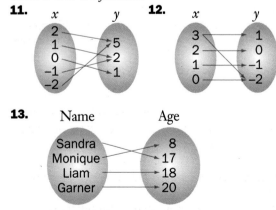

12.

13.

14. The function $f(x) = x^2 + 2$ has a domain {−2, −1, 0, 1, 2}.
a) Show the function as an arrow diagram.
b) Find the range of f.

234

Graph each function. The domain is R. Find the range of g.
15. $g(x) = 3x - 4$ **16.** $g(x) = x^2 - 2$

Applications and Problem Solving

17. Mach number An aircraft breaks the sound barrier when it flies at about 1200 km/h. This speed is known as Mach 1. The Mach number is given by the function $f(s) = \dfrac{s}{1200}$, where s is the speed of the aircraft in kilometres per hour.
a) What is the value of $f(2400)$? $f(3000)$?
b) In the function defined by the ordered pairs (speed, Mach number), identify the dependent variable and the independent variable. Explain your reasoning.

18. Measurement The volume of a cube is given by the function $f(x) = x^3$, where x is the length of an edge. Find
a) $f(4)$ **b)** $f(6)$

19. Sequences Find $f(1)$, $f(2)$, $f(3)$, and $f(4)$ to find the first 4 terms of the sequences defined by the following equations. In each case, state whether the sequence is an arithmetic sequence. Explain your reasoning.
a) $f(n) = 3n - 1$ **b)** $f(n) = 5n + 2$ **c)** $f(n) = n^2 - 3$

20. Canadian population a) Use the census data in the table to sketch a graph of the population of Canada as a function of the year.

Year	Population (millions)
1921	8.8
1931	10.4
1941	11.5
1951	14.0
1961	18.2
1971	21.6
1981	24.3
1991	27.3

b) Use the graph to estimate the population in 1958.
c) Use the graph to predict the year in which the population will reach 40 million.

21. Algebra If $f(x) = x^2 - 4$, what value(s) of x give each of the following values of $f(x)$?
a) 0 **b)** 12 **c)** −4 **d)** 7

State which of the following are graphs of functions.
Write the domain and range in each case.

22. **23.**

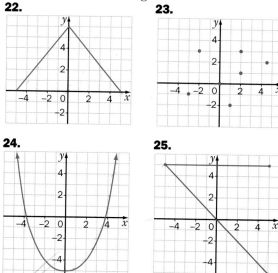

24. **25.**

26. **Vertical line test a)** For which of the graphs from questions 22–25 can you draw one vertical line through more than one point on the graph?
b) Use your findings to write a rule for identifying a relation that is not a function.
c) Make up 2 graphs of functions and 2 graphs of relations that are not functions. Use your graphs to test your rule.

27. **Fingerprints a)** Is the set of ordered pairs (n, f) a function, if n is a person's name and f is the person's fingerprints?
b) Reverse the terms of the ordered pairs so that the set of ordered pairs is (f, n). Is the new set of ordered pairs a function? Explain.

28. **Names of people** Is the set of ordered pairs (f, l) a function, if f is the first name of a person in your school and l is the last name of the person? Explain.

29. **Algebra a)** If $f(x) = 4x + 3$, write and simplify $f(2a)$.
b) If $f(x) = 2 - 3x$, write and simplify $f(n + 1)$.
c) If $g(x) = x^2 + 1$, write and simplify $g(m - 1)$.
d) If $f(x) = 2x^2 - 3$, write and simplify $f(2k + 1)$.
e) If $g(x) = x^2 + 4x - 1$, write and simplify $g(3t - 1)$.
f) If $f(x) = 3x^2 - 2x + 4$, write and simplify $f(3 - 2w)$.

30. If a linear relation is not a function, what can you state about the graph of the relation? Explain.

31. **Gym rental** The cost of renting a gym includes an initial fee, plus an additional fee for each hour or part of an hour. The rates are as shown in the table.

Time (h)	Cost ($)
Up to and including 1	120
Greater than 1; up to and including 2	200
Greater than 2; up to and including 3	280
Greater than 3; up to and including 4	360

The graph of the cost as a function of time is as shown.

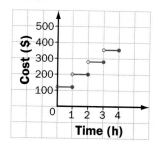

a) Explain why the open and closed dots are used on the graph.
b) The function represented by the graph is known as a **step function**. Explain why.
c) State the domain and range of the function.
d) Write 2 ordered pairs with the same cost value to represent points on the graph.
e) Write 2 ordered pairs with different cost values to represent points on the graph.
f) Is it possible to write 2 different ordered pairs with the same time value to represent points on the graph? Explain.
g) Represent the 4 ordered pairs from parts d) and e) in the form of an arrow diagram.

32. **Visualization** Use your research skills to write a rule used to calculate each of the following. Then, sketch the graph of each function.
a) the cost of mailing a first-class letter to an address in Canada as a function of the mass of the letter
b) the cost of a customer-dialed long-distance phone call from your home to a city of your choice in Canada
c) a taxi fare as a function of the distance travelled

235

5.6 Applications of Linear Functions

Because water expands when it freezes, the density of ice is less than the density of water. Therefore, ice floats on lakes and rivers, and icebergs float in the ocean.

Explore: Draw a Graph

The table shows the volume of ice produced when a given volume of water freezes.

Volume of Water (cm^3)	Volume of Ice (cm^3)
100	109
200	218
300	327
400	436
500	545

a) Graph the volume of ice as a function of the volume of water.
b) Is the function linear or non-linear?

Inquire

1. For each ordered pair in the table, divide the volume of ice by the volume of water. How are the ratios related?

2. Let I be the volume of ice and w be the volume of water. Write an equation of the form $I = \blacksquare w$ to describe the function, where ▒ represents a number.

3. Use the equation to find
a) the volume of ice formed when 800 cm^3 of water freezes
b) the volume of water needed to form 1308 cm^3 of ice

4. Large icebergs can be over 300 km long. Suppose that a large iceberg is 300 km long and 20 km wide. The height of the iceberg is 0.75 km, including the part under the water.
a) Calculate the volume of the iceberg in cubic kilometres. What assumption have you made?
b) Calculate the volume of water produced when the iceberg melts, to the nearest cubic kilometre.

The table shows the distances travelled in different lengths of time by an albatross gliding at a constant speed of 10 m/s. If the time is doubled, the distance is doubled. If the time is tripled, the distance is tripled. If the time is halved, the distance is halved. We say that the distance *varies directly* as the time. This type of function is known as a **direct variation**.

Time (s)	Distance (m)
1	10
2	20
3	30
4	40
5	50

In a direct variation, the ratio of corresponding values of the variables is a constant. In the case of the cruising albatross,

$$\frac{10}{1} = 10 \qquad \frac{20}{2} = 10 \qquad \frac{30}{3} = 10 \qquad \frac{40}{4} = 10 \qquad \frac{50}{5} = 10$$

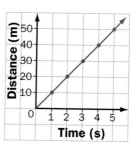

So, $\frac{d}{t} = k$, or $d = kt$, where k is the **constant of variation**.

The graph of distance as a function of time is shown.

In general, if y varies directly as x,
- $y \propto x$ or $y = kx$
- the graph is a straight line that passes through the origin

The symbol $\propto$ means "varies as."

Example 1   Hourly Rates of Pay
The amount that Olga earns varies directly as the number of hours she works. If she earns \$132 in 8 h, how much does she earn in 15 h?

Solution 1

$\frac{e}{h} = k$, where e represents earnings and h represents hours.

Find the constant, k.
$$\frac{132}{8} = k$$
$$16.5 = k$$

So, $\frac{e}{h} = 16.5$ or $e = 16.5h$

When $h = 15$,
$e = 16.5(15)$
$\quad = 247.5$

Estimate
$20 \times 15 = 300$

Solution 2
Solve a proportion.
$$\frac{132}{8} = \frac{x}{15}$$
$$15 \times \frac{132}{8} = 15 \times \frac{x}{15}$$
$$\frac{15 \times 132}{8} = x$$
$$247.5 = x$$

A proportion is a statement that two ratios are equal.

Estimate
$\frac{15 \times 132}{8} \doteq 2 \times 130$
$\doteq 260$

Olga earns \$247.50 in 15 h.

Note that, in Solution 2, above, both sides of the proportion equal the constant of variation.
$$\frac{132}{8} = 16.5 \qquad \frac{247.5}{15} = 16.5$$

For this reason, the constant of variation is sometimes called the *constant of proportionality*. Olga's earnings can be described as *directly proportional* to the hours she works.

The table shows the cost of a banquet at the Perogy Place. As the number of people increases, the cost increases. However, as the number of people is doubled, the cost is not doubled. So, this function is not an example of a direct variation.

Number of People	Cost ($)
10	700
20	1200
30	1700
40	2200
50	2700

The graph of the function is a straight line. However, extending the graph to the vertical axis shows that there is a fixed cost of $200 for every banquet. This amount covers such expenses as hydro and room rental, which do not change with the number of people at the banquet. We say that the cost *varies partially* as the number of people. This type of function is known as a **partial variation**.

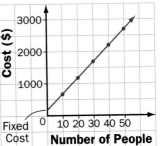

An equation for the function is

$$C = 50n + 200$$

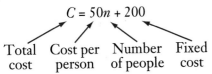

Total cost   Cost per person   Number of people   Fixed cost

In general, if y varies partially as x,
• the equation is of the form $y = mx + b$, $b \neq 0$
• the graph is a straight line that does not pass through the origin

Example 2 Base Salary Plus Commission

Ron sells clothing and shoes in a store. He earns $400/week, plus a commission of 5% of his sales.
a) Write the partial variation equation.
b) Calculate Ron's earnings in a week when his sales total $4600.

Solution

a) $E = 0.05S + 400$, where E represents earnings and S represents sales.

b) $E = 0.05S + 400$
$= 0.05(4600) + 400$
$= 230 + 400$
$= 630$

Estimate
$0.05 \times 5000 = 250$
$250 + 400 = 650$

Ron's earnings are $630.

Practice

Given that y varies directly as x, copy and complete each table of values. Then, graph each function.

1.

x	y
2	6
3	9
4	12
5	15
6	18

2.

x	y
3	-6
4	-8
5	-10
6	-12
7	-14

3.

x	y
1	5
2	10
3	15
4	20
5	25

Given that the first variable varies directly as the second, find
a) *the constant of variation*
b) *the equation that relates the variables*
4. when $p = 24$, $q = 8$
5. when $a = 10$, $b = 4$
6. when $s = 12$, $t = 20$
7. when $m = 11$, $n = 33$

8. If y varies directly as x, and $y = 24$ when $x = 3$, find y when $x = 7$. $y = 8 \; (7) \quad y = kx \quad 24 = k3$

9. If c varies directly as d, and $c = 12$ when $d = -3$, find c when $d = -10$.

10. If v varies directly as w, and $v = 4$ when $w = 8$, find w when $v = 16$.

11. If r varies directly as s, and $r = 39$ when $s = 26$, find s when $r = 21$.

12. a) Graph the function $y = 2x$.
b) Is this function a direct variation or a partial variation?

13. a) Graph the function $y = 2x + 1$.
b) Is this function a direct variation or a partial variation?

14. In the graph, y varies partially as x. State the equation that represents the function.

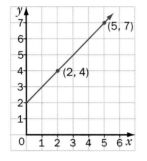

Applications and Problem Solving

15. Sequences Find $f(1), f(2), f(3), f(4)$, and $f(5)$ for each of the following arithmetic sequences. State whether $f(n)$ varies directly or partially with n for each sequence.
a) $f(n) = 3n$ **b)** $f(n) = 2n - 1$
c) $f(n) = -2n$ **d)** $f(n) = 2 - n$

16. Sequence The E-shapes are made from asterisks.

```
                              * * * * *
                              *
              * * * *         *
              *               * * * *
  * * *       * * *           *
  * *         *               *
  * * *       * * * *         * * * * *
Diagram 1         2               3
```

a) How many asterisks are in the fourth diagram? the fifth diagram?
b) Plot the number of asterisks versus the diagram number for the first 5 diagrams. Is the graph discrete or continuous?
c) Does the graph represent a direct variation or a partial variation?
d) Write an equation of the form $A = \blacksquare$ to represent the function, where A is the number of asterisks in the nth diagram.
e) Use the equation to find the number of asterisks in the 55th diagram.

17. Fitness The amount of energy burned when skipping rope varies directly as the time spent skipping. Paula burned 200 kJ of energy by skipping rope for 5 min.
a) Find the constant of variation and write the equation.
b) How much energy does Paula burn if she skips rope for 3.5 min?
c) For how long would Paula skip rope to burn 320 kJ of energy?

18. Bouncing ball The height that a ball bounces varies directly as the height from which it is dropped. A ball dropped from a height of 20 cm bounces 15 cm.
a) Find the constant of variation.
b) Find the height of the bounce if the ball is dropped from a height of 240 cm.
c) If the height of the bounce is 105 cm, from what height was the ball dropped?

19. Pizza toppings A large tomato sauce and cheese pizza with one extra topping costs $13.50. With two extra toppings, the cost is $15.00. With three extra toppings, the cost is $16.50.
a) Graph the cost as a function of the number of extra toppings.
b) What is the cost of a large pizza with no extra toppings?
c) Write the equation for the partial variation.

20. Advertising There is a fixed cost of $500 to write and design an advertising flyer. It costs $0.15 to print a flyer.
a) Write the partial variation equation that relates the total cost to the number of flyers printed.
b) Calculate the total cost of producing 60 000 flyers.
c) How many flyers can be produced for a total cost of $12 500?

21. Service calls
Rajiv is an electrician. The table shows how much he charges for service calls.

Cost ($)	Time (h)
110	1
160	2
210	3
260	4

a) Graph cost as a function of time.
b) What is the fixed cost?
c) Write the equation for the partial variation.
d) How much does Rajiv charge for a 7-h service call?
e) How long is a service call that costs $335?

22. Measurement What is the constant of variation for the direct variation of the circumference of a circle with
a) the diameter?   **b)** the radius?

23. Weight and mass
The weight of an object is the force of gravity acting on the object. Weight is measured in newtons (N). The table shows the weights of objects of known masses on the Earth's surface.

Mass (kg)	Weight (N)
5	49
10	98
15	147
20	196
25	245

a) Graph weight as a function of mass.
b) Is the variation direct or partial?
c) Write the equation.
d) What is the weight of a 75-kg object on the surface of the Earth?
e) What is the mass of an object that weighs 539 N on the surface of the Earth?
f) Could you use the equation from part c) to determine the weight of a 30-kg object on the surface of the moon? Explain.

24. Field trip It costs a total of $550 to take 25 students to an art gallery. The fixed cost of the bus is $300.
a) Write the equation for the partial variation.
b) The total cost is divided equally among the students attending. What is the difference in the cost per student if 15 students attend instead of 25?
c) Explain why there is a difference in part b).
d) How much would it cost to take 15 students to the art gallery?

25. Planning a trip Maylin is saving for a trip to Europe. She puts away the same amount at the end of every month. After 3 months, she still needs to save $3150. After 7 months, she still needs to save $1750.
a) Draw the graph of the amount to be saved versus time.
b) Write the equation for the partial variation.
c) How much does Maylin need altogether for the trip?
d) How long will she take to save this amount?
e) If she decided at the beginning that she needed to save an extra $700 for the trip, how would the graph change?

26. Travel times The table shows the time it takes to travel 100 km at different speeds.

Speed (km/h)	12.5	25	50	100
Time (h)	8	4	2	1

a) As the speed is doubled, what happens to the time?
b) The relationship between speed and time is known as an **inverse variation**. Explain why.
c) Graph time versus speed. Describe the shape of the graph.
d) Take the reciprocal of each speed. Then, graph time versus the reciprocal of the speed. What type of variation does the graph represent?

27. If $a \propto b$ and $b \propto c$, how is a related to c? Explain.

LOGIC POWER

XY and YZ are the diagonals of two faces of a cube.

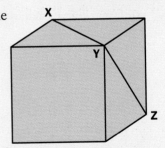

a) What is the measure of $\angle XYZ$?
b) Explain your reasoning.

Summer Olympics

Use the *Olympics* database, from the Computer Databank for ClarisWorks, to complete the following.

1 Timed Events

There are over 250 Summer Olympics events, many of which have taken place at each Summer Olympics since 1896. The 20 Summer Olympics events that are in this database are timed events where winning times are less than two minutes.

1. Find the winning time in 1996 that is over 60.00 s. For which event is it?

2. Find the *Swimming* records, and then sort them from fastest to slowest time. What is the fastest time? For which event and in which year is it? Who is the winning athlete?

3. Repeat question 2 for *Track and Field* records.

2 Canadian Winners

1. Find all the records where Canada is the winning country. How many records are displayed?

2. Canadians have won 45 gold medals at Summer Olympics from 1896 to 1996. What fraction of those are in this database?

3. In which events are Canada's most recent wins? Who are the athletes?

4. In which event and year was Canada's first win? Who was the athlete?

3 World Records

The *World Record, s* field tells what the world record was going into the Olympic event.

1. Find all the records for which *World Record, s* is available.

2. Do winning Olympic times usually beat world records? Explain.

4 Finding Trends

1. Create a table to display only these fields for all records.

Event   Year   Winning Time, s

2. Find all the records for *100-m Backstroke, Men,* and then sort them by year.

3. Copy and paste those records onto a spreadsheet.

4. Create a graph to show winning time as a function of year.

5. Describe any trends you observe.

6. Write a problem about your graph. Have a classmate solve your problem.

7. Repeat questions 2 to 6 for the following.
a) 100-m Freestyle, Women
b) 110-m Hurdles, Men
c) 4×100-m Relay, Women

5 Female and Male Athletes

There are usually separate competitions for female and male athletes. In the past, men's Olympic times have been faster than women's.

1. Work with classmates to compare men's and women's winning times in at least six events. Are women's winning times getting closer to men's winning times? Explain.

2. If the trend continues in each event from question 1, will the women's winning time ever be faster than the men's winning time? If so, estimate the first year in which the women's winning time will be faster.

3. Do you think that the trends found in question 1 will actually continue? Explain.

CONNECTING MATH AND SPORTS

The Mathematics of Punting a Football

1 Punting a Ball Nose First

The graph describes the flight of a football punted nose first at about 100 km/h when there is no wind. The graph shows the relationship between the following three variables.
- the angle at which the ball is kicked
- the distance it travels
- the time it spends in the air

For a ball punted at an angle of 30°, locate 30° on the graph. Read horizontally from 30° to find the distance the ball travels, which is about 55 m. Read vertically to find the time in the air, which is about 2.7 s.

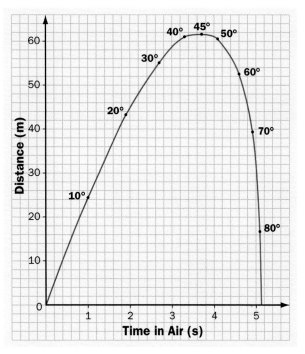

1. What are the distance and the time in the air for a ball punted at an angle of 70°? 10°?

2. If you wanted to punt a ball 30 m, what punt angles could you use? What would be the time in the air for each punt?

3. If the time in the air is 3 s, what are the punt angle and the distance for the punt?

4. What punt angle gives the maximum distance? What is the time in the air for this punt?

5. What punt angle would give the maximum time in the air? What would be the distance for this punt?

2 Punting a Ball End-Over-End

The graph describes the flight of a football punted end-over-end at a speed of about 100 km/h.

1. What are the distance and the time in the air for a ball punted at an angle of 20°? 60°?

2. If the time in the air is 4 s, what are the punt angle and the distance of the punt?

3. What punt angle gives the maximum distance? What is the time in the air for this punt?

4. What punt angle would give the maximum time in the air? What would be the distance of this punt?

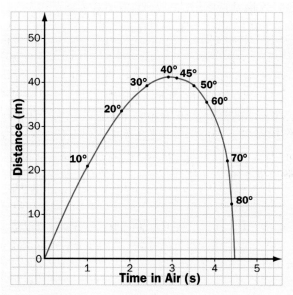

3 Making Comparisons

1. What is the difference in the maximum distances for the punts that are nose first and those that are end-over-end?

2. What is the difference in the maximum times in the air for the punts that are nose first and those that are end-over-end?

3. Why is the difference in the maximum distances so large while the difference in the maximum times in the air is so small?

4. You are the punter for a football team.
a) Do you choose to punt the ball end-over-end or nose first? Explain.
b) At what angle do you try to punt the football? Explain.

Review

5.1 *For each relation,*
   a) *write the domain and range*
   b) *draw the relation as an arrow diagram*
   c) *describe the relation in words*
   1. (3, 6), (4, 5), (7, 2), (9, 0)
   2. (8, 5), (7, 4), (5, 2), (4, 1)
   3. (0, 4), (0, 6), (0, 9), (0, 11)

For each arrow diagram,
   a) *write the relation as a set of ordered pairs*
   b) *write the domain and range*

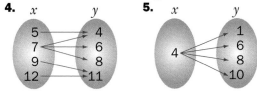

4. x   y   **5.** x   y

For each graph,
   a) *express the relation as a set of ordered pairs*
   b) *write the domain and range*
   c) *describe the relation in words*
   6.   **7.**

5.2 *Graph each relation. The domain is R.*
   8. $y = x + 5$ **9.** $y = 3x - 4$ **10.** $y = -x + 3$

Graph each relation. The domain is R.
   11. $x + y = 5$ **12.** $x - 2y = 4$ **13.** $2x + 3y = 6$

14. Grizzly bear
Over short
distances, a grizzly
bear has a top speed
of about 13 m/s.
a) Copy and
complete the table
of values for a grizzly bear at top speed.

Time (s)	Distance (m)
0	
1	
2	
3	
4	

b) Use the ordered pairs to plot a graph of distance versus time.
c) Use the graph to estimate the distances a grizzly bear can cover in 2.6 s and in 4.8 s.

15. Olympic times The women's 4×100-m relay was held for the first time in the 1928 Olympics. Canada won the gold medal with a time of 48.4 s. The table gives the winning times of the women's 4×100-m relay for 6 Olympic Games.

Year	Winning Time (s)	Year	Winning Time (s)
1928	48.4	1960	44.5
1936	46.9	1968	42.8
1952	45.9	1976	42.55

a) Plot winning time versus year for this relation.
b) Draw the line of best fit for the data.
c) Use your graph to predict what the winning time might have been in 1944.
d) Use your graph to predict the winning times in 1984 and in 1992. How close were your predictions to the actual winning times?

5.3 *Graph each relation using a graphing calculator, or complete the table of values for each relation and draw the graph on a grid. The domain is R.*
   16. $y = x^2 + 2$ **17.** $y = x^2 - 3$

x	y
0	
1	
-1	
2	
-2	
3	
-3	

x	y
0	
1	
-1	
2	
-2	
3	
-3	

5.4 18. Sailboat race A triangular race course for sailboats is shown, with the starting buoy marked S and the other buoys marked A and B.

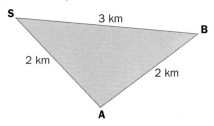

A boat sailed from S to A to B and back to S. Sketch a graph of
a) the boat's distance from S versus time
b) the boat's distance from A versus time
c) the boat's distance from B versus time

State whether each set of ordered pairs represents a function.

19. (2, 5), (4, 3), (6, 1), (8, −1), (9, −2)
20. (3, 2), (5, 6), (6, 8), (3, −2), (6, −4)
21. (2, 3), (2, 2), (2, 1), (2, 0), (2, −1)
22. (8, 1), (7, 1), (−3, 1), (−4, 1)

23. If $f(x) = 2x + 3$, find
a) $f(4)$ **b)** $f(8)$ **c)** $f(0)$
d) $f(−1)$ **e)** $f(−4)$ **f)** $f(0.5)$
g) $f(−0.1)$ **h)** $f(100)$ **i)** $f(5000)$

24. If $f(x) = 9 − 3x$, find
a) $f(2)$ **b)** $f(6)$ **c)** $f(−2)$
d) $f(0)$ **e)** $f(−5)$ **f)** $f(0.1)$
g) $f(−0.2)$ **h)** $f(10)$ **i)** $f(200)$

Graph each function. The domain is R.
Find the range of g.
25. $g(x) = 2x − 4$ **26.** $g(x) = x^2 − 1$

State whether each of the following is a graph of a function. Write the domain and range in each case.
27. **28.**

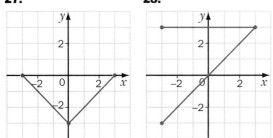

29. If a varies directly as b, and $a = 18$ when $b = 3$, find a when $b = 5$.

30. If x varies directly as y, and $x = −4$ when $y = 8$, find x when $y = 40$.

31. Driving distance Dave is driving from Cranbrook to Dawson Creek at a constant speed. After driving for 2 h, he has 480 km left to drive. After driving for 5 h, he has 240 km left to drive.
a) Draw a graph of remaining distance versus time.
b) How far is it from Cranbrook to Dawson Creek?
c) How long will Dave take to drive the whole distance?
d) Write the equation for the partial variation.

Exploring Math

Scheduling Games

A round robin is a way of scheduling games so that each team plays each of the other teams once.
1. Four basketball teams are entered in a round robin. A table can be used to decide which teams play each other in each round. The 1s in the table show that, in round 1, the Lions play the Bears, and the Eagles play the Tigers.

	Lions	Bears	Eagles	Tigers
Lions		1		
Bears	1			
Eagles				1
Tigers			1	

a) Copy the table and complete it using the numbers 2 and 3 to show the teams who play each other in round 2 and in round 3.
b) Keeping in mind that two teams play in each game, decide how many games will be played.

2. Three soccer teams are entered in a round robin.

	Giants	Sting	Colts
Giants		1	
Sting	1		
Colts			

Because there is an odd number of teams, in each round, one team will get a bye and not play. As the table shows, the Colts get a bye in the first round.
a) Copy and complete the table to show the other rounds of the tournament.
b) How many rounds are needed?
c) How many games will be played?

3. Draw up a schedule of games to be played in a 6-team round robin.

4. a) Copy and complete the table for round robins with different numbers of teams.

Number of Teams	Number of Rounds Needed	Number of Games Played
2		
3		
4		
5		
6		

b) How many rounds would be needed for 12 teams? 19 teams? 50 teams?
c) How many games would be played for 12 teams? 19 teams? 50 teams?

Chapter Check

For each relation,
a) *write the domain and range*
b) *draw the relation as an arrow diagram*
c) *describe the relation in words*
1. (8, 4), (6, 3), (4, 2), (2, 1)
2. (3, 2), (5, 4), (7, 6), (9, 8)
3. (3, 3), (3, 5), (3, 7), (3, 8)

For each arrow diagram,
a) *write the relation as a set of ordered pairs*
b) *write the domain and range*
c) *describe the relation in words*

4. x y **5.** x y

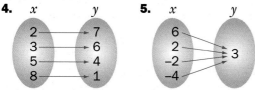

For each graph,
a) *express the relation as a set of ordered pairs*
b) *write the domain and range*
c) *describe the relation in words*

6.   **7.**

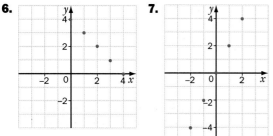

Graph each relation. The domain is R.
8. $y = x + 5$ **9.** $y = 4x - 3$

Graph each relation. The domain is R.
10. $x + y = 7$ **11.** $2x + y = 8$

State whether each set of ordered pairs represents a function.
12. (2, 4), (3, 5), (7, 9), (2, −5), (3, −7)
13. (5, 4), (4, 3), (3, 2), (2, 1), (1, 0)
14. (−1, 6), (0, −6), (1, −6), (2, −6)

15. If $f(x) = 6x - 3$, find
a) $f(5)$ **b)** $f(9)$ **c)** $f(-2)$
d) $f(0)$ **e)** $f(-7)$ **f)** $f(0.5)$
g) $f(-0.5)$ **h)** $f(1000)$ **i)** $f(0.1)$

Graph each relation using a graphing calculator, or complete the table of values for each relation and draw the graph on a grid. The domain is R.
16. $y = x^2 + 5$ **17.** $y = x^2 - 1$

x	y
0	
1	
−1	
2	
−2	
3	
−3	

x	y
0	
1	
−1	
2	
−2	
3	
−3	

Graph each function. The domain is R. Find the range of g.
18. $g(x) = 5x - 4$ **19.** $g(x) = x^2 + 1$

20. a) Graph the function $y = 3x + 2$.
b) Is this function a direct variation or a partial variation?

21. Olympic events The table gives the number of events held at the Summer Olympic games since 1968.

Year	Events	Year	Events
1968	172	1984	223
1972	196	1988	237
1976	199	1992	257
1980	200	1996	271

a) Plot events versus year for this relation.
b) Draw the line of best fit for the data.
c) Use your graph to predict the number of events for the games to be held in 2012.

22. Walking the dog The diagram shows the route Rosa travelled from her home, H, when she took her dog for a walk. Sketch a graph of Rosa's distance from home versus time for the walk.

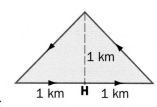

23. Keyboarding The number of words input varies directly as the time spent keyboarding. Janos can input 240 words in 5 min.
a) Find the constant of variation and write the equation.
b) How many words can Janos input in 12 min?
c) How long will it take him to input a 3000-word short story?

Using the Strategies

1. Four students have the first names Tim, Trip, Terry, and Thomas. Their middle names are Bob, Bill, Bevan, and Brooks. Their last names are Sol, Sand, Stone, and Silver. Each student has a different number of letters in each of his names. None of Terry's names has 6 letters. The name Bill does not belong to Tim or Sol. What is the full name of each student?

2. In a **descending number**, each digit is greater than the digit on the right. The number 743 is a descending number. How many descending numbers are there between 500 and 600?

3. A store sells bicycles, tricycles, and wagons. In total, they have 180 wheels and 60 pedals. The numbers of wagons and tricycles are equal. How many bicycles, tricycles, and wagons are in the store?

4. What is the total number of squares in this diagram?

5. What is the last digit of 17^{115}?

6. In how many different ways can you arrange 2 blue blocks and 2 red blocks in a straight line?

7. The mean or average of six numbers is 33. Four identical numbers are added, and the new mean of the ten numbers is 39. What is the value of each new number added?

8. The contestants in a chess tournament were numbered from 1 to 18. When the players were paired for the first game, the sum of the two numbers for each pair was a perfect square. What were the pairings for the first game?

9. What is the next number in the sequence?

21, 22, 23, 42, 25, 62, 27, 82, █

10. A bag contains one cube. The cube is red or green, but you do not know which. A red cube is put into the bag, and the bag is shaken. A cube is then taken from the bag. The cube is red. What is the chance that the cube left in the bag is red?

11. Copy the sentence. Then, fill in the blank with a number, written in words, to make the sentence true.

This sentence has █ letters.

12. A rectangular prism is made up of 1-cm cubes. The dimensions of the prism are 3 cm by 4 cm by 5 cm. If the faces of the prism are painted orange, how many of the cubes have orange paint on 1 face?

13. A restaurant menu includes orders of 6, 9, and 20 cheese sticks. To get 15 sticks, you can order 6 and 9 sticks. You cannot order 16 sticks, because no combination of 6, 9, or 20 adds to 16. What is the largest number of sticks that you *cannot* order?

14. One year, about 45 000 cyclists took part in the 71-km Tour de l'Ile de Montréal. If all the bicycles had been placed end to end, would they have stretched the entire length of the tour? Explain your reasoning and state your assumptions.

15. The diagram shows one way to cut the figure into 4 congruent pieces using straight lines. Find 3 other ways.

D A T A   B A N K

1. If x represents the annual number of hours of sunshine in Edmonton, write 3 different expressions in terms of x that represent the annual number of hours of sunshine in Churchill.

2. a) Write 7 different ordered pairs of the form (relative humidity, air temperature) that give humidex readings of 40°C.
b) Use the ordered pairs to plot a graph of air temperature versus relative humidity. Describe the relationship.
c) Find the air temperature at which the humidex is 40°C when the relative humidity is 50%.
d) Find the relative humidity at which the humidex is 40°C when the air temperature is 34°C.

CHAPTER 6

Coordinate Geometry

The source of the Mississippi River is Lake Itasca in Minnesota. The Mississippi flows to the Gulf of Mexico, entering it near New Orleans, Louisiana.

The source of the St. Lawrence River is Lake Ontario. The St. Lawrence flows to the Gulf of St. Lawrence, meeting it at Anticosti Island.

1. The elevation of Lake Itasca is 446 m. When the Mississippi passes through Minneapolis, the river has an elevation of about 214 m. The length of the river between Lake Itasca and Minneapolis is about 960 km. On average, how many centimetres does the river drop for every kilometre of its length between these two locations?

2. The length of the Mississippi between Minneapolis and the Gulf of Mexico is about 2810 km. On average, how many centimetres does the river drop for every kilometre of its length between these two locations?

3. On average over the whole length of the Mississippi, how many centimetres does the river drop for every kilometre of its length?

4. The elevation of Lake Ontario is 75 m. The length of the St. Lawrence is 3060 km. On average over the whole length of the St. Lawrence, how many centimetres does the river drop for every kilometre of its length?

5. On average over its whole length, which river is steeper — the Mississippi or the St. Lawrence? Explain.

GETTING STARTED

Street Geography and Geometry

1 Taxicab Geometry

In Euclidean geometry, the Pythagorean Theorem is used to calculate the distance between two points. The Euclidean distance, d_E, between the theatre and the arena in the diagram, is $\sqrt{41}$ blocks.

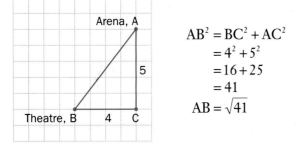

$$AB^2 = BC^2 + AC^2$$
$$= 4^2 + 5^2$$
$$= 16 + 25$$
$$= 41$$
$$AB = \sqrt{41}$$

The taxicab distance, d_T, between the theatre and the arena, is 9 blocks. This is the smallest number of blocks a taxi could drive or a person could walk from A to B.

1. Find d_T and d_E for each of the following pairs of points.
a) A and B
b) C and D
c) E and F
d) G and H
e) A and C
f) B and D

2. When is the taxicab distance between two points equal to the Euclidean distance?

3. a) Mark a point, A, on a sheet of grid paper. Then, mark a set of points so that the taxicab distance of each point from point A is 7 units. If you marked all the points in the set, what figure would you form?
b) If you repeated part a) but you made the Euclidean distance of each point from point A 7 units, what figure would you form?

2 School Boundaries

When a mathematical model is used to represent a real situation, assumptions may be made to simplify the model. In spite of the assumptions, information learned from the model can often be applied to the real situation.

In the following problems, all streets are assumed to be perpendicular and straight, and to have no width. All blocks are assumed to be the same length.

A community is built on a 10 by 10 coordinate grid, as shown. There are three schools, Applewood at (10, 8), Benton at (2, 7), and Crestview at (6, 1). Copy the grid and plot the schools on it.

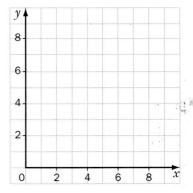

1. A student lives at every corner on the grid. Draw school boundaries to separate the areas served by each school, so that each student's walking distance to school is as short as possible. If a student lives the same distance from two schools, decide which one the student should attend.

2. There are plans to build a swimming pool so that
• students from Applewood have the shortest distance to walk from school to the pool;
• students from Crestview have the longest distance to walk from school to the pool;
• students from Crestview do not walk more than 9 blocks.
How many locations for the pool are possible?

3. A new school, Deerfield, is to be built at (1, 1). Draw the new school boundaries.

3 Neighbourhood Planning

The rectangular street grid for cities and towns has been around for centuries. The grid is easy to understand and can be extended in any direction.

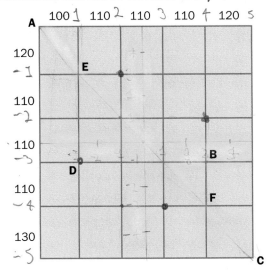

Today, town planners assume that the walking rate is 1.2 m/s. The dimensions on the street grid for the neighbourhood shown above are given in metres.

1. Assuming the streets are of negligible width, how long would it take to walk from A to B? from C to D?   *10.7 min*

2. Suppose that town planners decided to construct a diagonal street from E to F. Using all or part of the diagonal street, what is the shortest time it would take to walk from A to B? from C to D?   *8.9 min   5*

3. Instead of constructing the diagonal street from E to F, the town planners decided to use buses. To meet the needs of the people in the neighbourhood, the following rules were used by planners for bus stops.
• Bus stops must be located at corners.
• A corner without a bus stop must be within a five-minute walking distance of a bus stop.
a) How many bus stops are needed?
b) Where should they be placed?

Mental Math

Order of Operations

Evaluate.

1. $\dfrac{3+9}{2}$ **2.** $\dfrac{6+12}{2}$ **3.** $\dfrac{8+(-2)}{2}$

4. $\dfrac{-3+7}{2}$ **5.** $\dfrac{-1+(-5)}{2}$ **6.** $\dfrac{-6+(-4)}{2}$

Evaluate.

7. $\dfrac{6-2}{3-1}$ **8.** $\dfrac{10-1}{5-2}$ **9.** $\dfrac{5-(-1)}{8-2}$

10. $\dfrac{1-9}{2-4}$ **11.** $\dfrac{5-(-1)}{-3-(-1)}$ **12.** $\dfrac{-3-(-2)}{-2-(-1)}$

Multiplying by Multiples of 11

To multiply a two-digit number by 11, insert the sum of the two digits between the digits.
For 32×11, think $3 + 2 = 5$.
Inserting 5 between the 3 and the 2 gives 352.
So, $32 \times 11 = 352$.

If the sum of the two digits is greater than 9, add 1 to the first digit.
For 57×11, think $5 + 7 = 12$.
Inserting 2 and adding 1 to the 5 gives 627.
So, $57 \times 11 = 627$.

Calculate.

13. 18×11 **14.** 43×11 **15.** 52×11
16. 66×11 **17.** 78×11 **18.** 94×11

To multiply by 0.11, 1.1, 110, and so on, first multiply by 11, then place the decimal point.
Thus, $32 \times 1.1 = 35.2$, and $57 \times 110 = 6270$.

Calculate.

19. 14×110 **20.** 69×110 **21.** 23×1.1
22. 75×1.1 **23.** 36×0.11 **24.** 65×0.11

25. The expression $10x + y$ represents the value of a two-digit number, where x is the first digit and y is the second digit.
Multiplying by 11:
$$11(10x + y) = 110x + 11y$$
$$= 100x + 10x + 10y + y$$
$$= 100x + 10(x + y) + y$$

Use the expression $100x + 10(x + y) + y$ to explain why the rule for multiplying by 11 works.

INVESTIGATING MATH

Lengths of Horizontal and Vertical Line Segments

1 Lengths on a Number Line

1. What is the distance between A and B?

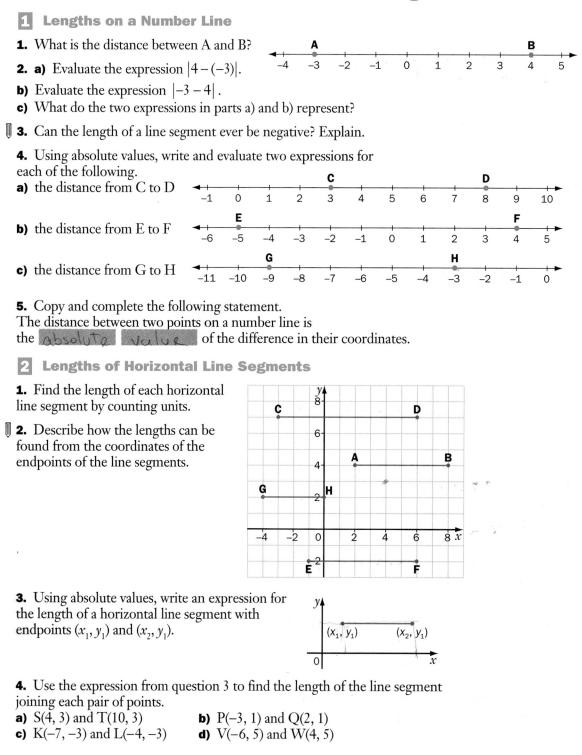

2. a) Evaluate the expression $|4-(-3)|$.

b) Evaluate the expression $|-3-4|$.

c) What do the two expressions in parts a) and b) represent?

3. Can the length of a line segment ever be negative? Explain.

4. Using absolute values, write and evaluate two expressions for each of the following.

a) the distance from C to D

b) the distance from E to F

c) the distance from G to H

5. Copy and complete the following statement.
The distance between two points on a number line is the ~~absolute~~ ~~value~~ of the difference in their coordinates.

2 Lengths of Horizontal Line Segments

1. Find the length of each horizontal line segment by counting units.

2. Describe how the lengths can be found from the coordinates of the endpoints of the line segments.

3. Using absolute values, write an expression for the length of a horizontal line segment with endpoints (x_1, y_1) and (x_2, y_1).

4. Use the expression from question 3 to find the length of the line segment joining each pair of points.
a) S(4, 3) and T(10, 3)
b) P(-3, 1) and Q(2, 1)
c) K(-7, -3) and L(-4, -3)
d) V(-6, 5) and W(4, 5)

252

3 Lengths of Vertical Line Segments

1. Find the length of each vertical line segment by counting units.

2. Describe how the lengths can be found from the coordinates of the endpoints of the line segments.

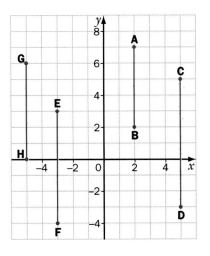

3. Using absolute values, write an expression for the length of a vertical line segment with endpoints (x_1, y_1) and (x_1, y_2).

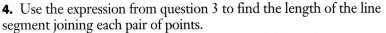

4. Use the expression from question 3 to find the length of the line segment joining each pair of points.

a) R(4, 6) and S(4, 1)   **b)** P(1, 2) and Q(1, −4)
c) V(−2, −7) and W(−2, −3)   **d)** T(−4, 0) and U(−4, −3)

4 Problem Solving

1. Plot the points and make a quadrilateral by joining them in order. Identify the type of quadrilateral, and find its perimeter and area.

a) A(2, 3), B(2, 7), C(6, 7), D(6, 3)
b) E(−3, 2), F(2, 2), G(2, −3), H(−3, −3)
c) K(1, 1), L(4, 1), M(4, 9), N(1, 9)
d) P(5, −2), Q(5, −5), R(−1, −5), S(−1, −2)

2. Three vertices of rectangle WXYZ are W(4, 3), X(−1, 3), and Y(−1, −1). What are the coordinates of Z?

3. Two adjacent vertices of a square have coordinates (−1, −3) and (−1, 0). What are the possible coordinates of the other two vertices?

4. Rectangle QRST has an area of 6 square units, and Q is at the origin. R lies on the *x*-axis, T lies on the *y*-axis, and the coordinates of the vertices are whole numbers. What are the possible sets of coordinates for the vertices of the rectangle?

6.1 Distance Between Two Points

The cost of a long distance call depends on the length of the call, in minutes, and the distance between the two locations.

There are thousands of cities and towns in North America. Instead of making a list of all the distances between them, telephone companies placed a coordinate grid over North America. The origin, (0, 0), is located off the northeast coast of Canada. Each North American city and town has a set of coordinates of the form (horizontal, vertical).

The coordinates for Edmonton are (7801, 4943).
The coordinates for Vancouver are (9023, 5976).

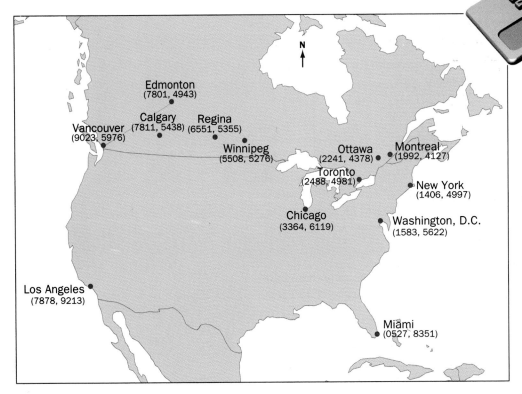

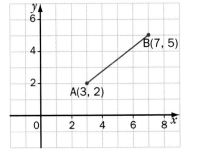

Explore: Develop a Method

a) Copy the graph of line segment AB onto grid paper.

b) Locate point C, below point B, so that △ABC is a right triangle, with AB the hypotenuse.

c) Write the coordinates of point C on the grid.

d) Find the lengths of legs AC and BC.

254

Inquire

1. Use the lengths of AC and BC and the Pythagorean Theorem to calculate the length of AB.

2. How can the length of AC be determined using the x-coordinates of A and B?

3. How can the length of BC be determined using the y-coordinates of A and B?

4. Would the length of AB change if the right triangle was formed by locating point C above point A? Explain.

5. If you are given the coordinates of the endpoints of any line segment, describe a method for finding the length of the segment without plotting the points.

6. Use the method from question 5 to find the length of the line segment joining each of the following pairs of points. Round each answer to the nearest tenth, if necessary.
a) C(2, 3) and D(10, 9) **b)** E(2, 4) and F(7, 1) **c)** G(−3, 1) and H(4, 5)

7. Calculate the distance between Edmonton and Vancouver, to the nearest telephone grid unit.

8. If each unit of length on the telephone grid is 0.51 km, what is the distance between Edmonton and Vancouver, to the nearest kilometre?

Example 1   Distance Between Two Points
Determine the distance between A(3, 1) and B(6, 7).

Solution
Draw the line segment AB.
Construct right triangle ABC.
The coordinates of C are (6, 1).

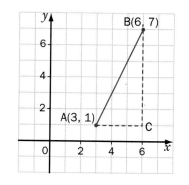

Length of AC $= |6 - 3|$
$\quad\quad\quad = 3$
Length of BC $= |7 - 1|$
$\quad\quad\quad = 6$
Use the Pythagorean Theorem.

$(AB)^2 = (AC)^2 + (BC)^2$
$AB = \sqrt{(AC)^2 + (BC)^2}$
$\quad = \sqrt{3^2 + 6^2}$
$\quad = \sqrt{9 + 36}$
$\quad = \sqrt{45}$
$\quad = 3\sqrt{5}$

The distance between A and B is $3\sqrt{5}$.

The distance formula may be generalized as follows.
The points $A(x_1, y_1)$ and $B(x_2, y_2)$ represent any two points on a grid.

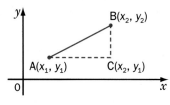

$$(AB)^2 = |\text{change in } x|^2 + |\text{change in } y|^2$$

Since the square of any number is always positive, it is not necessary to use absolute value, so we can write

$$(AB)^2 = (\text{change in } x)^2 + (\text{change in } y)^2$$
$$AB = \sqrt{(\text{change in } x)^2 + (\text{change in } y)^2}$$
$$AB = \sqrt{(x_2 - x_1)^2 + (y_2 - y_1)^2}$$
$$\text{or } l = \sqrt{(x_2 - x_1)^2 + (y_2 - y_1)^2}$$

Example 2   Using the Distance Formula
Find the length of the line segment joining $C(2, -4)$ and $D(-3, 5)$, to the nearest tenth.

Solution
Let $(x_1, y_1) = (2, -4)$ and $(x_2, y_2) = (-3, 5)$.
Then
$$CD = \sqrt{(x_2 - x_1)^2 + (y_2 - y_1)^2}$$
$$= \sqrt{(-3 - 2)^2 + (5 - (-4))^2}$$
$$= \sqrt{(-5)^2 + 9^2}$$
$$= \sqrt{25 + 81}$$
$$= \sqrt{106}$$
$$\doteq 10.3$$

Estimate
$\boxed{\sqrt{100} = 10}$

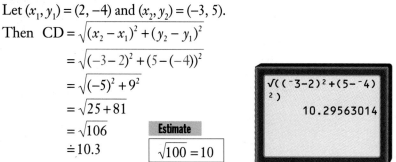

The length of the line segment joining C and D is 10.3, to the nearest tenth.

Practice

Determine the distance between each pair of points. Express each answer in radical form, if necessary.
1. $(2, 1)$ and $(3, 5)$
2. $(3, -5)$ and $(-6, 7)$
3. $(3, 0)$ and $(4, -1)$
4. $(-1, 2)$ and $(-6, -3)$
5. $(2, 1)$ and $(2, 9)$
6. $(4, -7)$ and $(11, -7)$

7. $(8.1, 3.7)$ and $(3.2, -5.4)$
8. $(0.1, 0.2)$ and $(-0.1, -0.2)$

Determine the radius of each circle, given its centre and a point on its circumference. Round each radius to the nearest tenth, if necessary.
9. centre $(0, 0)$, point $(1, 5)$
10. centre $(2, 3)$, point $(2, 7)$
11. centre $(-3, 4)$, point $(-4, -6)$
12. centre $(10, 0)$, point $(-11, 0)$

256

Applications and Problem Solving

13. Long distance calls Use the map on page 254 to find the distance between each pair of cities, to the nearest kilometre. Each unit of length on the telephone grid is 0.51 km.
a) Ottawa and Toronto
b) Montreal and Los Angeles
c) Calgary and Washington, D.C.
d) Winnipeg and Edmonton
e) Regina and Miami

14. Measurement A quadrilateral has vertices P(3, 5), Q(−4, 3), R(−3, −2), and S(5, −4). Find the lengths of the diagonals, to the nearest tenth.

15. Measurement The vertices of a right triangle are S(−2, −2), T(10, −2), and R(4, 5). Find the area of the triangle.

16. Coast guard A coast guard patrol boat is located 5 km east and 8 km north of the entrance to St. John's harbour. A tanker is 9 km east and 6 km south of the entrance. Find the distance between the two ships, to the nearest tenth of a kilometre.

17. Measurement Show that A(4, 2), B(−2, −2), and C(2, −8) are the vertices of an isosceles right triangle.

18. Measurement Show that C(−5, −1) is the midpoint of the line segment joining A(−2, 5) and B(−8, −7).

19. Measurement The coordinates of the endpoints of the diameter of a circle are (6, 4) and (−2, 0). Find the length of the radius of the circle.

20. Measurement Classify each triangle as equilateral, isosceles, or scalene. Then, find each perimeter, to the nearest tenth.
a) A(2, 5), B(−2, −1), C(6, −1)
b) D(−2, −5), E(−3, 2), F(1, 3)
c) P(2, 1), Q(5, 3), R(0, 4)
d) G(3, 0), H(0, 3$\sqrt{3}$), I(−3, 0)

21. Measurement Show that the quadrilateral with vertices K(3, 4), L(2, 0), M(−3, −2), and N(−1, 3) is a kite.

22. Measurement Show that the parallelogram with vertices W(1, 3), X(4, 1), Y(1, −1), and Z(−2, 1) is a rhombus.

23. Algebra a) Develop a formula for the distance between O(0, 0) and P(x_1, y_1).
b) Use the formula to find the distance between the origin and the point A(4, 3); the point B(5, −12); the point C(−6, −2).

24. Measurement Three or more points are **collinear** if they lie on the same straight line. Use lengths of line segments to show that the points P(3, 2), Q(0, 5), and R(−2, 7) are collinear. Explain your reasoning.

25. Algebra Write an expression for the distance between each pair of points.
a) (a, b) and (2a, 2b)
b) (x, 2y) and (2x, −y)
c) (m + 1, n − 1) and (m − 1, 2n − 1)
d) (2p, −p) and (4p, 3p)

26. Algebra The point (x, −1) is 13 units from the point (3, 11). What are the possible values of x?

27. List all the points that have whole-number coordinates and are
a) 5 units from the origin
b) 10 units from the point (2, 1)

LOGIC POWER

Assume that no small cubes are missing from the back of the stack.

1. How many small cubes are needed to complete the large cube?

2. The faces on the outside of the large cube must all be red. The faces hidden inside the large cube must all be yellow. Some of the small cubes you found in question 1 need to have 2 red faces and 4 yellow faces.
a) What other combinations of red and yellow faces are needed?
b) Find how many small cubes are needed with each combination of red and yellow faces.

INVESTIGATING MATH

Midpoint

1 Midpoints of Horizontal Line Segments

1. Graph each line segment. Count units to find the coordinates of M, the midpoint of each line segment.
a) The endpoints of AB are A(1, 4) and B(9, 4).
b) The endpoints of CD are C(–3, 2) and D(7, 2).
c) The endpoints of EF are E(0, –3) and F(6, –3).
d) The endpoints of GH are G(–7, –2) and H(–1, –2).
e) The endpoints of IJ are I(–2, 0) and J(10, 0).

2. Describe how the coordinates of the midpoint can be found from the coordinates of the endpoints of a horizontal line segment.

3. Given the endpoints of each line segment, find the coordinates of the midpoint without plotting the points.
a) K(2, 3) and L(6, 3)   **b)** P(–1, 1) and Q(5, 1)
c) R(–7, 5) and S(–3, 5)   **d)** V(0, –5) and W(8, –5)

4. AB is a line segment joining the points A(2, 3) and B(10, 3). Find the coordinates of the three points that divide AB into 4 equal parts.

2 Midpoints of Vertical Line Segments

1. Graph each line segment. Count units to find the coordinates of M, the midpoint of each line segment.
a) The endpoints of AB are A(2, 8) and B(2, 2).
b) The endpoints of CD are C(5, 6) and D(5, –2).
c) The endpoints of EF are E(–3, –1) and F(–3, –5).
d) The endpoints of GH are G(–2, –4) and H(–2, 0).
e) The endpoints of IJ are I(0, 8) and J(0, –2).

2. Describe how the coordinates of the midpoint can be found from the coordinates of the endpoints of a vertical line segment.

3. Given the endpoints of each line segment, find the coordinates of the midpoint without plotting the points.
a) S(4, 7) and T(4, 1)   **b)** P(1, 4) and Q(1, –6)
c) V(–2, –4) and W(–2, 4)   **d)** L(0, 9) and N(0, 1)

4. CD is a line segment joining the points C(3, 8) and D(3, –4). Find the coordinates of the three points that divide CD into 4 equal parts.

6.2 Midpoint of a Line Segment

When you look in a mirror, your reflection image is the same distance behind the mirror as you are in front of it. The line segment that connects your eyes and the image of your eyes is perpendicular to the mirror. The midpoint of this line segment lies on the mirror.

Explore: Use the Diagrams

Each diagram shows a green mirror line, a line segment joining a point and its reflection image, and the midpoint of the segment.

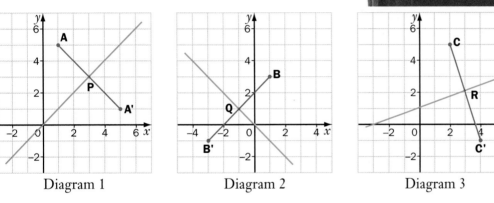

Diagram 1   Diagram 2   Diagram 3

a) In diagram 1, what are the coordinates of A? the image point A′? the midpoint P?
b) In diagram 2, what are the coordinates of B? the image point B′? the midpoint Q?
c) In diagram 3, what are the coordinates of C? the image point C′? the midpoint R?

Inquire

1. a) How is the x-coordinate of P related to the x-coordinates of A and A′?
b) How is the y-coordinate of P related to the y-coordinates of A and A′?

2. a) How is the x-coordinate of Q related to the x-coordinates of B and B′?
b) How is the y-coordinate of Q related to the y-coordinates of B and B′?

3. a) How is the x-coordinate of R related to the x-coordinates of C and C′?
b) How is the y-coordinate of R related to the y-coordinates of C and C′?

4. Write a rule for finding the coordinates of the midpoint of a line segment using the coordinates of the endpoints.

5. Find the coordinates of the midpoint of each of the following line segments, given the coordinates of the endpoints.
a) A(4, 9), B(2, 1)   **b)** C(0, −2), D(−2, 4)   **c)** E(−3, −4), F(−1, 6)

If A has coordinates (x_1, y_1) and B has coordinates (x_2, y_2), then the coordinates of the midpoint, M, of the segment AB are

$$\left(\frac{x_1 + x_2}{2}, \frac{y_1 + y_2}{2} \right)$$

Example 1   Midpoint of a Line Segment
Determine the coordinates of the midpoint, M, of the line segment with endpoints A(−2, −3) and B(4, 7).

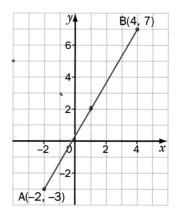

Solution
Substitute into the formula for midpoint.

$$M \left(\frac{x_1 + x_2}{2}, \frac{y_1 + y_2}{2} \right)$$

The midpoint of A(−2, −3) and B(4, 7) is

$$\left(\frac{-2+4}{2}, \frac{-3+7}{2} \right) = (1, 2)$$

The coordinates of the midpoint are (1, 2).

Example 2   Finding an Endpoint
For a line segment DE, one endpoint is D(−4, 5) and the midpoint is M(−1, 3). Find the coordinates of the endpoint E.

Solution
Let the coordinates of E be (x_1, y_1).
Since the coordinates of M are related to the coordinates of D and E by the formula for midpoint,

$$(-1, 3) = \left(\frac{x_1 + (-4)}{2}, \frac{y_1 + 5}{2} \right)$$

Solve for x_1.

$$-1 = \frac{x_1 - 4}{2}$$

Multiply both sides by 2:

$$2 \times (-1) = 2 \times \frac{x_1 - 4}{2}$$
$$-2 = x_1 - 4$$

Add 4 to both sides:

$$-2 + 4 = x_1 - 4 + 4$$
$$2 = x_1$$

Solve for y_1.

$$3 = \frac{y_1 + 5}{2}$$

Multiply both sides by 2:

$$2 \times 3 = 2 \times \frac{y_1 + 5}{2}$$
$$6 = y_1 + 5$$

Subtract 5 from both sides:

$$6 - 5 = y_1 + 5 - 5$$
$$1 = y_1$$

The coordinates of the endpoint E are (2, 1).

Practice

Determine the midpoint of each line segment from the given endpoints.

1. (5, 7) and (3, 9)
2. (4, 2) and (6, 8)
3. (−1, 0) and (1, −6)
4. (−2, −4) and (−2, 8)
5. (−3, −3) and (−1, −7)
6. (5, 4) and (−3, 4)
7. (0.2, 1.5) and (3.6, 0.3)
8. $\left(\dfrac{1}{2}, \dfrac{5}{2}\right)$ and $\left(\dfrac{3}{2}, \dfrac{-7}{2}\right)$
9. (200, 125) and (403, 174)
10. (a, b) and (c, d)

One endpoint and the midpoint of each line segment are given. Find the coordinates of the other endpoint.

	Endpoint	Midpoint
11.	(3, −5)	(7, 7)
12.	(6, 10)	(0, 0)
13.	(3, −4)	(3, 6)
14.	(−2, −6)	(1, 1)
15.	(4, 0)	(7, 2)
16.	(8, 4)	(2, 4)
17.	(1, −2.8)	(4.7, −1.3)
18.	(p, q)	(c, d)

Applications and Problem Solving

19. A diameter of a circle joins the points C(−7, −4) and D(−1, 10). What are the coordinates of the centre of the circle?

20. The endpoints of AB are A(10, 16) and B(−6, −12). Find the coordinates of the points that divide the segment into four equal parts.

21. Measurement Triangle RST has vertices R(4, 4), S(−2, 0), and T(4, −2).
a) Find the coordinates of A, the midpoint of RS.
b) Find the coordinates of B, the midpoint of RT.
c) Show that the length of AB is half the length of ST.

22. The centre of a circle has coordinates (−1, −3). One endpoint of a diameter of the circle has coordinates (−3, 0). What are the coordinates of the other endpoint of the diameter?

23. Measurement A quadrilateral has vertices P(0, 8), Q(−4, 4), R(2, −2), and S(6, 4). Find the perimeter of the figure whose vertices are the midpoints of the sides of the quadrilateral.

24. Measurement A quadrilateral ABCD has vertices A(4, −2), B(6, 6), C(0, 8), and D(−2, 4).
a) Plot the points on a grid.
b) Find the coordinates of the following points.
E, the midpoint of AB
F, the midpoint of BC
G, the midpoint of CD
H, the midpoint of AD
c) Show that the length of EF equals the length of GH.
d) Show that the length of FG equals the length of EH.

25. Algebra Write the coordinates of the midpoint between each of the following pairs of points.
a) (x, y) and (2x, 2y)
b) (4a, 3b) and (8a, −b)
c) (m + 1, n + 3) and (m − 1, n − 2)
d) (2t, 3t + 1) and (−4t, 1 − t)

26. Algebra The endpoints of PQ are P(3, −4) and Q(11, c). The midpoint of PQ is M(d, 3). Find the values of c and d.

27. Measurement A rectangle has vertices A(−4, 4), B(−4, −2), C(7, −2), and D(7, 4).
a) Find the perimeter of the rectangle.
b) Find the area.
c) Find the midpoint of each diagonal.
d) Show that the diagonals bisect each other.

28. For the line segment ST, S has coordinates (6, 2) and T is on the y-axis. The midpoint, M, of ST is on the x-axis. Find the coordinates of T and M.

29. Measurement Triangle RST has vertices R(4, 4), S(−6, 2), and T(2, 0).
a) Draw the triangle on a grid.
b) Find the lengths of the three medians.

30. Measurement A right triangle has vertices P(2, 3), Q(6, −5), and R(−6, −1). Show that the midpoint of the hypotenuse is equidistant from each vertex of the triangle.

261

31. Building a road Historic Fort Bragg is to be rebuilt for visitors. A map of the area around the fort is drawn on a coordinate grid. The coordinates of the fort are (5.2, 16.6). A straight road is to be constructed to the fort from a point with coordinates (12.8, 7.8) on a highway. The federal government has agreed to pay half the cost of the road. Construction costs are $790/m.
a) Find the coordinates of the midpoint of the road.
b) If one unit of length on the grid represents 1 km, use the midpoint to find half the length of the road.
c) Calculate the government's cost.
d) Describe a different method of solving the problem.

32. Algebra A line segment has endpoints $A(x_1, y_1)$ and $B(x_2, y_2)$. Find expressions in terms of $x_1, x_2, y_1,$ and y_2 for the coordinates of the points that divide AB into 4 equal parts.

33. The base of an isosceles triangle lies on the *x*-axis. The coordinates of the midpoints of the equal sides of the triangle are (2, 3) and (–2, 3). What are the coordinates of the vertices of the triangle?

34. The midpoints of the equal sides of an isosceles triangle have coordinates (4, 3) and (4, –1). What are the coordinates of the vertices of the triangle if
a) two vertices lie on the *y*-axis?
b) only one vertex lies on the *y*-axis?

35. Decide whether each statement is always true, sometimes true, or never true. Explain.
a) Two line segments with the same midpoint have the same length.
b) Two parallel line segments have the same midpoint.
c) The midpoint of a line segment is equidistant from the endpoints of the segment.
d) A point equidistant from the endpoints of a line segment is the midpoint of the segment.

36. Algebra Prove that M is the midpoint of each segment AB by using the distance formula to show that the lengths of AM and MB are equal.
a)

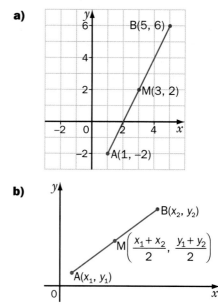

b)

37. Visualization Without using the word *midpoint*, describe to a classmate the meaning of the midpoint of a line segment in two different ways.

PATTERN POWER

48	25	39	82
32	10	27	16
15	35	63	
53	75	97	46

a) Describe how the numbers in the second row of the chart are related to the numbers in the first row.
b) Find the missing number.

6.3 Slope

The **slope** of a line is the measure of how steep the line is. The slope also describes the direction of the line.

Diagram 1 shows the steepness of two objects as you move up them. Diagram 2 shows the steepness of the same objects as you move down them. In each diagram, line I represents the median of a triangular face of the Great Pyramid of Cheops, line II represents Russian Hill, which is a street in San Francisco.

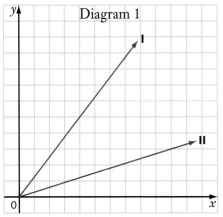

Diagram 1

The lines in diagram 1 rise from left to right. They have positive slopes.

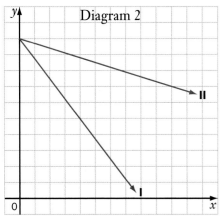

Diagram 2

The lines in diagram 2 fall from left to right. They have negative slopes.

The slope of a line is found by dividing the vertical change, called the **rise** (or fall), by the horizontal change, called the **run**.

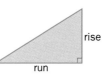

263

Explore: Develop a Method

Determine the rise and run for each line segment. Then, find the slope.

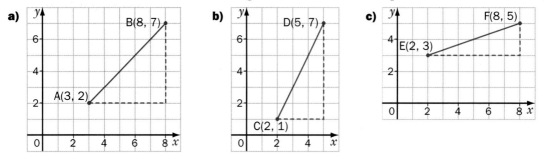

a) B(8, 7), A(3, 2)

b) D(5, 7), C(2, 1)

c) F(8, 5), E(2, 3)

Inquire

1. Which line segment is the steepest?

2. Which line segment has the greatest slope?

3. How can the rise of each line segment be calculated using the *y*-coordinates of the endpoints?

4. How can the run of each line segment be determined using the *x*-coordinates of the endpoints?

5. If you are given the coordinates of the endpoints of a line segment, describe a method for finding the slope without plotting the points.

6. Which has the greater slope — the face of the Great Pyramid or Russian Hill? Explain.

Example 1 Slopes of Line Segments

The coordinates of three points on a line are shown.
a) Calculate the slopes of segments AB, BC, and AC.
b) How are the slopes related?

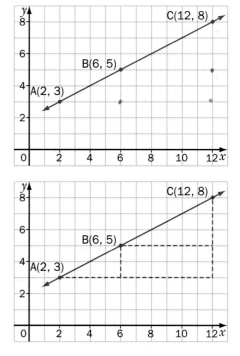

Solution

a) Find the rise and run for each segment by counting units.

For segment AB,
rise = 2
run = 4
$$\text{slope} = \frac{2}{4}$$
$$= \frac{1}{2}$$

For segment BC,
rise = 3
run = 6
$$\text{slope} = \frac{3}{6}$$
$$= \frac{1}{2}$$

For segment AC,
rise = 5
run = 10
$$\text{slope} = \frac{5}{10}$$
$$= \frac{1}{2}$$

b) The slopes are equal.

As Example 1 suggests, the slopes of all segments on a line are equal. Therefore, the slope of a line can be found using any two points on the line.

The slope, m, of a line containing the points $P(x_1, y_1)$ and $Q(x_2, y_2)$ is

$$m = \frac{\text{vertical change}}{\text{horizontal change}} \quad \text{or} \quad \frac{\text{rise}}{\text{run}}$$

$$= \frac{y_2 - y_1}{x_2 - x_1}, \quad x_2 \neq x_1$$

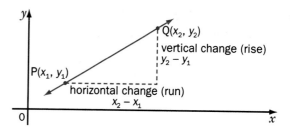

Example 2   Positive Slope of a Line

Find the slope of the line that passes through the points S(2, 1) and T(6, 8).

Solution

Let $(x_1, y_1) = (2, 1)$ and $(x_2, y_2) = (6, 8)$

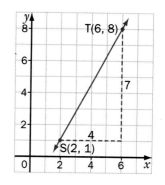

$$m_{ST} = \frac{y_2 - y_1}{x_2 - x_1}$$

Substitute known values: $= \dfrac{8 - 1}{6 - 2}$

Simplify: $= \dfrac{7}{4}$

The slope of the line that passes through the points S and T is $\dfrac{7}{4}$.

Note that, if a line rises from left to right, its slope is a positive number.

Example 3   Negative Slope of a Line

Find the slope of the line that passes through the points E(−5, 4) and F(3, −1).

Solution

Let $(x_1, y_1) = (-5, 4)$ and $(x_2, y_2) = (3, -1)$

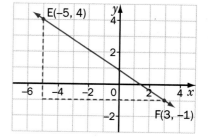

$$m_{EF} = \frac{y_2 - y_1}{x_2 - x_1}$$

Substitute known values: $= \dfrac{-1 - 4}{3 - (-5)}$

Simplify: $= \dfrac{-5}{8}$

The slope of the line that passes through the points E and F is $\dfrac{-5}{8}$.

Note that, if a line falls from left to right, its slope is a negative number.

Example 4   Slope of a Horizontal Line

Find the slope of the line that passes through the points J(2, 4) and K(−5, 4).

Solution

$$m_{JK} = \frac{y_2 - y_1}{x_2 - x_1}$$

Substitute known values: $= \dfrac{4 - 4}{-5 - 2}$

Simplify: $= \dfrac{0}{-7}$

$= 0$

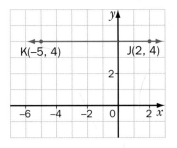

The slope of the line that passes through the points J and K is 0.

Note that the slope of a horizontal line is zero.

Example 5   Slope of a Vertical Line

Find the slope of the line that passes through the points V(3, 5) and W(3, −4).

$$m_{VW} = \frac{y_2 - y_1}{x_2 - x_1}$$

$y_2 - y_1 = -4 - 5$
$\qquad = -9$

$x_2 - x_1 = 3 - 3$
$\qquad = 0$

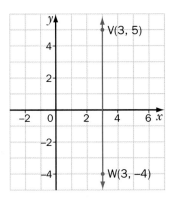

Since division by zero is undefined, the slope of the line that passes through V and W is not defined by a real number.

Note that the slope of a vertical line is undefined.

As Examples 2 to 5 suggest, the slopes of lines can be summarized as follows.

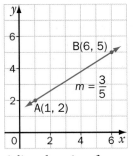

A line that rises from left to right has a positive slope.

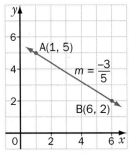

A line that falls from left to right has a negative slope.

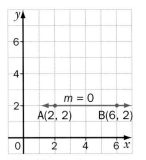

The slope of a horizontal line is zero.

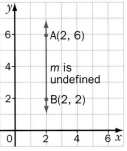

The slope of a vertical line is undefined.

266

Example 6   Finding a Coordinate

The slope of a line is 2. The line passes through $(-1, k)$ and $(4, 8)$. Find the value of k.

Solution

$$m = \frac{y_2 - y_1}{x_2 - x_1}$$

Substitute known values: $2 = \dfrac{8 - k}{4 - (-1)}$

Simplify: $2 = \dfrac{8 - k}{5}$

Multiply both sides by 5: $5 \times 2 = 5 \times \dfrac{8 - k}{5}$

$$10 = 8 - k$$

Subtract 8 from both sides: $10 - 8 = 8 - k - 8$

$$2 = -k$$

$$-2 = k$$

The value of k is -2.

Practice

a) *Without calculating the slope, state whether the slope of each line is positive, negative, zero, or undefined.*
b) *State the rise.*
c) *State the run.*
d) *Calculate the slope, where possible.*

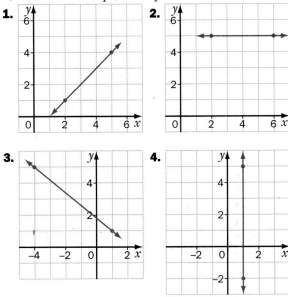

1.   **2.**

3.   **4.**

5.   **6.**

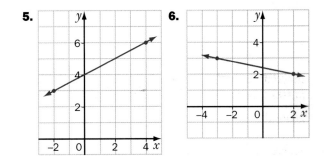

Find the slope of the line passing through the points.

7. $(0, 0)$ and $(2, 3)$ **8.** $(0, 0)$ and $(-2, -4)$
9. $(1, 3)$ and $(2, 7)$ **10.** $(-4, 5)$ and $(6, 5)$
11. $(5, -2)$ and $(-3, 4)$ **12.** $(0, 6)$ and $(4, 0)$
13. $(-5, 7)$ and $(-4, -2)$ **14.** $(-2, 5)$ and $(0, 8)$
15. $(-6, -5)$ and $(0, 0)$ **16.** $(-8, -7)$ and $(-4, -3)$
17. $(5, 7)$ and $(5, -3)$ **18.** $(-6, -1)$ and $(-2, 5)$

Find the slope of the line passing through the points.

19. $(3.4, 1.6)$ and $(5.4, 2.2)$
20. $(0, -1.7)$ and $(0.3, -3.8)$
21. $(11.9, -2.3)$ and $(15.4, 8.2)$
22. $\left(\dfrac{1}{2}, 4\right)$ and $(2, -6)$ **23.** $\left(\dfrac{1}{3}, 1\dfrac{1}{2}\right)$ and $\left(2, 3\dfrac{1}{2}\right)$

267

24. The slope of a line is 3. The line passes through $(2, k)$ and $(4, 1)$. Find the value of k.

25. The slope of a line is –2. The line passes through $(t, -1)$ and $(-4, 9)$. Find the value of t.

Write the coordinates of two points on a line that satisfies the given condition.

26. The line rises from left to right.

27. The line is horizontal.

28. The line falls from left to right.

29. The line is vertical.

Applications and Problem Solving

30. Visualization Given a point on the line and the slope, sketch the graph of the line.

a) $(2, 3)$, $m = 2$ **b)** $(-1, 1)$, $m = 3$

c) $(0, 4)$, $m = -2$ **d)** $(-3, 0)$, $m = \dfrac{1}{2}$

e) $(-3, -2)$, $m = \dfrac{2}{3}$ **f)** $(-3, 4)$, $m = \dfrac{-4}{3}$

g) $(4, -1)$, $m = 0$ **h)** $(-4, 5)$, m is undefined

31. Given the point $A(2, 2)$, find the coordinates of a point B so that the slope of AB is

a) 1 **b)** -1 **c)** 2 **d)** -2

e) 0 **f)** $\dfrac{1}{2}$ **g)** undefined

h) 7 **i)** -4 **j)** $-\dfrac{1}{3}$ **k)** $\dfrac{5}{2}$

32. A line passes through the point $(1, -3)$. State the coordinates of one other point on the line so that the slope of the line is

a) 0 **b)** 4 **c)** undefined

d) -1 **e)** $\dfrac{3}{4}$ **f)** $-\dfrac{1}{2}$ **g)** $\dfrac{5}{4}$

h) -8 **i)** 12 **j)** $\dfrac{7}{5}$ **k)** $\dfrac{-8}{3}$

33. Highway grade A sign on a highway indicates that the next hill has a grade of 10%. What is the slope of the hill?

34. Algebra A line of slope –3 passes through $(r, 3)$ and $(5, r)$. Find the value of r.

35. Algebra A line of slope $\dfrac{3}{2}$ passes through $(-1, 2)$ and $(s, s + 1)$. What are the possible values of s?

36. Quadrants
The quadrants on a coordinate grid are named as follows.

Point 1 and Point 2 lie on a line. Write the coordinates of the points so that the following conditions are satisfied.

	Location of Points		Slope of Line
	Point 1	**Point 2**	
a)	in 1st quadrant	in 1st quadrant	negative
b)	in 2nd quadrant	in 3rd quadrant	positive
c)	in 2nd quadrant	in 3rd quadrant	negative
d)	in 3rd quadrant	in 4th quadrant	negative
e)	in 3rd quadrant	in 4th quadrant	positive
f)	in 4th quadrant	in 3rd quadrant	zero
g)	in 3rd quadrant	on x-axis	negative
h)	in 3rd quadrant	on x-axis	positive
i)	on y-axis	on x-axis	positive
j)	on y-axis	on x-axis	negative
k)	in 1st quadrant	in 2nd quadrant	$\dfrac{1}{2}$
l)	in 1st quadrant	in 2nd quadrant	$-\dfrac{1}{2}$
m)	in 4th quadrant	in 3rd quadrant	$\dfrac{2}{3}$
n)	in 4th quadrant	in 3rd quadrant	$-\dfrac{2}{3}$
o)	in 3rd quadrant	in 2nd quadrant	$\dfrac{3}{2}$
p)	in 3rd quadrant	in 2nd quadrant	$-\dfrac{3}{2}$

37. Measurement Line segments AB and AC intersect at point A. The slope of AB is 0. The slope of AC is undefined. What is the measure of ∠BAC?

38. Find the coordinates of the vertices of a triangle so that the slopes of all three sides are positive.

39. Find the coordinates of the vertices of a quadrilateral so that the slopes of all four sides are negative.

40. Equations Given the equation of each line, find two points on the line and calculate the slope.
a) $x + y = 4$ **b)** $y = x + 2$
c) $x - y = 1$ **d)** $y = 5 - 2x$

41. Roof of a building
An A-frame building is 8 m wide and 6 m high. A cross section of the roof of the building can be drawn on the grid shown.
a) What is the slope of each half of the roof?
b) Is one half of the roof steeper than the other? Explain.
c) When comparing steepness, would it be reasonable to use absolute values of slopes? Explain.

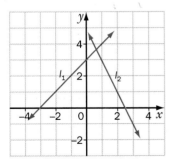

d) Which of the lines l_1 and l_2, shown on the following grid, is steeper? Explain.

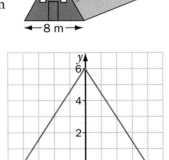

42. Three or more points are collinear if they lie on the same straight line.
a) Use slopes to determine whether the points A(6, 7), B(2, 1), and C(−2, −5) are collinear.
b) Use slopes to determine whether the points P(−2, 5), Q(1, 3), and R(5, 1) are collinear.
c) Give the coordinates of three collinear points if one point is in the first quadrant, one is in the second quadrant, and one is in the fourth quadrant.
d) If the points D(−2, −1), E(1, 1), and F(7, k) are collinear, find the value of k.

43. The equal sides of isosceles triangle ABC meet at A(5, 0). Vertices B and C lie on the y-axis. The slope of AB is 2. Find
a) the coordinates of B
b) the coordinates of C
c) the slope of AC

44. Archaeology The North Stone Pyramid in Egypt is 140 m high. The side length of its square base is 220 m.
a) If you climbed the pyramid, starting at the middle of one side of the base, what would be the slope of your climb, to the nearest hundredth?
b) If you climbed the pyramid along one of the edges, what would be the slope of your climb, to the nearest hundredth?

45. Stairways For comfortable stairways, the slope $\frac{3}{4}$ is often used. This is the maximum slope permitted by many building codes.
a) Why do building codes specify a maximum slope for stairways?
b) What is the slope of the stairways in your school?

LOGIC POWER

You have an empty 7-L pail, an empty 3-L pail, and a 10-L pail full of water. There are no volume markings on the pails. In one pouring, you can fill or empty a pail by pouring water from one pail to another. What is the smallest number of pourings needed to get 5 L of water in the 10-L pail and 5 L of water in the 7-L pail?

6.4 Slope as Rate of Change

The slope of a line is the rate of change between any two points on the line. The slope shows how much the vertical measure changes with respect to each unit change in the horizontal measure.

In applied problems, the slope of a line describes a constant or average rate of change. The rate can be found by dividing the change in the *y*-quantity by the change in the *x*-quantity.

Explore: Interpret the Graph

Marnie McBean and Kathleen Heddle won Olympic gold medals for Canada in the women's pairs rowing event at both the Barcelona and the Atlanta games. The graph shows their approximate times at the 500-m and 1500-m marks of their gold medal race in Atlanta. All Olympic rowing events are 2000 m long.

Find the slope of the graph, to the nearest tenth.

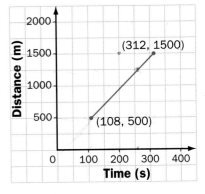

Inquire

1. The slope of the graph represents the average speed from the 500-m mark to the 1500-m mark. In what units should the speed be expressed? How do you know?

2. a) Would you expect that the average speed for the first 500 m of the race was higher or lower than the speed in question 1? Explain.
b) Determine the average speed for the first 500 m of the race.
c) On the same graph, which is steeper — a slope that represents a higher speed or a slope that represents a lower speed?

3. If the point (*k*, 1250) lies on the graph,
a) what is the value of *k*?
b) in what units is *k* expressed?
c) what does *k* represent?

Example   Internet Access

Christina's Internet provider charges a flat fee for the first 8 h of Internet access per month, plus an hourly rate for additional access. One month, Christina was on-line for 15 h and paid $25.88. The next month, she was on-line for 27 h and paid $49.76.
a) What is the hourly rate for hours above 8 h/month?
b) What is the flat fee for the first 8 h/month?

Solution

a) The hourly rate is the slope of the graph of cost versus time.

$$m = \frac{49.76 - 25.88}{27 - 15}$$

Estimate

$24 \div 12 = 2$

$$= 1.99$$

The rate for hours above 8 h/month is $1.99/h.

b) The flat fee for times up to 8 h can be found by determining the cost at a time of 8 h.

Let the ordered pair for this point be $(8, y_1)$.

Let (x_2, y_2) be the point $(15, 25.88)$.

$$m = \frac{y_2 - y_1}{x_2 - x_1}$$

Substitute known values:

$$1.99 = \frac{25.88 - y_1}{15 - 8}$$

$$= \frac{25.88 - y_1}{7}$$

Multiply both sides by 7:

$$7 \times 1.99 = 7 \times \frac{25.88 - y_1}{7}$$

$$13.93 = 25.88 - y_1$$

Subtract 25.88 from both sides:

$$13.93 - 25.88 = 25.88 - y_1 - 25.88$$

$$-11.95 = -y_1$$

$$11.95 = y_1$$

The flat fee for the first 8 h/month is $11.95.

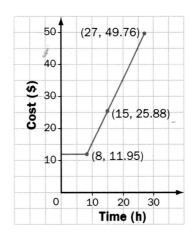

In part b) of the above solution, the ordered pair $(8, y_1)$ cannot be replaced with an ordered pair that includes a time below 8 h. At times below 8 h, points do not lie on the line of slope 1.99. The cost is constant up to 8 h/month, so a graph of cost versus time is horizontal for times up to 8 h.

Applications and Problem Solving

1. University students In 1971, there were 323 000 full-time university students in Canada. In 1995, there were 574 000. Find the average rate of change, to the nearest hundred students per year.

2. Televisions In 1970, 12.1% of Canadian households had colour televisions. In 1995, the figure was 98.5%. Find the average rate of change, to the nearest tenth of a percent per year.

271

3. Railways In 1947, Canadian railways logged 9 570 000 000 passenger-kilometres. In 1995, they logged 2 300 000 000 passenger-kilometres. Find the average rate of change, to the nearest million passenger-kilometres per year.

4. Running Bill jogged for 5 min, then sprinted 400 m in 64 s.
a) At what rate did he sprint?
b) Can you use the rate from part a) to find how far Bill jogged? Explain.

5. Thunderstorms Because light travels much faster than sound, you see lightning before you hear a thunderclap. If the storm is 960 m from you, the time interval between the flash and the thunderclap is 2.8 s. If the storm is 1680 m from you, the time interval is 4.9 s.
a) Determine the rate of change, to the nearest metre per second.
b) Describe the rate of change in words.
c) If the time interval is 3.7 s, determine your distance from the storm, to the nearest ten metres.
d) If you are 2500 m from the storm, what is the time interval, to the nearest tenth of a second?

6. Commission Mariko is paid a base salary plus commission for selling kitchen appliances. One week, her sales totalled $3800, and she earned $594. In a busier week, her sales totalled $5750, and she earned $652.50.
a) What rate of commission is Mariko paid? Express your answer as a percent.
b) What is her weekly base salary?
c) How much would she earn in a week if her sales totalled $4325?

7. Taxes Tax freedom day is the first day of the year on which average Canadians keep the money they earn. Before that day, everything they earn is paid in taxes to various levels of government. In 1961, tax freedom day was May 3. In 1996, tax freedom day was June 25.
a) Find the average rate of change in the tax freedom day, to the nearest tenth of a day per year.
b) Is there a limit to how long this average rate of change can continue? Explain.

8. Canada's population In 1871, Canada's population was 3 689 999. In 1996, it was 29 964 000.
a) Find the average rate of change, to the nearest thousand people per year.
b) Predict the year in which Canada's population reached 20 000 000. Then, use your research skills to check your prediction.
c) Predict the year in which Canada's population will reach 50 000 000.
d) Do you think that your prediction from part c) is valid? Explain.

9. Farming In 1931, the number of Canadians living on farms was 3 238 000. In 1991, the number was 807 000.
a) Find the average rate of change, to the nearest hundred people per year.
b) Predict the number of Canadians who lived on farms in 1966, to the nearest thousand. Then, use your research skills to check your prediction.
c) Use the data to predict the year in which no Canadians will be living on farms.
d) Do you think that your prediction from part c) is valid? Explain.

10. a) On a graph of y versus x using data you make up, is it possible to have a slope of 0? a positive slope? a negative slope? an undefined slope?
b) On a graph of distance versus time using real-world data, is it possible to have a slope of 0? a positive slope? a negative slope? an undefined slope? Explain.

11. Give three examples of rates of change that you use in your everyday life, and describe how you use them. Compare your examples with a classmate's.

NUMBER POWER

If $\dfrac{9^{28} - 9^{27}}{8} = 3^x$, what is the value of x?

Alpine Skiing in Canada

Use the *Skiing* database, from the Computer Database for ClarisWorks, to complete the following.

1 Vertical Drop

1. Find all the records for which *Summit Elevation, m* and *Base Elevation, m* are available.

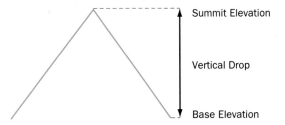

2. Add a calculation field called *Vertical Drop, m* for those records.

3. Sort the records from greatest to least vertical drop. In which province is the ski area with the greatest vertical drop?

2 Finding Slope

1. Find the records from those above for which *Longest Trail, km* is also available.

2. Add a calculation field called *Longest Trail, m* for those records.

3. Create a table for those records to display only the following fields.

Ski Area   Vertical Drop, m   Longest Trail, m

4. Outline a plan to calculate the slope of the longest trail using the two calculation fields — *Vertical Drop, m* and *Longest Trail, m*. Describe any other calculation fields you would add. What assumptions are you making?

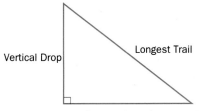

5. Compare your plan with a classmate's, and revise it if necessary.

6. Carry out your plan.

7. Sort the records from greatest to least slope of the longest trail. For which ski area does the longest trail have the greatest slope?

8. Explain, in terms of the assumptions you made, why the longest trail for that ski area might not actually have the greatest slope.

3 Ski Vacations

1. Consider the information provided in this database. What else would alpine skiers want to know about the ski areas when planning ski vacations? What fields would you add so that the database could be used to plan alpine ski vacations?

273

TECHNOLOGY

Programming a Graphing Calculator

1 Distance Between Two Points

The graphing calculator program shows how to calculate the distance between two points with coordinates (x_1, y_1) and (x_2, y_2).

```
PROGRAM:DISTANCE
:ClrHome
:Disp "X1="
:Input P
:Disp "Y1="
:Input Q
:Disp "X2="
:Input R
:Disp "Y2="
:Input S
:R–P→X
:S–Q→Y
:√(X²+Y²) →D
:Disp "DISTANCE"
:Disp "IS"
:Disp D
```

```
PROGRAM:DISTANCE
:ClrHome
:Disp "X1="
:Input P
:Disp "Y1="
:Input Q
:Disp "X2="
:Input R
```

1. Describe what each line of the program does.

2. Enter the program into your graphing calculator.
Use the program to find the distance between
a) (6, 7) and (1, 3), to the nearest tenth
b) (−4, −9) and (−2, 8), to the nearest tenth
c) (3.7, 16.2) and (11.8, −9.9), to the nearest hundredth

3. Write a computer program to find the distance between two points.

2 Midpoint of a Line Segment

1. Modify the distance program to find the coordinates of the midpoint of a line segment, given the coordinates of its endpoints.

2. Have a classmate check your program by using it to find the midpoint of a line segment with the following endpoints.
a) (4, 10) and (−6, 2)
b) (−3, −11) and (5, −1)
c) (3, −6) and (2, −1)
d) (2.4, −1.6) and (−1.8, −5.3)

3. Show that the diagonals of the parallelogram with vertices P(4, 1), Q(−2, −2), R(−9, 0), and S(−3, 3) bisect each other.

3 Endpoint of a Line Segment

1. Modify the midpoint program to find the coordinates of an endpoint of a line segment, given the coordinates of the midpoint and the other endpoint.

2. Given one endpoint and the midpoint, find the coordinates of the other endpoint.

	Endpoint	Midpoint
a)	(11, 8)	(7, 3)
b)	(−2, −1)	(−6, 5)
c)	(−4, 3)	(0.5, −1.5)

4 Slope of a Line

1. Modify the distance program to find the slope of a line, given the coordinates of two points on the line.

2. Have a classmate check your program by using it to find the slope of a line that passes through the following points.
a) (8, 14) and (2, 4)
b) (4, −1) and (16, −4)
c) (1.7, −3.8) and (−4.3, 9.7)

3. A line passes through the point (−1, 6) and has a slope of −3. Find one other point on the line. Use your slope program to check your answer.

4. The human heart pumps 18 L of blood in 3.6 min and 37.5 L in 7.5 min.
a) Determine the rate at which the heart pumps blood.
b) Determine the volume of blood the heart pumps in a day.

5. As the human heart grows, it beats more slowly. At one year of age, a person's normal pulse rate is 115 beats/min. At age 20, the normal rate is 75 beats/min.
a) Determine the average rate of change.
b) Predict the pulse rate at age 16.
c) Predict the age at which the pulse rate is 102 beats/min.
d) Predict the pulse rate of a newborn baby.
e) The pulse rate of a newborn is actually 140 beats/min. Does the rate of change you found in part a) apply before age 1? Explain.

5 Collinear Points

1. Program a calculator to determine whether three points are collinear, given the coordinates of the points.

2. Use the program to determine whether the following points are collinear.
a) (8, 5), (4, −1), and (−6, −16)   **b)** (−1, −1), (4, −11), and (−6, 8)

3. Two points on a line are given. Find one other point on the line. Use the program to check your answer.
a) (3, 4) and (5, 5)   **b)** (0, 3) and (3, −2)

Using Topographical Maps

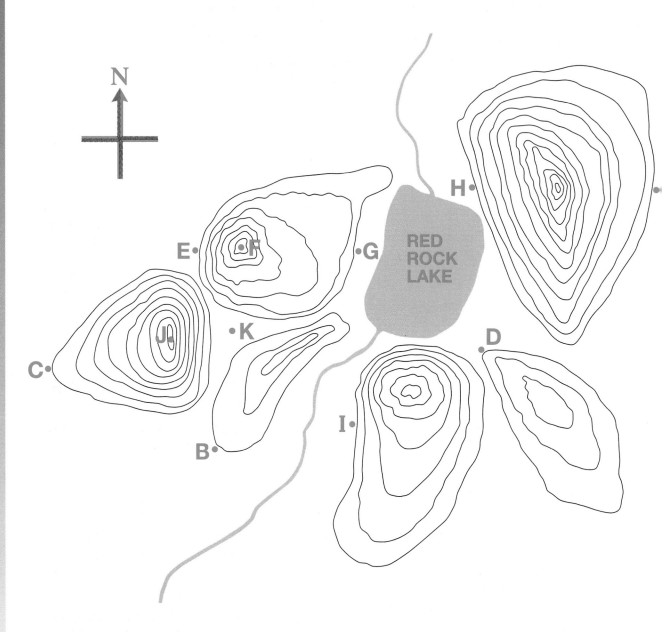

The map is called a topographical map. Like other maps, topographical maps show a top view of an area.

Topographical maps include contour lines, which are rings that represent changes in elevation. The change in elevation from one contour line to the next is given by the *contour interval* printed on the map. The contour interval on the above map is 10 m, so there is a change in elevation of 10 m from one contour line to the next.

Contour lines do not show whether a change in elevation is a rise or a drop. An easy way to decide is to look for lakes and streams. On the map, you can assume that the lake and the streams are at the lowest points. Therefore, in a group of contour lines, the innermost one shows the highest elevation relative to the surrounding area.

The scale for the map is 1:10 000. In other words, one unit of distance on the map represents 10 000 units on the Earth. Topographical maps are drawn with true north at the top of the map.

1 Sketching Views

If you were standing on the western shore of the lake looking east, this is the view you would see.

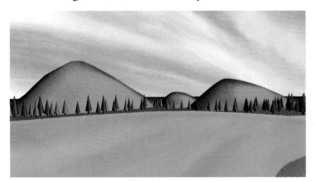

Sketch the view you would see if you were standing
a) on the eastern shore of the lake looking west
b) on the northern shore of the lake looking south
c) on top of the hill south of the lake looking north

2 Calculating Distances

1. How many metres on the Earth does 1 cm on the map represent?

2. What is the maximum straight-line distance across the lake, in metres?

3. How many metres above the lake is the top of the hill to the northeast of the lake?

4. What are the heights of the three hills to the west of the lake and the streams?

5. The easiest hiking trail between two points follows the shortest route over level ground.
a) Describe the easiest hiking trail from A to K. About how long is the trail?
b) Describe the easiest hiking trail from C to D. About how long is the trail?

6. a) What is the shortest distance from E to G if you go over the hill?
b) What is the distance from E to G if you go around the south side of the hill?
c) A rule of thumb for hiking is that the time taken to climb 10 m equals the time taken to hike 100 m on flat ground. The time taken to hike a certain distance downhill equals the time taken to hike the same distance on flat ground. What is the fastest way to get from E to G? How long would it take at 80 m/min?
d) What is the fastest way to get from A to H? How long would it take at 70 m/min?

3 Sketching Graphs

1. Sketch a graph of altitude versus distance if you took the most direct route from C to G.

2. Sketch a graph of altitude versus distance if you took the most direct route from A to I to E.

4 Using Slopes

1. Find the average slope for the climb from
a) E to F   **b)** G to F   **c)** C to J

2. If you walk the shortest distance from H to A, which is steeper — the climb or the descent?

3. Can you tell which way the stream to the south of Red Rock Lake is flowing? Explain.

4. If you hike from C to D along a trail that crosses no contour lines, do all parts of the trail have a slope of 0? Explain.

5. Suppose that one side of a hill is a vertical drop, but the other sides are sloping.
a) What is the slope of the vertical drop?
b) Sketch how the hill might be represented on a topographical map.

6.5 Linear Equations: Point-Slope Form

One way to find the age of a tree is to count the rings in the trunk. However, for a living tree, this method means cutting the tree down. For many types of trees, except for palms, firs, yews, horse chestnuts, and redwoods, the circumference of a tree gives its approximate age.

Explore: Develop a Method

The graph of the circumference of a tree, in centimetres, versus its age, in years, is linear.
One point on the line is (0, 0). The slope of the line is 2.5.
Let (x, y) represent any other point on the line.

a) Use the points (0, 0) and (x, y) to write and simplify an expression for the slope in the form $m = \dfrac{}{}$.

b) Substitute the given value of the slope for m.

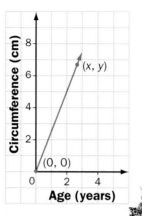

Inquire

1. a) Write the equation from b), above, in the form $y = \boxed{}\, x$, where $\boxed{}$ is a number.

b) How does the value of $\boxed{}$ compare with the slope?

2. Use the equation to find the circumference of a tree with an age of
a) 20 years   **b)** 85 years

3. Use the equation to find the age of a tree with a circumference of
a) 80 cm   **b)** 175 cm

4. Rewrite the equation in the form $\blacktriangle\, x + \bullet\, y = 0$, where $\blacktriangle$ and $\bullet$ are integers.

When the equation of a line is written in the form $Ax + By + C = 0$, the equation is in **standard form**.
For an equation in standard form, A, B, and C are integers, A and B are not both zero, and variables x and y represent real numbers.

The equation $2x + 3y - 7 = 0$ is in standard form.

The definition of slope can be used to find the equation of a line, given a point on the line and the slope of the line.

Example 1   Finding an Equation Given a Point and the Slope
a) Find an equation of the line that passes through $(1, 4)$ with slope $m = 3$.
b) Write the equation in standard form.
c) State the values of A, B, and C.

Solution
a) To find an equation that satisfies the given conditions, let (x, y) be any point on the line, other than $(1, 4)$. The slope of the line is 3.

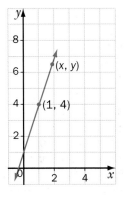

$$\frac{y_2 - y_1}{x_2 - x_1} = m$$

Substitute known values: $\dfrac{y - 4}{x - 1} = 3$

Multiply both sides by $x - 1$: $y - 4 = 3(x - 1)$

b) Simplify the equation by removing the brackets.
$$y - 4 = 3(x - 1)$$

Expand:
$$y - 4 = 3x - 3$$
Add 3 to both sides:
$$y - 4 + 3 = 3x - 3 + 3$$
$$y - 1 = 3x$$
Subtract $3x$ from both sides:
$$y - 1 - 3x = 3x - 3x$$
$$-3x + y - 1 = 0$$
$$\text{or } 3x - y + 1 = 0$$

For an equation in standard form, the value of A is usually not negative.

c) In the equation $3x - y + 1 = 0$, $A = 3$, $B = -1$, and $C = 1$.

In general, an equation for a line through (x_1, y_1) with slope m can be found as follows.

Let (x, y) be any point on the line, other than (x_1, y_1).

Then $\dfrac{y - y_1}{x - x_1} = m$

so $y - y_1 = m(x - x_1)$

Therefore, given a point on a line, (x_1, y_1), and the slope, m, an equation of the line may be expressed as
$y - y_1 = m(x - x_1)$

This equation is written in **point-slope form**.
The solution to Example 1 a), $y - 4 = 3(x - 1)$, is written in point-slope form.

Example 2   Using Point-Slope Form

Write an equation in standard form for the line through $(-4, -2)$ with slope $\dfrac{2}{3}$.

Solution

Use the point-slope form:

$(x_1, y_1) = (-4, -2)$ and $m = \dfrac{2}{3}$

Substitute known values:

Expand:

Multiply both sides by 3:

Expand:
Subtract 6 from both sides:

Subtract $3y$ from both sides:
Simplify:

$$y - y_1 = m(x - x_1)$$

$$y - (-2) = \frac{2}{3}(x - (-4))$$

$$y + 2 = \frac{2}{3}(x + 4)$$

$$y + 2 = \frac{2}{3}x + \frac{8}{3}$$

$$3 \times (y + 2) = 3 \times \left(\frac{2}{3}x + \frac{8}{3}\right)$$

$$3y + 6 = 2x + 8$$
$$3y + 6 - 6 = 2x + 8 - 6$$
$$3y = 2x + 2$$
$$3y - 3y = 2x + 2 - 3y$$
$$0 = 2x - 3y + 2$$

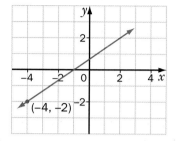

An equation of the line through $(-4, -2)$ with a slope of $\dfrac{2}{3}$ is $2x - 3y + 2 = 0$.

Example 3   Finding an Equation, Given Two Points

a) Write an equation in standard form for the line that passes through the points $(-2, 3)$ and $(2, -5)$.
b) Use the equation to find two other points on the line. Check the solutions.

Solution

a) To use the point-slope form to write the equation of a line, you need a point on the line and the slope.
Two points on the line are given. Use the two points to find the slope of the line.

$$m = \frac{y_2 - y_1}{x_2 - x_1}$$
$$= \frac{-5 - 3}{2 - (-2)}$$
$$= \frac{-8}{4}$$
$$= -2$$

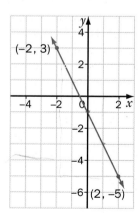

Use the point-slope form and either of the known points for (x_1, y_1).

Method 1
Using the point $(-2, 3)$ for (x_1, y_1),
$$y - y_1 = m(x - x_1)$$
$$y - 3 = -2(x - (-2))$$
$$y - 3 = -2(x + 2)$$
$$y - 3 = -2x - 4$$
$$2x + y + 1 = 0$$

Method 2
Using the point $(2, -5)$ for (x_1, y_1),
$$y - y_1 = m(x - x_1)$$
$$y - (-5) = -2(x - 2)$$
$$y + 5 = -2x + 4$$
$$2x + y + 1 = 0$$

An equation in standard form for the line through the points $(-2, 3)$ and $(2, -5)$ is $2x + y + 1 = 0$.

b) To find two other points on the line, it is convenient to solve the equation for y.
$$2x + y + 1 = 0$$
$$y = -2x - 1$$
When $x = 0$, $y = -2(0) - 1$
$$= -1$$
When $x = 1$, $y = -2(1) - 1$
$$= -2 - 1$$
$$= -3$$
Two other points on the line are $(0, -1)$ and $(1, -3)$.

Check:
Substitute the coordinates of the points into the equation $2x + y + 1 = 0$.

For $x = 0$ and $y = -1$,

L.S. $= 2x + y + 1$ **R.S.** $= 0$
$$= 2(0) + (-1) + 1$$
$$= 0$$

For $x = 1$ and $y = -3$,

L.S. $= 2x + y + 1$ **R.S.** $= 0$
$$= 2(1) + (-3) + 1$$
$$= 0$$

Example 4   Renting a Van

The U-Move-It company rents out vans for a daily charge, plus an amount for each kilometre driven. Renting a van and driving 150 km in a day costs $135. Renting a van and driving 220 km in a day costs $170.
a) Use the data to plot a graph.
b) Write an equation of the line in standard form.
c) Use the equation to find the cost of renting a van and driving 275 km in a day.

Solution

a) As the cost depends on the distance driven, plot the cost, in dollars, versus the distance, in kilometres. The ordered pairs are (150, 135) and (220, 170). Draw a line through the points. The line ends at the vertical axis, because it is impossible to drive a negative distance.

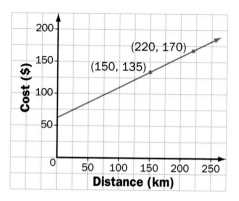

b) Replace x by d (distance) and y by c (cost). Find the slope of the line.

$$m = \frac{y_2 - y_1}{x_2 - x_1}$$
$$= \frac{c_2 - c_1}{d_2 - d_1}$$
$$= \frac{170 - 135}{220 - 150}$$
$$= \frac{35}{70}$$
$$= \frac{1}{2}$$

Use the point-slope form.

$$y - y_1 = m(x - x_1)$$
$$c - c_1 = m(d - d_1)$$

Substitute known values:
$$c - 135 = \frac{1}{2}(d - 150)$$

Multiply both sides by 2:
$$2c - 270 = d - 150$$

Write in standard form:
$$-d + 2c - 120 = 0$$
$$\text{or } d - 2c + 120 = 0$$

So, an equation of the line in standard form is $d - 2c + 120 = 0$.

281

c) Solve the equation for c:
$$d - 2c + 120 = 0$$
$$d + 120 = 2c$$
$$\frac{d + 120}{2} = c$$

Substitute the known value of d:
$$\frac{275 + 120}{2} = c$$
$$197.5 = c$$

The cost of renting a van and driving 275 km in a day is $197.50.

Note that, in Example 4, the slope of the line is $\frac{1}{2}$ or 0.5. The meaning of this rate of change is that the van rental company charges $0.50/km or 50¢ per kilometre driven.

Practice

Write an equation of the line that passes through the given point and has the given slope. Express the equation in standard form.

1. $(2, 3)$; $m = 4$
2. $(1, 4)$; $m = 3$
3. $(-5, 2)$; $m = 2$
4. $(3, -6)$; $m = -3$
5. $(-5, -1)$; $m = -2$
6. $(0, 7)$; $m = -1$
7. $(-6, 0)$; $m = 5$
8. $(5, 4)$; $m = 0$
9. $(1, -3)$; $m = 0$
10. $(2, 4)$; $m = \frac{1}{2}$
11. $(-3, 4)$; $m = -\frac{1}{3}$
12. $\left(-\frac{1}{2}, -5\right)$; $m = -\frac{1}{2}$
13. $\left(\frac{1}{2}, 6\right)$; $m = \frac{4}{3}$
14. $(-2, 1)$; $m = 1.5$

Write an equation of the line that passes through the given point and has the given slope. Use the equation to find two other points on the line. Check your solutions.

15. $(1, 5)$; $m = 1$
16. $(-2, 2)$; $m = 3$
17. $(4, 3)$; $m = -1$
18. $(3, -1)$; $m = -4$
19. $\left(1, \frac{1}{2}\right)$; $m = 2$
20. $\left(-\frac{3}{2}, 2\right)$; $m = -2$

Write an equation of the line that passes through the given points. Express the equation in standard form.

21. $(3, 4)$ and $(4, 6)$
22. $(-2, -4)$ and $(0, 6)$
23. $(3, 2)$ and $(6, -7)$
24. $(-5, -8)$ and $(-7, -9)$
25. $(-1, -2)$ and $(3, 0)$
26. $(3, -1)$ and $(9, -5)$
27. $(-1, 4)$ and $(3, 9)$
28. $(8, -7)$ and $(-6, -7)$
29. $(0, 4)$ and $(-5, 0)$
30. $(1.5, 2)$ and $(-1.5, -7)$

31. $(0.3, 0.4)$ and $(0.5, 0.7)$
32. $(3.4, -7.2)$ and $(2.2, -5.4)$
33. $\left(\frac{2}{3}, \frac{1}{4}\right)$ and $\left(\frac{1}{3}, \frac{1}{3}\right)$

Write an equation of the line that passes through the given points. Use the equation to find two other points on the line. Check your solutions.

34. $(1, 4)$ and $(3, 10)$
35. $(3, -3)$ and $(-6, 6)$
36. $(1, -4)$ and $(-3, 12)$
37. $(0, -1)$ and $(2, -2)$

Given the graph of a line, determine an equation of the line.

38.

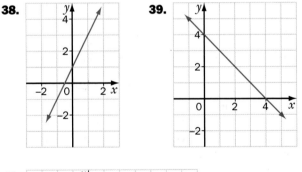

39.

40.

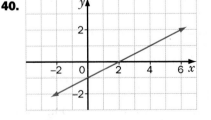

Applications and Problem Solving

41. Aviation The Strato 2C aircraft was built so that scientists could study the ozone layer. The number of hours the aircraft can fly depends on its altitude. The plane can cruise for 18 h at an altitude of 24 000 m, or for 48 h at an altitude of 18 000 m.
a) Plot a graph of time, in hours, versus altitude, in thousands of metres, for this relation.
b) Using a for altitude and t for time, find an equation of the line.
c) Use the equation to find the time for which the plane can fly at 20 000 m; at 26 000 m.

42. Find an equation of the line through $(a, 0)$ and $(0, b)$.

43. Show that the point $(7, 3)$ is on the line through the points $(3, 4)$ and $(-5, 6)$.

44. The three lines shown on the grid intersect to form triangle RST. Write the equation for each line in standard form.

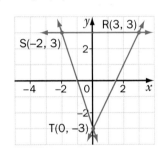

45. A rectangle has vertices K(2, 4), L(2, −6), M(−1, −6), and N(−1, 4). Write an equation in standard form for each diagonal of the rectangle.

46. The Mariana Trench The Mariana Trench in the Pacific Ocean is the deepest spot on Earth. The bottom of the trench is 11 000 m below the surface of the ocean. A winged submersible, called *Deep Flight*, allows scientists to explore the trench for the first time. Conventional submersibles rely on ballast to sink. *Deep Flight* is powered by thrusters to move it through the water. The ordered pairs (15, 1830) and (55, 6710) give the time, in minutes, that *Deep Flight* has been diving and its depth, in metres.
a) Calculate the slope of the line.
b) What rate of change does the slope represent?
c) Find an equation of the line.
d) Use the equation to determine how long it will take *Deep Flight* to reach the bottom of the Mariana Trench.

47. Space debris NASA scientists are worried about the amount of debris being put into Earth orbit, because collisions with particles as small as a paint chip can damage a space shuttle. The collisions happen at a speed of 35 000 km/h. The table gives the mass of space debris, in millions of kilograms, put into Earth orbit each year for four years.

Year	Mass Added (millions of kilograms)
1994	0.98
1995	1.02
1996	1.06
1997	1.1

a) Let 1994 = 1, 1995 = 2, and so on. Write the ordered pairs in the form (y, d), where y is the year and d is the mass of debris added in the year.
b) Calculate the slope of the line.
c) What rate of change does the slope represent?
d) Find an equation of the line.
e) Use the equation to find the mass of debris that will be put into orbit in the year 2010.

48. Temperature in the Earth The average temperature on the Earth's surface is 15°C. For every kilometre you go down into the Earth's crust, the temperature increases by 3.5°C.
a) Graph this relation.
b) Find an equation of the line.

49. Algebra An equation of a line is $kx - 5y + 6 = 0$. If the line passes through $(2, 4)$, what is the value of k?

50. Algebra An equation of a line is $8x + ky - 6 = 0$. If the line passes through $(1, -2)$, what is the value of k?

51. Algebra For the line $3x - 2y + 10 = 0$, find the coordinates of a point for which
a) the x-coordinate is 4 times the y-coordinate
b) the y-coordinate is 5 more than the x-coordinate

52. In the standard form of a linear equation, $Ax + By + C = 0$, A and B cannot both be zero. Explain why.

53. a) Can the point-slope form be used to write an equation of the vertical line through the point $(1, 3)$? Explain.
b) Write an equation of the line in standard form.

TECHNOLOGY

Graphing Calculators and Families of Lines

Lines that share a common characteristic are said to belong to a **family of lines**. For example, all five lines in the family shown in the graph pass through the point (1, 2).

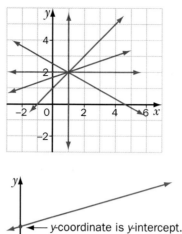

The y-coordinate of the point where a line crosses the y-axis is called the **y-intercept**. The x-coordinate of this point is 0.

y-coordinate is y-intercept.

1 Slope and y-Intercept

1. a) Use a graphing calculator to graph the equation $y = 2x - 1$.
b) Use the table feature of the calculator to display a table of values for the equation. Use the table to determine the y-intercept, that is, the value of y when $x = 0$. If your calculator does not have a table feature, find the y-intercept by substituting 0 for x in the equation $y = 2x - 1$.
c) Use the table of values or substitute a different integer for x to find the coordinates of another point on the line.
d) Use the coordinates from parts b) and c) to calculate the slope of the line.
e) Where do the values of the slope and the y-intercept appear in the equation $y = 2x - 1$?

2. State the slope and the y-intercept of each of the following lines. Use a graphing calculator to check your answers.
a) $y = 2x + 1$ **b)** $y = 0.5x + 2$ **c)** $y = -x - 1$

2 A Family of Lines

1. Graph the following equations so that they all appear in the same standard viewing window of your graphing calculator.
a) $y = 0.5x + 1$ **b)** $y = 0.5x$ **c)** $y = 0.5x - 2$

2. What is the y-intercept of each line?

3. What is the slope of each line?

4. Describe this family of lines in words by noting how the lines are the same and how they are different.

5. Write the equation of another line that belongs to this family. Use your graphing calculator to check your answer.

3 A Different Family of Lines

1. Graph the following equations so that they all appear in the same standard viewing window of your graphing calculator.
a) $y = 3x + 1$ **b)** $y = x + 1$ **c)** $y = -0.5x + 1$

2. What is the y-intercept of each line?

3. What is the slope of each line?

4. Describe this family of lines in words by noting how the lines are the same and how they are different.

5. Write the equation of another line that belongs to this family. Use your graphing calculator to check your answer.

4 Related Lines

1. Graph the following equations so that they appear in the same standard viewing window of your graphing calculator.

a) $y = \dfrac{3}{2}x + 3$ **b)** $y = \dfrac{3}{2}x + 1$

c) $y = \dfrac{3}{2}x$ **d)** $y = \dfrac{3}{2}x - 2$

2. What is the y-intercept of each line?

3. What is the slope of each line?

4. Do the four lines belong to a family? Explain.

5. a) Use the zoom square or equivalent feature on your calculator to adjust the viewing window so that one unit on the x-axis is equal in length to one unit on the y-axis.
Then, graph the equation $y = -\dfrac{2}{3}x - 1$ in this viewing window with the four graphs from question 1.
b) At what angle does the fifth line appear to intersect the other four?
c) What is the slope of the fifth line?
d) What is the product of the slope of the fifth line and the slope of each of the other four lines?

6. a) Graph the equations $y = 2x + 2$ and $y = -\dfrac{1}{2}x - 1$ in one viewing window, adjusted so that the units of length on the x- and y-axes are equal.
b) How do the lines appear to be related?
c) What is the slope of each line?
d) What is the product of the slopes?

6.6 Linear Equations: Slope and *y*-Intercept Form

As the altitude of an aircraft increases, the outside air temperature decreases. An equation that shows the relationship between temperature and altitude for a ground-level temperature of 20°C is $150t + a = 3000$, where t is the temperature in degrees Celsius and a is the altitude in metres.

Explore: Use the Graphs

The **y-intercept** of a relation is the *y*-coordinate of the point where the relation intersects the *y*-axis. In general, the coordinates of the *y*-intercept are $(0, b)$. The *y*-intercept is b. The *x*-coordinate has a value of zero at this point.

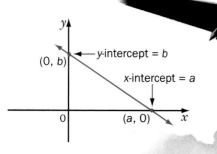

The **x-intercept** of a relation is the *x*-coordinate of the point where the relation intersects the *x*-axis. In general, the coordinates of the *x*-intercept are $(a, 0)$. The *x*-intercept is a. The *y*-coordinate has a value of zero at this point.

Copy and complete the following table using these steps.
• Solve each equation for *y*, if necessary.
• Graph each equation on a grid or a graphing calculator.
• Find the coordinates of two points on each line.
• Find the slope of each line using the coordinates of the two points.
• Find the *y*-intercept of each line.

	Equation	y = ▓▓ Form	Points	Slope	y-Intercept
a)	$y = 2x + 3$	$y = 2x + 3$	(1,5) (2,7)	2	3
b)	$x + y = 2$	$y = -x + 2$	(1,1) (2,0)	-1	2
c)	$y - x = 0$	$y = x + 0$	(1,1) (2,2)	1	0
d)	$3x + y + 1 = 0$	$y = x - 1/3$	(1,0) (2,½)	1/3	-1/3
e)	$2x + 2y + 6 = 0$	$y = -x - 3$	(1,-4) (2,-5)	-1	-3

Inquire

1. When an equation is solved for y, what is the relationship between
a) the numerical coefficient of x and the slope?
b) the constant term and the y-intercept?

2. Solve each of the following equations for y. Then, state the slope and y-intercept of each line.
a) $3y = 6x + 1$ **b)** $2x + y + 4 = 0$ **c)** $x - y = 2$ **d)** $x + 2y - 1 = 0$

3. The equation $y = mx + b$ is called the **slope and y-intercept form** of a linear equation. Explain why.

4. a) Write the equation for the relationship between the temperature outside an aircraft and the altitude in the slope and y-intercept form, $t = $ ▓. $t = 20 - \dfrac{a}{150}$
b) What is the value of t where the line crosses the t-axis? What does this value of t represent?
c) What is the slope of the line? What rate of change does the slope $-.13$ represent?

Example 1   Slope and y-Intercept Form
Write an equation of the line through $(0, -3)$ with slope -2.

Solution

Use the point-slope form: $y - y_1 = m(x - x_1)$
Substitute known values: $y - (-3) = -2(x - 0)$
Simplify: $y + 3 = -2x$
Write in slope and y-intercept form: $y = -2x - 3$

An equation of the line through $(0, -3)$ with slope -2 is $y = -2x - 3$.

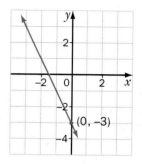

In general, an equation for a line through $(0, b)$ with slope m can be found as follows.

$y - y_1 = m(x - x_1)$
$y - b = m(x - 0)$
$y - b = mx$
$y = mx + b$

The graph of the equation of a line expressed in the form $y = mx + b$, where x and y are real numbers, has a slope m and a y-intercept b.

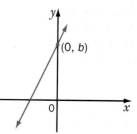

Example 2   Using the Slope and y-Intercept Form
A line has a slope of 3 and a y-intercept of 4. Write an equation of the line in standard form.

Solution

Use the slope and y-intercept form: $y = mx + b$
Substitute known values: $y = 3x + 4$
Write in standard form: $-3x + y - 4 = 0$
                 or $3x - y + 4 = 0$

An equation of the line with slope 3 and y-intercept 4 is $3x - y + 4 = 0$.

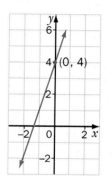

287

Example 3   Finding the Slope and y-Intercept

Determine the slope and y-intercept of the line $2x - 3y - 6 = 0$.

Solution

Write the equation in the form $y = mx + b$.

$$2x - 3y - 6 = 0$$

Isolate the y-term:

$$-3y = -2x + 6$$

Divide both sides by -3:

$$\frac{-3y}{-3} = \frac{-2x}{-3} + \frac{6}{-3}$$

$$y = \frac{2}{3}x - 2$$

$$y = mx + b$$

The slope $m = \dfrac{2}{3}$, and the y-intercept $b = -2$.

Practice

Find the slope and y-intercept of each line.

1. $y = 3x + 1$

2. $y = \dfrac{1}{2}x - 2$

3. $y = -4x + 3$

4. $x + y = 5$

5. $x + y - 7 = 0$

6. $y + 4 = 5x$

7. $y - 2x = 0$

8. $y = 3$

Find the slope and y-intercept of each line.

9. $4x + 2y = 3$

10. $x - y = 4$

11. $3x + 2y + 6 = 0$

12. $2y + 6 = 0$

13. $x - 3y - 9 = 0$

14. $5x + 2y = 10$

15. $22x + 0.5y - 1 = 0$

16. $0 = x - 2y - 4$

17. $6.4x = 0.8y$

18. $1.2x - 0.3y = 0.12$

Given the slope and y-intercept, write an equation of the line in the slope and y-intercept form. Then, write the equation in standard form.

19. $m = 2; b = 3$

20. $m = 3; b = -2$

21. $m = -4; b = 6$

22. $m = -5; b = -7$

23. $m = \dfrac{1}{2}; b = 1$

24. $m = \dfrac{2}{3}; b = -2$

25. $m = -0.5; b = 0$

26. $m = -\dfrac{2}{5}; b = -\dfrac{1}{3}$

Draw the graph of each line.

27. $y = 2x + 5$

28. $y = 3x - 1$

29. $y = -x + 4$

30. $y = \dfrac{2}{3}x - 2$

31. $y = -\dfrac{3}{4}x + 3$

32. $y = -0.5x$

Find the slope and y-intercept of each line. Then, write an equation of the line.

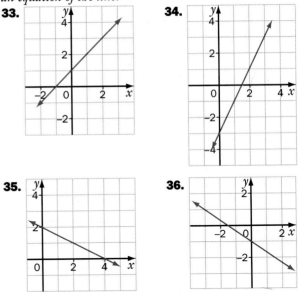

33.

34.

35.

36.

Applications and Problem Solving

37. Algebra An equation of a line is $y = 2x + b$. Find the value of b if the line passes through the point

a) $(4, 2)$ **b)** $(-3, 5)$ **c)** $(2, -6)$ **d)** $(-1, -3)$

38. Algebra An equation of a line is $y = mx + 3$. Find the value of m if the line passes through the point

a) $(2, 1)$ **b)** $(-4, 5)$ **c)** $(4, -5)$ **d)** $(-1, -6)$

288

39. Explain why these lines belong to a family.

$$y = \frac{3}{2}x - 1 \qquad\qquad 2x + y + 1 = 0$$
$$5x - 2y - 2 = 0 \qquad\qquad y + 1 = 0$$

40. **Box of crackers** The equation $m = 6n + 55$ relates the mass, in grams, of a box of crackers to the number of crackers, n, in the box.
a) Graph the line.
b) What is the mass of the empty box?
c) What is the mass of each cracker?
d) If the total mass of a box full of crackers is 355 g, how many crackers does the box hold?

41. **Driving distances** Virden, Manitoba, is 80 km west of Brandon along the Trans-Canada Highway. Marika drove west from Virden to Swift Current, Saskatchewan. Her average speed was 75 km/h.
a) Plot a graph of her distance, d kilometres, from Virden versus her driving time, t hours. Write an equation of the line in the form $d = $ ▨ .
b) Repeat part a), but use the distance from Brandon instead of the distance from Virden.
c) If Marika drove for a total of 7 h, what is the driving distance to Swift Current from Virden? from Brandon?

42. **Dog years** For dogs that are two or more years old, the equation $5x - y + 12 = 0$ compares dog years to human years. In the equation, x represents the actual age of a dog, and y is the age of the dog in human years.
a) Write the equation in the form $y = mx + b$.
b) What is the age in human years of a 5-year-old dog?
c) What is the slope of the line?
d) What rate of change does the slope represent?

43. **School trip** A group of students is travelling by bus to visit a planetarium. The equation $13n - 2c + 500 = 0$ relates the total cost of the trip, c dollars, to the number of students, n. The total cost includes the bus rental charge and admission to the planetarium.
a) Graph the total cost versus the number of students.
b) What is the cost of renting the bus?
c) What is the admission cost per student to the planetarium?

44. **Treasure hunting** Finding treasure on sunken ships can be a very profitable business. One important consideration for treasure-hunting divers is the pressure they experience. The pressure is a function of depth. At a depth of 4 m, the pressure is 140 kPa (kilopascals). At a depth of 9 m, the pressure is 190 kPa.
a) Plot the points on a grid in the form (d, p), where d is the depth, in metres, and p is the pressure, in kilopascals. Join the points with a line.
b) Find the equation of the line and write it in the form $p = md + b$.
c) What rate of change does the slope represent?
d) What is the value of the p-intercept, and what does it represent?
e) At what depth is the pressure double the pressure on the surface?
f) The pirate Blackbeard's ship sank off the coast of North Carolina around 1718. It was discovered in 1997, at a depth of 6 m. What was the pressure on the divers at this depth?

45. **Algebra** The equation of a line is $x + py - q = 0$. The slope is $-\frac{1}{2}$, and the y-intercept is 3. What are the values of p and q?

46. Is it possible to write the equation of the line $x = -2$ in the slope and y-intercept form? Explain.

47. Do lines with equations of the form $y = mx$ all belong to a family? Explain.

48. **Measurement** The sum, s degrees, of the interior angles of an n-sided polygon is given by the equation $s = 180(n - 2)$.
a) Write the equation in the slope and y-intercept form.
b) What is the value of the s-intercept?
c) Does the s-intercept have any meaning for a real polygon? Explain.
d) What is the smallest value n can have?

NUMBER POWER

Find four consecutive whole numbers so that the sum of the cubes of the three smallest numbers equals the cube of the largest number.

CAREER CONNECTION

Oceanography

Oceans and seas, with a total surface area of 3.6×10^8 km², cover about 70% of the Earth's surface. Their physical aspects, such as salinity, temperature, and movement of water, are studied by oceanographers. The distribution, population, and life cycles of plants and animals of the oceans are also the concern of oceanographers.

1 Antarctic Ice

International teams of oceanographers study the waters of Antarctica. This continent is covered with frozen fresh water that is thousands of metres thick. Around its perimeter, there is another type of ice, called sea ice, which is formed from salty ocean water.

The graph shows how the area of sea ice increases and decreases during a year.

1. What is the area of the sea ice, in square kilometres, for the first 60 days of the year?

2. a) By how much does the area of the sea ice increase between day 60 and day 240?
b) How does this increase compare with the area of Canada?

3. a) What is the slope of the line segment between day 60 and day 240?
b) What rate of increase does the slope represent?
c) How fast does the area of the sea ice increase each hour? each minute?

4. What happens to the sea ice between day 240 and day 270?

5. What happens to the sea ice between day 270 and day 360?

6. a) What is the slope of the line segment between day 270 and day 360?
b) What rate of decrease does the slope represent?
c) How fast does the area of the sea ice decrease each hour? each minute?

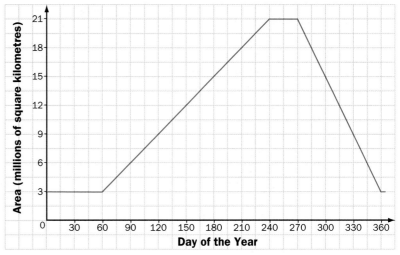

7. How does the slope you found in question 6 compare with the slope you found in question 3?

2 Antarctic Krill

Krill are crustaceans that grow under the Antarctic sea ice. They feed the millions of penguins, millions of seals, and thousands of whales in the Antarctic. The total mass of krill is estimated to be 1.5 billion tonnes.

1. Estimate the total mass, in tonnes, of all the humans on Earth.

2. How does the total mass of krill compare with the total mass of humans?

3 Locating Information

Use your research skills to find information about the education needed for a career in oceanography. Share your information with your classmates.

6.7 Parallel and Perpendicular Lines

A laser produces a narrow beam of very bright light. One application of a laser is to create holograms, which are three-dimensional images. Part of the process of forming a hologram involves splitting a laser beam into two perpendicular beams. Some aircraft use a hologram for projecting data onto a see-through screen above the instrument console.

Explore: Use the Graphs

Parallel Lines

a) Place a ruler on a grid so that each edge passes through at least two points whose coordinates are integers.

b) Draw a line along each edge of the ruler. Label the parallel lines AB and CD, so that the points A, B, C, and D have integer coordinates.

c) Use the labelled points to calculate the slope of each line. Compare the slopes.

d) Repeat the steps for a different pair of parallel lines.

e) Compare your results with a classmate's.

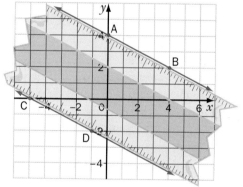

Perpendicular Lines

a) Place the right-angled corner of a piece of paper, a plastic triangle, or a set square on a grid at a point with integer coordinates. Label the point P.

b) Rotate the paper or triangle about P until each arm of the right angle passes through at least one point with integer coordinates. The arms should not be horizontal or vertical. Label the two points A and B. Draw the perpendicular lines PA and PB.

c) Calculate the slope of each line. Compare the slopes.

d) Repeat the steps for a different pair of perpendicular lines.

e) Compare your results with a classmate's.

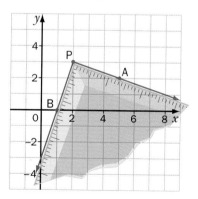

Inquire

1. How are the slopes of parallel lines related?

2. How are the slopes of perpendicular lines related?

3. What is the product of the slopes of two perpendicular lines?

4. Find the slope of each line by writing the equation in the form $y = mx + b$.
a) $2x - y = 2$　**b)** $x + y = -2$　**c)** $x - y = 1$　**d)** $y - 2x = -3$

5. Which lines in question 4 are parallel? Explain.

6. Which lines in question 4 are perpendicular? Explain.

7. Suppose that a laser beam has been split into two beams to create a hologram.
a) If a grid were superimposed on the two beams, how would their slopes be related? Explain.
b) Could the beams point in directions in the coordinate plane that would make the answer to part a) untrue? Explain.

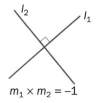

Parallel Lines
• Two non-vertical lines are parallel if they have the same slope.
• All vertical lines, which have undefined slopes, are parallel.

Perpendicular Lines
• Two lines that are not vertical or horizontal are perpendicular if the product of their slopes is –1.
• A vertical line is perpendicular to a horizontal line.

Example 1　Parallel Lines
Write an equation of the line parallel to $5x - 8y + 12 = 0$ and through the point $(-2, 3)$.

Solution
An equation of a line can be found from the slope of the line and a point on the line. The slope of the new line will equal the slope of $5x - 8y + 12 = 0$, because the lines are parallel.
To find the slope of $5x - 8y + 12 = 0$, write the equation in the form $y = mx + b$.

$$5x - 8y + 12 = 0$$

Isolate the y-term:　　$-8y = -5x - 12$

Divide both sides by –8:　　$y = \dfrac{-5x}{-8} - \dfrac{12}{-8}$

$$= \dfrac{5x}{8} + \dfrac{3}{2}$$

The slope of $5x - 8y + 12 = 0$ is $\dfrac{5}{8}$.

The slope of a line parallel to $5x - 8y + 12 = 0$ is also $\dfrac{5}{8}$.

So, for the required line, $m = \dfrac{5}{8}$ and $(x_1, y_1) = (-2, 3)$.

Use the point-slope form: $\qquad\qquad\qquad y - y_1 = m(x - x_1)$

Substitute known values: $\qquad\qquad\quad y - 3 = \dfrac{5}{8}(x - (-2))$

$$y - 3 = \dfrac{5}{8}(x + 2)$$

Expand: $\qquad\qquad\qquad\qquad\qquad y - 3 = \dfrac{5}{8}x + \dfrac{5}{4}$

Multiply both sides by 8: $\qquad\quad 8 \times (y - 3) = 8 \times \left(\dfrac{5}{8}x + \dfrac{5}{4} \right)$

Expand: $\qquad\qquad\qquad\qquad\qquad 8y - 24 = 5x + 10$

Write in standard form: $\qquad\qquad -5x + 8y - 34 = 0$

$$\text{or } 5x - 8y + 34 = 0$$

An equation of the line is $5x - 8y + 34 = 0$.

Example 2   Perpendicular Lines

Two perpendicular lines intersect on the y-axis. An equation of one of the lines is $y = 2x + 4$. Find an equation of the second line.

Solution

The two lines intersect on the y-axis.
The line $y = 2x + 4$ has a y-intercept of 4.
So, the lines intersect at $(0, 4)$.

If the equation is not in the slope and y-intercept form, you can find the y-intercept by substituting 0 for x.

The slope of the line $y = 2x + 4$ is 2.
The slope, m, of the line perpendicular to $y = 2x + 4$ is given by $2 \times m = -1$

$$m = -\dfrac{1}{2}$$

Here are two methods to find the required equation.

Method 1
Use the point-slope form.

$m = -\dfrac{1}{2}$ and $(x_1, y_1) = (0, 4)$

$$y - y_1 = m(x - x_1)$$
$$y - 4 = -\dfrac{1}{2}(x - 0)$$
$$y - 4 = -\dfrac{1}{2}x$$
$$2y - 8 = -x$$
$$x + 2y - 8 = 0$$

Method 2
Use the slope and y-intercept form.

$m = -\dfrac{1}{2}$ and $b = 4$

$$y = mx + b$$
$$y = -\dfrac{1}{2}x + 4$$
$$2y = -x + 8$$
$$x + 2y - 8 = 0$$

An equation of the second line is $x + 2y - 8 = 0$.

Practice

Given the slopes of two lines, determine whether the lines are parallel, perpendicular, or neither.

1. $m_1 = \dfrac{2}{3}$, $m_2 = \dfrac{3}{2}$

2. $m_1 = \dfrac{1}{2}$, $m_2 = \dfrac{4}{8}$

3. $m_1 = -3$, $m_2 = \dfrac{1}{3}$

4. $m_1 = \dfrac{2}{5}$, $m_2 = -\dfrac{2}{5}$

5. $m_1 = 1$, $m_2 = -1$

6. $m_1 = 2$, $m_2 = -2$

7. $m_1 = \dfrac{20}{25}$, $m_2 = \dfrac{4}{5}$

8. $m_1 = \dfrac{1}{4}$, $m_2 = -\dfrac{3}{12}$

9. $m_1 = -0.5$, $m_2 = 2$

10. $m_1 = -0.1$, $m_2 = 1$

Find the slope of a line perpendicular to a line with the given slope.

11. 3   **12.** $\dfrac{3}{4}$ **13.** $-\dfrac{1}{8}$

14. $-\dfrac{5}{2}$ **15.** -6 **16.** undefined

State the slope of a line
a) *parallel to each line*
b) *perpendicular to each line*

17. $y = 2x - 4$ **18.** $y = -x - 2$

19. $y = \dfrac{1}{3}x + 5$ **20.** $y = -\dfrac{2}{7}x$

21. $y + 4x = 4$ **22.** $3x + y = -2$
23. $2x - y = 8$ **24.** $5x + 3 = 2y$
25. $2x + 3y - 1 = 0$ **26.** $3x - 5y + 2 = 0$

27. Given the coordinates of the following 8 points, determine whether each pair of lines is parallel, perpendicular, or neither.
A(3, 2), B(6, 4), C(–8, –2), D(–2, 2), E(8, 1), F(2, 0), G(11, 3), H(6, –3)
a) AB and CD   **b)** DF and CH
c) CD and BE   **d)** AH and GH
e) DF and AF   **f)** BG and AE
g) FC and DG   **h)** BE and AB

The following are slopes of parallel lines. Find the value of the variable.

28. $2, -\dfrac{6}{m}$ **29.** $-3, \dfrac{w}{4}$

30. $-\dfrac{2}{3}, -\dfrac{n}{9}$ **31.** $\dfrac{z}{3}, \dfrac{1}{2}$

32. $-\dfrac{x}{3}, -\dfrac{2}{5}$ **33.** $\dfrac{2}{k}, -\dfrac{4}{5}$

The following are slopes of perpendicular lines. Find the value of the variable.

34. $3, \dfrac{m}{6}$ **35.** $2, \dfrac{2}{q}$

36. $-\dfrac{1}{2}, \dfrac{4}{w}$ **37.** $\dfrac{2}{3}, \dfrac{x}{4}$

38. $\dfrac{z}{9}, -\dfrac{3}{5}$ **39.** $\dfrac{4}{t}, \dfrac{9}{2}$

Identify whether each pair of lines is parallel, perpendicular, or neither.
40. $x - y + 1 = 0$ and $4x + 4y + 1 = 0$
41. $3x - 2y + 12 = 0$ and $-2x - 3y - 12 = 0$
42. $2x + 5y - 13 = 0$ and $2x - 5y + 23 = 0$
43. $x + 9y + 1 = 0$ and $9x + y + 1 = 0$

Determine an equation of each of the following lines.
44. the line parallel to $2x - 3y + 1 = 0$ and passing through the point (1, 2)
45. the line perpendicular to $x - 5y + 2 = 0$ and passing through the point (–2, 5)
46. the line parallel to $x + 3 = 0$ and passing through (–6, –7)
47. the line perpendicular to $y - 4 = 0$ and passing through (–1, 6)
48. the line parallel to $x + 9y - 2 = 0$ and having the same x-intercept as the line $2x - 9y + 27 = 0$
49. the line perpendicular to $3x - 12y + 16 = 0$ and having the same y-intercept as $14x - 13y - 52 = 0$

50. Two perpendicular lines intersect on the x-axis. An equation of one line is $y = 3x + 1$. Find an equation of the other line.

51. Plot the points R(5, 3), S(–1, 0), and T(3, –1) on a grid. Determine an equation of
a) the line through R and parallel to ST
b) the line through S and perpendicular to RT

Applications and Problem Solving

52. Explain why these lines belong to a family.

$$y = -\frac{1}{2}x + 3 \qquad\qquad x + 2y + 1 = 0$$

$$3x + 6y - 7 = 0 \qquad\qquad y - 2 = -\frac{1}{2}(x - 3)$$

Plot and join the points in order on a grid. Classify each figure as a square, rectangle, parallelogram, or trapezoid. Give reasons for each answer.

53. P(–2, 5), Q(–4, 3), R(4, –5), and S(5, –2)
54. E(–1, 0), F(3, 1), G(2, 5), and H(–2, 4)
55. A(1, 3), B(–2, –3), C(2, –5), and D(5, 1)
56. U(–1, 2), V(–4, –1), W(–1, –4), and Z(2, –1)
57. J(3, –1), K(2, 2), L(–4, 4), and M(–3, 1)

58. Write an equation of the line that is perpendicular to the y-axis and has a y-intercept of –2.

59. Comparing journeys Hugo drove 255 km from Calgary to Crowsnest Pass at 85 km/h. He left Calgary at 09:00. Kayla drove along the same route at the same speed, but she left Calgary at 11:00.
a) Graph the distance driven versus the time of day for both journeys on the same set of axes.
b) How are the lines related? Why are they related in this way?

60. Appliance repairs Kate and Marius repair electrical appliances, such as washing machines and stoves. Kate charges a fee of $50 for a service call, whereas Marius charges $40. In addition, they each charge $35.00/h for completing the repairs.
a) On the same set of axes, graph the total cost of a service call versus the time taken to complete the repairs for Kate and for Marius.
b) For jobs that take the same length of time, which person always charges more — Kate or Marius? Explain why.

61. Algebra Find the value of k if the lines $3x - 2y - 5 = 0$ and $kx - 6y + 1 = 0$ are
a) parallel   **b)** perpendicular

62. Algebra a) Find the values of k for which the lines $kx - 2y - 1 = 0$ and $8x - ky + 3 = 0$ are parallel.
b) Are there any values of k that would make the two lines perpendicular? Explain.

63. Given the equations $2y = 3x + 4$ and $8y = 12x + 16$,
a) what are the slopes?
b) are the lines parallel? Explain.

64. State whether the two lines on each grid are perpendicular. Explain.

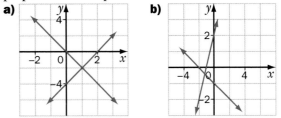

65. Write the equations of lines that form the sides of a square, so that no sides are vertical or horizontal. Have a classmate check your equations by graphing the square.

LOGIC POWER

Copy the grid, with the letters A, E, I, and O placed as shown.

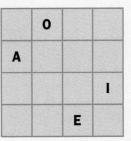

Place the letters B, C, D, F, G, H, J, K, L, M, N, and P in the empty squares of the grid, according to the following directions.

- H and B are in the top row.
- F is in the left column.
- G, M, and N are in corners.
- P is in the third column but not in the third row.
- L is in the same row as E.
- C is in the same column as I.
- No row or column contains two consecutive letters of the alphabet.

6.8 Graphing Linear Equations

Imagine that the Great Lakes were drained of water. Then, imagine pouring all the world's oil reserves, known and estimated, into the lakebeds. The world's oil reserves would fill less than 5% of Lake Superior. The other lakes would be empty.

Explore: Draw a Graph

Oil geologists use an equation of a line to estimate the world's remaining oil reserves, starting at the year 2000. The equation is $b + 20n = 1300$, where b represents the world's oil reserves, in billions of barrels, and n represents the number of years after the year 2000, starting with the year 2000 equal to 0.

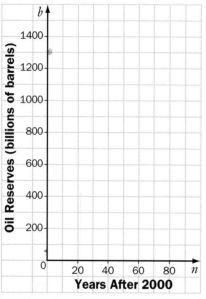

a) If $n = 0$, what is the value of b?
b) If $b = 0$, what is the value of n?
c) Use the solutions to a) and b) to write two ordered pairs in the form (n, b).
d) Draw axes like the ones shown and plot the ordered pairs. Join the ordered pairs with a straight line.

Inquire

1. What does the b-intercept represent?
2. What does the n-intercept represent?
3. Use the graph to find the reserves in the year 2030; 2045.
4. Find the slope of the line.
5. What rate of change does the slope represent?
6. What world conditions might change the slope?

It is often convenient to use the x- and y-intercepts to draw the graph of a line.

Example 1 Graphing Using the Intercepts

a) Use the intercepts to graph $3x - 4y = 12$. The domain is R. State the range.
b) Find the slope.

Solution

a) To find the x-intercept, let $y = 0$.

$$3x - 4y = 12$$
$$3x - 4(0) = 12$$
$$3x = 12$$
$$x = 4$$

One point on the line is $(4, 0)$.

To find the y-intercept, let $x = 0$.

$$3x - 4y = 12$$
$$3(0) - 4y = 12$$
$$-4y = 12$$
$$y = -3$$

Another point on the line is $(0, -3)$.
Plot the points on a grid.
Because the domain is R, draw a line through the points.
The range is also R.

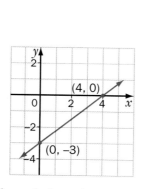

b) Use the points $(4, 0)$ and $(0, -3)$ to find the slope.

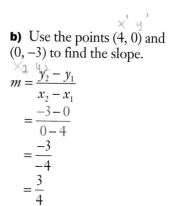

$$m = \frac{y_2 - y_1}{x_2 - x_1}$$
$$= \frac{-3 - 0}{0 - 4}$$
$$= \frac{-3}{-4}$$
$$= \frac{3}{4}$$

Example 2 Graphing Using Different Methods

Graph the equation $y = \frac{1}{2}x + 1$, where x is a real number.

Solution 1
Use any two points.

$$y = \frac{1}{2}x + 1$$

When $x = 2, y = 2$.
When $x = 4, y = 3$.
Plot $(2, 2)$ and $(4, 3)$.
Draw the graph.

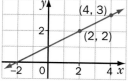

Solution 2
Use the intercepts.

$$y = \frac{1}{2}x + 1$$

When $x = 0, y = 1$.
When $y = 0, x = -2$.
Plot $(0, 1)$ and $(-2, 0)$.
Draw the graph.

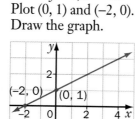

Solution 3
Use the slope and the y-intercept.

$$y = \frac{1}{2}x + 1$$

The y-intercept is 1. The slope is $\frac{1}{2}$.

Plot the point $(0, 1)$. Use the slope to locate another point on the line. Draw the graph.

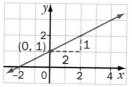

Solution 4
Use a graphing calculator.

Practice

Use the x- and y-intercepts to graph each line.

1. $3x + 2y = 6$ **2.** $4x + 3y = 12$
3. $x + 2y = 8$ **4.** $2x + y - 4 = 0$
5. $2x - 5y = 10$ **6.** $5x - 4y + 20 = 0$
7. $x - y - 3 = 0$ **8.** $x - 3y + 6 = 0$

Graph each equation using the slope and y-intercept.

9. $y = 3x - 2$ **10.** $y = 0.5x + 4$

11. $y = -4x - 1$ **12.** $y + 3 = \frac{1}{3}x$

13. $2x + y = 0$ **14.** $3x + 2y - 6 = 0$

Graph using a method of your choice. Find the intercepts, slope, and range for each line. The domain is R.

15. $y = 3x - 9$ **16.** $5x + y + 5 = 0$

17. $y = 2(x - 3)$ **18.** $y + 4 = \frac{1}{2}x$

19. $y = -\frac{1}{3}x + 2$ **20.** $y + 2 = -(x + 1)$

21. $\frac{x}{3} + \frac{y}{2} = 1$ **22.** $3x - 4y = -24$

Find an equation for each of the following lines.

23.

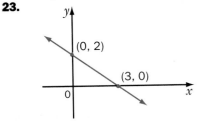

24.

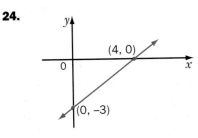

25.

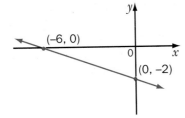

26.

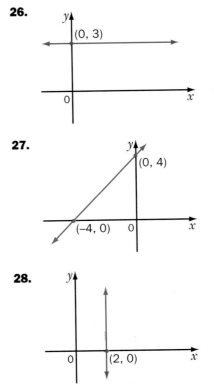

27.

28.

Applications and Problem Solving

29. Write an equation of a line whose graph does not have an x-intercept.

30. Write an equation of a line whose graph does not have a y-intercept.

31. Write an equation of a line whose x- and y-intercepts are the same, but not 0.

32. Write an equation of a line whose x- and y-intercepts are both 0.

33. Calculate the distance between the x-intercept of the line $3x - 2y + 10 = 0$ and the y-intercept of the line $3x + 7y + 21 = 0$.

34. If the x- and y-intercepts of a line are equal and not zero, what is the slope of the line? Explain.

35. Air travel A plane is flying from Montreal to Rome. An equation that relates the distance from Rome, d kilometres, to the flying time, t hours, is $d = 6600 - 825t$.

a) What is the d-intercept and what does it represent?

b) What is the t-intercept and what does it represent?

c) What is the slope and what does it represent?

d) What is the distance from Rome when the plane has been flying for 5.2 h?

e) For how many hours has the plane been flying when it is 3465 km from Rome?

36. University degrees The percent of Canadians with university degrees was about 4.5% in 1971, 8% in 1981, and 11.5% in 1991. Let 1971 represent year 0.

a) Express the data as ordered pairs in the form (year, percent).

b) Plot the points on a grid and join them with a straight line.

c) Find the slope of the line. What does the slope represent?

d) Find an equation of the line.

e) Use your equation to predict the percent of Canadians who will have degrees in the year 2001; the year 2010.

f) Will the equation be valid forever? Explain.

37. Describe each of the following lines in terms of its slope, intercepts, domain, and range.

a) $y = 7$ **b)** $x = -2$ **c)** $y = 0$
d) $x = 5$ **e)** $x = 0$ **f)** $y = -4$

38. Algebra Describe each of the following lines in terms of its slope, intercepts, and range. The domain is R.

a) $x = a$ **b)** $y = b$
c) $y = x$ **d)** $y = mx + b$

39. Algebra Find the general equation of the line with x-intercept a and y-intercept b.

40. Reaction time The stopping distance of a car has two components — the distance the car travels during the time you take to move your foot from the gas pedal to the brake pedal, and the distance the car travels after the brakes are applied. At a speed of 40 km/h, a car travels 8 m during the time your foot moves from the gas pedal to the brake. At 70 km/h, the car travels 14 m before your foot hits the brake.

a) Plot the points in the form (speed, distance) on a grid. Join them with a straight line.

b) Find the slope of the line. What does the slope represent?

c) Find an equation of the line.

d) Use the equation to find the distance a car travels while your foot is moving between the gas pedal and the brake, if the car has a speed of 50 km/h; 100 km/h.

e) What are the x- and y-intercepts? What meaning do they have? What do they suggest about the domain and range of the relation? Explain.

41. Algebra The standard form of an equation is $Ax + By + C = 0$.

a) Write the equation in the form $y = mx + b$.

b) Determine the rules that connect A, B, and C to the slope and the x- and y-intercepts.

42. Use the rules you found in question 41 to find the slope and intercepts for each of the following.

a) $2x + 3y - 6 = 0$ **b)** $3x - 4y + 12 = 0$
c) $x + y = -2$ **d)** $5x - 8y - 20 = 0$

43. The standard form of an equation is $Ax + By + C = 0$. Describe the graph of the line when

a) $A = 0, B \neq 0, C \neq 0$
b) $B = 0, A \neq 0, C \neq 0$
c) $C = 0, A \neq 0, B \neq 0$
d) $A = 0, C = 0, B \neq 0$
e) $B = 0, C = 0, A \neq 0$

LOGIC POWER

A box contains 5 coloured cubes and an empty space the size of a cube.

Use moves like those in checkers. In one move, one cube can slide to an empty space or jump over one cube to an empty space. Find the smallest number of moves needed to reverse the order of the cubes.

Review

Find the distance between each pair of points.
1. A(7, 9) and B(1, 1)
2. W(4, 5) and X(–2, 3)
3. E(–2, 8) and F(–5, 5)
4. R(–10, 5) and T(4, –1)
5. U(1.2, –0.4) and V(–0.8, 3.6)

6. **Boating** Two pleasure boats set out from Grand Beach on Lake Winnipeg. After an hour, one boat is 10 km west and 6 km south of Grand Beach. The other boat is 5 km west and 8 km north of Grand Beach. How far apart are the two boats, to the nearest tenth of a kilometre?

6.2 *Find the coordinates of the midpoint of each line segment, given the endpoints.*
7. P(2, –7) and Q(–3, 5)
8. S(6, –2) and T(2, 2)
9. M(2, –5) and N(5, –1)
10. G(–2, 0) and H(–4, 3)
11. V(2.9, 3.2) and W(3.1, –4.2)
12. $A\left(3\frac{1}{2}, \frac{1}{2}\right)$ and $B\left(-2\frac{1}{2}, 1\frac{1}{2}\right)$

13. The coordinates of an endpoint of a line segment are (–4, 3), and its midpoint is (1, –2). What are the coordinates of the other endpoint?

6.3 *Find the slope of the line passing through the points.*
14. E(3, 5) and F(5, 9)
15. W(–2, 7) and X(–5, 8)
16. S(–3, 7) and T(6, –2)
17. P(0, –9) and Q(1, –6)
18. K(1.3, –5.4) and L(0.3, –0.6)
19. A(3, –1) and B(–2, –1)
20. U(1, 2) and V(1, –6)

Given a point on the line and the slope, sketch the graph of the line.
21. (3, 5); $m = 3$
22. (0, –4); $m = \frac{2}{3}$
23. (2, –6); $m = -1$
24. (4, 2); $m = -\frac{3}{2}$
25. (5, 0); m is undefined

6.4 26. **Population statistics** Statistics Canada projects that the percent of the Canadian population who will be at least 65 years old will be 14% in the year 2011 and 17.5% in the year 2021.
a) Find the rate of change, in percent per year.
b) Predict the year in which all Canadians will be at least 65 years old.
c) Is your prediction from part b) valid? Explain.

6.5 *Write an equation in standard form for the line that passes through the given point and has the given slope.*
27. (4, 5); $m = 2$
28. (–1, 3); $m = -3$
29. (2, –6); $m = -1$
30. (–4, –3); $m = \frac{1}{2}$
31. (–3, 0); $m = -0.2$

Write an equation in standard form for the line that passes through the given points.
32. (2, 5) and (4, 9)
33. (–2, 7) and (–1, 4)
34. (–4, 7) and (–2, 6)
35. (1, –6) and (3, –9)
36. (0, –3) and (4, –2)

Write an equation in standard form for the line that passes through the given points. Use the equation to find two other points on the line. Check your solutions.
37. (2, 5) and (4, 13)
38. (1, –5) and (2, –6)

6.6 *Find the slope and y-intercept of the following lines.*
39. $y = 4x + 6$
40. $4x + 5y – 20 = 0$
41. $x – 2y = -4$
42. $2x – 12 = 3y$

Given the slope and y-intercept, write an equation of the line in standard form.
43. $m = 5$ and $b = 2$
44. $m = -4$ and $b = -3$
45. $m = \frac{4}{3}$ and $b = 0$

46. **Clothing sales** Rolf earns a salary plus commission for selling clothes in a store. The equation $e = 480 + 0.05s$ relates his weekly earnings, e, to his weekly sales, s.

a) Graph the line.
b) What is Rolf's weekly salary, excluding commission?
c) What is the rate of commission, expressed as a percent?
d) If Rolf earned $600 one week, what were his sales that week?

6.7 *State the slope of the line*
a) *parallel to each line*
b) *perpendicular to each line*
47. $y = 5x + 4$
48. $y = \dfrac{1}{4}x - 4$

49. $y = -\dfrac{3}{5}x - 7$

50. $8x + 4y = 11$
51. $x - 3y - 1 = 0$

Write an equation of each of the following lines.
52. a line parallel to $3x + 2y - 4 = 0$ and passing through the point (2, 3)
53. a line perpendicular to $x - 2y + 3 = 0$ and passing through the point (4, –1)

6.8 *Find an equation for each of the following lines.*
54.

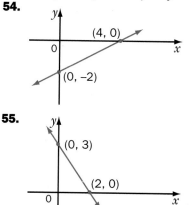

55.

Use the x- and y-intercepts to graph each line.
56. $x + 5y = 10$
57. $7x - 2y - 14 = 0$
58. $y = -6x + 12$

Graph each equation. If the domain is R, state the range, slope, and intercepts.
59. $y - 3 = 2(x + 1)$
60. $4x + 2y - 5 = 0$

Exploring Math

Patterns in a Table

Assume that the pattern continues.

Row	Column A	B	C	D	E
1	1	2	3	4	5
2	6	7	8	9	10
3	11	12	13	14	15
4	16	17	18	19	20

1. What numbers are in row 10? row 50?

2. Write an equation in the form $N =$ ▨ to express a number, N, in terms of the row number, r, for the numbers in
a) column A   **b)** column B   **c)** column C
d) column D   **e)** column E

3. a) If you plotted the points for each equation from question 2, how would the five sets of points be related? Explain.
b) Would you be justified in drawing a straight line through each set of points in part a)? Explain.

4. Use the equations from question 2 to find the numbers in row 211; row 507.

5. What is the number in
a) row 256, column C?   **b)** row 333, column B?

6. Which row will include
a) the number 2354?   **b)** the number 3926?

7. Which column will include
a) the number 822?   **b)** the number 4443?

8. Show that the sum of *any* number in column A and *any* number in column B will be found in column C.

9. In which row and which column will you find the sum of the number from row 22, column A, and the number from row 37, column B?

10. Show that the sum of *any* number in column A and *any* number in Column D will be found in column E.

11. In which row and which column will you find the sum of the number from row 17, column A, and the number from row 33, column D?

Chapter Check

Find the distance between each pair of points.
1. $(4, -1)$ and $(1, 3)$ **2.** $(2, 0)$ and $(4, -5)$

3. Measurement The endpoints of a diameter of a circle are $(-1, -3)$ and $(3, 6)$. Calculate the radius of the circle, to the nearest tenth.

Determine the coordinates of the midpoint of each line segment, given the endpoints.
4. $(3, 4)$ and $(5, -2)$ **5.** $(-2, -5)$ and $(-8, 3)$

Find the slope of the line that passes through each pair of points.
6. $(3, 1)$ and $(3, -5)$ **7.** $(0, 5)$ and $(4, 7)$
8. $(-1, 2)$ and $(2, -3)$ **9.** $(-1, -3)$ and $(-5, -3)$

Write an equation in standard form for the line that passes through the given point and has the given slope.

10. $(2, -1)$; $m = 3$ **11.** $(-1, -2)$; $m = -\dfrac{2}{5}$

Write an equation in standard form for the line that passes through the given points.
12. $(2, 5)$ and $(7, 10)$
13. $(3, 1)$ and $(9, -2)$

Find the slope and y-intercept of each line.
14. $y = 2x - 3$
15. $x + 2y + 8 = 0$

16. How are the two lines in questions 14 and 15 related? Explain.

Use the slope and y-intercept to write an equation of each line in standard form.
17. $m = 2$ and $b = -1$

18. $m = -\dfrac{1}{4}$ and $b = 2$

19. Write an equation of a line parallel to $2x + y - 2 = 0$ and passing through the point $(4, 5)$.

Use a method of your choice to graph each line. If the domain is R, state the range, the slope, and the intercepts.
20. $y = 3 + 4x$
21. $2x - 5y + 10 = 0$

22. Prairie cities Saskatoon is about 700 km from Winnipeg. The elevation of Saskatoon is 484 m. The elevation of Winnipeg is 232 m.
a) Write two ordered pairs of the form (d, e) to represent the data, where d is the distance from Winnipeg, in kilometres, and e is the elevation, in metres.
b) Graph the ordered pairs and connect them with a line segment.
c) Find the slope and state the units in which it is expressed.
d) What rate of change does the slope represent?
e) Write an equation of the line in slope and y-intercept form.
f) Use the equation to predict the elevation of a point between Winnipeg and Saskatoon, and 250 km from Winnipeg.
g) Do you think that your prediction in part f) is valid? Explain.

Using the Strategies

1. Each letter in the box represents a different number. The sums of the columns and two of the rows are given. Find the missing sums.

A	B	A	B	36
A	C	A	C	34
B	B	B	C	?
D	E	F	A	?
37	40	41	32	

2. The school yearbook committee is meeting in a room where there are three-legged stools and four-legged chairs. A person is seated on every chair and every stool. The total number of legs in the room is 39. How many stools, chairs, and people are in the room?

3. Four friends — Alexi, Kala, Lisa, and Jake — are sitting at a round table. The person wearing the red shirt, who is not Alexi or Kala, is sitting between Lisa and the person wearing the blue shirt. The person wearing the green shirt is sitting between Kala and the person wearing the yellow shirt. What colour shirt is each person wearing?

4. The shape is to be divided into 6 congruent parts by cutting along the grid lines. What shapes can the congruent parts have?

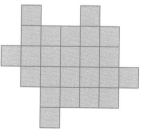

5. Here are three of the Platonic solids.

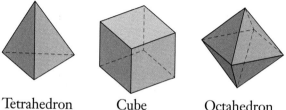

Tetrahedron   Cube   Octahedron

You can trace the edges of one of the solids without lifting your pencil and without tracing any edge twice. Which one?

6. A telephone keypad is shown.

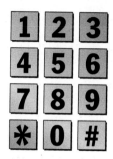

The distance between the centre of the 1 button and the centre of the 2 button is 2 cm. The distance between the centre of the 1 button and the centre of the 4 button is 2 cm. What is the minimum distance your finger would have to travel to enter the following number, to the nearest centimetre? 1-726-567-2194

7. Two glasses, C and D, have exactly the same size and shape. Glass C is empty, and glass D has some water in it. Half the water in D is then poured into C. This process is repeated two more times. Each time, half the remaining water in D is poured into C. After three pourings, C is half full. What fraction of a glassful is now left in D?

8. About how many boxes of breakfast cereal are sold each year in Canada?

1. a) What time is it at the North Pole? Explain using time zones.
b) In what way is the time zone map on pages 404—405 misleading?

2. At cruising speed, a Boeing 767 aircraft can fly from Calgary to Montreal in 3.75 h and from Saskatoon to Toronto in 2.75 h.
a) Plot a graph of distance travelled versus time for this aircraft.
b) Find the slope of the graph. What does this slope represent?
c) Describe a way to find the value that the slope represents without plotting a graph.

CHAPTER 7

Measurement

A place on the Earth is located using its latitude and longitude. The city of Nanking, China, has a latitude of about 35°N and a longitude of about 120°E. If you travelled from Nanking through the centre of the Earth and out the other side, you would arrive close to the city of Buenos Aires, Argentina. Buenos Aires has a latitude of about 35°S and a longitude of about 60°W.

A location's antipodal point is the place on Earth directly opposite it. Nanking is the antipodal point to Buenos Aires. Two other cities related in this way are Lima, Peru, and Phnom Penh, Cambodia. Lima is located at about 10°S, 75°W. Phnom Penh is located at about 10°N, 105°E.

1. For each pair of cities, what is the relationship between the latitudes?

2. For each pair of cities, what is the relationship between the longitudes?

3. To the nearest degree, what are the latitude and the longitude of the place where you live?

4. Find the latitude and the longitude of the antipodal point to where you live. Does anyone live there?

5. If two places are at antipodal points on the Earth,
a) what is the straight-line distance between them?
b) what is the shortest distance you can travel on the Earth from one place to the other? State any assumptions you made.

GETTING STARTED

Enlargements and Reductions

1 Scale Factors

1. Figures 1 and 2 are similar. Figures 3 and 4 are similar. Explain why.

2. The ratio of the side lengths in an image figure to the side lengths in the original figure is known as the scale factor. If Figure 2 is the image of Figure 1, what is the scale factor for this enlargement?

3. If Figure 4 is the image of Figure 3, what is the scale factor for this reduction?

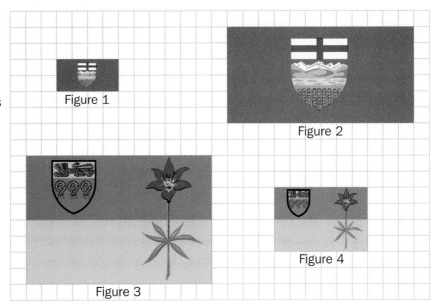

Figure 1

Figure 2

Figure 3

Figure 4

2 A Statue

In Flin Flon, Manitoba, the main industry is mining for zinc, copper, gold, and silver. Just outside the city stands a statue of a fictional character called Josiah Flintabbatey Flonatin. The total height of the statue and the base on which it stands is 7.5 m. Flin Flon was named after Josiah because of his connection with gold mining. He discovered gold by inventing a submarine and going down into the Earth through a lake. His statue was designed by Al Capp, a cartoonist who was famous for his comic strip *Li'l Abner*.

1. a) Determine the total length of the image of the statue and the base in the photograph. Find the scale factor for the reduction.
b) To enlarge the photograph of the statue and the base to the same height as the statue and the base, what is the scale factor?
c) How are the scale factors in parts a) and b) related? Explain why they are related in this way.

2. a) In the photograph, measure the greatest width of Josiah's foot. Then, calculate the greatest width of the foot on the statue.
b) Measure the greatest width of one of your feet. What scale factor would reduce the width of the foot on the statue to the width of one of your feet?

306

3 Dido's Challenge

A Greek legend tells the story of Dido, who founded the ancient city of Carthage in North Africa. In the *Aeneid*, Virgil described how Dido and her companions bought land to build the city.

Here they bought ground; they used to call it Byrsa,
That being a word for bull's hide; they bought only
What a bull's hide could cover.

Dido's challenge was to enclose as much land as possible with a bull's hide. She cut the hide into strips and attached them to form a long piece of rawhide. Dido used the seashore as one boundary of the city and used the rawhide to mark the other boundaries.

Assume that Dido's piece of rawhide had a length of 100 units.

1. Suppose that Dido tried a rectangular shape for the city.

a) What dimensions would have given the maximum area?
b) What was the maximum area?

2. Find the area of the city for each of the following shapes.
a) equilateral triangle

b) half of a regular hexagon

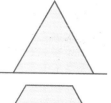

3. a) For half of a regular octagon, the area would have been 1510 square units. Use this information and your results from questions 1 and 2 to predict the shape that Dido used to give the maximum area.

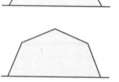

b) What were the dimensions of the shape?
c) What was the maximum area?

Mental Math

Order of Operations

Evaluate.

1. $5 \times 11 \times 4$ **2.** $10 \times 12 \times 8$
3. $\frac{1}{2} \times 8 \times 9$ **4.** $\frac{1}{2} \times 15 \times 6$
5. $\frac{1}{3} \times 6 \times 6 \times 4$ **6.** $\frac{1}{3} \times 2 \times 4 \times 9$
7. $3.14 \times 10 \times 10 \times 3$ **8.** $3.14 \times 5 \times 5 \times 4$

Evaluate.

9. $\sqrt{6^2 + 8^2}$ **10.** $\sqrt{5^2 + 12^2}$
11. $180 - (48 + 42)$ **12.** $180 - (37 + 63)$
13. $6^2 + 4 \times \frac{1}{2} \times 6 \times 5$ **14.** $3^2 + 4 \times \frac{1}{2} \times 6 \times 6$

Squaring Numbers Ending in 5

To calculate 75^2, multiply 7 by the next whole number, 8; then, affix 25.

$7 \times 8 = 56$
Affixing 25 to 56 gives 5625.
So, $75^2 = 5625$

Calculate.

1. 15^2 **2.** 25^2 **3.** 55^2 **4.** 45^2
5. 85^2 **6.** 95^2 **7.** 35^2 **8.** 105^2

To calculate 6.5^2 or 650^2, think 65^2; then, place the decimal point.

$65^2 = 4225$
So, $6.5^2 = 42.25$, and $650^2 = 422\,500$.

Calculate.

9. 2.5^2 **10.** 3.5^2 **11.** 1.5^2
12. 4.5^2 **13.** 7.5^2 **14.** 8.5^2
15. 150^2 **16.** 350^2 **17.** 550^2
18. 750^2 **19.** 250^2 **20.** 850^2
21. 3500^2 **22.** 5500^2 **23.** 7500^2

24. The expression $10n + 5$, where n is a whole number, represents any number ending in 5.
a) Expand $(10n + 5)^2$.
b) Explain why the rule for squaring numbers ending in 5 works.

7.1 Reviewing Surface Area and Volume

The **surface area** of a polyhedron is the sum of the areas of all its faces.

Explore: Use the Data

Edmonton's Muttart Conservatory is a botanical garden. The plants are grown in four greenhouses that are square-based pyramids. For the two largest greenhouses, the side length of the base is about 25 m. The height of each triangular glass face is about 27 m. For each pyramid, find
a) the area of the square base
b) the area of each triangular face

Inquire

1. What is the total area of the glass faces of the pyramid?

2. What is the surface area of the pyramid?

3. For the two smallest greenhouses at the conservatory, the side length of the base is about 20 m. The height of each triangular glass face is about 21 m. Express the area of glass in each pyramid as a percent of the area of glass in one of the largest pyramids, to the nearest percent.

4. Suppose that the side lengths of the bases and the heights of the triangular faces were unchanged, but all the pyramids were triangular instead of square.
a) Would your answer to question 1 change? Explain.
b) Would your answer to question 2 change? Explain.
c) Would your answer to question 3 change? Explain.

The volume of a right pyramid is given by the formula $V = \frac{1}{3}Bh$, where B is the area of the base and h is the height.

Example 1   Volume of a Pyramid

Calculate the volume of the pyramid.

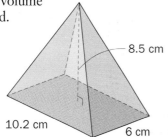

Solution

$$V = \frac{1}{3}Bh$$

$$= \frac{1}{3}(10.2)(6)(8.5)$$

$$= 173.4$$

Estimate

$10 \times 6 \times 10 = 600$
$600 \div 3 = 200$

The volume of the pyramid is 173.4 cm³.

The volume of a right prism is given by the formula $V = Bh$, where B is the area of the base and h is the height.

Example 2 Volume of a Triangular Prism
Calculate the volume of the triangular prism.

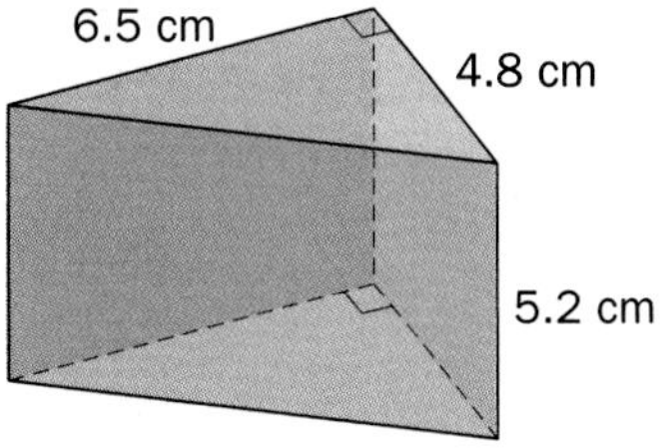

Solution
$V = Bh$
For the triangular base,

$B = \dfrac{1}{2}(4.8)(6.5)$
$= 15.6$

Estimate
$\dfrac{1}{2} \times 5 \times 6 = 15$

$V = (15.6)(5.2)$
$= 81.12$

Estimate
$16 \times 5 = 80$

The volume of the triangular prism is 81.12 cm^3.

For a right cone, the volume, V, and surface area, SA, are given by the formulas

$V = \dfrac{1}{3}\pi r^2 h$ and $SA = \pi r^2 + \pi rs$

where r is the radius, h is the height, and s is the slant height.

Example 3 Volume and Surface Area of a Cone
A solid cone has a radius of 5 cm and a height of 12 cm. Find
a) the volume of the cone, to the nearest cubic centimetre
b) the surface area of the cone, to the nearest square centimetre

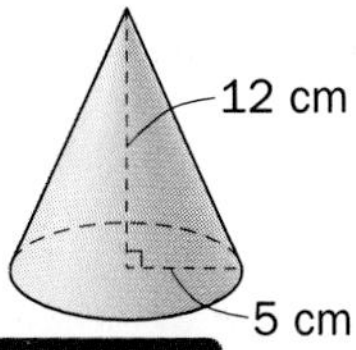

Solution
a) $V = \dfrac{1}{3}\pi r^2 h$

$= \dfrac{1}{3}(\pi)(5)^2(12)$

$\doteq 314$

Estimate
$25 \times 12 = 300$

```
1/3π5²12
            314.1592654
```

The volume of the cone is 314 cm^3, to the nearest cubic centimetre.

b) Find the slant height, which is the hypotenuse of a right triangle. Using the Pythagorean Theorem,

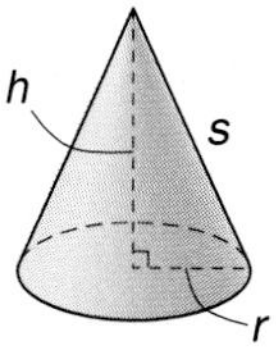

$s = \sqrt{r^2 + h^2}$
$= \sqrt{5^2 + 12^2}$
$= \sqrt{25 + 144}$
$= \sqrt{169}$
$= 13$

$SA = \pi r^2 + \pi rs$
$= \pi(5)^2 + \pi(5)(13)$
$\doteq 283$

Estimate
$3 \times 25 = 75$
$3 \times 70 = 210$
$75 + 210 = 285$

```
π5²+π5*13
            282.7433388
```

The surface area of the cone is 283 cm^2, to the nearest square centimetre.

For a right cylinder, the volume, V, and surface area, SA, are given by the formulas
$$V = \pi r^2 h \qquad \text{and} \qquad SA = 2\pi r^2 + 2\pi rh$$
where r is the radius and h is the height.

Example 4 Surface Area of a Cylindrical Container

A cylindrical container for a perfume bottle has a radius of 3 cm and a height of 16 cm. What is the surface area of the container, to the nearest square centimetre?

3 cm

16 cm

Solution

$$\begin{aligned} SA &= 2\pi r^2 + 2\pi rh \\ &= 2\pi(3)^2 + 2(\pi)(3)(16) \\ &\doteq 358 \end{aligned}$$

Estimate

$2 \times 3 \times 10 = 60$
$2 \times 10 \times 16 = 320$
$60 + 320 = 380$

2π3²+2π3*16
358.1415625

The surface area of the can is 358 cm², to the nearest square centimetre.

Example 5 Finding the Radius of a Cylindrical Container

A new style of blue jeans was marketed in a cylindrical can. An engineering design company designed a can with a volume of 8000 cm³. The designer decided that the can should be 40 cm high. What was the radius of the can, to the nearest centimetre?

40 cm

Solution

$$V = \pi r^2 h$$

Substitute for V and h:
$$8000 = \pi r^2(40)$$

Divide both sides by 40π:
$$\frac{8000}{40\pi} = \frac{\pi r^2(40)}{40\pi}$$

$$\frac{8000}{40\pi} = r^2$$

Take the square root of both sides:
$$\sqrt{\frac{8000}{40\pi}} = r$$

$$8 \doteq r$$

Estimate

$8000 \div 40 = 200$
$200 \div 3 \doteq 70$
$\sqrt{70} \doteq 8$

√(8000/40/π)
7.978845608

The radius of the can was 8 cm, to the nearest centimetre.

Another way to solve Example 5 is to solve the formula $V = \pi r^2 h$ for r. The resulting formula is $r = \sqrt{\dfrac{V}{\pi h}}$. Substituting values of V and h into this formula gives the value of r. This method is very useful when the calculation of r is repeated using different values of V and h.

Practice

Find the volume and surface area of each figure.

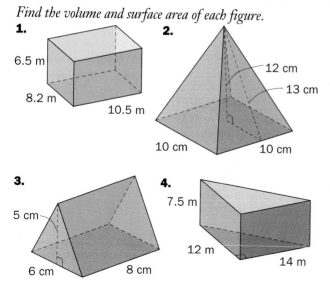

1.
6.5 m, 8.2 m, 10.5 m

2.
12 cm, 13 cm, 10 cm, 10 cm

3.
5 cm, 6 cm, 8 cm

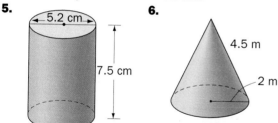

4.
7.5 m, 12 m, 14 m

For each solid, calculate
a) *the surface area, to the nearest square unit*
b) *the volume, to the nearest cubic unit*

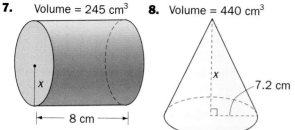

5.
5.2 cm, 7.5 cm

6.
4.5 m, 2 m

Applications and Problem Solving

Calculate the missing dimension, x, to the nearest tenth of a centimetre.

7. Volume = 245 cm³
x, 8 cm

8. Volume = 440 cm³
x, 7.2 cm

9. NASA The Vehicle Assembly Building at Complex 39 of the John F. Kennedy Space Center in Florida is in the shape of a rectangular prism. The building is 218 m long, 158 m wide, and 160 m high. Calculate the volume of the building.

10. The Louvre The Louvre, in Paris, is one of the world's finest art museums. The entrance is a square-based glass pyramid of altitude 21.6 m and base dimensions 34.2 m by 34.2 m.
a) Determine the height of each triangular face, to the nearest tenth of a metre.
b) Find the total area of the glass used to make the 4 triangular faces.

11. Giant teepee The frame of a giant teepee, originally built for the Calgary Winter Olympics, is now in Medicine Hat, Alberta. The teepee commemorates the role of Aboriginal peoples in the history of Southern Alberta. The teepee is 65 m in height and has a base diameter of 50 m. What area of covering, to the nearest square metre, would be needed to cover the curved surface of the teepee?

12. Road salt A cone-shaped pile of road salt is covered with tarpaulins to keep it dry. The radius of the base of the pile is 9.5 m, and the height of the pile is 7.2 m.
a) Calculate the volume of salt in the pile, to the nearest cubic metre.
b) Allowing 15% extra for overlap, what is the area of the tarpaulins, to the nearest square metre?

13. Engines One factor affecting the power of a gasoline engine is its displacement. In a gasoline engine, each piston moves within a cylinder. The displacement of each cylinder is the volume of space through which the piston travels. The diameter of the cylinder is called the bore; the distance the piston travels is called the stroke. Calculate each of the following, to the nearest cubic centimetre.

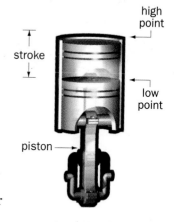

high point
stroke
low point
piston

a) the displacement of a cylinder with a 9.2 cm bore and a 7.2 cm stroke
b) the total displacement of a 6-cylinder engine in which each cylinder has a 10 cm bore and a 7.9 cm stroke

14. Composite solid Find the volume and the surface area of this solid.

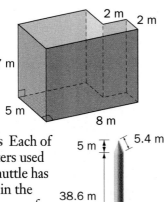

15. Rocket boosters Each of the solid rocket boosters used to launch the space shuttle has dimensions as shown in the sketch. Find the total area of the curved surfaces of the cylinder and the cone, to the nearest square metre.

16. Newsprint The *Halifax Chronicle-Herald* uses newsprint in rolls of diameter 1.03 m around an inside tube of diameter 10 cm. The width of the roll is 1.37 m. What is the volume of newsprint in the roll, to the nearest hundredth of a cubic metre?

17. Birdhouse The inside dimensions of a birdhouse suitable for a bluebird's nest are as shown. There is just one entrance hole, which has a diameter of 4 cm.
a) Determine the inside volume of the birdhouse.
b) Determine the area of the inside walls of the birdhouse.

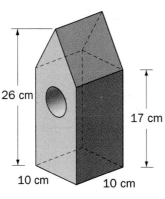

18. $2 coin A Canadian $2 coin is 28 mm in diameter. The outer ring is made of a nickel alloy. The inner core has a diameter of 16 mm and is made of a copper alloy. The thickness of the coin is 1.8 mm. Calculate the volume of nickel alloy in the coin
a) in cubic millimetres **b)** in cubic centimetres

19. Packing crate What is the largest number of rectangular boxes of dimensions 30 cm by 50 cm by 20 cm, that can fit into a crate whose dimensions are 2 m by 1.5 m by 0.5 m?

20. A wooden cone is cut as shown to make a cone that is only half as tall.
a) How many times greater is the surface area of the original cone than the surface area of the smaller cone?
b) How many times greater is the volume of the original cone than the volume of the smaller cone?

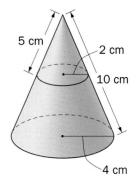

21. Algebra The edges of a rectangular prism are $2y$, $5y$, and $8y$, units long. The volume of the prism is 2160 cubic units. What is the value of y?

22. Algebra The edges of a rectangular prism are $2x$, $3x$, and $4x$ units long. If the total surface area is 1300 cubic units, find the value of x.

23. Algebra a) The expression representing the surface area of a cylinder is $2\pi r^2 + 2\pi rh$. Write this expression in factored form.
b) For a cylinder with radius 5 cm and height 8 cm, use both the original form and the factored form to calculate the surface area. Which form of the expression is more convenient? Why?
c) In a similar way, factor the expression for the surface area of a cone. Use both forms to find the surface area of a cone with radius 5 cm and height 8 cm.

24. A cone and a cylinder have equal bases and equal volumes. Which is taller and by what factor?

25. Give examples of the following and compare your examples with your classmates'. Limit yourself to whole-number dimensions of 10 units or less.
a) two rectangular prisms with equal volumes but different surface areas
b) two rectangular prisms with equal surface areas but different volumes

7.2 Volume and Surface Area of a Sphere

The sphere is a common shape. Planets, grapefruit, snapping turtle eggs, and the balls used in many sports are among the many objects that are approximately spherical.

Explore: Use the Diagram

The sphere of radius r just fits inside the cylinder.
Use the diagram to state the following in terms of r.
a) the radius of the top and bottom of the cylinder
b) the height of the cylinder

Inquire

1. Write a formula for the volume of the cylinder in terms of r.

2. The volume of the sphere is two thirds of the volume of the cylinder. Write a formula for the volume of the sphere in terms of r.

3. A snapping turtle can lay up to 80 eggs at a time. The radius of a snapping turtle egg is about 1.4 cm.
a) Use the formula from question 2 to calculate the volume of a snapping turtle egg, to the nearest tenth of a cubic centimetre.
b) Calculate the total volume of the eggs a snapping turtle can lay at one time, to the nearest cubic centimetre.

Example 1   Surface Area of a Soccer Ball
A soccer ball has a radius of about 11 cm. Calculate the surface area, to the nearest square centimetre.

11 cm

Solution
Assume that a soccer ball is a sphere. For a sphere with radius r, the surface area is given by the formula

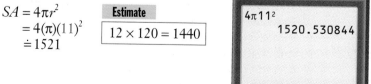

$SA = 4\pi r^2$
$\quad = 4(\pi)(11)^2$
$\quad \doteq 1521$

Estimate
$12 \times 120 = 1440$

$4\pi 11^2$
$\qquad 1520.530844$

The surface area of a soccer ball is 1521 cm², to the nearest square centimetre.

313

Example 2   Diameters and Volumes of Basketballs

The basketballs used by men and women are different sizes. A women's basketball has a diameter of about 23 cm. A men's basketball has a volume that is about 870 cm^3 greater than the volume of a women's basketball. Calculate

a) the volume of a women's basketball, to the nearest cubic centimetre

b) the diameter of a men's basketball, to the nearest centimetre

Solution

a) For a sphere with radius r, the volume is given by the formula

$$V = \frac{4}{3}\pi r^3.$$

For a women's basketball, the radius is $\frac{23}{2}$ or 11.5 cm.

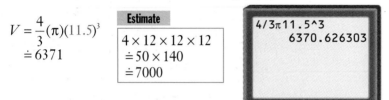

$$V = \frac{4}{3}(\pi)(11.5)^3$$
$$\doteq 6371$$

Estimate
$4 \times 12 \times 12 \times 12$
$\doteq 50 \times 140$
$\doteq 7000$

```
4/3π11.5^3
          6370.626303
```

The volume of a women's basketball is 6371 cm^3, to the nearest cubic centimetre.

b) The volume of a men's basketball is
$$6371 + 870 = 7241$$
Work backward from this volume to find the radius.

$$V = \frac{4}{3}\pi r^3$$

Substitute for V:
$$7241 = \frac{4}{3}\pi r^3$$

Multiply both sides by 3:
$$3 \times 7241 = 3 \times \frac{4}{3}\pi r^3$$

$$21\ 723 = 4\pi r^3$$

Divide both sides by 4π:
$$\frac{21\ 723}{4\pi} = \frac{4\pi r^3}{4\pi}$$

$$\frac{21\ 723}{4\pi} = r^3$$

Take the cube root of both sides:
$$\sqrt[3]{\frac{21\ 723}{4\pi}} = r$$
$$12 \doteq r$$
$$2 \times 12 \doteq d$$
$$24 \doteq d$$

```
³√(21723/4/π)
          12.00153086
```

The diameter of a men's basketball is 24 cm, to the nearest centimetre.

Practice

For each sphere, find
a) *the surface area, to the nearest square centimetre*
b) *the volume, to the nearest cubic centimetre*

1.   **2.**

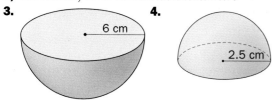

For each solid hemisphere, find
a) *the surface area, to the nearest square centimetre*
b) *the volume, to the nearest cubic centimetre*

3.   **4.**

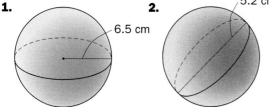

5. If a sphere has a surface area of 100 m², what is its radius, to the nearest tenth of a metre?

6. If a sphere has a volume of 250 m³, what is its radius, to the nearest tenth of a metre?

Applications and Problem Solving

7. World's fair Expo 67 was a world's fair held in Montreal to mark the hundredth anniversary of Confederation. The American pavilion was a sphere 76 m in diameter. What was the surface area of the sphere, to the nearest 10 m²?

8. Canada and the moon Canada is the world's second-largest country, with an area of about 10 000 000 km². The diameter of the moon is about 3500 km. How does the surface area of the moon compare with the area of Canada?

9. Butane tank North America's largest butane storage tank is at Point Tupper, Nova Scotia. The tank is a sphere with an inside radius of 12.8 m. What volume of butane can the tank hold, to the nearest cubic metre?

10. Planetarium A planetarium in Miyazaki, Japan, has a hemispherical dome with an inside diameter of 27 m. What is the volume of the interior of the dome, to the nearest cubic metre?

11. Weather balloons Balloons carry scientific instruments that record information about weather and the atmosphere. Some of the larger balloons have volumes of about 85 000 m³.
a) What is the radius of a spherical balloon of this size, to the nearest metre?
b) What is the surface area of a balloon of this size, to the nearest hundred square metres?

12. Earth's atmosphere The radius of the Earth is about 6400 km. The Earth's atmosphere extends about 1600 km above the surface of the Earth. What is the volume of the Earth's atmosphere?

13. Water tanks A spherical water tank has a radius of 5 m.
a) Find its volume, to the nearest tenth of a cubic metre.
b) If a cubic tank has the same volume, what is the length of each edge, to the nearest hundredth of a metre?

14. Find the surface area of each figure, to the nearest square centimetre. The rounded ends are hemispheres.

a)   **b)**

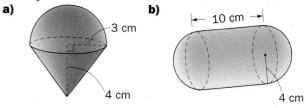

15. Find the volume of each figure, to the nearest tenth of a cubic metre. The rounded ends are hemispheres.

a)   **b)**

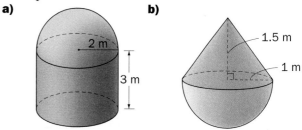

16. Golf and tennis The volume of a golf ball is about 98.5 cm³ less than the volume of a tennis ball. If the diameter of a tennis ball is 6.4 cm, what is the diameter of a golf ball, to the nearest tenth of a centimetre?

17. Baseball and softball The surface area of a softball is about 118 cm^2 greater than the surface area of a baseball. The radius of a baseball is 3.7 cm.
a) What is the radius of a softball, to the nearest tenth of a centimetre?
b) How many times greater is the volume of a softball than the volume of a baseball, to the nearest tenth?

18. Packing boxes A lighting company ships large spherical light bulbs of diameter 40 cm in cubic boxes with 45 cm edges. Styrofoam is used to fill the box and protect the bulb from breakage. What volume of styrofoam is needed for a shipment of 100 light bulbs? Round your answer to the nearest 100 cm^3.

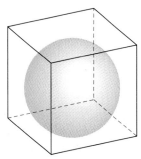

19. Packaging a puzzle A container with a volume of 1000 cm^3 is required to hold a jigsaw puzzle.
a) Find dimensions for a container with this volume if the container was a cube; a rectangular prism other than a cube; a cylinder; a sphere.
b) Which shape of container would take the least material to make?
c) Why might your answer to part b) not be the best shape for the container?

20. Racquet balls Two racquet balls of diameter 5.6 cm just fit inside a cylindrical container.
a) What is the volume of each racquet ball, to the nearest cubic centimetre?
b) What is the volume of the cylindrical container, to the nearest cubic centimetre?
c) What percent of the volume of the container is empty, to the nearest percent?
d) Describe how you could complete part c) without knowing the radius of a racquet ball.

21. If the volume of a sphere in cubic metres is numerically equal to its surface area in square metres, what is the radius of the sphere?

22. The Earth Aside from the thin crust on which we live, the Earth has three layers. The inner core is a solid metallic sphere, about 1216 km in radius. The outer core is liquid and has a thickness of about 2270 km. The mantle is a rocky layer with a thickness of about 2885 km. Calculate the volume of each of the three layers in cubic kilometres. Write each answer in scientific notation, expressing the decimal part to two decimal places.

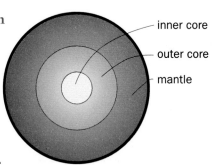
inner core
outer core
mantle

23. The edge of a solid cube, the diameter of a solid sphere, and the diameter and height of a solid cylinder are all equal. Which solid has
a) the greatest volume?
b) the least volume?
c) the greatest surface area?
d) the least surface area?

24. Visualization A solid sphere of radius r is tightly wrapped in a cylindrical piece of paper open at both ends. The height of the cylinder equals the diameter of the sphere.
a) How does the surface area of the sphere compare with the surface area of the paper cylinder?
b) Use your finding from part a) to suggest a way of using a piece of paper, a pencil, and a ruler to find the approximate surface area of a sphere.

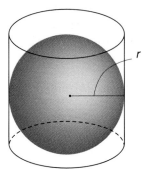
r

Cartography

Cartography, or the making of maps and charts, is a very old profession. The earliest surviving maps, dating from about 2300 B.C., are on clay tablets found in Iraq. Aboriginal people drew maps of parts of Canada long before Europeans arrived.

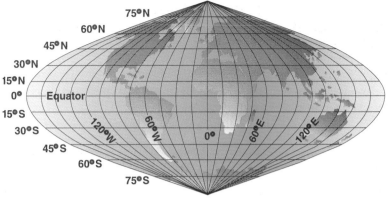

In the 2nd century A.D., an Egyptian named Ptolemy wrote a geography book that included maps of the known world. Vast parts of the world, including the Americas, were omitted. When Columbus sailed from Europe to the Americas in 1492, he thought that he had sailed to Asia! It was not until the 16th century that Gerardus Mercator's maps revolutionized navigation.

Cartography has changed dramatically. Maps, which were drawn by hand for centuries, are now created on computers. Information, often obtained from aerial photography or satellite imaging, is converted into digital form. Cartographers work with the digital information to decide colour, lines, lettering, and scales. A laser plotter draws maps directly onto photographic film for printing.

1 Map Projections

A map projection is the depiction of the three-dimensional Earth on a two-dimensional surface. It is impossible to do this without distortion. The projection a cartographer uses depends upon the purpose of the map, its scale, and the region of the Earth being mapped. Equal-area projections keep areas correct. Equidistant projections keep distances correct. Conformal projections keep shapes correct.

1. Using several different atlases, find
a) how many different projections are used
b) which projections are used most

2. To represent certain parts of the world, are some projections used more than others? If so, give possible reasons.

2 Area of the Earth

1. Assume that both the equatorial and polar radii of the Earth are 6400 km. Calculate the following, to the nearest whole number.
a) the surface area of the Earth
b) the equatorial circumference
c) the semi-circumference from pole to pole
d) the area that seems to be represented by the following Mercator projection, using dimensions from parts b) and c)

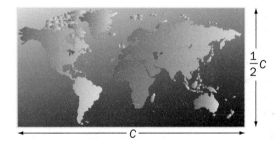

2. How many times as great as the surface area of the Earth is the area that seems to be represented by this Mercator projection?

3. Which parts of the Earth are most distorted by a Mercator projection? Explain.

3 Locating Information

1. Use your research skills to find more information about
a) the Mercator projection, and why it revolutionized navigation
b) other projections and what they distort

2. Find out how cartographers are trained and where they work.

317

INVESTIGATING MATH

Similar Figures

Use models or diagrams to complete the following.

1 Areas of Rectangles

Each small square on the grid has a side length of 1 cm. The diagrams show three rectangles with dimensions 2 cm × 1 cm, 4 cm × 2 cm, and 6 cm × 3 cm.

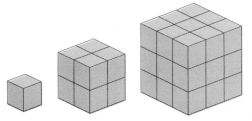

1. State the area of each rectangle.

2. By what factor is the area multiplied when both side lengths are multiplied by 2? Express your answer as a power of 2.

3. By what factor is the area multiplied when both side lengths are multiplied by 3? Express your answer as a power of 3.

4. By what factor is the area multiplied when both side lengths are multiplied by a scale factor k?

5. Check your answer to question 4 by drawing two pairs of similar rectangles with whole-number side lengths and finding how the areas are related. Describe your findings.

2 Surface Areas and Volumes of Cubes

Each small cube has an edge of 1 cm. The diagrams show three cubes with edges of 1 cm, 2 cm, and 3 cm.

1. Copy and complete the table for the three cubes.

Edge (cm)	Surface Area (cm^2)	Volume (cm^3)
1		
2		
3		

2. By what factor is the surface area multiplied when the edge is multiplied by 2? Express your answer as a power of 2.

3. By what factor is the surface area multiplied when the edge is multiplied by 3? Express your answer as a power of 3.

4. By what factor is the surface area multiplied when the edge is multiplied by a scale factor k?

5. By what factor is the volume multiplied when the edge is multiplied by 2? Express your answer as a power of 2.

6. By what factor is the volume multiplied when the edge is multiplied by 3? Express your answer as a power of 3.

7. By what factor is the volume multiplied when the edge is multiplied by a scale factor k?

8. Extend the table from question 1 to include cubes with edges of 4 cm, 5 cm, and 10 cm.

3 Surface Areas and Volumes of Rectangular Prisms

The rectangular prism has dimensions 3 cm × 2 cm × 1 cm.

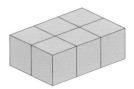

1. For the rectangular prism, determine
a) the surface area   **b)** the volume

2. Multiply each dimension of the rectangular prism by 2. For the new prism, determine
a) the surface area   **b)** the volume

3. Multiply each dimension of the original rectangular prism by 3. For the new prism, determine
a) the surface area   **b)** the volume

4. As the dimensions of the rectangular prism change, do the surface area and volume change in the same ways you found for cubes? Explain.

5. State the surface area and volume of rectangular prisms with the following dimensions.
a) 15 cm × 10 cm × 5 cm
b) 30 cm × 20 cm × 10 cm

4 Surface Areas and Volumes of Composite Solids

The composite solid is made from three 1-cm cubes.

1. For the composite solid, determine
a) the surface area
b) the volume

2. Multiply each dimension of the composite solid by 2. For the new solid, determine
a) the surface area   **b)** the volume

3. Multiply each dimension of the original composite solid by 3. For the new solid, determine
a) the surface area   **b)** the volume

4. As the dimensions of the composite solid change, do the surface area and volume change in the same ways you found for cubes? Explain.

5. Repeat questions 1–4 for composite solids of your own design made from
a) four 1-cm cubes   **b)** five 1-cm cubes

319

7.3 Surface Areas and Volumes of Similar Figures

In *Gulliver's Travels*, by Jonathan Swift, Gulliver's first voyage was to Lilliput. The inhabitants, known as Lilliputians, were very tiny. Gulliver described how his height and food requirements compared with those of the Lilliputians.

"...his Majesty's Mathematicians, having taken the Height of my Body...and finding it to exceed theirs in the proportion of Twelve to One...concluded from the Similarity of the Bodies that mine must contain at least 1728 of theirs, and consequently would require as much Food as was necessary to support that Number of Lilliputians."

Explore: Use the Diagrams

The diagrams show 4 square pyramids, A, B, C, and D.

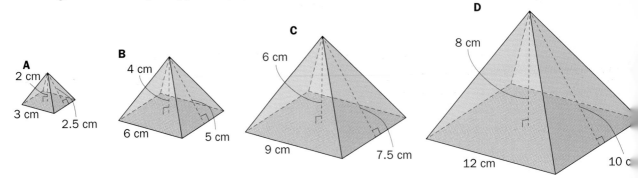

a) How many times as great are the dimensions of pyramid B as those of pyramid A?
b) How many times as great are the dimensions of pyramid C as those of pyramid A?
c) How many times as great are the dimensions of pyramid D as those of pyramid A?

Inquire

1. The pyramids are all similar. Explain why.

2. Calculate the surface area and volume of each pyramid. Copy and complete the table.

Pyramid	Surface Area (cm²)	Volume (cm³)
A		
B		
C		
D		

3. Use the data in the table to answer the following.
a) If the dimensions of a three-dimensional figure are multiplied by a scale factor k, by what factor is its surface area multiplied?
b) If the dimensions of a three-dimensional figure are multiplied by a scale factor k, by what factor is its volume multiplied?

4. Draw a diagram to show the dimensions of a pyramid similar to pyramid A and with 36 times the surface area of pyramid A.

5. Draw a diagram to show the dimensions of a pyramid similar to pyramid A and with 125 times the volume of pyramid A.

6. a) Explain the reasoning of Lilliputian mathematicians when they used Gulliver's height to decide that he needed as much food as 1728 Lilliputians.
b) If the mathematicians had calculated how many times as much soap Gulliver needed for hand washing and showers as the average Lilliputian, what number would they have found? Explain.

Example 1   Volumes of Cubes
Find the ratio of the volume of the larger cube to the volume of the smaller cube.

a)   **b)**

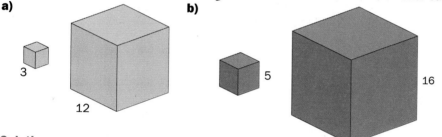

3

12

5

16

Solution
The ratio of the volumes of two similar figures equals the cube of the ratio of the lengths of any pair of corresponding segments.

a) The ratio of the lengths of corresponding segments is $\dfrac{12}{3}$ or $\dfrac{4}{1}$.

$$\text{Ratio of volumes} = \left(\frac{4}{1}\right)^3$$
$$= \frac{64}{1}$$

The ratio of the volume of the larger cube to the volume of the smaller cube is 64:1.

b) The ratio of the lengths of corresponding segments is $\dfrac{16}{5}$.

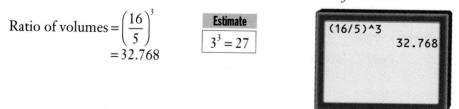

$$\text{Ratio of volumes} = \left(\frac{16}{5}\right)^3$$
$$= 32.768$$

Estimate
$3^3 = 27$

(16/5)^3   32.768

The ratio of the volume of the larger cube to the volume of the smaller cube is 32.768:1.

Example 2 Enlarging a Photograph

A photo has an area of 40 cm². By what factor must each dimension be multiplied to increase the area by 80 cm²? Round your answer to the nearest hundredth.

Solution

The photo and its enlargement are similar.

The ratio of the areas of two similar figures equals the square of the ratio of the lengths of any pair of corresponding segments. Therefore, the ratio of the lengths of any pair of corresponding segments equals the square root of the ratio of the areas.

If the area increases by 80 cm², the final area is 40 + 80 or 120 cm².

The ratio of the areas is $\dfrac{120}{40}$ or $\dfrac{3}{1}$.

Ratio of lengths $= \sqrt{\dfrac{3}{1}}$

$\doteq \dfrac{1.73}{1}$

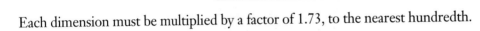

√(3)
                1.732050808

Each dimension must be multiplied by a factor of 1.73, to the nearest hundredth.

Example 3 Cans of Pineapple

A can that holds 540 mL of pineapple chunks is 8.5 cm in diameter and 11.5 cm in height. If the dimensions of the can were increased using a scale factor of 1.5,
a) what quantity of pineapple chunks would the larger can hold?
b) what would be the area of the label on the larger can, to the nearest 10 cm²?

Solution

a) The ratio of the volumes of two similar objects equals the cube of the ratio of the lengths of any pair of corresponding segments.

Ratio of lengths $= 1.5$ or $\dfrac{3}{2}$

Ratio of volumes $= \left(\dfrac{3}{2}\right)^3$

$= \dfrac{27}{8}$

Solve a proportion to find the unknown quantity.

$$\dfrac{27}{8} = \dfrac{x}{540}$$

$$540 \times \dfrac{27}{8} = 540 \times \dfrac{x}{540}$$

$$1822.5 = x$$

The larger can would hold 1822.5 mL of pineapple chunks.

b) Assume that the label covers the whole of the curved surface of each can. For the smaller can, the area of the curved surface is

$$2\pi rh = 2(\pi)\left(\frac{8.5}{2}\right)(11.5)$$
$$= 97.75\pi$$

The ratio of the surface areas of two similar objects equals the square of the ratio of the lengths of any pair of corresponding segments.

Ratio of areas of curved surfaces $= \left(\frac{3}{2}\right)^2$

$$= \frac{9}{4}$$

For the larger can, the area of the curved surface would be

$$\frac{9}{4} \times 97.75\pi \doteq 690$$

Estimate
$2 \times 100 \times 3 = 600$

The area of the label on the larger can would be 690 cm², to the nearest 10 cm².

Practice

Determine whether the figures in each pair are similar.

1.

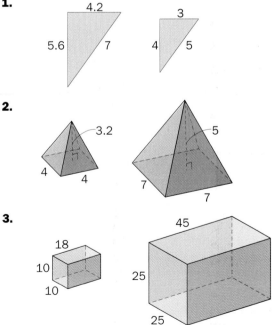

2.

3.

Calculate the surface area and volume of each solid. Round to the nearest square or cubic unit. Then, using the scale factor provided, calculate the surface area and volume of a solid similar to the original solid.

4.

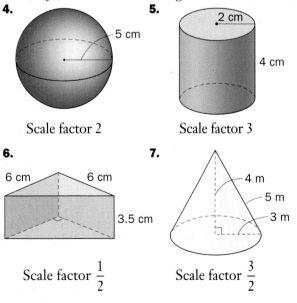

5 cm

Scale factor 2

5.

2 cm

4 cm

Scale factor 3

6.

6 cm   6 cm

3.5 cm

Scale factor $\frac{1}{2}$

7.

4 m

5 m

3 m

Scale factor $\frac{3}{2}$

8. If the dimensions of a rectangle are increased by a factor of 4, by what factor does each of the following increase?
a) perimeter   **b)** area

9. If the dimensions of a cone are decreased by a factor of 3, by what factor does each of the following decrease?
a) volume   **b)** surface area
c) area of base   **d)** circumference of base

Applications and Problem Solving

10. The ratio of the volumes of two spheres is 27:8. What is the ratio of the radii of the spheres?

11. The ratio of the surface areas of two cubes is 64:81. What is the ratio of the edges of the cubes?

12. If the ratio of the surface areas of two similar cones is 4:9, what is the ratio of their volumes?

13. If the ratio of the volumes of two similar cylinders is 125:27, what is the ratio of their surface areas?

14. **Egyptian pyramids** The pyramids at Giza are roughly similar. The dimensions of the pyramid of Khafre are about double the dimensions of the pyramid of Menkaure. How do the masses of the two pyramids compare? Explain.

15. **Polar bear** A statue of a polar bear in Cochrane, Ontario, symbolizes the town's northerly location and the Polar Bear Express train, which runs north from the town to Moosonee. The length of the statue is about 5 m. A real polar bear has a length of about 2.5 m.
a) What scale factor was used to build the statue?
b) How does the width of the nose on the statue compare with the width of a polar bear's nose?
c) How does the area of an eye on the statue compare with the area of a polar bear's eye?
d) How does the volume of the statue compare with the volume of a polar bear?

16. **Floor plan** A floor plan of an apartment is drawn using a scale of 1:70. The plan measures 18 cm by 10 cm.
a) Explain the meaning of "a scale of 1:70."
b) Show two ways to calculate the actual area of the apartment, in square metres.

17. **Earth and moon** The moon has a diameter of about 3480 km. The Earth has a diameter of about 12 760 km. Write each of the following ratios as a fraction in lowest terms.
a) the diameter of the moon to the diameter of the Earth
b) the surface area of the moon to the surface area of the Earth
c) the volume of the moon to the volume of the Earth

18. *Hamlet* **set** The technical crew built a scale model of the set for a school production of *Hamlet*. Renée painted the scale model and used 140 mL of paint. If the scale of the model was 1:15, how much paint was needed for the real set?

19. **Model rocket** A model of a rocket was built to a scale of 1:40.
a) The length of the model was 75 cm. What was the length of the actual rocket, in metres?
b) The area of metal used to cover the curved surfaces of the model was 2750 cm^2. What area of metal covered the curved surfaces of the actual rocket, in square metres?
c) The nose cone of the rocket had a volume of 96 m^3. What was the volume of the nose cone of the model, in cubic centimetres?

20. A triangular pyramid has a height of 36 cm and a volume of 1728 cm^3. A similar pyramid has a volume of 1000 cm^3. What is the height of the second pyramid?

21. **Moons of Saturn** Titan and Rhea are the two largest moons of Saturn. The ratio of the surface area of Titan to the surface area of Rhea is about 34:3. If the diameter of Rhea is about 1530 km, what is the diameter of Titan, to the nearest ten kilometres?

22. **Comparison shopping** Five 30-cm diameter pizzas cost the same as three 40-cm diameter pizzas with the same toppings. Which is the better buy? Explain your answer and state any assumptions you make.

23. A triangle has an area of 20 cm^2. By what factor must the dimensions of the triangle be multiplied to give a similar triangle and increase the area by each of the following? Round each answer to the nearest hundredth, if necessary.

a) 60 cm^2 **b)** 160 cm^2
c) 20 cm^2 **d)** 80 cm^2

24. A rectangular prism has a volume of 1000 cm^3. By what factor must the dimensions of the prism be multiplied to give a similar prism and reduce the volume by
a) 875 cm^3?   **b)** 992 cm^3?

25. Architecture The design of a new office building included a central courtyard.
a) If the architects decided to increase each dimension of the courtyard by 15%, what was the percent increase in the area of the courtyard?
b) If the architects decided to keep the courtyard the same shape but increased the area by 21%, what was the percent increase in each dimension of the courtyard?

26. Lewis Carroll An Oxford mathematician named Charles Dodgson, writing under the pen name Lewis Carroll, wrote *Alice's Adventures in Wonderland*. In this novel, Alice ate a special cookie and grew to 10 times her original height.
a) What was the effect of the change in height on the area of her skin? on the volume of her body? on the mass of her body?
b) What was the effect of the change in height on the strength of her bones? Explain.
c) Would the bones of the larger body be strong enough to support the larger mass? Explain.
d) Describe the effects of the change in height on other biological systems of the body, including the circulatory system. Is it likely that a real person could have survived the change in height?

27. Decide whether each statement is always true, sometimes true, or never true. Explain.
a) Two rectangles are similar.
b) A scalene triangle and an isosceles triangle are similar.
c) Two circles are similar.
d) Two spheres are similar.
e) Two cones are similar
f) Two cubes are similar.
g) Two cylinders are similar.

28. Decide whether each statement is always true. Explain, using examples.
a) If two rectangular prisms are similar, and corresponding edges are in the ratio 2:1, the volumes are in the ratio 8:1.
b) If the volumes of two rectangular prisms are in the ratio 8:1, the two prisms are similar.

29. Write a problem that involves the surface area and volume of two similar figures. Have a classmate solve your problem.

PATTERN POWER

The difference $10^2 - 10^1$ equals 90, when expressed in standard form.

1. Express each of the following differences in standard form.
a) $10^3 - 10^1$ **b)** $10^4 - 10^1$
c) $10^3 - 10^2$ **d)** $10^4 - 10^3$
e) $10^5 - 10^2$ **f)** $10^6 - 10^4$

2. For the difference $10^m - 10^n$ in standard form, where m and n are positive integers and $m > n$,
a) how many 9s are there?
b) how many 0s are there?

3. Use the pattern you found in question 2 to write each of the following differences in standard form.
a) $10^9 - 10^6$ **b)** $10^{11} - 10^{10}$ **c)** $10^{16} - 10^{12}$

4. Find and describe the pattern for the difference $10^a - 10^b$ in standard form, where a and b are negative integers and $a > b$.

5. Use the pattern you found in question 4 to write each of the following differences in standard form.
a) $10^{-5} - 10^{-6}$ **b)** $10^{-3} - 10^{-8}$
c) $10^{-1} - 10^{-7}$ **d)** $10^{-9} - 10^{-14}$

CONNECTING MATH AND DESIGN

Architectural and Engineering Drawings

There are two common types of drawings that architects, designers, and engineers use to display an object.

An **isometric view** is an attempt to represent a three-dimensional object in a two-dimensional drawing. An isometric view can also be called a corner view of the object. When drawing an isometric view, it is helpful to use isometric dot paper.

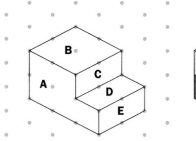

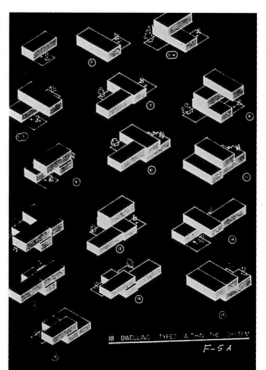

An **orthographic view** represents two dimensions of an object, as it appears from the top, front, and right side. When drawing orthographic views, it is helpful to use grid paper or square dot paper.

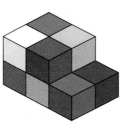

In an orthographic view, any hidden edges of an object are shown as dashed lines. In an isometric view, any hidden edges are not shown.

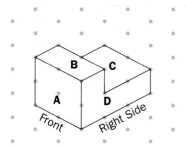

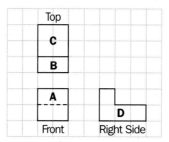

Habitat, designed by Canadian architect Moshe Safdie

1 Orthographic Views

Use the isometric view of each object to draw its orthographic views.

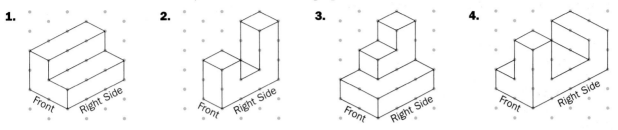

1. Front Right Side

2. Front Right Side

3. Front Right Side

4. Front Right Side

2 Isometric Views

Use the orthographic views of each object to draw its isometric view.

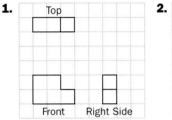

1. Top Front Right Side

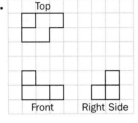

2. Top Front Right Side

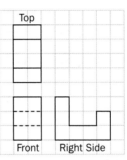

3. Top Front Right Side

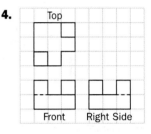

4. Top Front Right Side

3 Base Designs

For an object built from linking cubes, the **base design** shows the top view of the object and the number of cubes in each column.

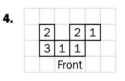

	2	
2	1	1

Base Design

Use the following base designs to draw the orthographic and isometric views of each object.

1.
2	3
1	2

Front

2.
1	2	2
1	1	2

Front

3.

3	2	1
1		1

Front

4.
2		2	1
3	1	1	

Front

5. A given object has the base design shown:

1	1

Front

a) Which of these base designs represents an object that is similar to the given object?

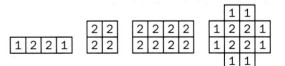

2	2

1	2	2	1

2	2
2	2

2	2	2	2
2	2	2	2

	1	1	
1	2	2	1
1	2	2	1
	1	1	

b) Draw isometric and orthographic views of the two similar objects.

6. A given object has the base design shown:

1	2	1

Front

a) Draw the base design of an enlargement of the object using the scale factor 2.
b) Draw isometric and orthographic views of the two similar objects.

INVESTIGATING MATH

Trigonometric Ratios

1 Sine, Cosine, and Tangent

In the right triangle ABC,

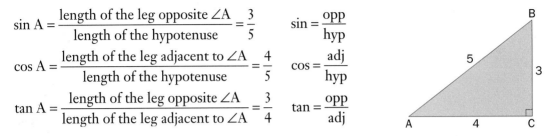

$$\sin A = \frac{\text{length of the leg opposite } \angle A}{\text{length of the hypotenuse}} = \frac{3}{5} \qquad \sin = \frac{\text{opp}}{\text{hyp}}$$

$$\cos A = \frac{\text{length of the leg adjacent to } \angle A}{\text{length of the hypotenuse}} = \frac{4}{5} \qquad \cos = \frac{\text{adj}}{\text{hyp}}$$

$$\tan A = \frac{\text{length of the leg opposite } \angle A}{\text{length of the leg adjacent to } \angle A} = \frac{3}{4} \qquad \tan = \frac{\text{opp}}{\text{adj}}$$

1. Find sin B, cos B, and tan B. Leave each answer in fraction form.

2. In each of the following triangles, find the three trigonometric ratios for each acute angle. Leave each answer in fraction form. Do not rationalize radical denominators.

a) **b)**

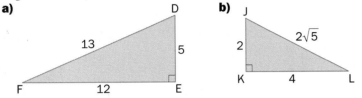

2 Special Right Triangles

The 45°–45°–90° triangle is an isosceles right triangle. If each leg has a length of 1 unit, the length of the hypotenuse is $\sqrt{2}$ units.

Drawing the altitude of an equilateral triangle with a side length of 2 units gives the 30°–60°–90° triangle with the side lengths shown.

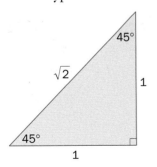

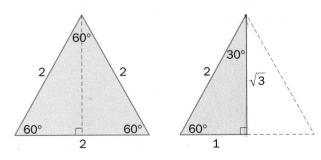

In each of the following, leave each answer in fraction form. Do not rationalize radical denominators.

1. Use the 45°–45°–90° triangle to find
a) sin 45°   **b)** cos 45°   **c)** tan 45°

2. Use the 30°–60°–90° triangle to find
a) sin 30°   **b)** cos 30°   **c)** tan 30°

3. Use the 30°–60°–90° triangle to find
a) sin 60°   **b)** cos 60°   **c)** tan 60°

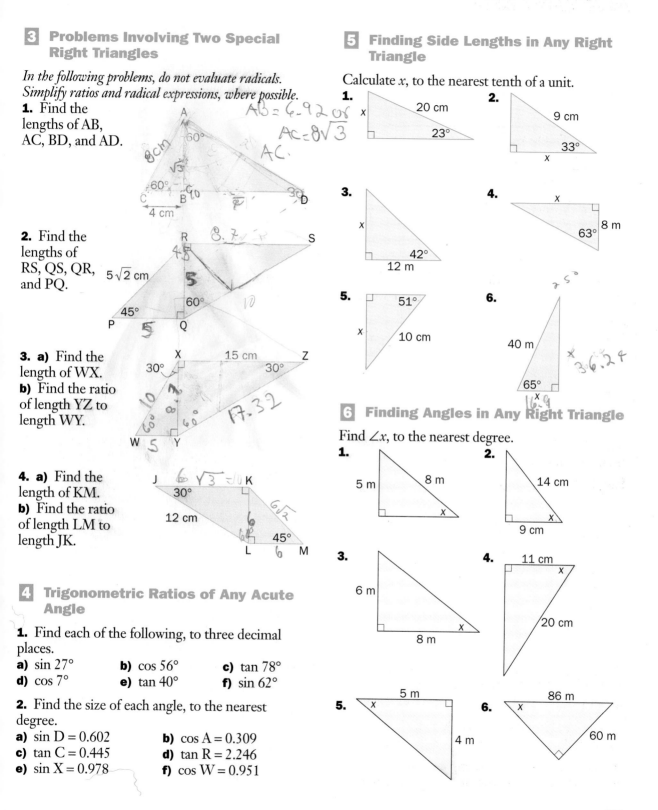

3 **Problems Involving Two Special Right Triangles**

In the following problems, do not evaluate radicals. Simplify ratios and radical expressions, where possible.

1. Find the lengths of AB, AC, BD, and AD.

$AB = 6.92$ or
$AC = 8\sqrt{3}$
AC.

A
60°
8 cm
$\sqrt{3}$
60°
C   B   30   D
4 cm

2. Find the lengths of RS, QS, QR, and PQ.

R   8.7   S
45
$5\sqrt{2}$ cm   5   10
45°   60°
P   5   Q

3. a) Find the length of WX.
b) Find the ratio of length YZ to length WY.

X   15 cm   Z
30°   30°
10   8.7   17.32
60   60
W   5   Y

4. a) Find the length of KM.
b) Find the ratio of length LM to length JK.

J   6$\sqrt{3}$ =10 K
30°
12 cm   6   6$\sqrt{2}$
L   6   M
45°

4 **Trigonometric Ratios of Any Acute Angle**

1. Find each of the following, to three decimal places.
a) sin 27°   **b)** cos 56°   **c)** tan 78°
d) cos 7°   **e)** tan 40°   **f)** sin 62°

2. Find the size of each angle, to the nearest degree.
a) sin D = 0.602   **b)** cos A = 0.309
c) tan C = 0.445   **d)** tan R = 2.246
e) sin X = 0.978   **f)** cos W = 0.951

5 **Finding Side Lengths in Any Right Triangle**

Calculate *x*, to the nearest tenth of a unit.

1.
x   20 cm
23°

2.
9 cm
33°
x

3.
x
42°
12 m

4.
x
63°   8 m

5.
51°
x   10 cm

6.
$7.5°$
40 m
65°
x   3.6.24
16.9

6 **Finding Angles in Any Right Triangle**

Find ∠*x*, to the nearest degree.

1.
5 m   8 m
x

2.
14 cm
x
9 cm

3.
6 m
x
8 m

4.
11 cm
x
20 cm

5.
5 m
x
4 m

6.
86 m
x
60 m

7.4 Reviewing Trigonometric Ratios

In the short story *The Musgrave Ritual*, Sherlock Holmes found the solution to a mystery at a certain point. To find the point, he had to start near the stump of an elm tree and take 20 paces north, then 10 paces east, then 4 paces south, and finally 2 paces west.

Explore: Draw a Diagram

Let E be the point where Holmes started pacing and S be the point where he stopped. Draw a diagram of his path. Join ES. Draw another line segment so that ES is the hypotenuse of a right triangle.

Inquire

1. What are the lengths of the two legs of the right triangle?

2. What trigonometric ratio can you use to find ∠E from the lengths of the legs? $\frac{16}{8}$

3. Find ∠E, to the nearest degree. $63°$

4. What methods could you use to find the length of ES, in paces?

5. What is the length of ES, to the nearest pace? Pythagorean Theorem $a^2 + b^2 = c$

6. In what direction, and for how many paces, could Holmes have walked in order to go directly from E to S? toward NE 17.9

The primary trigonometric ratios are

$$\text{sine } \theta = \frac{\text{opposite}}{\text{hypotenuse}}$$

$$\text{cosine } \theta = \frac{\text{adjacent}}{\text{hypotenuse}}$$

$$\text{tangent } \theta = \frac{\text{opposite}}{\text{adjacent}}$$

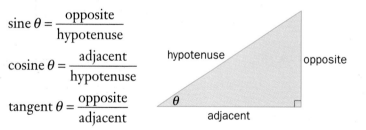

Example 1   Solving a Triangle
In △DEF, ∠E = 90°, d = 7.4 m, and f = 6.5 m. Solve the triangle by finding
a) the unknown angles, to the nearest tenth of a degree
b) the unknown side, to the nearest tenth of a metre

Solution
a) From the diagram,

$$\tan D = \frac{7.4}{6.5}$$

$$\angle D \doteq 48.7°$$

$$\angle F = 90° - 48.7°$$

$$= 41.3°$$

Your calculator must be in the degree mode.

330

b) From the diagram, $\quad \sin 48.7° = \dfrac{7.4}{e}$

$$e \times \sin 48.7° = 7.4$$

$$e = \dfrac{7.4}{\sin 48.7°}$$

$$\doteq 9.9$$

In $\triangle DEF$, $\angle D = 48.7°$, $\angle F = 41.3°$, and $e = 9.9$ m.

```
7.4/sin(48.7)
        9.850064272
```

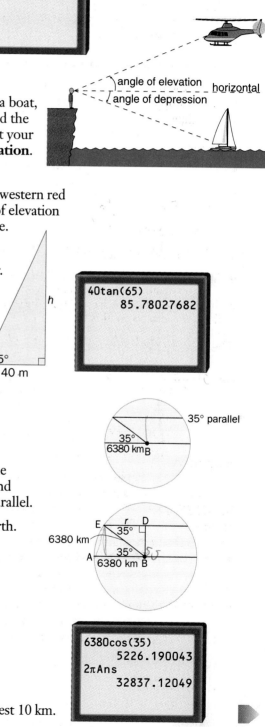

angle of elevation   horizontal
angle of depression

If you are standing on a cliff beside a river, and you look down at a boat, the angle that your line of sight makes with the horizontal is called the **angle of depression**. If you look up at a helicopter, the angle that your line of sight makes with the horizontal is called the **angle of elevation**.

Example 2   Western Red Cedars

Cathedral Grove, on Vancouver Island, is a rain forest of firs and western red cedars. From a point 40 m from the foot of one cedar, the angle of elevation of the top is 65°. Find the height of the cedar, to the nearest metre.

Solution

Draw and label a diagram. Let h represent the height of the cedar.

$$\dfrac{h}{40} = \tan 65°$$

$$h = 40 \tan 65°$$

$$\doteq 86$$

The cedar is 86 m tall, to the nearest metre.

```
40tan(65)
        85.78027682
```

h

$65°$
40 m

Example 3   Parallel of Latitude

Find the length of the 35° parallel of latitude, to the nearest 10 km. Assume that the radius of the Earth is 6380 km.

35° parallel

35°
6380 km B

Solution

In the diagram, B is the centre of the Earth, and A is a point on the equator. D is the centre of the circle defined by the 35° parallel, and E is a point on its circumference. DE is the radius, r, of the 35° parallel.

$\angle BDE$ is a right angle. $BA = BE$, because both are radii of the Earth.
$\angle DEB = \angle ABE$ (alternate angles)

In $\triangle DEB$, $\quad \dfrac{r}{6380} = \cos 35°$

$$r = 6380 \cos 35°$$

$$\doteq 5226$$

The length of the 35° parallel of latitude is its circumference, C.

$C = 2\pi r$

$\doteq 2\pi(5226)$

$\doteq 32\ 840$

E r D
35°
6380 km
A   35°
6380 km B

```
6380cos(35)
        5226.190043
2πAns
        32837.12049
```

The length of the 35° parallel of latitude is 32 840 km, to the nearest 10 km.

331

Practice

Solve each triangle. Round each side length to the nearest unit and each angle to the nearest degree.

1.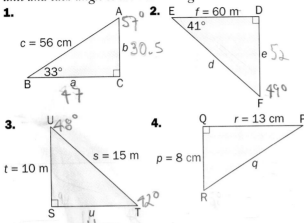

2.

3.

4.

Solve each triangle. Round each side length to the nearest tenth of a unit and each angle to the nearest tenth of a degree.

5.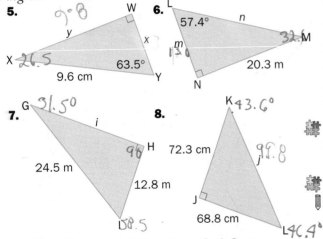

6.

7.

8.

Applications And Problem Solving

9. Prince Shoal lighthouse The light on the Prince Shoal lighthouse is 25 m above the water level. From a position beside the light, the angle of depression of a sailboat is 12°. How far is the sailboat from the lighthouse, to the nearest metre?

10. Two CN Towers Edmonton's CN Tower is a highrise office building. From a point 35 m from the base of the building and level with the base, the angle of elevation of the top is 72.5°.
a) Find the height of Edmonton's CN Tower, to the nearest metre.

b) Toronto's CN Tower is a tourist attraction, with a height of 555 m. How many times as tall as Edmonton's CN Tower is Toronto's CN Tower?
c) Edmonton's CN Tower has 27 storeys. If an office building were the height of Toronto's CN Tower, how many storeys would you expect it to have?

11. Coast guard A coast guard patrol boat is 14.8 km east of the Rose Blanche lighthouse. A disabled yacht is 7.5 km south of the lighthouse.
a) How far is the patrol boat from the yacht, to the nearest tenth of a kilometre?
b) At what angle south of due west, to the nearest tenth of a degree, should the patrol boat travel to reach the yacht?

12. Highway signs A sign shows that a hill has a grade of 9%. What angle does the hill make with the horizontal, to the nearest tenth of a degree?

13. Stairs Comfortable stairs have a slope of $\frac{3}{4}$. What angle do the stairs make with the horizontal, to the nearest degree?

14. Latitude Find the length of the 40° parallel of latitude, to the nearest 10 km.

15. Arctic Circle Find the length of the Arctic Circle, which is at 66.55° north.

16. Show that the length of any parallel of latitude is equal to the length of the equator times the cosine of the latitude angle.

17. Use right triangles ABC and DEF to complete the table. Leave all ratios in fraction form.

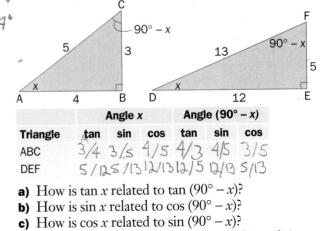

Triangle	Angle x			Angle (90° − x)		
	tan	sin	cos	tan	sin	cos
ABC	3/4	3/5	4/5	4/3	4/5	3/5
DEF	5/12	5/13	12/13	12/5	12/13	5/13

a) How is tan x related to tan (90° − x)?
b) How is sin x related to cos (90° − x)?
c) How is cos x related to sin (90° − x)?
d) Explain the relationships in parts a), b), and c).

7.5 Problems Involving Two Right Triangles

A popular tourist attraction in Ottawa is the Peace Tower, which rises from the Centre Block of the Parliament buildings. A bronze mast, which flies the Canadian flag, stands on top of the tower. From a point 25 m from the foot of the tower, the angle of elevation of the top of the tower is 74.8°. From the same point, the angle of elevation of the top of the mast is 76.3°.

Explore: Use a Diagram

To find the height of the mast, draw a diagram. Mark the given information on the diagram. DC is the height of the mast, and BC is the height of the tower.

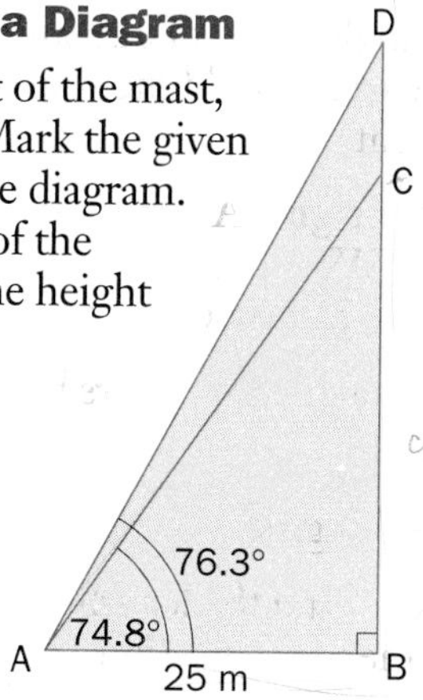

Inquire

1. a) Use $\triangle ABC$ and write a trigonometric equation for BC in terms of $\angle BAC$ and AB.
b) Substitute the values of $\angle BAC$ and AB into the equation.
c) Find BC, to the nearest tenth of a metre.

2. a) Use $\triangle BAD$ and write a trigonometric equation for BD in terms of $\angle BAD$ and AB.
b) Substitute the values of $\angle BAD$ and AB into the equation.
c) Find BD, to the nearest tenth of a metre.

3. What is the height of the mast, to the nearest tenth of a metre?

4. From a point 50 m from the foot of the tower, what is the angle of elevation, to the nearest tenth of a degree?
a) the top of the tower
b) the top of the mast

Example 1   Cloud Height at Night
Many pilots flying from small airports follow visual flight rules. Under these rules, an aircraft can take off only if the visibility is 5 km or greater, and the cloud height is 300 m or higher. Airport managers determine the cloud height visually during the day. To determine the cloud height at night, many small airports have a spotlight that shines on the clouds.

The angle the light beam makes with the ground is 70°. An observer, located on the ground 300 m from the light, measures the angle of elevation of the point where the light shines on the clouds. Suppose the angle of elevation of this point is 60°, and the light and the observer are on opposite sides of the point. Find the cloud height, to the nearest metre.

Recall the Steps

Understand the Problem

Think of a Plan

Carry Out the Plan

Look Back

Solution
Draw and label a diagram. Let h represent the cloud height.

Mark the given data on the diagram. Also include the measures of $\angle ADB$ and $\angle CDB$, which can be determined from the given data.

Since $x + y = 300$, use the two right triangles to find expressions for x and y in terms of h.

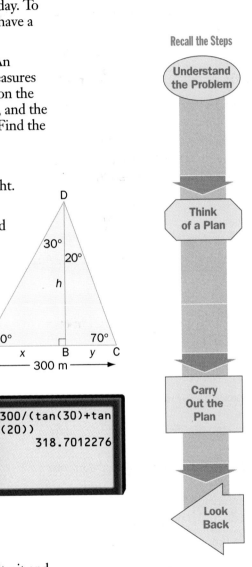

In $\triangle ABD$,

$$\frac{x}{h} = \tan 30°$$
$$x = h \tan 30°$$

In $\triangle BCD$,

$$\frac{y}{h} = \tan 20°$$
$$y = h \tan 20°$$

But $x + y = 300$
so $h \tan 30° + h \tan 20° = 300$
and $h(\tan 30° + \tan 20°) = 300$

$$h = \frac{300}{\tan 30° + \tan 20°}$$
$$h \doteq 319$$

```
300/(tan(30)+tan
(20))
          318.7012276
```

The cloud height is 319 m, to the nearest metre.

Does the answer seem reasonable?

Example 2   Confederation Bridge
The Confederation Bridge spans the Northumberland Strait and joins New Brunswick and Prince Edward Island. From one point in the strait, the angle of elevation of the highest point on the bridge is 26.6°. From a point 100 m closer to the bridge, the angle of elevation is 71.7°. How high is the bridge, to the nearest tenth of a metre?

Solution

Draw and label a diagram. Let h represent the height of the bridge. Mark the given data on the diagram. Also include the measures of $\angle BAD$ and $\angle CAD$, which can be determined from the given data.

Since $y - x = 100$, use the two right triangles to find expressions for x and y in terms of h.

In $\triangle ABD$,

$$\frac{y}{h} = \tan 63.4°$$
$$y = h \tan 63.4°$$

In $\triangle ACD$,

$$\frac{x}{h} = \tan 18.3°$$
$$x = h \tan 18.3°$$

But
$$y - x = 100$$
so $h \tan 63.4° - h \tan 18.3° = 100$
and $h(\tan 63.4° - \tan 18.3°) = 100$

$$h = \frac{100}{\tan 63.4° - \tan 18.3°}$$
$$h \doteq 60.0$$

The height of the bridge is 60.0 m, to the nearest tenth of a metre.

```
100/(tan(63.4)-t
an(18.3))
        60.0155269
```

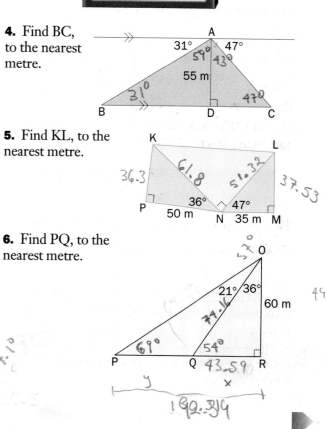

Practice

1. Find YZ, to the nearest tenth of a metre.

2. Find EF, to the nearest tenth of a metre.

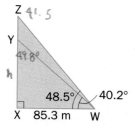

3. Find UV, to the nearest tenth of a metre.

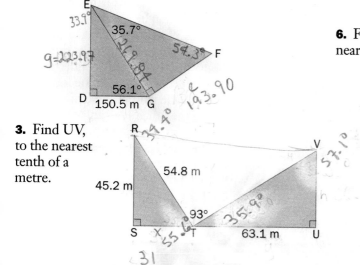

4. Find BC, to the nearest metre.

5. Find KL, to the nearest metre.

6. Find PQ, to the nearest metre.

335

7. Find RS, to the nearest metre.

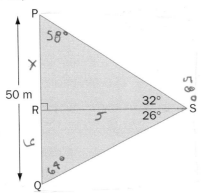

8. Find TU, to the nearest tenth of a metre.

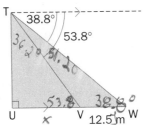

Applications and Problem Solving

9. Height of a building From the window of one building, Sam finds the angle of elevation of the top of a second building is 41° and the angle of depression of the bottom is 54°. The buildings are 56 m apart. Find, to the nearest metre,
a) the height of the second building
b) the height Sam is above the ground

10. Horseshoe Falls From the bridge of a boat on the Niagara River, the angle of elevation of the top of the Horseshoe Falls is 64°. The angle of depression of the bottom of the falls is 6°. If the bridge of the boat is 2.8 m above the water, calculate the height of the Horseshoe Falls, to the nearest tenth of a metre.

11. Capilano Suspension Bridge The Capilano Suspension Bridge in North Vancouver is the world's largest and highest footbridge of its kind. The bridge is 140 m long. From the ends of the bridge, the angles of depression of a point on the river under the bridge are 41° and 48°. How high is the bridge above the river, to the nearest metre?

12. Surveying A surveyor uses an instrument called a theodolite to measure angles vertically and horizontally. To measure the height of an inaccessible cliff, a surveyor sets up the theodolite at E and measures ∠DEF to be 56.7°. Then, the baseline EG is laid off perpendicular to EF. The length of EG is 74.6 m, and ∠EGF measures 49.1°.

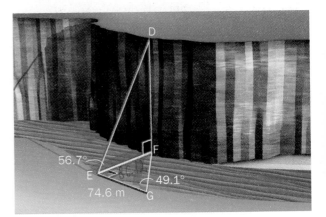

a) Find DF, to the nearest tenth of a metre.
b) If the theodolite is 1.6 m tall, what is the height of the cliff?

13. Whooping crane The whooping crane is Canada's tallest bird. When the angle of elevation of the sun decreases from 30° to 25°, the length of a whooping crane's shadow increases by 62 cm. How tall is the whooping crane, to the nearest centimetre?

14. Lovett Lookout In heavily forested areas, a fire-watch system is usually maintained during dry periods. The system consists of a chain of lookout towers, often supplemented by small aircraft and by satellite photographs. The system is linked by radio to a central office, which can dispatch fire fighting crews. Lovett Lookout is in Northern Alberta and has a height of 24 m. From the top of the lookout, a ranger sees two fires. One is at an angle of depression of 6° and the other at an angle of depression of 4°. The fires and the tower are in a straight line. Find the distance between the fires, to the nearest metre, if they are
a) on the same side of the tower
b) on opposite sides of the tower

15. Pont du Gard aqueduct The Pont du Gard is an aqueduct that crosses the River Gard near Nîmes in France. From one point on the river, the angle of elevation of the top of the aqueduct is 43.6°. From a point 30 m farther away from the aqueduct, the angle of elevation is 31.2°. Find the height of the aqueduct, to the nearest metre.

16. Satellites The angle of elevation of an orbiting satellite from tracking station Alpha is 63° and from tracking station Bravo is 58°. The two tracking stations are 1300 km apart on the Earth.
a) What is the height of the satellite, to the nearest kilometre, if it is is position S_1?
b) What is the height of the satellite, to the nearest kilometre, if it is in position S_2?

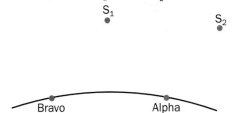

c) What assumptions have you made when calculating the heights?

17. Aircraft speed An Airbus A320 aircraft is cruising at an altitude of 10 000 m. The aircraft is flying in a straight line away from Chandra, who is standing on the ground. If she sees the angle of elevation of the aircraft change from 70° to 33° in one minute, what is its cruising speed, to the nearest kilometre per hour?

18. Pendulum When a pendulum swings 40° from the vertical, the bob moves 20 cm horizontally and 7.3 cm vertically. What is the length of the pendulum, to the nearest centimetre?

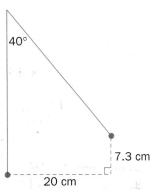

19. Measurement In a regular octagon, AB is a diagonal and CD joins the midpoints of two opposite sides. The side length of the octagon is 4 cm.

To the nearest tenth of a centimetre, find
a) AB
b) CD

20. Write a problem that involves the use of two right triangles. Have a classmate solve your problem.

LOGIC POWER

Opposite faces of the large cube are shaded in the same way. The interior is filled with small white cubes.

How many of the small cubes have
a) 3 blue faces?
b) 2 blue faces?
c) 1 blue face?
d) 0 blue faces?

7.6 Angles in Standard Position

The trigonometric ratios have been defined in terms of the sides and acute angles of right triangles. Trigonometric ratios can also be defined for angles in **standard position** on a coordinate grid. In the diagram, angle θ is in standard position.

The vertex of angle θ is at the origin. One ray, called the **initial arm**, is fixed on the positive x-axis. The other ray, called the **terminal arm**, rotates about the origin. The measure of the angle is the amount of rotation from the initial arm to the terminal arm.

The clapper board used in TV or movie production makes angles that look like angles in standard position. The lower part of the board is used to record such information as the name of the movie or TV program, the director, and the scene and take numbers. At the start of each take, the hinged arm is clapped onto the lower part of the board. In the editing stage, the loud sound and the visual image of the arm hitting the board are used to synchronize the moving pictures with the sound track.

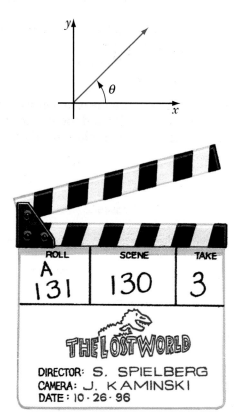

Explore: Look for Patterns

Copy the table. Complete it using a calculator. Round each calculated value to four decimal places.

Angle A	sin A	cos A	Angle A	sin A	cos A
0°	0	1	105°	0.9769	-0.3698
15°	0.3698	1.1769	120°	0.9760	-0.5
30°	0.5	0.9760	135°	0.7071	-0.7071
45°	0.7171	0.7171	150°	0.5	-0.9760
60°	0.9760	0.5	165°	0.3698	-1.1769
75°	.9659	0.3698	180°	0	-1
90°	1	0			

Inquire

1. How does the value of sin A change as the measure of ∠A goes from 0° to 90°? from 90° to 180°? Increase as then starts decreasing at 1

2. a) What pairs of angles have equal sine values?
b) How are the angles in each pair related?

3. If sin A = sin 40°, and ∠A is between 0° and 180°, what are the possible measures of ∠A?

4. If sin B = sin 130°, and ∠B is between 0° and 180°, what are the possible measures of ∠B?

5. How does the value of cos A change as the measure of ∠A goes from 0° to 90°? from 90° to 180°?

6. a) How are cos 30° and cos 150° the same? How are they different?
b) How are cos 165° and cos 15° the same? How are they different?

7. a) Measure the angle made by the clapper board in the photo, to the nearest degree.   013

b) State the measure of an obtuse angle whose sine equals the sine of the angle from part a).

c) Is there an obtuse angle whose cosine equals the cosine of the angle from part a)? Explain.

Let (x, y) be a point on the terminal arm of an angle θ in standard position. The side opposite θ is y and the side adjacent to θ is x. The hypotenuse, r, can be found using the Pythagorean Theorem.

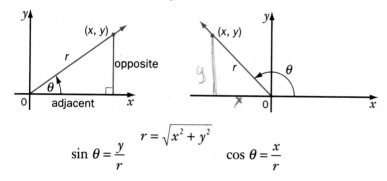

$$r = \sqrt{x^2 + y^2}$$

$$\sin \theta = \frac{y}{r} \qquad \cos \theta = \frac{x}{r}$$

Example 1 Sine and Cosine

The point $(-4, 3)$ lies on the terminal arm of an angle θ in standard position. Find $\sin \theta$ and $\cos \theta$.

Solution

Sketch the angle in standard position. Make a triangle by drawing a perpendicular from the point $(-4, 3)$ to the x-axis.

$$r^2 = x^2 + y^2$$
$$r = \sqrt{x^2 + y^2}$$
$$= \sqrt{(-4)^2 + 3^2}$$
$$= \sqrt{16 + 9}$$
$$= \sqrt{25}$$
$$= 5$$

$$\sin \theta = \frac{y}{r} \qquad \cos \theta = \frac{x}{r}$$
$$= \frac{3}{5} \qquad\quad = \frac{-4}{5} \text{ or } -\frac{4}{5}$$

For an angle θ in standard position, the **reference angle** is the acute angle formed by the terminal arm and the x-axis. For an obtuse angle θ, the reference angle measures $180° - \theta$.

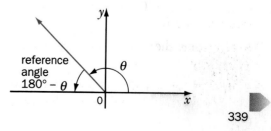

Example 2   Sine and Cosine of Special Obtuse Angles
Find sin 120° and cos 120°.

Solution
Sketch the angle 120° in standard position. The reference angle is 180° – 120° or 60°. Make a triangle by drawing a perpendicular to the x-axis from a point P(x, y) on the terminal arm. The triangle is a 30°–60°–90° triangle. Mark the appropriate lengths on the triangle. The length, r, of the terminal arm is always positive. If P is chosen so that $r = 2$, then $x = -1$, and $y = \sqrt{3}$.

$$\sin 120° = \frac{y}{r} \qquad \cos 120° = \frac{x}{r}$$

$$= \frac{\sqrt{3}}{2} \qquad\qquad = \frac{-1}{2} \text{ or } -\frac{1}{2}$$

Example 3   Finding Special Angles
If $0° \le A \le 180°$, find the measure(s) of $\angle A$ when

a) $\sin A = \dfrac{1}{\sqrt{2}}$ **b)** $\cos A = \dfrac{1}{\sqrt{2}}$ **c)** $\cos A = -\dfrac{1}{\sqrt{2}}$

Solution
a) The sine of an angle is positive when the terminal arm is in the first or second quadrant.

Draw an angle in standard position in each quadrant. Draw a perpendicular from each terminal arm to the x-axis to make two triangles.

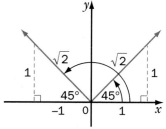

Since $\sin A = \dfrac{y}{r}$ and $\sin A = \dfrac{1}{\sqrt{2}}$, then $y = 1$ if $r = \sqrt{2}$.

Both triangles must be 45°–45°–90° triangles.

Both reference angles are 45°.

So, for $\sin A = \dfrac{1}{\sqrt{2}}$, $\angle A = 45°$ and 135°.

b) The cosine of an angle is positive when the terminal arm is in the first quadrant but not in the second quadrant, so $\angle A = 45°$.

c) The cosine of an angle is negative in the second quadrant, so $\angle A = 135°$.

Note from Examples 2 and 3 that sin 120° = sin 60° and sin 135° = sin 45°. In general, sin (180° – θ) = sin θ.

For example, sin 146° = sin 34°.

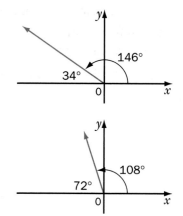

From Examples 2 and 3, cos 120° = –cos 60° and cos 135° = –cos 45°.

In general, cos (180° – θ) = –cos θ.

For example, cos 108° = –cos 72°.

You can use a calculator to find the sine or cosine of an obtuse angle directly, without considering the reference angle.

Example 4   Sine and Cosine of Any Obtuse Angle

Evaluate, to four decimal places.

a) sin 133°   **b)** cos 119.7°

Solution

a)

$\sin 133° \doteq 0.7314$

b)

$\cos 119.7° \doteq -0.4955$

Example 5   Finding Any Obtuse Angle

Find ∠A, to the nearest tenth of a degree, if $0° \le A \le 180°$ and

a) sin A = 0.3214   **b)** cos B = −0.5804

Solution

a)

b)

The sine of an angle is positive when the terminal arm is in the first or second quadrants.
So, ∠A ≐ 18.7° or ∠A = 180° − 18.7°
                 = 161.3°

The cosine of an angle is negative when the terminal arm is in the second quadrant.
So, ∠A ≐ 125.5°

Practice

1. The point (5, 12) lies on the terminal arm of an angle A in standard position. Find sin A and cos A.

2. The point (−3, 4) lies on the terminal arm of an angle θ in standard position. Find sin θ and cos θ.

If $0° \le A \le 180°$, find the measure(s) of ∠A.

3. $\sin A = \dfrac{1}{2}$

4. $\cos A = \dfrac{1}{2}$

5. $\cos A = -\dfrac{1}{2}$

Determine, without using a calculator.

6. sin 150° $\frac{1}{2}$ **7.** cos 150° $-\frac{\sqrt{3}}{2}$

8. sin 135° $\frac{1}{\sqrt{2}}$ **9.** cos 135° $-\frac{1}{\sqrt{2}}$

Evaluate, to four decimal places.

10. $\cos 144°$　　　　　　**11.** $\sin 105°$
12. $\sin 167°$　　　　　　**13.** $\cos 92°$
14. $\cos 134.7°$　　　　　**15.** $\sin 121.3°$
16. $\sin 178.8°$　　　　　**17.** $\cos 113.1°$

Find $\angle A$, to the nearest tenth of a degree, if $0° \le A \le 180°$.

18. $\sin A = 0.2568$
19. $\cos A = 0.4561$
20. $\cos A = -0.5603$

Applications and Problem Solving

21. If two angles in standard position are supplementary and have terminal arms that are perpendicular, what are the measures of the angles?

22. If $\sin A = \sin 125°$, what are the possible values of A?

23. If $\sin \theta = \dfrac{\sqrt{3}}{2}$, what is $\cos \theta$ if angle θ is

a) acute?　　　　　　**b)** obtuse?

24. If $\cos \theta = -\dfrac{\sqrt{3}}{2}$ for an obtuse angle θ, what is $\sin \theta$?

25. Algebra The point $P(k, 24)$ is 25 units from the origin. If P lies on the terminal arm of an angle in standard position, find, for each value of k,
a) the sine of the angle　**b)** the cosine of the angle

26. Book pages If you assume that the pages of this book are flat, the bottom edges of the two pages facing you form a 180° angle. Describe the changes in the sine of the angle and the cosine of the angle as you close the book.

27. Pendulum The amplitude, A, of a pendulum is the maximum horizontal distance of the bob from its central position. The displacement, x, is the actual horizontal distance of the bob from its central position at a certain time. The displacement

is related to the amplitude by the equation

$$x = A \sin \frac{360t}{T}$$

In this equation, t is the time since the bob passed through the central position, and T is the time the pendulum takes for a complete back-and-forth swing.

a) A pendulum has an amplitude of 10 cm and takes 2 s to complete one back-and-forth swing. To the nearest centimetre, how far is the bob from the central position when t is 0 s, 0.25 s, 0.5 s, 0.75 s, and 1 s?

b) Use your results from part a) to describe the motion of the pendulum from $t = 0$ to $t = 1$.

28. Number operations For obtuse angles A and B, $\cos A - \cos B = \dfrac{\sqrt{3} - \sqrt{2}}{2}$. If $\angle B = 150°$, what is the measure of $\angle A$?

29. a) Find three pairs of angles for which the sine of the first angle equals the cosine of the second angle.
b) How are the angles in each pair related?

LOGIC POWER

Nine playing cards are in a 3 by 3 arrangement. There are at least two aces, two kings, two queens, and two jacks. The cards border on each other horizontally and vertically, but not diagonally. Every jack borders on a king and a queen. Every queen borders on a king and an ace. Every king borders on an ace. Find an arrangement for the cards.

342

Construction

Many structures that we see from day to day — including houses, office buildings, bridges, and so on — have resulted from the skills of construction workers. Even the simpler aspects of construction work, such as the ramps used to make buildings wheelchair accessible, require a knowledge of mathematics.

Construction tradespeople, such as carpenters, electricians, and plumbers, may be college educated, but many learn their trade through on-the-job training as apprentices.

1 Wheelchair Ramps

The standard slope of a wheelchair ramp is $\frac{1}{12}$.

1. What is the measure of the indicated angle, to the nearest tenth of a degree?

2. For each of the following heights that a wheelchair ramp rises, find the length it runs, and the length of the ramp itself, to the nearest centimetre.

a) 50 cm   **b)** 42 cm
c) 76 cm   **d)** 1.05 m

2 Raked Stages

To improve sightlines for theatre audiences, set designers often include rakes, or sloping sections, in their sets. Because plywood and lumber are sold in 244-cm lengths, set designers often design rakes to be multiples of this length.

1. A rake is built with a length of 488 cm and a rise of 60 cm.

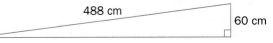

What angle does the rake make with the horizontal, to the nearest tenth of a degree?

2. Suppose the angle with the horizontal of the 488-cm rake were changed to 5.3°. To the nearest tenth of a centimetre, what would be the change in
a) the rise?   **b)** the run?

3 Locating Information

Use your research skills to find information about

1. a construction career that interests you

2. a construction project that is under way in your community or province

7.7 The Law of Sines

The Joint Astronomy Centre operates two telescopes on the summit of Mauna Kea in Hawaii. They are the United Kingdom Infrared Telescope and the James Clerk Maxwell Telescope. Canada, the Netherlands, and the United Kingdom share the use and the costs of the telescopes.

Explore: Interpret the Diagram

The classical way to find the height of a mountain uses a theodolite, which measures horizontal and vertical angles. To find the height of Mauna Kea, a surveyor finds the length of a baseline AB. A and B are at the same altitude, and, in this case, AB is 6 km. C is the summit of the mountain. D is directly below the summit and is at the same altitude as A and B.

Standing at A, the surveyor points the theodolite at B and then at C. In this case, the theodolite gives the horizontal angle BAD as 42° and the angle of elevation CAD as 33°.

Standing at B, the surveyor points the theodolite at A and then at C. In this case, the theodolite gives the horizontal angle ABD as 53° and the angle of elevation CBD as 37.8°.

Inquire

1. Why are △CDA and △CDB right triangles?

2. A triangle that is not right-angled is called an **oblique triangle**. Is △ABD right-angled or oblique?

3. In △ABD, is each of the following lengths greater than, less than, or equal to the length AB? Explain.
a) length x
b) length y

4. How could the height h be calculated
a) if the length x was known?
b) if the length y was known?

In an oblique triangle, when two angles and a side are given, the other sides can be found using the law of sines, which can be developed as follows.

In △ABC, draw AD perpendicular to BC. AD is the altitude or height, h, of △ABC.

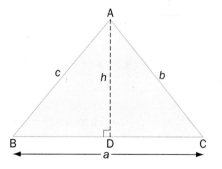

In $\triangle$ABD, $\dfrac{h}{c} = \sin B$ In $\triangle$ACD, $\dfrac{h}{b} = \sin C$

$h = c \sin B$ $h = b \sin C$

Then $c \sin B = b \sin C$

Divide both sides by bc: $\dfrac{c \sin B}{bc} = \dfrac{b \sin C}{bc}$

Simplify: $\dfrac{\sin B}{b} = \dfrac{\sin C}{c}$

Drawing the altitude from C and repeating the steps gives

$\dfrac{\sin A}{a} = \dfrac{\sin B}{b}$

Combining the results gives two forms of the law of sines.

$\dfrac{\sin A}{a} = \dfrac{\sin B}{b} = \dfrac{\sin C}{c}$ or $\dfrac{a}{\sin A} = \dfrac{b}{\sin B} = \dfrac{c}{\sin C}$

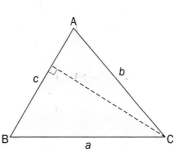

Example 1   Two Angles and a Side

In $\triangle$RST, $\angle S = 82.6°$, $r = 53$ m, and $\angle T = 25.9°$. Solve the triangle. Round each side length to the nearest metre.

Solution

First, determine the measure of $\angle R$.

$\angle R = 180° - 82.6° - 25.9°$

$\quad = 71.5°$

Use the law of sines to find s and t.

$\dfrac{s}{\sin S} = \dfrac{r}{\sin R}$

$\dfrac{s}{\sin 82.6°} = \dfrac{53}{\sin 71.5°}$

$s = \dfrac{53 \sin 82.6°}{\sin 71.5°}$

$s \doteq 55$

```
53sin(82.6)/sin(
71.5)
          55.42260948
```

$\dfrac{t}{\sin T} = \dfrac{r}{\sin R}$

$\dfrac{t}{\sin 25.9°} = \dfrac{53}{\sin 71.5°}$

$t = \dfrac{53 \sin 25.9°}{\sin 71.5°}$

$t \doteq 24$

```
53sin(25.9)/sin(
71.5)
          24.41201868
```

In $\triangle$RST, $\angle R = 71.5°$, $s = 55$ m, and $t = 24$ m.

Example 2   Area of a Triangle

In △RST, $\angle R = 24°$, $\angle S = 61°$, and $t = 200$ m.
Find the area of the triangle.

Solution

$\angle T = 180° - 24° - 61°$
$= 95°$

Use the law of sines to find s.

$$\frac{s}{\sin S} = \frac{t}{\sin T}$$

$$\frac{s}{\sin 61°} = \frac{200}{\sin 95°}$$

$$s = \frac{200 \sin 61°}{\sin 95°}$$

$$s \doteq 175.6$$

Draw an altitude, h, from T.

$$\frac{h}{s} = \sin 24°$$

$$\frac{h}{175.6} = \sin 24°$$

$$h = 175.6 \sin 24°$$

$$h \doteq 71.4$$

Area of △RST $= \dfrac{1}{2} bh$

$= \dfrac{1}{2} \times 200 \times 71.4$

$= 7140$

The area of △RST is 7140 m².

```
200sin(61)/sin(9
5)
        175.5921225
```

```
175.6sin(24)
        71.42295452
```

Example 3   Height of Mauna Kea

The diagram shows the measurements a surveyor
used to calculate the height of Mauna Kea.
a) Find h, to the nearest metre.
b) If the elevation of A, B, and D is 1077 m,
what is the height of Mauna Kea?

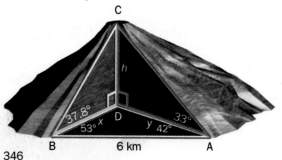

Solution
a) Find the measure of $\angle BDA$.
$\angle BDA = 180° - 53° - 42°$
$= 85°$

Use the law of sines to find y.

$$\frac{y}{\sin DBA} = \frac{AB}{\sin BDA}$$

$$\frac{y}{\sin 53°} = \frac{6000}{\sin 85°}$$

$$y = \frac{6000 \sin 53°}{\sin 85°}$$

$$\doteq 4810$$

In $\triangle ACD$, $\dfrac{h}{4810} = \tan 33°$

$h = 4810 \tan 33°$

$\doteq 3124$

```
6000sin(53)/sin(
85)
        4810.117008
Ans*tan(33)
        3123.726509
```

So, h is 3124 m, to the nearest metre.

b) Since the elevation of A, B, and D is 1077 m, the height of Mauna Kea is $3124 + 1077$ or 4201 m.

Practice

Find the length of the indicated side, to the nearest tenth.

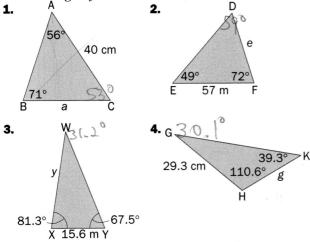

1.

A
56°
40 cm
71° 53°
B a C

2.

D
50°
e
49° 72°
E 57 m F

3.

W 31.2°
y
29.3 cm
81.3° 67.5°
X 15.6 m Y

4.

G 30.1°
39.3° K
110.6° g
H

Find the indicated quantity, to the nearest tenth.

5. In $\triangle ABC$, $\angle B = 38.2°$, $\angle C = 65.6°$, and $b = 54$ cm. Find c.

6. In $\triangle DEF$, $\angle D = 117.4°$, $d = 7.4$ m, and $\angle E = 30.2°$. Find f.

7. In $\triangle GHK$, $\angle G = 44.1°$, $k = 9.5$ cm, and $\angle H = 29.4°$. Find h.

In questions 8–10, solve each triangle. Round each answer to the nearest tenth, if necessary.

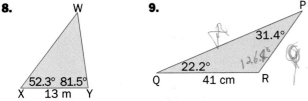

8.

W
52.3° 81.5°
X 13 m Y

9.

P
31.4°
22.2° 126.4°
Q 41 cm R

10. In $\triangle JKL$, $\angle J = 31.9°$, $j = 20.5$ cm, and $\angle K = 75.4°$.

Find the area of each triangle, to the nearest square unit.

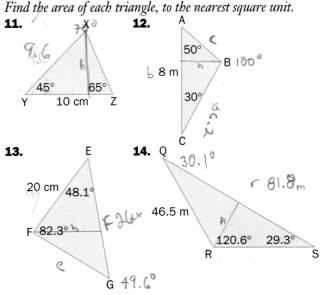

11.

X 70°
9.6
h
45° 65°
Y 10 cm Z

12.

A
50° c
b 8 m B 100°
h
30°
C 36.2

13.

E
20 cm 48.1°
F 82.3° h 26 m
e
G 49.6°

14.

Q 30.1°
81.8 m
46.5 m
A
120.6° 29.3°
R S

Applications and Problem Solving

15. Nile River Two museums are located on the same side of the Nile River in Thebes. The museums are on the bank of the river and are 1600 m apart. Across the river is the dock for the two ferries that bring tourists to the museums. The angles made by the river bank and the lines drawn from the museums to the dock, D, are shown.

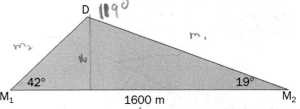

D 119°
m₂ m₁
w
42° 19°
M₁ 1600 m M₂
d

a) How far is each museum from the dock, to the nearest metre?

b) What is the width of this part of the Nile, to the nearest metre?

16. Surveying One of the many things surveyors do is determine the area of a plot of land. A field is in the shape of a triangle. Label the triangle ABC. A surveyor determines that the length of side AB is 620 m. Using a theodolite, the surveyor finds that ∠A is 56.8° and ∠B is 60.4°. Find
a) the perimeter of the field, to the nearest metre
b) the area of the field, to the nearest square metre

17. Measurement
Find the volume of the triangular solid, to the nearest cubic centimetre.

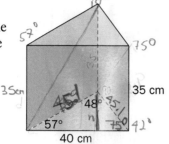

35 cm

40 cm

18. Inaccessible cliff To measure the height, AB, of an inaccessible cliff, a surveyor records the data shown.

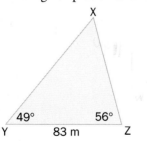

Find the height of the cliff, to the nearest metre.

19. Heron's formula A surveyor records the data shown for the triangular plot of land XYZ.

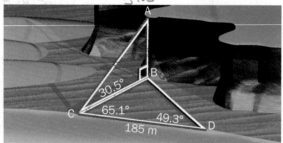

a) Find y and z, to the nearest metre.

b) Heron's formula states that the area, A, of a triangle is given by $A = \sqrt{s(s-a)(s-b)(s-c)}$ where the semiperimeter, $s = \frac{1}{2}(a+b+c)$.

Use Heron's formula to find the area of the plot of land, to the nearest square metre.

20. Measurement A diagonal, AC, of the parallelogram ABCD is 8.5 cm long. The diagonal makes angles of 43° and 32° with sides AB and AD.

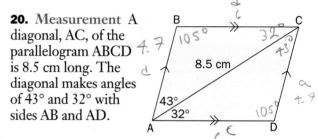

Find the side lengths of the parallelogram, to the nearest tenth of a centimetre.

21. Use the law of sines to show that the two equal sides of an isosceles triangle are opposite the two equal angles.

22. Saturn V spacecraft The Saturn V spacecraft that took astronauts to the moon was much bigger than the space shuttle. From one point on the ground, the angle of elevation of the top of Saturn V was 57.7°. From a point 20 m closer, the angle of elevation of the top was 65.7°.
a) Use the law of sines to find the height of the Saturn V, to the nearest metre.
b) Compare this method of finding a height with the one used in Example 2 in Section 7.5. Which method do you prefer? Explain why.

WORD POWER

1. Unscramble the letters to make words. Identify the word that does not belong to the set. Explain.

tiffy   neevel   veens   tyneevs

fryot   thige   hyfritt   velwet

2. Write a similar problem using words from algebra, geometry, or statistics.

7.8 The Law of Cosines

Around 700 B.C., engineers on the Greek island of Sámos constructed a tunnel through Mount Castro to bring water from one side of the mountain to a city on the other side. This feat was amazing because the digging teams started at opposite ends and met at the centre, with only a small error. Archaeologists discovered the tunnel in 1882.

Explore: Use the Diagram

A way for a surveyor to find the length of the tunnel, XY, is to select a point Z and measure $\angle Z$ and the lengths x and y.

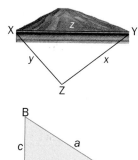

Inquire

1. In $\triangle ABC$, if $\angle A = 90°$, what is the relationship between a^2 and $b^2 + c^2$?

2. a) Suppose lengths b and c remain constant, but $\angle A$ is increased to make an obtuse angle. Does length a increase or decrease?
b) What is the relationship between a^2 and $b^2 + c^2$ when $\angle A$ is obtuse?

3. a) Suppose lengths b and c remain constant, but $\angle A$ is decreased to make an acute angle. Does length a increase or decrease?
b) What is the relationship between a^2 and $b^2 + c^2$ when $\angle A$ is acute?

4. A surveyor examining the tunnel through Mount Castro selected a point 5000 m from one end of the tunnel and 5500 m from the other end.

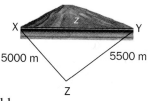

a) If $\angle Z$ was 90°, what would be the length of the tunnel, to the nearest metre?
b) The surveyor found that $\angle Z$ was actually 10°. Is the length of the tunnel greater than or less than the length you found in part a)?
c) Estimate the length of the tunnel.

5. Can you use the law of sines to find the actual length of the tunnel? Explain.

349

Some oblique triangles that cannot be solved using the law of sines can be solved using the law of cosines, which can be developed as follows.

In $\triangle ABC$, draw CD perpendicular to AB.
CD is the altitude, h, of $\triangle ABC$.
Let $AD = x$.
Then, $BD = c - x$.

In $\triangle ACD$, $b^2 = h^2 + x^2$

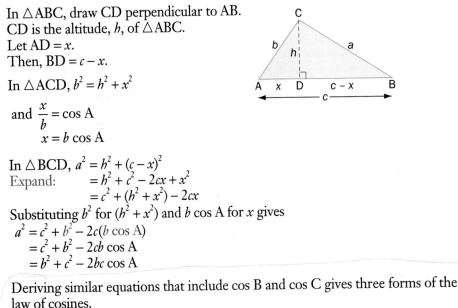

and $\dfrac{x}{b} = \cos A$

$x = b \cos A$

In $\triangle BCD$, $a^2 = h^2 + (c - x)^2$
Expand: $\quad = h^2 + c^2 - 2cx + x^2$
$\quad = c^2 + (h^2 + x^2) - 2cx$
Substituting b^2 for $(h^2 + x^2)$ and $b \cos A$ for x gives
$a^2 = c^2 + b^2 - 2c(b \cos A)$
$\quad = c^2 + b^2 - 2cb \cos A$
$\quad = b^2 + c^2 - 2bc \cos A$

Deriving similar equations that include $\cos B$ and $\cos C$ gives three forms of the law of cosines.
$a^2 = b^2 + c^2 - 2bc \cos A$
$b^2 = a^2 + c^2 - 2ac \cos B$
$c^2 = a^2 + b^2 - 2ab \cos C$

Example 1   Two Sides and the Contained Angle
In $\triangle ABC$, $\angle B = 42°$, $a = 52$ m, and $c = 36$ m.
Find b, to the nearest metre.

Solution
Use the law of cosines.
$b^2 = a^2 + c^2 - 2ac \cos B$
$\quad = 52^2 + 36^2 - 2(52)(36) \cos 42°$
$b \doteq 35$
The length b is 35 m, to the nearest metre.

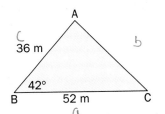

The law of cosines can also be used to find the measure of an angle of a triangle when the lengths of the three sides are known. To do this, the formula can be solved for the cosine value.

$$a^2 = b^2 + c^2 - 2bc \cos A$$

Rearrange: $\quad 2bc \cos A = b^2 + c^2 - a^2$

Divide both sides by $2bc$: $\quad \cos A = \dfrac{b^2 + c^2 - a^2}{2bc}$

Similarly $\quad \cos B = \dfrac{a^2 + c^2 - b^2}{2ac}$

and $\quad \cos C = \dfrac{a^2 + b^2 - c^2}{2ab}$

Example 2   Three Sides

In $\triangle DEF$, $d = 62$ cm, $e = 51$ cm, and $f = 48$ cm.
Find each angle of the triangle, to the nearest degree.

Solution

Use the law of cosines to find $\angle F$.

$$\cos F = \frac{d^2 + e^2 - f^2}{2de}$$
$$= \frac{62^2 + 51^2 - 48^2}{2(62)(51)}$$
$$\angle F \doteq 49°$$

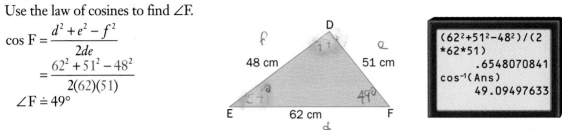

```
(62²+51²-48²)/(2
*62*51)
          .6548070841
cos⁻¹(Ans)
          49.09497633
```

Use the law of cosines or the law of sines to find $\angle D$.

Using the law of cosines,

$$\cos D = \frac{e^2 + f^2 - d^2}{2ef}$$
$$= \frac{51^2 + 48^2 - 62^2}{2(51)(48)}$$
$$\angle D \doteq 77°$$

```
(51²+48²-62²)/(2
*51*48)
          .2167075163
cos⁻¹(Ans)
          77.48427716
```

Using the law of sines,

$$\frac{\sin D}{d} = \frac{\sin F}{f}$$
$$\frac{\sin D}{62} = \frac{\sin 49°}{48}$$
$$\sin D = \frac{62 \sin 49°}{48}$$
$$\angle D \doteq 77°$$

```
62sin(49)/48
          .9748332078
sin⁻¹(Ans)
          77.11849488
```

$$\angle E = 180° - 49° - 77°$$
$$= 54°$$

In $\triangle DEF$, $\angle D = 77°$, $\angle E = 54°$, and $\angle F = 49°$, to the nearest degree.

Example 3   Tunnel Through Mount Castro

Find the length of the tunnel, XY, to the nearest metre.

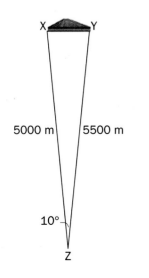

Solution

Use the law of cosines to find z.

$$z^2 = x^2 + y^2 - 2xy \cos Z$$
$$= 5500^2 + 5000^2 - 2(5500)(5000) \cos 10°$$
$$z \doteq 1042$$

```
5500²+5000²-2*55
00*5000cos(10)
          1085573.584
√(Ans)
          1041.908626
```

The tunnel is 1042 m long, to the nearest metre.

Practice

Find the missing side length, to the nearest tenth of a unit.

1.

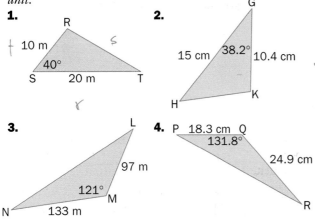

2.

3.

4.

Find the indicated angle, to the nearest tenth of a degree.

5.

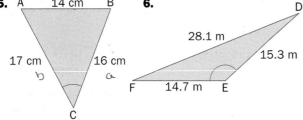

6.

Find the indicated quantity, to the nearest tenth of a unit.

7. In $\triangle PQR$, $p = 15.3$ m, $q = 18.2$ m, and $\angle R = 70°$. Find r.

8. In $\triangle UVW$, $u = 10.3$ cm, $v = 11.4$ cm, and $w = 12.5$ cm. Find $\angle V$.

9. In $\triangle BCD$, $b = 35.6$ cm, $c = 22.1$ cm, and $d = 22.1$ cm. Find $\angle B$.

10. In $\triangle MNO$, $m = 36.5$ m, $o = 51.4$ m, and $\angle N = 126°$. Find n.

In questions 11–14, solve each triangle. Round each calculated value to the nearest tenth of a unit, if necessary.

11.

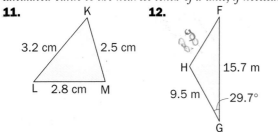

12.

13. In $\triangle IJK$, $i = 10.5$ m, $j = 20.1$ m, and $k = 12.5$ m.

14. In $\triangle TUW$, $w = 25.4$ cm, $u = 34.2$ cm, and $\angle T = 43.1°$.

Applications and Problem Solving

15. Golf At the Fisherville Golf Club, the shortest distance to the eighth hole from the tee is 348 m. On his first shot, Mario drives the ball so that it lands 195 m from the tee. However, he slices the ball so that the line connecting the tee and the ball makes an angle of 12° with the line connecting the tee and the hole. After his first shot, how far is the ball from the hole, to the nearest metre?

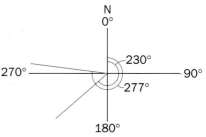

16. Space shuttle The tops of the solid rocket boosters used to launch the space shuttle are cones of diameter 3.7 m and slant height 5.4 m. Find the angle that the curved surface of the cone makes with a diameter, to the nearest tenth of a degree.

17. Navigation Two ships left Estevan Point, British Columbia, at the same time. One travelled at 16 km/h on a course of 277°. The other travelled at 14 km/h on a course of 230°. How far apart were the ships after two hours, to the nearest tenth of a kilometre?

18. Measurement Given the side lengths in △RST, find the area of the triangle, to the nearest square metre.

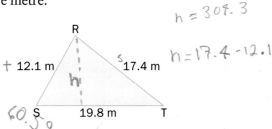

$h = 309.3$

$h = 17.4 - 12.1$

19. Triathlon The three phases of a triathlon involve swimming, cycling, and running, in that order. The distances for each phase can vary. For the triathlon held in Hawaii each year, competitors swim in the ocean, bicycle 112 km, and run 41.8 km. In the diagram, S is the start of the swim, and F is the finish. A surveyor used the dimensions shown to calculate the length SF across the bay.

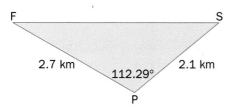

Find the distance the athletes swim in the Hawaiian triathlon, to the nearest metre. What assumptions have you made?

20. Treasure hunting There are hundreds of known shipwrecks off the coast of Vancouver Island. To try to locate the wreck of the *North Star*, which sank in 1892, a ship leaves Pachena Point and sails on a bearing of S20°W for 18 km, then N85°E for 6 km.
a) How far are the treasure hunters from Pachena Point?
b) In what direction, to the nearest degree, should they sail to return to Pachena Point?

21. Measurement A parallelogram has side lengths of 14.2 cm and 16.3 cm. One diagonal has a length of 23.5 cm. Find the angle measures in the parallelogram, to the nearest tenth of a degree.

22. Measurement Find the volume of the prism, to the nearest cubic metre.

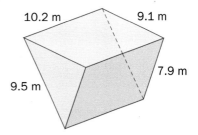

23. Coordinate geometry Solve △ABC with vertices A(2, 1), B(7, 4), and C(1, 5). Round each calculated value to the nearest tenth of a unit.

24. Explain whether you can use the law of sines to find p in △PQR when given ∠P = 47°, q = 9 m, and r = 11 m.

25. Explain whether you can use the law of cosines to find r in △RST when given ∠T = 53°, t = 15 m, and s = 14 m.

26. The Pythagorean Theorem can be thought of as a special case of the law of cosines. Explain why.

27. Algebra Use the law of cosines to show that each angle of an equilateral triangle must measure 60°.

28. Try to solve a triangle with side lengths 3 cm, 8 cm, and 4 cm using the law of cosines. What do you find? Explain why.

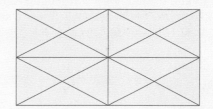

LOGIC POWER

How many triangles are in this figure?

TECHNOLOGY

Graphing Calculator Programs

1 The Law of Sines

The graphing calculator program shows how to calculate the length of a side of a triangle using the law of sines.

A is the unknown side length.
B is the given length.
X is the measure of the angle opposite B.
Y is the measure of the angle opposite A.

1. Describe what each line of the program does.

2. Enter the program into your graphing calculator. Use the program to find the unknown side length, A, in each of the following triangles, to the nearest tenth of a unit.

PROGRAM:SINES
:ClrHome
:Disp "B="
:Input B
:Disp "X="
:Input X
:Disp "Y="
:Input Y
:Bsin(Y)/sin(X)→A
:Disp "A="
:Disp A

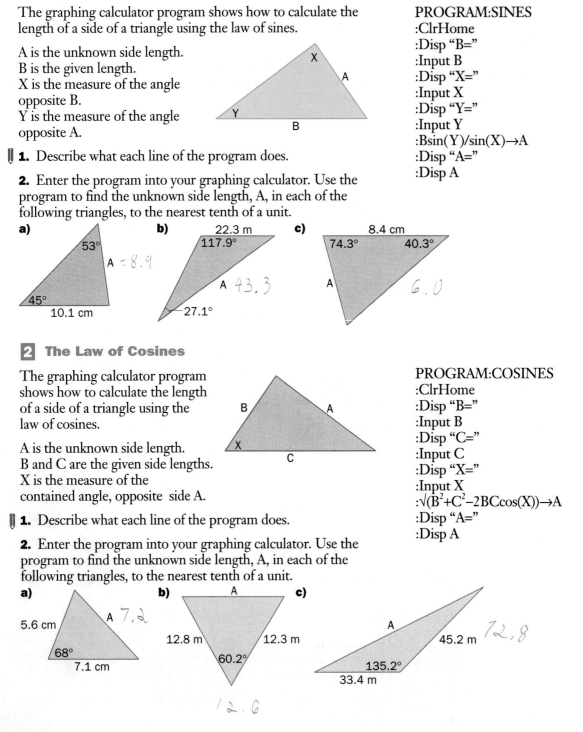

a)
53°
A = 8.9
45°
10.1 cm

b)
22.3 m
117.9°
A 43.3
27.1°

c)
8.4 cm
74.3°   40.3°
A   6.0

2 The Law of Cosines

The graphing calculator program shows how to calculate the length of a side of a triangle using the law of cosines.

A is the unknown side length.
B and C are the given side lengths.
X is the measure of the contained angle, opposite side A.

1. Describe what each line of the program does.

2. Enter the program into your graphing calculator. Use the program to find the unknown side length, A, in each of the following triangles, to the nearest tenth of a unit.

PROGRAM:COSINES
:ClrHome
:Disp "B="
:Input B
:Disp "C="
:Input C
:Disp "X="
:Input X
:√(B²+C²−2BCcos(X))→A
:Disp "A="
:Disp A

a)
5.6 cm
A 7.2
68°
7.1 cm

b)
A
12.8 m   12.3 m
60.2°

12.6

c)
A
45.2 m   12.8
135.2°
33.4 m

3. Suppose that, in the triangle, side lengths A, B, and C are given, but angle measure X is unknown. Modify the program to find the unknown angle measure, X.

4. Have a classmate check your program by using it to find the unknown angle measure, X, in each of the following triangles, to the nearest tenth of a degree.

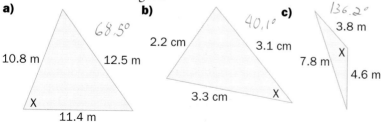

a)

68.5°

10.8 m   12.5 m

X

11.4 m

b)

2.2 cm   3.1 cm

3.3 cm   X

40.1°

c)

136.2°

3.8 m

7.8 m   X

4.6 m

3 Heron's Formula

The graphing calculator program shows how to calculate the area of a triangle using Heron's formula.

1. Describe what each line of the program does.

2. Enter the program into your graphing calculator. Use the program and the given side lengths to find the area of each of the following triangles, to the nearest tenth of a square unit.

a) 4.3 m, 3.9 m, 6.4 m
b) 18.2 cm, 29.8 cm, 25.3 cm

3. Try to find the area of a triangle, given side lengths of 4.2 cm, 10.3 cm, and 5.6 cm. What do you observe? Why?

PROGRAM:HERON
:ClrHome
:Disp "A="
:Input A
:Disp "B="
:Input B
:Disp "C="
:Input C
:(A+B+C)/2→S
:√(S(S−A)(S−B)(S−C))→X
:Disp "AREA="
:Disp X

Review

7.1 Calculate the volume and surface area of each figure.

1.

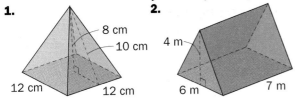

2.

Calculate
a) *the surface area, to the nearest square unit*
b) *the volume, to the nearest cubic unit*

3.

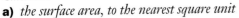

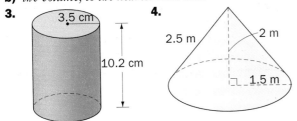

4.

5. Channel Tunnel The Channel Tunnel links Britain and France under the English Channel. There are two parallel rail tunnels, each connected to a central service tunnel. Each rail tunnel is a cylinder 7.6 m in diameter and 50 km long. What is the volume of each rail tunnel, to the nearest thousand cubic metres?

7.2 For each sphere or hemisphere, calculate
a) *the surface area, to the nearest square centimetre*
b) *the volume, to the nearest cubic centimetre*

6.

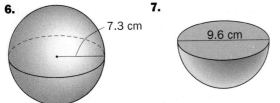

7.

8. Sputnik 1 The first artificial satellite to orbit the Earth was Russia's *Sputnik 1*. The satellite was a sphere of diameter 58 cm. Calculate
a) its surface area, to the nearest ten square centimetres
b) its volume, to the nearest hundred cubic centimetres

7.3
9. A rectangle has an area of 400 cm². By what factor must the dimensions be multiplied to give a similar rectangle and decrease the area by 300 cm²?

10. Calculate the surface area of a pyramid similar to the given pyramid but enlarged using a scale factor of $\frac{5}{2}$.

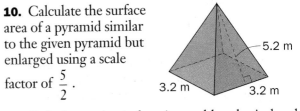

11. Raindrops A raindrop is roughly spherical and has a radius of about 0.1 mm. Each raindrop is formed from cloud droplets, which combine until the raindrop is large enough to fall. A cloud droplet has a radius of about 0.005 mm. How many cloud droplets combine to form one raindrop?

7.4 Solve each triangle. Round each side length to the nearest tenth of a unit and each angle to the nearest tenth of a degree.

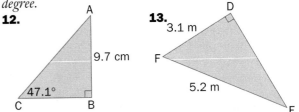

12.

13.

14. One Canada Square The tallest building in London, England, has the address "One Canada Square." From the top of the building, the angle of depression of a point on the ground 50 m from the foot of the building is 78.4°. Find the height of the building, to the nearest metre.

7.5
15. Find AB, to the nearest centimetre.

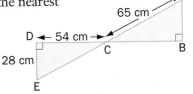

16. Lethbridge's High Level Bridge The High Level Bridge, which crosses the Oldman River in Lethbridge, Alberta, is over 1 km long. From one point on the river, the angle of elevation of the top of the bridge is 62.6°. From a point 20 m closer to the bridge, the angle of elevation of the top of the bridge is 72.8°. How high is the bridge above the river, to the nearest metre?

7.6 *Evaluate, to four decimal places.*
   17. sin 92°   **18.** cos 100°
   19. sin 129.3°   **20.** cos 163.7°

If 0° ≤ A ≤ 180°, find ∠A, to the nearest tenth of a degree.
   21. sin A = 0.7531   **22.** cos A = −0.3412

   23. a) If sin A = 0.5, and angles A and B are complementary, what is the measure of angle B?
   b) If cos C = −0.5, and angles C and D are supplementary, what is the measure of angle D?

7.7 *Find the length of the indicated side, to the nearest tenth.*
   24. In △KLM, ∠K = 54.3°, ∠L = 61.3°, and k = 23.2 m. Find l.

   25. In △PQR, ∠Q = 33.3°, ∠R = 45.9°, and p = 28.3 cm. Find r.

Find the area of each triangle, to the nearest square unit.

26.

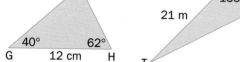

27.

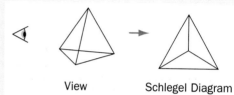

28. Covered bridge The world's longest covered bridge crosses the Saint John River at Hartland, New Brunswick. From two points, X and Y, 100 m apart on the same side of the river, the lines of sight to the far end of the bridge, Z, make angles of 85.6° and 79.8° with the river bank, as shown. What is the length of the bridge, b, to the nearest ten metres?

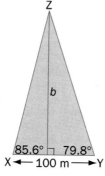

.8 *Find the indicated quantity, to the nearest tenth of a unit.*
   29. In △ABC, a = 14.3 m, c = 16.4 m, and ∠B = 97.4°. Find b.

   30. In △CDE, c = 4.8 cm, d = 5.3 cm, and e = 4.5 cm. Find ∠D.

   31. Boating Two pleasure boats left Churchill, Manitoba, at the same time. One travelled at 10 km/h on a course of 47°. The other travelled at 8 km/h on a course of 79°. How far apart were the boats after 45 min, to the nearest tenth of a kilometre?

Exploring Math

Schlegel Diagrams

Schlegel diagrams are named after the German mathematician Victor Schlegel, who invented them in 1883. Schlegel diagrams are two-dimensional representations of polyhedra, which are three-dimensional shapes. Schlegel diagrams look like three-dimensional objects that have been deflated or flattened.

A Schlegel diagram shows the edges of a polyhedron as they appear from a position just outside the centre of one face. The edges of this face appear as a frame around the diagram. All the other edges appear inside the frame. The following Schlegel diagram represents a regular tetrahedron.

View   Schlegel Diagram

1. Does the Schlegel diagram show all the vertices, edges, and faces of the tetrahedron?

2. Are the lengths of the edges of the tetrahedron represented accurately in the Schlegel diagram? Explain.

3. Are the shapes of the faces of the tetrahedron represented accurately in the Schlegel diagram? Explain.

4. If you used a model of a tetrahedron to help you draw its Schlegel diagram, would a solid, a skeleton, or a shell be the most useful model? Explain.

5. Draw a Schlegel diagram of each of the following regular polyhedra.
   a) cube
   b) octahedron

Chapter Check

Calculate
a) *the surface area, to the nearest square unit*
b) *the volume, to the nearest cubic unit*

1.

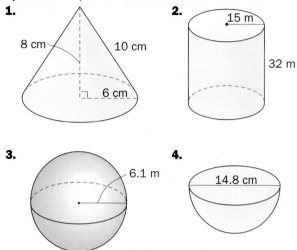

8 cm · 10 cm · 6 cm

2. 15 m · 32 m

3. 6.1 m

4. 14.8 cm

5. Triangles ABC and DEF are similar. The base, BC, of △ABC has a length of 4.2 cm. The height of △ABC is 3.6 cm. The area of △ABC is 9 times the area of △DEF.

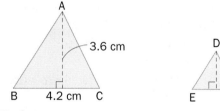

A · 3.6 cm · B · 4.2 cm · C · D · E · F

Find
a) the base length, EF   **b)** the height of △DEF

6. Find XY, to the nearest tenth of a metre.

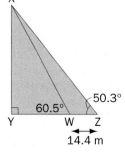

X · Y · 60.5° · 50.3° · W · Z · 14.4 m

Evaluate, to four decimal places.
7. sin 82.3°   **8.** sin 149.5°
9. cos 19.9°   **10.** cos 159.2°

358

11. If the cosine of an angle is −0.4527, can the angle be acute? Explain.

12. If the sine of an angle is 0.3892, can the angle be obtuse? Explain.

Solve each triangle. Round each answer to the nearest tenth, if necessary.

13.

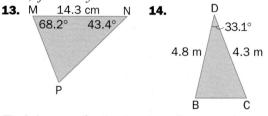

M · 14.3 cm · N · 68.2° · 43.4° · P

14.

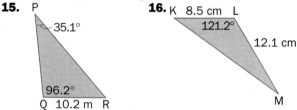

D · 33.1° · 4.8 m · 4.3 m · B · C

Find the area of each triangle, to the nearest square unit.

15. P · 35.1° · 96.2° · Q · 10.2 m · R

16. K · 8.5 cm · L · 121.2° · 12.1 cm · M

17. Hailstone The largest hailstone known to have fallen in Canada was found at Cedoux, Saskatchewan. The hailstone had a diameter of 11.4 cm. Calculate
a) the surface area of the hailstone, to the nearest square centimetre
b) the volume of the hailstone, to the nearest cubic centimetre

18. Moons of Uranus Two of the 15 known moons of Uranus are called Titania and Umbriel. The ratio of the volume of Titania to the volume of Umbriel is about 64:27. If the diameter of Titania is about 1600 km, what is the diameter of Umbriel?

19. Lion's Gate Bridge
The span of a bridge is the distance between its supports. A helicopter is hovering above the longest span of the Lion's Gate Bridge in Burrard Inlet, British Columbia.

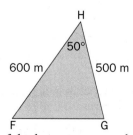

H · 50° · 600 m · 500 m · F · G

The distances to the ends of the longest span and the angle between the lines of sight are as shown. What is the length of the longest span of the bridge, FG, to the nearest ten metres?

Using the Strategies

1. The glass container shown is a cylinder, open at the top. How could you fill it exactly half-way with water, without using any other measuring tools?

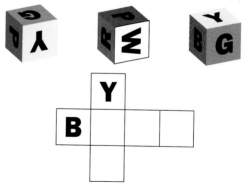

2. A clock's minute hand is 10 cm long and its hour hand is 7 cm long. Find the total distance moved by the tips of both hands in a 24-h period.

3. A regular tetrahedron is to be made from the net shown. Design another net for a tetrahedon, using faces in the same four colours, so that the arrangements of colours in the two tetrahedra can be distinguished.

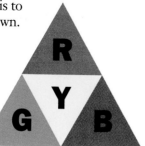

4. Using six different colours for the faces of a cube, how many distinguishable arrangements of colours are possible?

5. A rectangular solid is to be made using 60 blocks, each one a 1-cm cube.
a) How many rectangular solids are possible?
b) How many different surface areas are possible?

6. A valuable gold coin that had been stolen was recovered, along with seven counterfeit copies that had been placed in the same box. If the counterfeit coins are all identical and are slightly lighter than the real coin, how can the real coin be found with only two weighings?

7. Each edge of a cube is taped with either brown tape or clear tape. If every face of the cube has brown tape on at least one edge, what is the smallest possible number of brown edges?

8. A certain three-digit number is equal to the cube of the sum of its digits. What is the number?

9. Three views of a cube are shown. Copy the net and label it with the correct colours.

10. On a clockface, the hours 10, 12, 3, and 8 form a quadrilateral. What is the measure of each angle of the quadrilateral?

11. At one stall at the Bargain Flea Market, every item costs a dollar or less, including tax. If you have 3 quarters, 2 dimes, and 5 pennies, how many different prices could you pay with exact change?

12. Estimate the number of trees within a 2-km radius of your school.

13. a) About how many times could you handwrite the following sentence on pages with a total area that equals the area of Canada?

> Canada is the world's second-largest country, after Russia.

b) About how many years would you take to write the sentence this many times?

DATA BANK

1. The moon has a diameter of 3476 km. How does the surface area of the moon compare with the surface area of the smallest planet in the solar system? What assumption have you made?

2. How many times as great is the volume of the sun as the combined volumes of all the planets in the solar system?

CHAPTER 8

Statistics and Probability

bout 100 lightning bolts blast the Earth every second.

1. How many lightning bolts blast the Earth
a) in a year? **b)** in a day?

2. The area of Canada is about 2% of the area of the Earth. If the lightning bolts were evenly distributed on the Earth, how many times would Canada be hit by lightning
a) in a year? **b)** in a day?

3. The table gives the area of each province as a percent of the area of Canada. What is the expected number of times lightning will strike your province

Province	Percent of Canada's Area
Alberta	6.6
British Columbia	9.5
Manitoba	6.5
New Brunswick	0.7
Newfoundland	4.1
Nova Scotia	0.6
Ontario	10.7
Prince Edward Island	0.1
Quebec	15.5
Saskatchewan	6.5

a) in a year? **b)** in a day?

4. Why might the expected numbers from question 3 not be close to the actual numbers of lightning strikes in your province?

GETTING STARTED

Games

1 Scrabble®

The table gives the number of tiles and the points value for each letter in the game of SCRABBLE®.

Letter	Number	Points	Letter	Number	Points	Letter	Number	Points
A	9	1	J	1	8	S	4	1
B	2	3	K	1	5	T	6	1
C	2	3	L	4	1	U	4	1
D	4	2	M	2	3	V	2	4
E	12	1	N	6	1	W	2	4
F	2	4	O	8	1	X	1	8
G	3	2	P	2	3	Y	2	4
H	2	4	Q	1	10	Z	1	10
I	9	1	R	6	1			

There are also 2 blank tiles, for a total of 100 tiles.

The game starts with all tiles face down. Each player draws one tile to see who will play first. The player drawing a letter nearest the beginning of the alphabet plays first. A blank tile supersedes all other tiles.

1. If you are the first person to pick one tile, find
a) the probability of picking an A-tile; a Z-tile; a blank tile

b) a letter for which the probability of being picked is $\frac{1}{50}$

c) a letter that has twice the probability of being picked as the letter you named in part b)

2. If you are the first person to pick one tile, what is the probability that your first tile is
a) an X-tile?   **b)** a vowel?
c) a consonant?   **d)** a vowel or a Y-tile?
e) a tile with a value of 1 point?
f) a tile with a value of 2 points?
g) a tile with a value of 10 points?
h) a tile with a value of 9 points?
i) a tile with a value greater than 4 points?
j) a tile in the first half of the alphabet?
k) a vowel, a consonant, or a blank tile?

3. After the players decide who plays first, the tiles they chose are returned to the set, and the set is scrambled face down. Each player then draws 7 tiles to play the game. Suppose that you are the first person to draw 7 tiles. If the first tile you pick is an E-tile, and it is not returned to the set, what is the probability that the second tile you pick is

a) an E-tile?
b) a T-tile?
c) a vowel?
d) not a vowel?
e) a consonant?
f) a tile with a value of 1 point?
g) a tile with a value of 3 points?
h) a tile with a value greater than 5 points?

2 Stone, Scissors, Paper

In a *fair game*, each player has the same probability of winning. In an *unfair game*, players do not have the same probability of winning.

Stone, Scissors, Paper is a game for two players. At the same time, each player uses a hand to represent a stone, a pair of scissors, or a piece of paper.

| Stone | Scissors | Paper |

For each possible pair of objects, points are scored using the following rules.
• If each player shows the same object, no points are scored.
• Because stone breaks scissors, stone gets 1 point.
• Because scissors cut paper, scissors gets 1 point.
• Because paper covers stone, paper gets 1 point.

1. Play 15 rounds of *Stone, Scissors, Paper*. Record the points scored for each player.

2. The table shows two possible outcomes of the game.

Player A	Player B
stone	stone
stone	paper
⋮	⋮

Copy the table. Complete it by including the other possible outcomes. How many possible outcomes are there?

3. Use the table to find the probability that
a) no points are scored
b) player A wins
c) player B wins

4. Is the game fair? Explain.

Mental Math

Order of Operations With Rational Numbers

Simplify.

1. $\dfrac{4+5+(-3)}{3}$ **2.** $\dfrac{6+6+(-1)+(-3)}{4}$

3. $\dfrac{1}{2}\times4+\dfrac{1}{2}\times12$ **4.** $\dfrac{1}{2}\times6+\dfrac{1}{2}\times(-8)$

5. $\dfrac{1}{4}\times8+\dfrac{2}{5}\times(-10)+\dfrac{1}{2}\times(-4)$

Evaluate.

6. $10\times0.1+20\times0.2$ **7.** $25\times0.6+100\times0.4$
8. $8\times0.5+12\times0.25$ **9.** $14\times0.1+3\times0.2$
10. $6\times0.3+8\times0.2+5\times0.4$

Multiplying by Multiples of 9

To multiply a number by 9, first multiply the number by 10, then subtract the number itself. For 23×9,
multiply by 10: $23\times10=230$
subtract 23: $230-23=207$
So, $23\times9=207$

Calculate.

1. 19×9 **2.** 28×9 **3.** 55×9
4. 88×9 **5.** 110×9 **6.** 125×9
7. 4.5×9 **8.** 2.6×9 **9.** 5.7×9

To multiply by 0.9, 90, 900, and so on, first multiply by 9, as shown above, then place the decimal point.
Thus, $23\times0.9=20.7$, and $23\times90=2070$.

Calculate.

10. 13×90 **11.** 27×90 **12.** 41×90
13. 15×900 **14.** 37×900 **15.** 35×900
16. 4.9×90 **17.** 6.4×90 **18.** 10.5×90
19. 16×0.9 **20.** 38×0.9 **21.** 42×0.9

22. Explain why the rule for multiplying by 9 works.

23. Modify the rule for multiplying by 9 to write a rule for multiplying by 11.

Calculate using your rule from question 23.

24. 36×11 **25.** 48×11 **26.** 53×1.1
27. 85×1.1 **28.** 21×110 **29.** 66×110

8.1 Sampling Techniques

Does photo radar reduce the incidence of traffic accidents? Does one brand of battery last longer than another? Which party will win the next election? When someone poses such questions, the answers may be found in statistics, which is the science of collecting, organizing, and interpreting data.

Explore: Make Predictions

Work in groups of four or five and come to a group decision for the answers to the following questions.

a) What percent of students in grade 10 in your school know the capitals of the Canadian provinces?

b) What percent of the students in your school put on their right shoe before their left shoe in the morning?

c) What is the favourite weeknight TV show of high school students in your province, and what percent of the students choose this show as their favourite?

Inquire

1. Compare your group's answers to those of the other groups in the class. Is there a consensus in your class about the answers?

2. What groups of students should you survey next to improve the accuracy of the answers?

For some questions, the answer can be found using a **survey**, which is a way of collecting data. It is usually not possible to carry out a **census** by collecting data from all the people who could be surveyed, or the **population**. When a census is not possible, the answer is found using a part, or **sample**, of the population.

Example 1   Identifying Populations and Choosing Samples

Identify the population and suggest a sample that could be used to answer each of the following questions.

a) How many hours of homework does a grade 10 student in your school have each week?

b) What percent of the teenagers in your city or town wear a chain around their necks?

c) Should drivers in your province have to retake the driving test after they get a ticket for a driving infraction?

Solution

a) The population includes all the grade 10 students in your school. A sample might include all the students in your class.

b) The population includes all the teenagers in your city or town. A sample might include all the students in your school.

c) Highway safety affects most people in the province, including people who use public transit and do not drive cars. The population might include all people of voting age in the province. Another possible population is all people of legal driving age in the province. A sample might include the students and teachers who drive to your school.

The procedure used for collecting information from the sample is called the sampling technique. To collect reliable data, statisticians use techniques designed to eliminate **bias**. Bias is the difference between the results obtained by sampling and the truth about the whole population.

Sampling techniques can be divided into two categories, **probability sampling** and **non-probability sampling**. Probability sampling involves the random selection of units from a population. One probability sampling technique is called **simple random sampling**. Units are drawn at random from the population and every unit has the same probability of being selected.

Example 2   Simple Random Sampling

Design a simple random sampling procedure to select students in your school to study the following question.

Should the school year be changed so that there are 4 blocks of 13 weeks of classes, with 10 weeks of classes and 3 weeks vacation in each block?

Solution

The population is all the students in your school. Suppose there are 800 students. Let the sample size be 8% of the population.

8% of 800 = 64

The simple random sampling procedure might include the following steps:

• Obtain an alphabetical list of all the students.

• Assign each student a different number from 1 to 800.

• Generate 64 random numbers from 1 to 800 using a calculator or computer.

• The sample is the 64 students whose numbers were generated. Ask each of the 64 students the question and record the answers.

A second probability sampling technique is called **systematic sampling**. Units are selected from a list using a selection interval, k. Every kth element on the list is sampled.

Example 3   Systematic Sampling

Design a systematic sampling procedure to study the following question among students in your school.

If you had the chance, would you travel in space?

Solution

The population is all the students in your school. Suppose there are 500 students. Sample 10% of them.

10% of 500 = 50

Suppose you wanted to cover all age groups fairly. The systematic sampling procedure might include the following steps:

• Sort a list of all the students in the school in order of age.

Find the sampling interval, k. $k = \dfrac{500}{50}$
$$= 10$$

• Choose a random number, x, from 1 to 10 to be the starting point of the systematic sample. If $x = 4$, start with the 4th student in the sorted list.

• Sample every 10th student in the sorted list. If you start with the 4th student, also sample the 14th, 24th, and 34th students, and so on.

• Ask each sampled student the question and record the answers.

A third probability sampling technique is called **stratified sampling**. The population is divided into groups, or strata, from which random samples are taken.

Example 4   Stratified Sampling

Design a stratified sampling procedure to study the following question among students in your school.

Should school start one hour earlier and end one hour earlier each school day?

Solution

The population is all the students in your school. Suppose there are 660 students. Sample 15% of them.

15% of 660 = 99

The stratified sampling procedure might include the following steps:

• Divide the population into two groups — female and male.

• Survey 15% of the females and 15% of the males. If the school population includes 340 females and 320 males,

$n_f = 0.15 \times 340$ $\qquad\qquad$ $n_m = 0.15 \times 320$
$\quad\; = 51$ $\qquad\qquad\qquad\quad = 48$

• From the list of all students prepare an alphabetical list of female students and an alphabetical list of male students.

• Use simple random sampling to identify 51 females and 48 males.

• Ask each of the 99 students the question and record the answers

▌ Describe another way to stratify the student population in Example 4.

A fourth probability sampling technique is **clustered sampling**, which involves choosing a random sample from one group within a population.

Example 5   Clustered Sampling
Design a clustered sampling procedure to determine the most popular magazine among students in your school.

Solution
The population is all students in the school. As some students do not regularly read magazines, you might survey only those who do. These students make up a clustered sample of the population. The sampling procedure might include the following steps:
• If there are 1000 students in the school, randomly select 25%, or 250, of them to find out who reads at least one magazine per month. Suppose 40% of the selected students, or 100 students, regularly read magazines.
• Survey 50% of the clustered sample of students who regularly read magazines. 50% of 100 = 50
• Prepare an alphabetical list of the 100 students. Survey 50 of them by asking the first, third, and fifth students, and so on, to name their favourite magazine. Record the results.

Non-probability sampling techniques generally give less reliable results than probability sampling techniques. The main advantages of non-probability sampling techniques are that they are cheap and convenient. Two non-probability sampling techniques are called **convenience sampling** and **sampling of volunteers**.

An example of convenience sampling occurs when television reporters sample people on the street to determine public opinion. Little planning goes into the selection of the sample.

An example of sampling of volunteers occurs when a researcher studying the effects of sleep deprivation asks for volunteers to participate in the study.

Example 6   Non-Probability Sampling
Describe a non-probability sampling technique that might be used to study the following two questions.
a) Does being good in mathematics affect a person's musical ability?
b) What percent of people are vegetarians?

Solution
a) A researcher could try sampling volunteers by inviting members of the general public to participate in the study. The researcher might advertise in a local newspaper and offer payment to those who volunteer.
b) A student could use the convenience sampling technique by surveying some students in the cafeteria at lunch.

Practice

Identify the population implied in each statement.

1. *Hockey Night in Canada* is watched by 23% of people.

2. Two out of three people prefer to use *Toothtaste* to brush their teeth.

3. Today's youth prefer comfort over style in fashion.

4. Water-skiing is the most popular summer water activity.

5. A politician has the support of the party.

Describe how simple random sampling could be used to answer each of the following questions.

6. What is the average time spent by shoppers on each visit to a mall?

7. What are the favourite vacation destinations of Canadians?

8. What is the average mass of the salmon in a certain river?

Describe how the list of units could be organized if the systematic sampling technique were used to answer each of the following questions.

9. How many days per year does the average skier spend on the slopes?

10. Should humanity spend money on exploring the planets?

11. Should Canadian high school students take a Canadian history course every year?

12. What is the average speed of the cars on a section of highway?

Suggest appropriate stratifications of the population for each of the following questions.

13. Which school-related extra-curricular activity is the most popular?

14. What restrictions should be placed on young drivers?

15. Should the area of Canada's national and provincial parks be increased?

16. Should the federal government control fishing quotas?

Applications and Problem Solving

17. Mineral water A mineral water company wants to find out why people choose a particular brand of mineral water. Name each of the following sampling techniques that could be used. Identify the population in each case.
a) Attach a mail-in response card to the neck of each bottle of mineral water the company sells.
b) Phone every 200th person who mails in a coupon as part of a money-back offer on the purchase of mineral water.
c) Phone 400 numbers chosen randomly from all the residential phone numbers in a province.
d) Mail questionnaires to 200 men and 200 women chosen randomly from a list of men and a list of women compiled from a market research company's mailing list.
e) Ask the first 20 people seen carrying mineral water in a park on a summer day.

18. Car ownership A survey of car ownership among mall shoppers was conducted in a shopping mall. Passers-by were asked how many cars were owned by all the people in their household.
a) Name the sampling technique used in this case.
b) Identify the population.
c) Design a probability sampling technique that would be appropriate in this study.

19. Stranded airline travellers A television reporter interviewed travellers stranded at an airport during a snow storm about the efficiency of Canadian air travel.
a) Name the sampling technique used.
b) Design a probability sampling technique that could be used to study the efficiency of Canadian air travel.

20. Mosquito repellent A researcher was studying a new mosquito repellent. The study required people to apply the repellent to part of an arm and then expose it in a closed environment infested with mosquitos.
a) Name the sampling technique required in this case.
b) Suggest a possible source of sample units.

21. Soup choice The school cafeteria wants to introduce a new soup, to be chosen from three possibilities. How would you conduct a survey to decide which soup should be the new one?

22. Movie viewing A newspaper has asked you to find out how many times people 15 years old and older go a movie theatre in a year in your town or city. The data must be organized in the table shown.

Number of Times	Percent of People
0	
1 to 4	
5 to 8	
9 to 12	
13 +	

a) What sampling technique would you use?
b) How would you collect the data?

23. Shopping habits To get a sense of people's shopping habits in a mall, a survey was conducted on a Friday morning. The first 50 customers were asked where they would spend most of their money that day — in a department store, music store, clothing store, or restaurant. The results are shown in the table.

Place	Percent of People
Department Store	54
Music Store	12
Clothing Store	26
Restaurant	8

a) What generalization, if any, can be made from this survey?
b) Does the sample adequately represent the shopping habits of mall shoppers? Explain.
c) Design a more reliable method of obtaining information about the shopping habits of mall shoppers. Include the details you would put on the questionnaire. Describe the method you would use to select the sample.

24. There were 250 girls and 275 boys in a school. A random sample of 30 students was surveyed to decide on social activities for the following year.
a) Is it possible that the sample contained only boys?
b) Would a sample containing only boys represent all the students? Explain.

25. Decide whether each of the following methods will produce a random sample of the students in your class. Explain.
a) One student lists the names of 6 students from memory.

b) Five letters of the alphabet are chosen randomly. Students whose surnames begin with these letters form the sample.
c) Choose the five oldest students in the class.
d) The name of each student is written on a piece of paper, and all the pieces of paper are placed in a hat. The names of eight students are drawn.
e) The sample consists of the first six students to arrive in class in the morning.

26. Library books To find the average number of pages in a book, you could use all the books in a public library as a population. How would you select 30 books to obtain each of the following types of samples?
a) simple random   **b)** stratified
c) convenience   **d)** systematic
e) clustered

27. School vacations a) Design a sampling technique that could be used to survey opinions on the following statement.
The two-month vacation students have in July and August should be changed to January and February. That way, the cost of heating schools would be reduced.
b) What difficulties do you anticipate with this survey? How will you get around these difficulties?
c) Conduct the survey on your school population.
d) Might there be a better choice of population for the survey? Explain.

28. Jury duty How are people chosen to appear for possible jury duty in the area where you live? Is a sampling technique used? Are some people excluded from jury duty?

29. How Canadians spend their time Statistics Canada has found that all Canadian adults averaged, over a seven-day period, 3.6 h a day on paid work, 3.6 h a day on unpaid work (domestic chores, volunteering), 5.7 h a day of free time, 8.1 h a day sleeping, and 2.4 h a day on personal care, including eating. The remaining 0.5 h in the day was spent on education.
a) What sampling technique was probably used?
b) Design a survey in which you analyze the day of the students at your school.
c) Conduct the survey and conclude with a statement similar to the one above.

8.2 Inferences and Bias

Surveying samples of a population is an attempt to make generalizations about the population. For inferences to be reliable, a survey should use a suitable probability sampling technique and be as unbiased as possible. However, an absolute lack of bias is rarely possible.

Explore: Analyze the Questions

Work in groups. Analyze the four questions to decide what response is being encouraged by the wording of each question.

a) Without commercials, there would be no private radio stations. Are commercials a small price to pay for being able to listen to the radio for free?

b) In-line skaters who skate on sidewalks endanger the safety of pedestrians. Should in-line skaters be allowed to skate on sidewalks?

c) Shakespeare's plays are written in a style that is hard to understand and, besides, they are old. Should high school students be required to study one of Shakespeare's plays?

d) Do you really think fast food is nutritious?

Inquire

1. What influence might the information provided in the first sentence of questions a), b), and c) have on the responses?

2. How could questions a), b), and c) be reworded so that a certain response is not encouraged?

3. How does the wording of question d) direct the response?

4. How could you reword question d) so that it is unbiased?

5. Give two examples of these types of forced response questions you have seen or heard in advertising.

Bias that arises from the phrasing or construction of a survey question, as in questions a) to d) above, is called **response bias**.

Bias can also occur within the sampling technique itself. One example is to exclude one or more groups of people from the sample. This type of bias is called **selection bias**.

Example 1 Selection Bias

A newspaper conducted a survey to identify the sports its readers wanted to have highlighted in the sports section. The newspaper obtained lists of members of golf and tennis clubs. A simple random sample was taken from these lists.

a) Identify the population that was of interest in this survey.

b) What form of bias appeared in the sample design? Explain.

c) Design a different sampling method that would more accurately have represented the population.

Solution

a) The population of interest was all readers of the newspaper's sports section.

b) This sampling technique had selection bias, because some members of the population were excluded from the sample. Not all of the newspaper's readers were members of the golf or tennis clubs. For example, people who could not afford the club fees or who were not interested in these sports were not included in the survey.

c) The newspaper could have chosen a simple random sample from its list of subscribers and conducted a phone survey. This method would have excluded fewer people from the sample and removed some of the bias that was in the original sampling method.

A second form of bias within the sampling technique may arise if a large number of the people selected for the sample do not complete the survey. This type of bias is called **non-response bias.** For example, after a polling firm has selected the people in a sample, the firm still has to get their opinions. This process can be difficult. If a large number of the people selected do not respond to a questionnaire, interview, or telephone call, the results of the poll may be distorted.

Example 2   Non-Response Bias

A magazine publisher used a systematic sampling method to get the opinions of readers on a new feature for a magazine. The company took its list of subscribers and mailed a questionnaire to every five hundredth subscriber. Of the readers who received the mailing, 20% responded, and 60% of the respondents were in favour of the new feature.

a) What inferences about the population can be drawn from this information? Are the inferences reliable?

b) How could the sampling method be improved to reduce the possibility of a non-response bias?

Solution

a) Because 60% of the respondents were in favour of the new feature, it could be inferred that the readers approve of it. However, the inference may not be reliable, because only 20% of the people in the sample responded. The fact that 80% did not respond may mean that there is a non-response bias in the result.

b) To reduce the possibility of a non-response bias, the company could offer an incentive, such as a gift, to anyone who returns a completed survey. Another approach would be to contact each subscriber in the sample by telephone and conduct a personal interview. A typical response rate for personal interviews is 75%, compared with 25% for mailed questionnaires.

To give reliable results, polling companies must know how different types of bias can affect the results of a survey. The companies should make every effort to reduce the influence of bias on the results.

Practice

Identify the type(s) of bias that are most likely to influence the results of each of the following surveys.

1. A traffic survey was taken at a major intersection between the hours 13:00 and 16:00 to study the traffic volume in the central core of a city.

2. A survey to determine the types of voluntary work people were willing to do was conducted by leaving a stack of questionnaires on a counter at a recreation centre. They were to be filled out and dropped into a collection box.

3. A survey question is worded as follows.
Should there be national examinations for all grade 10 students? In what subjects?

4. A stratified sample, using the two strata male and female, was chosen in the workplace, to determine the attitude of the public to government welfare.

5. A survey question is worded as follows.
Is it really fair that young people are not allowed to drive until they are 16?

6. A travel agency obtained a list of department store credit card holders and surveyed a random sample of them as to their preference for either air or ground transportation when they went on vacation. The survey was conducted by telephone interview.

Reword the following questions to remove the response bias.
7. Why does Cola A taste better than Cola B?

8. Do you think that small, yappy dogs make good pets?

9. Deforestation of the tropical rainforest in Brazil occurs at a rate of about 14 000 km^2/year. Should logging be restricted in Canada?

10. If you cannot understand a painting, is it still art?

11. Do long, challenging projects make school work more interesting?

12. Is hydro-electric power less damaging to the environment than nuclear power, with its radioactive waste?

13. Does the thrill of skiing make it Canada's favourite winter sport?

Applications and Problem Solving

14. School diplomas A newspaper reported that thousands of Canadian high school dropouts are returning to school to get high school diplomas. The newspaper determined the students' reasons for dropping out and returning by surveying 892 students who returned to school.
a) What organizations would be interested in the results of this type of survey?
b) What type of sampling technique could be used for this survey?
c) Design a questionnaire that could be used in this survey.
d) Suppose the results of the survey suggested that most dropouts returned to school because there was a high unemployment rate for high school dropouts. What follow-up survey could be conducted to test this suggestion?

15. Highway bypass The Ministry of Transportation is studying the possibility of building a highway bypass around a town of 5000 people. The ministry plans to survey 320 town residents about the bypass, using a telephone survey.
a) How is the ministry attempting to control non-response bias?
b) Describe fully how you would choose the sample of 320 people. Should they be drawn from the whole population of the town or from the people most affected by the bypass?
c) Design a sequence of questions that could be used in the telephone survey.

16. Banking Many women own businesses in Canada. These businesses are a significant market for banks. A study found that the percent of business women reporting one or more problems in working with the banks has dropped by 15%, to about the same level as for men. The study was based on a survey of 800 businesses.
a) What inference can be drawn from the information?
b) What type of sampling method is most appropriate for this study?
c) The study was conducted by the Foundation for Women Business Owners. What do you think was the motivation for the study?

d) How would you conduct the survey to try to eliminate non-response bias?

e) Design a questionnaire that could be used in this study.

17. Phone-in poll To get the public reaction to a government decision announced on the news, a television station asked viewers to dial one phone number if they agreed with the decision and another if they disagreed. Identify the potential sources of bias in such a poll.

18. Federal election To predict the winner of a federal election, a national magazine compiled a list of 10 000 names from its subscription list, telephone books, and lists of homeowners. The questionnaire was mailed to all 10 000 people, and 1203 people returned it. The magazine used the 1203 responses to make its prediction. Describe the potential sources of bias in this survey.

19. The Internet A radio station posts a questionnaire on the Internet to get responses to a possible switch in format from country music to all-talk radio. What are the potential sources of bias in this survey?

20. Suggestion box If you set up a suggestion box in the school cafeteria and asked for suggested improvements to the menu,

a) in what way might the results be biased? Explain.

b) how could you gain a more accurate impression of students' opinions?

21. Census In a census of a population, can the following types of bias exist? Explain.

a) response bias

b) selection bias

c) non-response bias

22. Advertising A company that makes headache tablets advertises that three out of four doctors prefer their product.

a) Analyze this statement for its accuracy in terms of population, possible sampling method, and bias.

b) What inference does the company hope you will draw from its claim?

23. Destructive sampling In some situations, sampling methods may be **destructive**, which means that the sample cannot be returned to the population. For a firecracker manufacturer, testing the quality of firecrackers by setting off a sample of them is an example of destructive sampling.

a) Give two other examples of situations in which you think destructive sampling might be used.

b) When carrying out destructive sampling, what are the advantages and disadvantages of using a small number of units in the sample? a large number of units in the sample?

24. Media generalizations Use various media, such as newspapers, magazines, radio, and television, to find three examples of data from samples being used to make generalizations about populations. Explain why you agree or disagree with the generalizations. Share your findings with your classmates.

PATTERN POWER

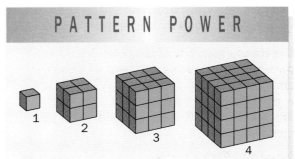

In diagram 1, there is one $1 \times 1 \times 1$ cube, so the total number of cubes is 1. In diagram 2, there are eight $1 \times 1 \times 1$ cubes and one $2 \times 2 \times 2$ cube, so the total number of cubes is 9.

1. What is the total number of cubes in

a) diagram 3?   **b)** diagram 4?

2. Describe the pattern in the total numbers of cubes.

3. Use the pattern to predict the total number of cubes in

a) diagram 5   **b)** diagram 6

INVESTIGATING MATH

Reviewing Probability

1 Theoretical Probability

Find the theoretical probability of each of the following
outcomes for the spinner.

1. $P(6)$
2. P(even number)
3. P(blue)
4. P(prime number)
5. P(less than 3)
6. P(greater than 7)
7. $P(11)$
8. P(number from 1 to 10)
9. P(red or blue)
10. P(4, 9, or 10)
11. P(multiple of 4)
12. P(not red)

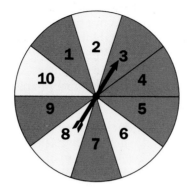

2 Independent Events — Tossing Coins

When a penny and a nickel are tossed, the possible outcomes,
or **sample space**, can be shown with a **tree diagram**. There are
four possible outcomes in the sample space, all equally likely.

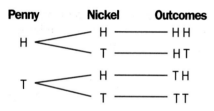

Penny	Nickel	Outcomes

From the tree diagram, $P(\text{HH}) = \dfrac{1}{4}$

$P(\text{HH})$ can also be calculated as follows.

For the penny, $P(\text{H}) = \dfrac{1}{2}$

For the nickel, $P(\text{H}) = \dfrac{1}{2}$

For all independent events, the probability of an outcome is the
product of the probabilities of the individual events.

So, $P(\text{HH}) = \dfrac{1}{2} \times \dfrac{1}{2}$

$\qquad\qquad = \dfrac{1}{4}$

The probability of each of the other three outcomes is also $\dfrac{1}{4}$.

1. Draw a tree diagram to show the possible outcomes when a penny, nickel, and dime are tossed.

2. What is the probability of tossing
a) 3 tails?   **b)** 2 heads and a tail?   **c)** 3 tails or 3 heads?

3. When a penny, dime, nickel, and quarter are tossed, what is the probability of tossing
a) 4 heads?   **b)** 3 tails?   **c)** 2 tails?   **d)** 1 head or 1 tail?
e) the same number of heads and tails?   **f)** at least 1 head?

3 Independent Events — Rolling Dice

1. Draw a tree diagram to show the possible outcomes when a red die, then a green die are rolled.

2. What is the probability of rolling
a) a total of 9?
b) a total of 7?
c) a total less than 4?
d) an even number followed by an odd number?
e) a 2 followed by a 6?
f) a number less than 5 followed by a 4?

4 Experimental Probability — A Folded Square

For many events, such as rolling a die, the probability of an outcome can be determined mathematically. For some other events, the probabilities of the outcomes can only be determined by experiment.

Cut off one end of an index card to make a square. Fold the square along a diagonal to make two triangular faces at an angle of about 45° to each other. When the folded card is tossed, assume that there are two ways it can land — on one of its triangular faces or on two of its edges.

1. Estimate the probability of the card landing in each of the two positions.

2. Toss the card 20 times and record the number of times it lands in each position.

3. Combine your results with your classmates'.

4. Use the combined results to find the experimental probability of the card landing in each position.

5. Compare the experimental probabilities with your estimates.

TECHNOLOGY

Graphing Calculators, Simulations, and Sampling

1 Generating Random Numbers

The following list of random numbers was generated by a scientific calculator.

0.538	0.383	0.826	0.755
0.017	0.404	0.304	0.385
0.003	0.089	0.554	0.859

The following list of random numbers was generated by a graphing calculator.

.6485679703	.6976612527	.0151786158
.1018231622	.7916268018	.3688383045
.5069113153	.4866367322	.9172764946

1. How are the random numbers generated by each calculator the same? How are they different?

2. Describe how you can generate random numbers with your calculator.

3. Random numbers generated by a calculator are never 0, 1, or negative. Calculators are designed to generate random numbers, r, in the interval $0 < r < 1$.
a) Use your calculator to multiply the random numbers it generates by 2.
b) If $2r$ represents the resulting numbers, write their interval in the form $\bullet < 2r < \blacksquare$, where $\bullet$ and $\blacksquare$ are numbers.

4. a) Use your calculator to multiply the random numbers it generates by 2, and then add 1.
b) If $2r + 1$ represents the resulting numbers, write their interval in the form $\blacktriangle < 2r + 1 < \blacktriangledown$, where $\blacktriangle$ and $\blacktriangledown$ are numbers.

5. The greatest integer function on a graphing calculator gives the greatest integer less than or equal to a number. For example, int(7.98) = 7 and int(−5.41) = −6. If the greatest integer function was applied to the numbers generated in question 4, what integers would result?

2 Simulating Coin Tosses and Rolling Dice

1. To simulate tossing a coin, enter the command int(2rand)+1 into your graphing calculator.
a) What are the possible outputs each time ENTER is pressed?
b) Describe how you could use your calculator to simulate a coin toss experiment.

2. a) Modify the command from question 1 so that only the integers from 1 to 6 can appear when you press ENTER.
b) Describe how you could use your calculator to simulate an experiment that involves rolling a single die.

3 Sampling

1. A way to select a random sample of students in a school is to list the students alphabetically, number each student, and generate random numbers to define the sample. Suppose the 750 students in a school are numbered from 1 to 750.
a) What command would you enter into your graphing calculator to generate random integers from 1 to 750?
b) If you wanted a sample of 60 students and generated 60 random integers, what would you do if the same integer were generated twice?

2. In a school, there are 460 male students, numbered from 1 to 460, and 420 female students, numbered from 1 to 420. A stratified sample of students is to contain 23 male students and 21 female students. Describe how you would use your graphing calculator to choose the students for the sample. Specify any commands you would use.

3. A school has 250 grade 10 students, 220 grade 11 students, and 200 grade 12 students. A sample stratified by grade is to include a total of 67 students. Describe how you would use your graphing calculator to choose the students for the sample. Specify any commands you would use.

8.3 Expected Values

Making predictions with probabilities often involves multiplying a probability by another number. For example, if you observed a rabbit at random times of the day and night, you would find that the probability of it being asleep is about $\frac{5}{12}$. You can predict the number of hours a rabbit sleeps in a day by multiplying this probability by the number of hours in a day.

$$\frac{5}{12} \times 24 = 10$$

So, a rabbit sleeps about 10 h/day.

Explore: Analyze the Game

To practice the addition of integers, some students play the following game. They play in pairs and take turns rolling a die. Rolling an odd number wins that number of positive points. Rolling an even number wins that number of negative points. The first player to reach a total of −10 points is the winner.

Suppose a player rolled each number once in 6 rolls.

a) Copy and complete the table to show the number of points won in each roll.

b) What is the total number of points won in 6 rolls?

Number Rolled	Points Won
1	
2	
3	
4	
5	
6	

Inquire

1. From question b), above, what is the average number of points won per roll?

2. What is the probability of rolling an odd number, $P(\text{odd})$?

3. What is the average number of points won when an odd number is rolled? Call this number the *odd payoff*.

4. What is the probability of rolling an even number, $P(\text{even})$?

5. What is the average number of points won when an even number is rolled? Call this number the *even payoff*.

6. Substitute the values found in questions 2, 3, 4, and 5 into this expression.
$P(\text{odd}) \times (\text{odd payoff}) + P(\text{even}) \times (\text{even payoff})$

7. How does the number you found in question 6 compare with the number you found in question 1?

In question 1, above, you found the average number of points won per roll. This is known as the **expected value** of a roll.

When you spin the spinner and land on a colour, you win the number of points shown for that colour. The number of points for each colour is called the **payoff** for that colour.

The outcomes red, yellow, blue, and green are equally likely. So, you would expect to get each colour once in every 4 spins, if you played many times. If you get each colour once in 4 spins, the total points won are $3 + (-4) + 9 + (-6) = 2$.

Since you expect to win 2 points in 4 spins, you expect the average winnings per spin to be $\dfrac{2}{4} = 0.5$ points/spin

This value of 0.5 points/spin is the expected value or **mathematical expectation** per spin.

Suppose the division is carried out before the points are totalled. If E is the expected value, then

$$E = \frac{3 + (-4) + 9 + (-6)}{4}$$

$$= \frac{3}{4} + \frac{-4}{4} + \frac{9}{4} + \frac{-6}{4}$$

$$= \frac{1}{4} \times 3 + \frac{1}{4} \times (-4) + \frac{1}{4} \times 9 + \frac{1}{4} \times (-6)$$

where $\dfrac{1}{4}$ is the probability of getting each colour, and 3, −4, 9, and −6 are the payoffs for each colour.

So, $E = \dfrac{3}{4} - 1 + \dfrac{9}{4} - \dfrac{6}{4}$

$$= \frac{2}{4}$$

$$= \frac{1}{2}$$

Thus, the expected value can be found by multiplying probability × payoff for each event, and then adding the results. Another way to calculate the expected value is as follows.

The probability of getting red or blue is $\dfrac{1}{2}$.

The average payoff for red or blue is $\dfrac{9+3}{2} = 6$.

The probability of getting yellow or green is $\dfrac{1}{2}$.

The average payoff for yellow or green is $\dfrac{(-4)+(-6)}{2} = -5$

$E = P(\text{red or blue}) \times (\text{red or blue payoff}) + P(\text{yellow or green}) \times (\text{yellow or green payoff})$

$$= \frac{1}{2} \times 6 + \frac{1}{2} \times (-5)$$

$$= 3 - 2.5$$

$$= 0.5$$

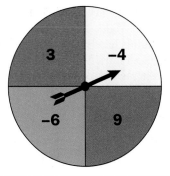

Colour	Points Won
red	3
yellow	−4
blue	9
green	−6

Example 1   Tossing Coins

Rosa and Malik play a game in which they each toss a coin. If one or both coins lands heads, Rosa wins 1 point and Malik loses 1 point. If both coins land tails, Malik wins 3 points and Rosa loses 3 points.
a) What is Rosa's mathematical expectation?
b) What is Malik's mathematical expectation?
c) Is this a fair game? Explain.

Solution

The sample space for this game is HH, HT, TH, and TT.
a) Rosa wins with the outcomes HH, HT, and TH, and her payoff for each win is 1 point. Rosa loses with TT, and her payoff is −3.

$$E = P(\text{HH}) \times (\text{HH payoff}) + P(\text{HT}) \times (\text{HT payoff}) + P(\text{TH}) \times (\text{TH payoff}) + P(\text{TT}) \times (\text{TT payoff})$$

$$= \frac{1}{4} \times 1 + \frac{1}{4} \times 1 + \frac{1}{4} \times 1 + \frac{1}{4} \times (-3)$$

$$= \frac{1}{4} + \frac{1}{4} + \frac{1}{4} - \frac{3}{4}$$

$$= 0$$

Or $E = P(1 \text{ or } 2 \text{ heads}) \times (\text{HH, HT, TH average payoff}) + P(\text{TT}) \times (\text{TT payoff})$

$$= \frac{3}{4} \times 1 + \frac{1}{4} \times (-3)$$

$$= \frac{3}{4} - \frac{3}{4}$$

$$= 0$$

So, Rosa's mathematical expectation is 0.

b) Malik wins with the outcome TT, and his payoff is 3. Malik loses with the outcomes HH, HT, and TH, and his payoff is −1 for each.

$$E = P(\text{TT}) \times (\text{TT payoff}) + P(\text{HH}) \times (\text{HH payoff}) + P(\text{HT}) \times (\text{HT payoff}) + P(\text{TH}) \times (\text{TH payoff})$$

$$= \frac{1}{4} \times 3 + \frac{1}{4} \times (-1) + \frac{1}{4} \times (-1) + \frac{1}{4} \times (-1)$$

$$= \frac{3}{4} - \frac{1}{4} - \frac{1}{4} - \frac{1}{4}$$

$$= 0$$

Or $E = P(\text{TT}) \times (\text{TT payoff}) + P(1 \text{ or } 2 \text{ heads}) \times (\text{HH, HT, TH average payoff})$

$$= \frac{1}{4} \times 3 + \frac{3}{4} \times (-1)$$

$$= \frac{3}{4} - \frac{3}{4}$$

$$= 0$$

So, Malik's mathematical expectation is 0.

c) The game is fair because the mathematical expectation of both players is the same. In this case, the mathematical expectation of each player is 0.

Example 2   Dice Game

The rules for a dice game are as follows.
- Each player takes a turn rolling a pair of dice and adding the results.
- Two points are scored if the sum is greater than or equal to 9.
- Three points are scored if the sum is less than or equal to 4.
- One point is lost if the sum is greater than or equal to 5 and less than or equal to 8.
- The first player to reach 10 points wins.

a) Determine the expected value for a roll of the dice.

b) How many rolls of the dice would you expect a player to take to reach 10 points?

Solution

a) The outcomes for rolling a pair of dice can be organized in a table. Each entry in the table represents the sum of the numbers rolled. Each of the 36 entries is equally likely. The probability of rolling the different points on a turn can be found as follows.

$$P(2 \text{ points}) = P(\text{sum} \geq 9)$$
$$= \frac{\text{number of outcomes} \geq 9}{\text{total number of outcomes}}$$
$$= \frac{10}{36}$$
$$= \frac{5}{18}$$

$$P(3 \text{ points}) = P(\text{sum} \leq 4)$$
$$= \frac{6}{36}$$
$$= \frac{1}{6}$$

$$P(-1 \text{ point}) = P(5 \leq \text{sum} \leq 8)$$
$$= \frac{20}{36}$$
$$= \frac{5}{9}$$

The expected value for a roll of the dice is given by

$$E = P(\text{sum} \geq 9) \times 2 + P(\text{sum} \leq 4) \times 3 + P(5 \leq \text{sum} \leq 8) \times (-1)$$
$$= \frac{5}{18} \times 2 + \frac{1}{6} \times 3 + \frac{5}{9} \times (-1)$$
$$= \frac{10}{18} + \frac{1}{2} - \frac{5}{9}$$
$$= \frac{10}{18} + \frac{9}{18} - \frac{10}{18}$$
$$= \frac{9}{18}$$
$$= 0.5$$

b) Since the expected value per roll is 0.5 points, you would expect a player to take 20 rolls of the dice to reach 10 points.

	●	⦂	⸪	⸬	⁙	⸭
●	2	3	4	5	6	7
⦂	3	4	5	6	7	8
⸪	4	5	6	7	8	9
⸬	5	6	7	8	9	10
⁙	6	7	8	9	10	11
⸭	7	8	9	10	11	12

	●	⦂	⸪	⸬	⁙	⸭
●	2	3	4	5	6	7
⦂	3	4	5	6	7	8
⸪	4	5	6	7	8	9
⸬	5	6	7	8	9	10
⁙	6	7	8	9	10	11
⸭	7	8	9	10	11	12

There are 10 outcomes where sum ≥ 9.

	●	⦂	⸪	⸬	⁙	⸭
●	2	3	4	5	6	7
⦂	3	4	5	6	7	8
⸪	4	5	6	7	8	9
⸬	5	6	7	8	9	10
⁙	6	7	8	9	10	11
⸭	7	8	9	10	11	12

There are 6 outcomes where sum ≤ 4.

	●	⦂	⸪	⸬	⁙	⸭
●	2	3	4	5	6	7
⦂	3	4	5	6	7	8
⸪	4	5	6	7	8	9
⸬	5	6	7	8	9	10
⁙	6	7	8	9	10	11
⸭	7	8	9	10	11	12

There are 20 outcomes where 5 ≤ sum ≤ 8.

Practice

Determine the expected value of a spin for each spinner.

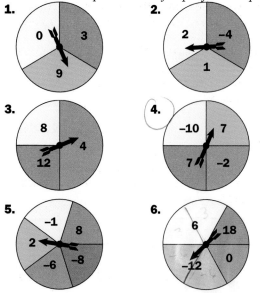

1. 0 3 9

2. 2 −4 1

3. 8 4 12

4. −10 7 7 −2

5. −1 8 2 −6 −8

6. 6 18 −12 0

Applications and Problem Solving

7. Coin toss game Carla and Sam play a game in which they each toss a coin at the same time. Carla wins 1 point if both coins are heads or both are tails. Sam wins 1 point for a head and a tail.
a) Draw a tree diagram to identify all the possible outcomes.
b) Is the game fair? Explain.

8. Coin toss game A coin toss game has the following rules.
• Each player takes turns tossing two coins.
• Five points are scored if both coins land heads.
• Two points are scored if one coin is a head and the other is a tail.
• Seven points are lost if both coins land tails.
• The first player to accumulate 10 points wins.
a) Determine the expected value of a turn.
b) Is the game fair? Explain.

9. Dice game Suppose you play a game in which you roll a die. If you roll an odd number, you win the number of points on the die. If you roll any even number, you lose 4 points. The player to reach −5 points first is the winner.
a) What is the expected value for a roll?
b) How many rolls would you expect a player to take to reach −5 points?

10. Scratch-and-save sale A store runs a sale in which each customer receives a discount by scratching a coupon at the checkout. Of every 10 coupons, 1 gives a 40% discount, 2 give a 30% discount, 3 give a 20% discount, and 4 give a 10% discount.
a) What is the expected value of the discount on one coupon?
b) If the average customer purchases goods with a regular price of $55 during the sale, what is the expected saving per customer?

11. a) Find the expected value for one roll of a single die.
b) Use your answer from part a) to predict the expected value for one roll of a pair of dice.
c) Check your prediction from part b) by calculating the expected value for one roll of a pair of dice.

12. Card game Each player takes turns dropping three playing cards, one after another. Each card is held between the thumb and the forefinger, so that the card is perpendicular to the floor. The card is then dropped from shoulder height with the arm extended. The other rules are as follows.
• Fifty points are scored if all the cards land face up or face down.
• Twenty points are scored if exactly 2 cards land face up.
• Forty points are lost if only 1 card lands face up.
• The first player to reach 100 points wins.
a) Draw a tree diagram to identify all of the possible outcomes.
b) What is the expected value of a turn?
c) How many turns would you expect to take to reach 100 points?

13. Fair game A game is played by rolling a single die. The player wins $12 if a 2 turns up, wins $24 if a 4 turns up, and loses $18 if a 6 turns up. The player loses a certain amount for rolling a 1, 3, or 5. What is this amount if the game is fair? Explain.

14. Bridge In the game of bridge, each of the four players is dealt a hand of 13 cards. The suit that contains the most cards is a player's longest suit. For example, in a hand that includes 6 hearts, 4 clubs, 2 spades, and 1 diamond, the longest suit

382

is hearts. The probabilities of being dealt various numbers of cards in the longest suit are shown.

Cards in Longest Suit	Probability
13	0.000 000 000 006
12	0.000 000 003
11	0.000 000 4
10	0.000 02
9	0.000 4
8	0.005
7	0.04
6	0.17
5	0.44
4	0.35
3	0

a) Determine the expected number of cards in the longest suit in a bridge hand.
b) Why is the probability zero that the longest suit has 3 cards?

15. If the expected value for a spin of the spinner is 4, how many points are in the fourth sector?

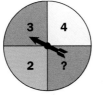

16. Drawing cards In a game for 2 players, you draw one card at random from a deck of 52 playing cards. If you draw an ace, you get 30 points. If you draw a king, queen, or jack, you get 20 points. If you draw any other card, your opponent gets a certain number of points. How many points should your opponent get for any other card if the expected value of a draw is 0?

17. Algebra If the expected value for a spin of the spinner is 3, what is the value of x?

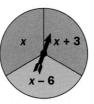

18. Algebra If the expected value for a spin of the spinner is −4, what is the value of x?

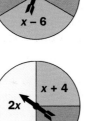

19. Dice game It costs $10 to roll a pair of dice. If no 1 turns up on either die, the player gets the $10 back and wins the sum of the numbers rolled, in dollars. The player loses the $10 if 1 turns up on either die or both dice.
a) What is the expected value, in dollars, of a roll?
b) Is the game fair? Explain.
c) What would be the expected value, in dollars, if it cost $28 to roll the dice?
d) Is the $28 game fair? Explain.
e) What would be the expected value if, in the $10 game, the 1 was changed to a 6?

20. Design each of the following spinners so that each sector has a different integer value. Compare your spinners with your classmates'.
a) 4 equally likely sectors; expected value 2 for a spin
b) 5 equally likely sectors; expected value −1 for a spin
c) 6 equally likely sectors; expected value $\frac{1}{2}$ for a spin

21. Platonic solids While the most common dice are cubes, a die with equally likely outcomes can also be made in the shape of each of the other Platonic solids.
a) Suppose one die is made in the shape of each of the five Platonic solids, and the faces are numbered with whole numbers, starting with 1, 2, 3, and so on. What is the expected value for a roll of each die?
b) How could you renumber the faces of an octahedral die so that each face is numbered with a different integer, and the expected value is 2 for one roll? Compare your answer with a classmate's.

LOGIC POWER

Suppose intercity buses travel from Montreal to Toronto and from Toronto to Montreal, leaving each city on the hour every hour from 06:00 to 20:00. Each trip takes $5\frac{1}{2}$ h. All buses travel at the same speed on the same highways. Your driver waves at each of her colleagues she sees driving an intercity bus in the opposite direction. How many times would she wave during the journey if your bus left Toronto at
a) 14:00?   **b)** 18:00?   **c)** 06:00?

8.4 Making Decisions Using Probability

Explore: Use the Data

The Adventure Wilderness Park is a wildlife park, with lions, tigers, giraffes, rhinos, and many other species. Visitors can drive through the park or take a tram with a guide. Admission is $12 per person. The average number of visitors on any day, and the number of park staff needed, depends on the weather. However, if the park is open, at least 15 staff are required.

Weather	Visitors	Staff
Sun	1000	30
Cloud	800	27
Rain	30	15

On a Friday, the weather office gave the following probabilities for the next day.

Weather	Probability
Sun	0.1
Cloud	0.2
Rain	0.7

Inquire

1. Calculate the expected number of visitors for Saturday.

2. What is the expected revenue from ticket sales?

3. Calculate the expected number of staff required for Saturday. Round your answer up to the nearest whole number.

4. Each staff member earns $150 a day. What is the expected cost of staff salaries for the day?

5. The park owner has a fixed cost of $1000 a day, excluding salaries, whether the park is open or not. Can the owner expect to make a profit on Saturday? Explain.

6. Should the owner close the park for the day? Give reasons for your answer.

Many decisions you need to make are based on statistics and probabilities.

Example 1   Lotteries

There were 5 000 000 tickets sold for a lottery, and each ticket cost $1. There were five prize categories. The number of tickets that won each prize are listed in the table.

Prize ($)	Number of Winning Tickets
500 000	1
50 000	9
5 000	90
500	900
50	9000

a) What is the expected value of a ticket?
b) Would you expect to win or lose money by buying a ticket? Explain.

Solution

a) To calculate the expected value of a ticket, you need the probability of winning each prize.
The total number of tickets sold is 5 000 000.

Prize ($)	Number of Winning Tickets	Probability of Winning Prize
500 000	1	1 ÷ 5 000 000 = 0.000 000 2
50 000	9	9 ÷ 5 000 000 = 0.000 001 8
5 000	90	90 ÷ 5 000 000 = 0.000 018
500	900	900 ÷ 5 000 000 = 0.000 18
50	9000	9000 ÷ 5 000 000 = 0.001 8

The expected value of a ticket is the sum of
(prize value) × (probability of winning that prize) for all possible prizes.
$E = 500\ 000 \times 0.000\ 000\ 2 + 50\ 000 \times 0.000\ 001\ 8 + 5000 \times 0.000\ 018$
$\qquad + 500 \times 0.000\ 18 + 50 \times 0.0018$
$\quad = 0.1 + 0.09 + 0.09 + 0.09 + 0.09$
$\quad = 0.46$
The expected value of a ticket is $0.46.

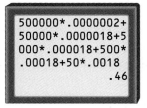

```
500000*.0000002+
50000*.0000018+5
000*.000018+500*
.00018+50*.0018
                .46
```

b) Since it costs $1 to purchase a ticket, you would expect to lose $1 – $0.46, or $0.54, on each ticket you purchase. It is possible that you would buy a winning ticket, but the probability is that you would lose money.

Example 2   House Insurance

An insurance company has analyzed its claims per policy on the loss or theft of household contents. Each policy is based on no deductible, that is, when the claim is made, the policyholder does not have to pay any amount.

Annual Claim Amount ($)	Probability
0 (no claim)	0.95
1 000	0.04
5 000	0.002
10 000	0.007
25 000	0.001

a) Determine the expected amount of money to be paid out per policy in a year.
b) What annual premium should the insurance company charge per policy, if the company adds a 33% markup to the expected amount?
c) Suppose the deductible was $200 per claim, that is, the policyholder pays the first $200 of the claim. What should the change in the annual premium be for the policyholder?

Solution

a) The expectation, E, is the sum of claim amount $\times$ probability of claim for all possible claims.
$E = 0 \times 0.95 + 1000 \times 0.04 + 5000 \times 0.002 + 10\ 000 \times 0.007 + 25\ 000 \times 0.001$
$\quad = 0 + 40 + 10 + 70 + 25$
$\quad = 145$
The company can expect to pay out $145 per policy in a year.

b) The premium, with a 33% markup added to the expected amount, is
$E + (0.33)E = 145 + (0.33)145$
$\qquad\qquad = 192.85$

> **Estimate**
>
> $\dfrac{1}{3} \times 150 = 50$
> $50 + 145 = 195$

The company would charge an annual premium of $192.85 per policy.
c) The policyholder pays the first $200, so each claim is reduced by $200.
The table becomes

Annual Claim Amount ($)	Probability
0 (no claim)	0.95
800	0.04
4 800	0.002
9 800	0.007
24 800	0.001

The new expectation is
$E = 0 \times 0.95 + 800 \times 0.04 + 4800 \times 0.002 + 9800 \times 0.007 + 24\ 800 \times 0.001$
$\quad = 0 + 32 + 9.6 + 68.6 + 24.8$
$\quad = 135$

The new premium would be $E + (0.33)E = 135 + (0.33)135$
$\qquad\qquad\qquad\qquad\qquad = 179.55$

> **Estimate**
>
> $\dfrac{1}{3} \times 135 = 45$
> $45 + 135 = 180$

So, raising the deductible to $200 reduces the premium from $192.85 to $179.55.
The change in the annual premium for the policyholder is a reduction of $13.30.

Example 3 Stock Market Investments

A stockbroker advises a client to keep a stock that is increasing in value as long as there is at least a 75% probability that it will continue to rise. Otherwise, the broker's advice is to sell. If probability of the stock falling on any given day is 0.05, on which day should the client sell the stock?

Solution

Make a partial tree diagram to find the pattern in the probability of the stock continuing to rise.

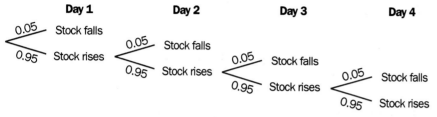

Use the pattern to calculate the probabilities that the stock will rise for different numbers of days.

Number of Days	Probability of Stock Rising
1	0.95
2	$(0.95)(0.95) = (0.95)^2 = 0.9025$
3	$(0.95)(0.95)(0.95) = (0.95)^3 \doteq 0.8574$
4	$(0.95)(0.95)(0.95)(0.95) = (0.95)^4 \doteq 0.8145$
5	$(0.95)(0.95)(0.95)(0.95)(0.95) = (0.95)^5 \doteq 0.7738$
6	$(0.95)(0.95)(0.95)(0.95)(0.95)(0.95) = (0.95)^6 \doteq 0.7351$

The probability of the stock rising stays above 75% for up to 5 days in a row. The probability of the stock rising for 6 days in a row is below 75%. Therefore, the client should sell the stock on day 5.

Applications and Problem Solving

1. Lotteries There were 2 000 000 lottery tickets sold. Copy and complete the table.

Prize	Number of Winning Tickets	Probability of Winning Prize
1st	1	0. 000 0605
2nd	2	0.000 001
3rd	50	0. 000 025
4th	800	0.000 4
5th	10 000	0.005

2. Summer jobs Axel has been guaranteed one of two part-time summer jobs. He cannot do both of them. One job pays $5400, and the other pays $4000. His probability of getting the $5400 job is 0.3, and his probability of getting the $4000 job is 0.7. What are Axel's expected earnings?

3. Contract proposal Jenna prepared a proposal for a construction contract that was worth $500 000. It cost her $10 000 to prepare the proposal, and her probability of getting the contract was 0.3. Calculate Jenna's expected gain.

4. Wave pool Tickets to get into the wave pool cost $8. The table shows how the average number of tickets sold and the number of staff required depend on the weather.

Weather	Tickets Sold	Staff
Sun	800	10
Cloud	700	8
Rain	100	3

For tomorrow's weather, the probability of sun is 0.7, of cloud is 0.2, and of rain is 0.1.
a) Calculate the expected number of tickets that will be sold tomorrow.
b) What is the expected revenue from ticket sales?
c) Calculate the expected number of staff required. Round your answer up the the nearest whole number.
d) Staff members earn $110 a day. What is the expected cost of staff salaries for the day?
e) The pool owner's fixed cost for a day is $300, excluding salaries. How much profit can the owner expect to make?

5. Lotteries There were 4 000 000 tickets sold for a lottery, and each ticket cost $1. There were five prize categories. The number of winning tickets for each category is shown in the table.

Prize ($)	Number of Winning Tickets
400 000	1
40 000	6
5 000	50
500	120
50	5000

a) What is the expected value of each ticket?
b) For every $100 spent on tickets, what is the expected loss?

6. Stock market An investor decides to keep a stock that is increasing in value as long as there is a probability of at least 0.6 that its value will continue to increase. The probability of the value falling is 0.07 in any week. In which week should the investor sell the stock?

7. Household insurance The table shows an insurance company's analysis of its claims per household contents policy. Each policy is based on no deductible.

Annual Claim Amount ($)	Probability
0 (no claims)	0.94
1 000	0.05
5 000	0.003
10 000	0.006
25 000	0.001

a) Determine the expected amount of money paid out per policy in a year.
b) What annual premium should the insurance company charge for each policy, if the company adds a 35% markup to the expected amount?
c) If the deductible was $300 per claim, what should the change in the annual premium be for the policyholder?

8. Multiple-choice test There are 50 questions on a multiple-choice test, and 5 possible choices for each question. You know 35 of the answers, and you have the option of guessing at the other 15. Use your expected score to explain whether it is a good idea to make guesses, if there is 1 point for each correct answer and
a) no penalty for each wrong answer
b) −1 point for each wrong answer
c) $-\dfrac{1}{4}$ point for each wrong answer

9. Renting videos The rental cost of a videotape is $2.99. The value of the tape is $40. The tape protection per rental costs $0.25. If the probability of the tape breaking is 0.01, is it worthwhile to take out the tape protection? Explain.

10. Renting skis Ski equipment can be rented at a resort for $20. Insurance on the rental is $1 for a day. The value of the skis and bindings is $150. If you are a novice skier, the probability of breaking the skis is approximately 0.02. If you are an intermediate skier, the probability of breaking the skis is 0.005. If you are an expert skier, the probability of breaking the skis is 0.01. Which type(s) of skier would you advise to buy the insurance? Explain.

11. Real estate Kwan is deciding when to list her house for sale. Based on the real estate market over the past two years, she thinks that there is a probability of 0.9 that the price of her house will go up each month. Kwan decides not to list her house for sale as long as the probability of the price continuing to rise remains above 0.65. How many months should she wait before listing the house for sale?

12. Scratch-and-win contest A scratch-and-win box on the underside of a lid is used in a promotion for jars of peanut butter. The consumer removes the lid and scratches the box to reveal the prize. The table lists the prizes and the number of lids winning each prize.

Prize ($)	Number of Winners
10 000	3
5 000	10
1 000	25
100	50
10	1000

There are 200 000 jars of peanut butter in the promotion.
a) Calculate the probability of winning each prize.
b) What is the expected value of the prize money for each purchase of a jar of peanut butter?
c) A jar of peanut butter usually sells for $2.79. To cover some of the costs of the promotion, the company raises the price by one half of the expected value of the prize money paid per jar. What is the new price of a jar of peanut butter?
d) Do you think that this promotion is an effective way to increase sales of peanut butter? Explain.

13. Weather The probability that no rain will fall on a day in June in Vancouver is $\frac{2}{3}$.

a) If you visit Vancouver for a week in June, what is the probability of it raining at least once in the week? Round your answer to the nearest hundredth.
b) How might your answer to part a) help you pack for your trip?

14. Charity lottery A school holds a lottery to raise funds for a charity. The rules are as follows.
• The lottery runs for 3 weeks.
• Tickets cost $1 each, and 300 are sold.
• One winner is drawn each week and wins $50.
• The winning ticket is included in the next week's draw.
What is the expected value of one ticket?

15. Collision insurance Tara's collision insurance for her car costs $360 per year. The average claim paid by insurance companies for collision damage is $980. The probability that Tara will have an accident resulting in collision damage is 0.2 per year.
a) From the insurance company's point of view, what is the expected value of Tara's collision insurance?
b) From Tara's point of view, what is the expected value of her collision insurance?
c) Explain why your answers to parts a) and b) compare in the way they do.
d) What factors should Tara consider when deciding whether to purchase the collision insurance? Explain.

16. Give explanations for each of the following. Compare your explanation with a classmate's.
a) Why might it be reasonable for the owner of a $135 000 home to pay $600 a year to insure the house against fire, even though the probability of a house fire is only 0.003 per year?
b) Why might it not be reasonable for a bank to loan someone $50 000 at 8% interest, if the probability of the loan being repaid is 0.95?

LOGIC POWER

There are two red disks and two blue disks on the grid. Each disk can move in an L-shape. No two disks can occupy the same square at the same time. Therefore, the disk in the top left corner can move either one square to the right and two squares down or one square down and two squares to the right. Find the minimum number of moves it takes to switch the positions of the red disks with those of the blue disks.

CONNECTING MATH AND ENGLISH

Keyboarding

The diagram on the left shows part of the QWERTY keyboard, which was invented by Christopher Sholes in 1872. The diagram on the right shows part of the Dvorak keyboard, named for August Dvorak who invented it in 1932. Dvorak's design was intended to improve typing speed and decrease errors.

Home Row

The red keys are for the right hand and the blue keys for the left hand.

The table gives the average number of times that each key is used in 100 letters of written English.

A	8.2	H	5.3	O	8	V	0.9
B	1.4	I	6.5	P	2	W	1.5
C	2.8	J	0.1	Q	0.1	X	0.2
D	3.8	K	0.4	R	6.8	Y	2
E	13	L	3.4	S	6	Z	0.05
F	3	M	2.5	T	10.5		
G	2	N	7	U	2.5		

1 Comparing Keyboards

1. The average length of an English word is 4.5 letters. In typing a 1000-word passage, what is the expected number of times you would use
a) the E-key?   **b)** the Q-key?   **c)** the R-key?

2. The more you can keep your fingers on the home row, the faster you will type and the fewer errors you will make. What percent of the key strokes are on the home row of
a) the QWERTY keyboard?   **b)** the Dvorak keyboard?

3. Explain why the letters not used often should be located on the bottom row of a keyboard.

4. What percent of the 6 least-used letters appear on the bottom row of
a) the QWERTY keyboard?   **b)** the Dvorak keyboard?

2 Popular Words

As a fraction of the words written, one third of all writing in English is made up of the following 25 words.

a, and, at, but, for, had, has, have, he, I, in, is, it, me,
my, not, of, on, she, that, the, too, was, with, you

1. If you select a word at random from material written in English, what is the probability that the word is *not* one of the 25 words listed?

2. What is the total number of times you would expect to type any of the 25 words in a 1000-word passage?

3. a) What percent of the 25 words can be typed without leaving the home row of a QWERTY keyboard?
b) How many words is that in a 1000-word passage?

4. a) What percent of the 25 words can be typed without leaving the home row of a Dvorak keyboard?
b) How many words is that in a 1000-word passage?

5. Whenever the right hand and the left hand type alternate letters, typing speed is increased. One hand can get into position for the next letter, while the other hand is typing the previous one. Twenty-three of the listed words include 2 or more letters. For what percent of the 23 words do the right hand and the left hand type alternate letters on
a) the QWERTY keyboard? **b)** the Dvorak keyboard?

3 Keyboard Trivia

1. Christopher Sholes' first keyboard used piano keys, with the letters in a single row and in alphabetical order from left to right. Which part of the QWERTY keyboard shows that it originated as an alphabetical arrangement?

2. The very first QWERTY keyboard did not have the letter R in the top row. The manufacturer put the letter R in the top row, so that salespeople could show off the machine by typing the word *typewriter* very quickly. Why did moving the letter R help?

3. It has been shown that it takes three times as long to learn to type on a QWERTY keyboard as on a Dvorak keyboard. Also, when typists use a QWERTY keyboard, instead of a Dvorak keyboard, they type half as fast, move their fingers 20 times as far, and make twice as many errors. Why do you think the Dvorak keyboard has not replaced the QWERTY keyboard?

4 Letter Game

Suppose you play a game in which each letter of the alphabet is assigned a numerical value. Each turn of the game involves choosing a letter at random from a passage written in English.

1. What is the probability of choosing each letter of the alphabet?

2. If each vowel has a numerical value of 2 and each consonant has a numerical value of 1, what is the expected value for a turn of the game?

3. The winner of the game is the first person to reach a total score of 30. How many turns would you expect to take to reach this total?

4. If the rules were changed so that each vowel still had a value of 2 but each consonant had a value of −1,
a) would you expect to reach a total score of 30? Explain.
b) is it possible that you could reach a total score of 30? Explain.

CAREER CONNECTION

Insurance Underwriting

The modern insurance industry began in the coffee houses of London, England, in the 17th century. Merchant ship owners would post a description of a future voyage. People willing to insure the ship and its cargo would write their names under the description. Therefore, insuring a ship came to be called "underwriting." Today, the people who decide whether something can be insured are called insurance underwriters.

Underwriters work for insurance companies, which charge premiums for taking on risks. Buying insurance is a way to share the risks of everyday living. For example, if an insured house is destroyed by fire, the rebuilding cost is paid by the insurance company out of premiums paid by all the people who bought insurance. An underwriter checks that the risks an insurance company takes on are likely to result in a profit for the company.

Many underwriters have completed university or college business programs. Underwriters may be certified by the Insurance Institute of Canada. Their work is based on the use of statistics and probabilities.

1 Car Liability Insurance

The table shows information about liability insurance for passenger vehicles in a Canadian province for one year, for drivers 25 years of age or older.

Vehicle Use	Number of Vehicles Insured	Number of Claims	Average Cost per Claim ($)
Business	81 560	4 095	5223
Commuting > 15 km/way	217 053	9 691	4924
Commuting < 16 km/way	1 076 308	46 690	4758
Pleasure	1 437 495	51 043	4811

1. Use the number of vehicles insured and the number of claims to calculate the probability of a claim being made in a year for each use. Round each probability to the nearest ten thousandth.

2. Use the probabilities from question 1 and the average cost per claim to determine the expected value of a claim in a year for each use. Round each expected value to the nearest dollar.

3. Suppose the annual premium for $1 000 000 in liability coverage for business use is $420. Decide a fair annual premium for the same coverage for each of the other three vehicle uses. Explain your reasoning.

2 Locating Information

1. Use your research skills to locate information about another type of insurance, such as house, property, business, life, disability, travel, or health insurance. Find out how much insurance costs for different amounts of coverage, the deductible (if any), and the risks insured by the policy. Share your findings with your classmates.

2. Which types of insurance do you think are the most important to have? Explain.

Car Insurance Claims

Use the *Insurance* database, from the Computer Data Bank for ClarisWorks, to complete the following. All the records are for third party liability or collision coverage in Ontario or Alberta.

1 Premiums and Claim Cost

1. Find all the records for either coverage and either province, for example, collision coverage in Alberta.

2. Explain why the following expression gives the premiums per vehicle insured.

$$\frac{\text{premiums}}{\text{number of vehicles insured}}$$

3. Explain why the following expression gives the claim cost per vehicle insured or the expected value of a claim.

$$\frac{\text{claim frequency}}{100} \times \text{average cost per claim}$$

4. Add two calculation fields called *Premiums per Vehicle Insured, $* and *Claim Cost per Vehicle Insured, $*, rounding the amounts to the nearest dollar.

2 Loss Ratio

Use all the records you found in *Premiums and Claim Cost*.

1. The loss ratio per vehicle insured is the ratio of claim cost per vehicle insured to premiums per vehicle insured. Does a ratio of less than 1 or greater than 1 result in a loss for the insurance companies? Explain.

2. Add a calculation field called *Loss Ratio per Vehicle Insured*, rounding the ratios to 2 decimal places.

3. Sort the records from greatest to least loss ratio per vehicle insured. Which record has the greatest loss ratio?

3 Expected Value

Use all the records you found in *Premiums and Claim Cost*.

1. The expected value per vehicle insured is found from the subtraction (premiums per vehicle insured) – (claim cost per vehicle insured).

a) Is a high or low expected value per vehicle insured worse for the insurance companies? Explain.

b) What does a negative expected value indicate?

2. Add a calculation field called *Expected Value per Vehicle Insured, $*.

3. Sort the records from least to greatest expected value per vehicle insured. Which record has the least expected value?

4 Paying Premiums

1. Based on your results in *Loss Ratio* and *Expected Value*, which category of driver by years licensed, driver training, driver's gender, and driver's age should pay the highest premium?

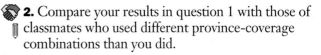 **2.** Compare your results in question 1 with those of classmates who used different province-coverage combinations than you did.

393

Review

8.1 **1.** Identify the population and suggest a sampling procedure that could be used to answer the following questions.
a) What is the favourite radio station of the students in your school?
b) How many households in your city or town own a dog?
c) What musical instrument is the most popular choice of amateur musicians?

2. School teams Design a simple random sampling procedure to select students in your school to study the following question.
Should all students be required to pay $10 a year to support school teams?

3. Part-time jobs Design a systematic sampling procedure to select high school students in your city or town to study the following question.
What is the best part-time job for high school students?

4. Running shoes Design a stratified sampling procedure to study the following question.
What percent of the people in your province own running shoes but do not run?

5. Extra-terrestrials Describe a non-probability sampling technique to study the following question.
Has the Earth had visitors from outer space?

8.2 *In questions 6–8, identify the type(s) of bias in each of the following situations.*
6. Swimming pool A neighbourhood group trying to get support for a new swimming pool puts petition forms in all variety stores in the neighbourhood.

7. Teen drivers A survey question is worded as follows.
Teenagers drive too fast. Should the cars they drive have speed regulators put on them?

8. Mail delivery A survey mailed to every hundredth person in the telephone book asked the following question.
Should there be mail delivery twice a day?

9. Radio listening times A radio station conducted a survey to determine the time of day most people listened to the radio. A simple random sample was taken from the telephone book. People were surveyed by telephone between 09:00 and 16:00 on weekdays in May.
a) Identify the population that is of interest in this survey.
b) Describe the bias that appears in the sample design.
c) Design another sampling method that would more accurately represent the population.

10. Study habits Suppose you studied the following question in your school.
What are the study habits of high school students?
a) Describe a stratified sampling method.
b) What population would you be targeting?
c) Design a questionnaire that you could use. Try to avoid any response bias.
d) What is non-response bias? How would you try to control it in this study?
e) If the study was conducted within a city or town, instead of being limited to one school, what would be a good stratification?

11. Advertising Suppose that an advertisement claimed that three out of five teenagers prefer a certain brand of cola.
a) What inferences would you be expected to draw from the advertisement?
b) Do you know that the claim is reliable? Explain.

8.3 *What is the expected value of a spin for each spinner?*

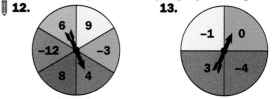

12. **13.**

14. If 10 number tiles, numbered 1 to 10, are placed in a bag and a tile is drawn at random, what is the expected value of a draw?

15. Coin toss game Suppose a game is played by tossing two coins. If the outcome is two heads, you win $10. If the outcome is a head and a tail, you win $5. If the outcome is two tails, you lose $k.
a) What is the value of k if the game is fair?
b) Devise a fair game that uses three coins.

16. Lottery A service club raises funds to support its programs by running a lottery. The grand prize is a $200 000-home. There are two second-place prizes of $40 000, three third-place prizes of $2000, four fourth-place prizes of $700, and five fifth-place prizes of $200. There are 4000 tickets, and they cost $100 each. Every prize must be awarded, even if all the tickets are not sold.
a) How much money will be raised if all the tickets are sold?
b) What is the expected value of each ticket?
c) If only 3000 tickets are sold, what is the expected value of a ticket? How much money will be raised in this case?
d) How many tickets must be sold in order for the service club to break even? What will be the expected value of each ticket in this case?

17. Insurance An insurance company analyzed the claims per policy for loss or damage caused by vandalism. Each policy is based on no deductible.

Annual Claim Amount ($)	Probability
0 (no claims)	0.96
200	0.02
400	0.01
600	0.005
800	0.003
1000	0.002

a) What is the expected amount of money to be paid out per policy in a year?
b) Suppose the deductible is changed to $100 per claim. What is the expected amount of money to be paid out per policy in a year?

18. Weather The probability of there being no snowfall on a day in January in Winnipeg is about 0.6. What is the minimum number of days you would need to spend in Winnipeg in January to have at least a 90% probability of there being a snowfall?

Exploring Math

Travelling Salesperson Problem

Some real-world problems cannot be solved by standard means on a computer, because of the time it takes to get a solution. One example, called the "travelling salesperson problem," is found in the trucking and airline industries. It entails finding the shortest route between a number of different cities. To solve this problem, the number of possible routes is found, then the shortest one is calculated.

For 2 cities, there are 2 routes.
A to B   B to A

For 3 cities, there are 6 routes.
A to B to C   A to C to B
B to A to C   B to C to A
C to A to B   C to B to A

1. a) How many possible routes are there for 4 cities?
b) Copy and complete the table.

Number of Cities	Number of Routes	Calculation
2	2	2×1
3	6	$3 \times 2 \times 1$
4		
5		

2. Another way to write $3 \times 2 \times 1$ is 3!, which is read "three factorial." Write the number of possible routes for 4 cities and 5 cities in factorial notation.

3. Suppose there are 10 cities.
a) Write the number of different routes in factorial notation.
b) Use your calculator to find the number of different routes.

4. Suppose a computer can solve a 10-city problem in 5 s. About how many years would it take to solve a 20-city problem? From your answer, you will see that computer programmers must find a different solution to a 20-city problem. Using the above method, some problems would take a computer 10 million years to solve. Finding more efficient solution methods is one of the challenges computer programmers face.

Chapter Check

1. Describe the difference(s) between probability sampling and non-probability sampling.

2. Describe the differences between simple random sampling, systematic sampling, and stratified sampling.

3. Favourite videos Describe how you could use each of the following sampling methods to find out the favourite type of video among students in your school.
a) simple random sampling
b) systematic sampling
c) stratified sampling

4. Politics a) Describe a non-probability sampling technique that might be used to study the following question.
What percent of Canadians would like to be a Member of Parliament?
b) If you drew inferences from the results in part a), how reliable would the inferences be? Explain.

5. Milk consumption a) If the following question were studied, what stratification of the sample might be appropriate?
How much milk do high school students in your city or town drink?
b) How could a random sample be chosen within each stratum?

Identify the type(s) of bias in each situation.
6. Advertising A survey question reads as follows.
Research shows that most people would rather watch television than listen to the radio or read a newspaper. Is television the best place to advertise?

7. Education A newspaper reporter asked people entering a bank on a Wednesday afternoon the following question.
Should elementary school students be allowed to use calculators in math class?

8. When you spin the spinner and land on a colour, you get the number of points shown for that colour. What is the expected value for a spin?

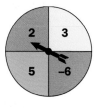

9. Dice game Nyla and Ken played a game in which a single die was rolled. If a 1 or a 2 was rolled, Nyla won 1 point. If a 3 was rolled, Nyla won 2 points. If a 4 was rolled, Ken won 2 points. If a 5 or a 6 was rolled, Ken won 1 point.
a) What was each player's expected value for one roll?
b) Was the game fair? Explain.
c) Predict the number of rolls needed for either player to win a total of 10 points.

10. Fund-raising A college is holding a lottery to raise money. There are 5000 tickets at $100 each. The first prize is $50 000, the second prize is $40 000, the third prize is $30 000, the fourth prize is $20 000, and the fifth prize is $10 000. All prizes must be awarded, no matter how many tickets are sold.
a) If all the tickets are sold, what is the expected value of each ticket?
b) How much money will the college raise if all the tickets are sold?
c) If 4000 tickets are sold, what is the expected value of a ticket?
d) How many tickets will have to be sold for the college to break even?

11. Insurance premium An insurance company analyzed its claims per policy on the theft of household contents. The results are shown in the table. Each policy is based on no deductible.

Annual Claim Amount ($)	Probability
0 (no claims)	0.96
1 000	0.03
5 000	0.006
10 000	0.003
20 000	0.001

a) What is the expected amount of money to be paid out per policy in a year?
b) What annual premium should the insurance company charge each policyholder, if it adds a 40% margin to the expectation?

Using the Strategies

1. Calculate the area of the figure, in square units, if each grid square represents 1 square unit.

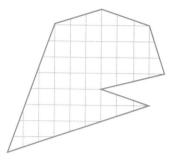

2. Find three different prime numbers whose sum is 40.

3. The figure is made up of squares. Square B has dimensions 4 cm by 4 cm. Square C has dimensions 5 cm by 5 cm. What is the total area of the figure, in square centimetres?

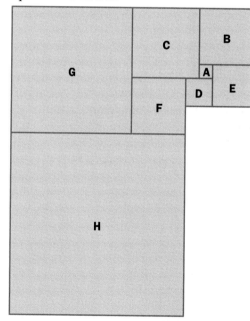

4. A two-digit whole number is "odd looking" if all the digits are odd. How many two-digit whole numbers are "odd looking"?

5. One way to write 96 as the difference of squares is $25^2 - 23^2$. Find the three other ways.

6. Three identical rectangles are placed as shown to form a large rectangle. The area of the large rectangle is 1536 m^2.

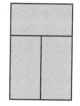

a) Find the dimensions of each of the small rectangles.
b) Find the area of the square that has the same perimeter as the large rectangle.

7. Each letter represents a different digit from 0 to 9 in this addition. Find the value of each letter.

BRUNCH
+LUNCH
DINNER

8. Here are two ways to put 4 counters on the grid shown, so that the counters lie on the vertices of a square.

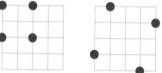

What is the largest number of counters that can be put on the grid, at the intersections of the grid lines, so that no 4 counters lie on the vertices of a square?

9. About how many kilometres will you travel on school property while you are in grade 10?

DATA BANK

1. Suppose you visited Churchill, Manitoba, on random days of the year and at random times of the day and night. What would be the probability that the sun was shining in Churchill when you visited? Round your answer to the nearest hundredth.

2. Use information from the Data Bank on pages 404 to 413 to write two problems. Have a classmate solve your problems.

CUMULATIVE REVIEW, CHAPTERS 5–8

Chapter 5

For each relation,
a) *write the domain and range*
b) *draw the relation as an arrow diagram*
c) *describe the relation in words*
1. (5, –2), (4, –2), (–1, –2), (0, –2)
2. (1, 4), (2, 8), (4, 16), (11, 44)
3. (12, 4), (6, 2), (3, 1), (–9, –3)

Graph each relation. The domain is R.
4. $y = x - 2$ **5.** $y = -3x + 1$
6. $4x + y = 3$ **7.** $x - 2y = 0$

Graph each relation using a graphing calculator, or complete the table of values for each relation and draw the graph on a grid. The domain is R.
8. $y = x^2 - 2$ **9.** $y = x^2 + 5$

x	y
0	
1	
–1	
2	
–2	
3	
–3	

x	y
0	
1	
–1	
2	
–2	
3	
–3	

State whether each set of ordered pairs represents a function.
10. (3, 5), (4, 6), (7, 9), (11, 13), (16, 18)
11. (9, 1), (9, 3), (9, 5), (9, 7), (9, 9)
12. (1, 7), (3, 7), (2, 7), (5, 8), (6, 7)

13. If $f(x) = 3x - 2$, find
a) $f(2)$ **b)** $f(9)$ **c)** $f(1)$
d) $f(0)$ **e)** $f(-3)$ **f)** $f(0.4)$
g) $f(-0.1)$ **h)** $f(10)$ **i)** $f(100)$

14. Graph the function g defined by $g(x) = 5x - 9$. The domain is R. Find the range of g.

15. If a varies directly as b, and $a = 9$ when $b = 2$, find a when $b = 10$.

16. Sales earnings Sathi sells sound systems in a store. She earns a weekly salary of $500, plus 4% commission on her sales.
a) Write the partial variation equation that relates her total weekly earnings to her sales.
b) Calculate her weekly earnings if her sales are $5500.
c) What are her sales if her weekly earnings are $975?

Chapter 6

Find the distance between each pair of points. Express each answer in simplest radical form.
1. M(3, 7) and A(1, 3)
2. E(–2, 9) and T(3, 1)
3. B(5, –1) and C(–3, 4)
4. D(–10, 0) and Y(0, –10)

Find the coordinates of the midpoint of each line segment, given the endpoints.
5. K(2, 5) and L(4, 3)
6. M(1, 13) and N(9, –3)
7. W(7, –2) and Z(–2, –3)
8. P(8.4, 2.5) and Q(–1.6, –1.1)

Find the slope of the line passing through the points.
9. G(1, 5) and H(5, 1)
10. R(–3, 8) and S(–8, 3)
11. F(0, –6) and A(1, –2)
12. J(2.3, 5.1) and T(–5.7, 3.5)

13. Maple products The value of Canadian maple products increased from $34.1 million in 1980 to $90.6 million in 1995.
a) Find the average rate of change to the nearest $100 000/year.
b) Predict the value of Canadian maple products in 1990.

Given a point on the line and the slope, sketch the graph of the line.
14. (4, 3); $m = 2$ **15.** (–5, –2); $m = -1$
16. (3, –3); m is undefined

17. Write an equation in standard form for the line that passes through the point (5, –3) and has a slope of –2.

18. Write an equation in standard form for the line that passes through (–2, 2) and (2, 0). What is the slope of the line? Use the equation to find two other points on the line.

19. State the slope of a line parallel to the line $y = 5x - 4$. What is the slope of a line perpendicular to the given line?

20. Write an equation of a line parallel to $3x - y = 5$ and passing through the point (1, –1).

21. Use the x- and y-intercepts to graph the line $2x + 3y = 12$.

Chapter 7

1. Calculate the volume of the sphere, to the nearest cubic centimetre.

6.4 cm

2. Dome-shaped lid The lid on a can of foam cleanser is in the shape of a dome or hemisphere. What is the surface area of the outside of the lid, to the nearest square centimetre?

5.2 cm

3. Two rectangular prisms are similar. Their side lengths are in the ratio 2.5:1. What is the ratio of
a) their surface areas?   **b)** their volumes?

4. Find AD, to the nearest tenth of a kilometre.

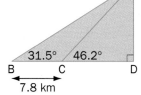
31.5°   46.2°
B   C   D
7.8 km
A

5. If cos A = – cos 25°, what is the measure of obtuse ∠A?

Solve each triangle. Round each answer to the nearest tenth, if necessary.
6. In △ABC, ∠A = 95.3°, ∠B = 44.4°, and b = 33.9 cm.
7. In △RST, ∠R = 64.5°, ∠T = 72.4°, and s = 24.4 cm.
8. In △DEF, f = 19.6 m, d = 15.3 m, and ∠E = 70.4°.
9. In △XYZ, x = 8.2 cm, y = 9.4 cm, and z = 7.5 cm.

10. In △JKL, ∠K = 52 °, ∠L = 72°, and j = 120 m. Find the area of the triangle, to the nearest square metre.

11. Air ambulance An ambulance helicopter flies 78 km from its base to an accident scene, then flies 134 km to a hospital. The angle between those legs of the trip is 124°. If the helicopter flies directly from the hospital back to its base, what is the total distance flown, to the nearest kilometre?

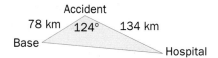
Accident
78 km   124°   134 km
Base
Hospital

Chapter 8

1. Identify four probability sampling techniques. Then, use an example to explain how any one of them can be used.

2. Health care A newspaper reporter asked four people on the street how they believed they would be affected by proposed changes to health care. The newspaper printed the responses, along with pictures of the people. Classify the sampling technique. Then, design a probability sampling technique that would be appropriate.

3. Local issue Identify an issue in your community. Write an unbiased survey question to ask about the issue.

4. Driving age Suppose a survey carried out using a simple random sample of students at your school asked the following question.
To make our highways safer, should the age for obtaining a driver's learning permit be 21?
a) Identify the types of bias in the survey.
b) How could you try to eliminate these types of bias from the survey?

5. Spinning green results in 4 points. Spinning yellow results in 6 points. Spinning blue or red results in 20 points.

a) Find the expected value of a spin.
b) How many spins would you expect to take to reach 100 points? Explain.

6. Charity lottery A charity sells 5000 tickets at $100 each. All expenses associated with the lottery are paid through donations. The grand prize is $10 000, and two second prizes are $5000 each.
a) How much money does the charity raise?
b) What is the expected value of a ticket?
c) How would you decide whether to buy one of these tickets?

7. Stock market An investor decides to keep a stock that is increasing in value as long as the probability of its value continuing to increase is at least 0.7. The probability of the value falling is 0.06 on any given day. On which day should the investor sell the stock?

Name the sets of numbers to which each number belongs.

1. $\sqrt{0.36}$ **2.** $-\sqrt{11}$ **3.** $\sqrt{144}$ **4.** $\sqrt{\dfrac{9}{25}}$

Evaluate.

5. $|-7|$ **6.** $|1-6|$ **7.** $|2|-|-3|$

Simplify.

8. $\sqrt{28}$ **9.** $\dfrac{\sqrt{40}}{\sqrt{8}}$ **10.** $\dfrac{4\sqrt{12}}{2\sqrt{2}}$

11. $2\sqrt{3}\times 3\sqrt{5}$ **12.** $4\sqrt{2}+7\sqrt{2}$

13. $9\sqrt{5}-4\sqrt{5}$ **14.** $\sqrt{20}-\sqrt{5}+\sqrt{40}$

15. Without using a calculator, arrange the following in order from least to greatest.
$$2\sqrt{32},\ 3\sqrt{16},\ 7\sqrt{3},\ 9\sqrt{2}$$

Expand and simplify.

16. $\sqrt{3}(2\sqrt{6}-4\sqrt{15})$

17. $(4\sqrt{3}+7\sqrt{2})(\sqrt{3}-\sqrt{2})$

18. $(\sqrt{3}+\sqrt{7})(\sqrt{3}-\sqrt{7})$

19. $(\sqrt{2}-2\sqrt{5})^2$

Simplify.

20. $\dfrac{10}{\sqrt{5}}$ **21.** $\dfrac{2}{\sqrt{6}-\sqrt{3}}$

Simplify.

22. $(2x^3y^{-1})^3$ **23.** $(-3x^3y^5)(4x^{-1}y^2)$

24. $(-9a^5b^2c)\div(-3abc)$ **25.** $\dfrac{48x^2y^5}{(2xy^{-3})(8xy)}$

Evaluate.

26. $81^{\frac{1}{2}}$ **27.** $\left(\dfrac{8}{27}\right)^{\frac{1}{3}}$ **28.** $64^{-\frac{1}{3}}$

29. $(-0.027)^{\frac{2}{3}}$ **30.** $16^{\frac{5}{4}}$ **31.** $\sqrt{\sqrt[3]{64}}$

32. Measurement The area of a rectangle is 20. One dimension is $\sqrt{5}+\sqrt{3}$. What is the other?

Find the amount of each investment.

33. $15\ 000 invested for 2 years at an annual interest rate of 4% compounded annually

34. $8000 invested for 3 years at an annual interest rate of 8% compounded quarterly

35. Vacation package Calculate the total price of a $2795 vacation package if the GST is 7%, the PST is 9%, and the PST is calculated
a) only on the price
b) on the price plus the GST

Given the general term, state the first 5 terms and the tenth term of each sequence.

36. $t_n = 3n + 2$ **37.** $t_n = 2n^2 - 1$

Write the first 5 terms determined by each recursion formula. Then, graph t_n versus n.

38. $t_1 = 7;\ t_n = t_{n-1} - 2$
39. $t_1 = -15;\ t_n = t_{n-1} + 4$

Find the indicated terms for each arithmetic sequence.

40. $-1, 4, 9, \ldots;\ t_5$ and t_{20}
41. $29, 18, 7, \ldots;\ t_{10}$ and t_{15}

Find the number of terms in each arithmetic sequence.

42. $13, 17, 21, \ldots, 277$
43. $59, 45, 31, \ldots, -95$
44. $-8, -5, -2, \ldots, 64$

45. Find three arithmetic means between 8 and 60.

46. Find the sum of the arithmetic series.
$$15 + 24 + 33 + \ldots + 96$$

Find the indicated sum for each arithmetic series.

47. S_{20} for $1 + 3 + 5 + \ldots$
48. S_{15} for $20 + 17 + 14 + \ldots$

49. The first term of an arithmetic series is -10. The eighth and last term of the series is 32. Find the sum of the series.

50. Pattern The first three diagrams in a pattern made using squares are shown. Find the total number of squares used to make the first 10 diagrams in the pattern.

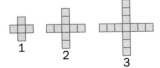

1   2   3

Simplify.
51. $(4x - 2y) - (5x + y)$
52. $(3a^2 + 5ab - 8b^2) + (2a^2 - ab + 9b^2)$
53. $(6xyz^3)(-7x^2yz)$
54. $\dfrac{36a^2b^3c}{-9abc}$

Expand and simplify.
55. $5(u+7)+9(u-2)$
56. $3t(4t^2-3t+7)-11t(2t^2+t-8)$
57. $(y+1)(y+3)+(y-1)(y+6)$
58. $(3a+2)(a^2+a-1)$
59. $(2x-1)^2$
60. $(6q+7)(6q-7)$
61. $3(g+1)^2$
62. $(5c-2)^3$

Factor, if possible.
63. $7d+49$
64. $t^2+7t+12$
65. x^2+x-6
66. k^2+k+1
67. n^2-9
68. m^2+9
69. $3x^2+6x+3$
70. $2y^2-2y+1$
71. $4m^2+4m+1$
72. $9t^2-1$

Factor fully.
73. $3x^2-12$
74. $8x^2-18y^2$
75. $2x^2-12xy+18y^2$
76. $4d^2-2d-6$
77. $12k^2+33k-9$

Divide. State any restrictions on the variables.
78. $\dfrac{9x^3-3x^2+6x}{3x}$
79. $\dfrac{12x^2y^4-8xy}{-2xy}$
80. $(x^2+12x+36)\div(x+6)$
81. $(4x^2-3x-1)\div(x-1)$
82. $(10r^2+r-2)\div(2r+1)$

Divide. Write each answer in the form $P=DQ+R$.
83. $(x^2+7x+13)\div(x+4)$
84. $(6y^2+y-3)\div(2y-1)$
85. $(12m^3+17m^2+3m)\div(3m+2)$
86. $(8n^3-10n^2-12n+12)\div(2n^2-3)$

Simplify. State any restrictions on the variables.
87. $\dfrac{m^2-3m+2}{m^2+2m-3}$
88. $\dfrac{3n^2+n}{9n^2+6n+1}$
89. $\dfrac{5b^2+3b-2}{5b-2}$
90. $\dfrac{4t^2-1}{6t^2-13t+5}$

Simplify. State any restrictions on the variables.
91. $\dfrac{8m^3n^2}{-2p^2q^2}\times\dfrac{-6p^3q^2}{12mn}$
92. $\dfrac{-7xy^3}{65x^2y^2}\div\dfrac{28x^4y}{-26y^4}$
93. $\dfrac{b^2-25}{b^2+12b+35}\times\dfrac{b^2+5b-14}{b+5}$
94. $\dfrac{2a^2-5a-12}{a^2+10a+25}\div\dfrac{3a^2-4a-32}{3a^2+23a+40}$

Simplify. State any restrictions on the variables.
95. $\dfrac{2a+1}{3}+\dfrac{5a}{2}$
96. $\dfrac{5-9x}{4}-\dfrac{x+3}{12}$
97. $\dfrac{t}{3t-6}-\dfrac{5}{2t-4}$
98. $\dfrac{3}{n^2+4n+4}+\dfrac{1}{n^2-4}$

Solve.
99. $7d-4=3+5d$
100. $3(r+5)=9(1-r)$
101. $\dfrac{c}{4}-\dfrac{5c}{8}=3$
102. $\dfrac{2}{3x-1}=\dfrac{5}{5x+3}$

103. National parks The combined area of Banff National Park and Kluane National Park is about 28 600 km². The area of Banff National Park is 1100 km² more than one quarter the area of Kluane National Park. Find the area of each park.

For each relation,
a) *write the domain and range*
b) *describe the relation in words*
104. $(4,-2),(3,-1),(1,1),(0,2)$
105. $(10,7),(8,5),(3,0),(1,-2)$
106. $(-2,-4),(2,4),(3,6),(6,12)$

The domain of each of the following relations is $\{-2,-1,0,1,2\}$. Complete a table of values. Then, graph each relation.
107. $y=x+2$
108. $y=-3x+4$
109. $4x-y=7$
110. $3x+2y=0$

Graph each of the following. The domain is R. For each graph, find the values of x when $y=5.2$ and $y=-3.7$. Round each value to the nearest tenth.
111. $y=3x+11$
112. $3y=-x-4$
113. $4x+2y=-1$
114. $2x-3y-6=0$

Graph each of the following. The domain is R. Find the range.

115. $y = x^2 - 3$ **116.** $y = |x| - 1$

State whether each set of ordered pairs represents a function.

117. $(-3, 5), (8, -2), (7, 0), (8, 13), (14, -8)$
118. $(-9, 1), (9, 1), (12, 1), (7, 1), (-13, 1)$
119. $(1, -7), (6, 1), (7, -4), (-2, 9), (-3, -2)$

120. If $f(x) = 3x + 1$, find
a) $f(2)$ **b)** $f(0)$ **c)** $f(-3)$
d) $f(0.4)$ **e)** $f(-0.5)$ **f)** $f(13)$

121. If c varies directly as d, and $c = 7$ when $d = 12$, find c when $d = 60$.

122. Find $f(1), f(2), f(3), f(4)$, and $f(5)$, for each of the following arithmetic sequences. State whether $f(n)$ varies directly or partially with n for each sequence.
a) $f(n) = 2n$ **b)** $f(n) = n + 3$ **c)** $f(n) = -5n$

123. Driving distances Medicine Hat is 100 km east of Lethbridge. Roy left his home in Medicine Hat and drove east at 90 km/h.
a) Write the partial variation equation that relates Roy's distance from Lethbridge to his driving time.
b) After driving for 2.5 h, how far was Roy from Lethbridge?
c) For how many hours had Roy been driving when he reached a point 424 km from Lethbridge?

Find the distance between each pair of points. Express each answer in radical form, if necessary.

124. $(2, 5)$ and $(-2, 2)$
125. $(-1, -3)$ and $(-2, -4)$
126. $(0, 3)$ and $(-6, 0)$
127. $(-1.6, 1.5)$ and $(2.4, -0.5)$

Find the coordinates of the midpoint of each line segment, given the endpoints.

128. $M(-2, 2), P(4, 8)$ **129.** $A(3, 3), J(-9, 3)$

130. The midpoint of AB is $M(-2, 1)$. If one endpoint is $A(-7, 3)$, what are the coordinates of B?

131. Measurement A rectangle has vertices $Q(5, 6), R(5, -3), S(1, -3)$, and $T(1, 6)$. Find
a) the perimeter   **b)** the area

Find the slope of the line passing through the points.

132. $E(7, -5), F(7, 1)$ **133.** $R(-6, 4), S(1, 5)$

134. The slope of a line is 2. The line passes through $(a, 5)$ and $(-3, -5)$. Find the value of a.

135. Population growth In 1991, Canada's population was 27.3 million. The population is projected to reach 42.3 million in 2021.
a) Find the average rate of change in the population.
b) Predict Canada's population in the year 2015.

Given a point on the line and the slope, sketch the graph of the line.

136. $(-5, 3); m = 0$
137. $(7, -2); m = -3$
138. $(0, -3); m$ is undefined

139. Write an equation in standard form for the line that has a slope of 3 and passes through the point $(4, -3)$.

140. Write an equation in standard form for the line that passes through $(-5, 0)$ and $(2, -4)$. Use the equation to find two other points on the line.

141. State the slope of a line parallel to the line $2y + x = -4$. What is the slope of a line perpendicular to the given line?

142. Write an equation of a line parallel to $2x - y = 3$ and passing through $(8, -5)$.

143. Use the x- and y-intercepts to graph the line $3x - 2y = 12$.

144. Tennis ball A tennis ball has a diameter of 6.5 cm. Calculate
a) the surface area, to the nearest square centimetre
b) the volume, to the nearest cubic centimetre

145. Silo A cylindrical silo with a hemispherical top has the dimensions shown. Calculate the volume of the silo, to the nearest cubic metre.

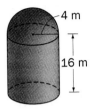

146. A square-based pyramid has a height of 16 cm and a volume of 1000 cm^3. A similar pyramid has a height of 24 cm. What is the volume of the second pyramid?

147. Royal Centre Tower From a point 35 m from the base of the Royal Centre Tower in Vancouver, and level with the base, the angle of elevation is 76°. What is the height of the tower, to the nearest metre?

148. Find BD, to the nearest metre.

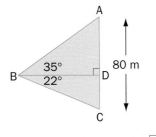

149. If $0° \leq A \leq 180°$ and $\sin A = \dfrac{\sqrt{3}}{2}$, what are the possible values of A?

Solve each triangle. Round each answer to the nearest tenth.

150. In $\triangle ABC$, $\angle B = 54.5°$, $\angle C = 76.3°$, and $a = 10.5$ m.

151. In $\triangle DEF$, $f = 20.8$ cm, $\angle E = 118.2°$, and $\angle F = 35.4°$.

152. In $\triangle FGH$, $h = 8.2$ m, $g = 10.1$ m, and $\angle F = 52.3°$.

153. In $\triangle RST$, $t = 4.5$ cm, $r = 3.9$ cm, and $s = 4.1$ cm.

154. Step angle We do not walk by placing one foot directly in front of the other. Our path is more of a zigzag.

The distance of a step from one foot to the next is a pace. The distance that one foot moves is a stride.

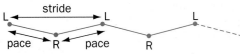

The angle formed between paces is the step angle. Find this step angle, to the nearest degree.

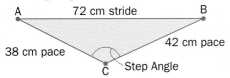

155. In $\triangle PQR$, $p = 100$ m, $\angle Q = 43°$, and $\angle R = 62°$. Find the area of the triangle, to the nearest square metre.

156. Athletes' salaries Design a systematic sampling procedure to study the following question among students in your school.
Are professional athletes paid too much?

157. Bridge toll A town council has proposed charging drivers each time they use a bridge. In this way, only those who use the bridge will pay for its upkeep. Residents who agree with the proposal have been asked to sign a petition by a certain date. Copies of the petition have been placed in all public libraries and postal stations in the town.
a) Identify the sources of bias in this method of collecting information.
b) Describe how you would try to eliminate the sources of the bias.

158. What is the expected value of a spin for this spinner?

159. Coin-die game A coin is tossed, and a die is rolled. If the outcome is a head and an even number, a player wins that even number of points. If the outcome is a tail and an odd number, a player loses that odd number of points. Other outcomes are ignored, and the player repeats the turn until points are won or lost. The first player to score at least 5 points wins.
a) Draw a tree diagram to identify all possible outcomes.
b) What's the expected value of a turn?
c) How many turns would a player be expected to take to win? Explain.

160. Lottery A lottery was held to raise money for improvements to a children's summer camp. There were 1000 tickets sold at $20 each. The first prize was $5000. There were 5 second prizes of $500 each, and 10 third prizes of $100 each. What was the expected value of a ticket?

Time Zones

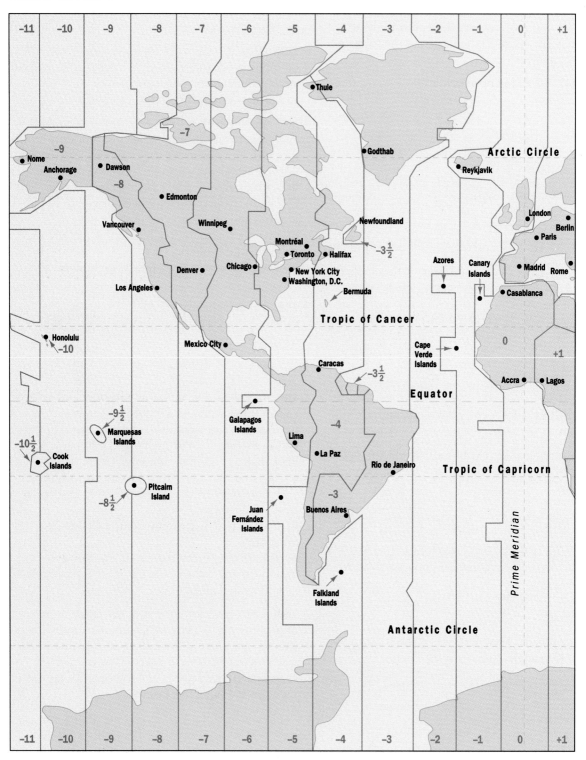

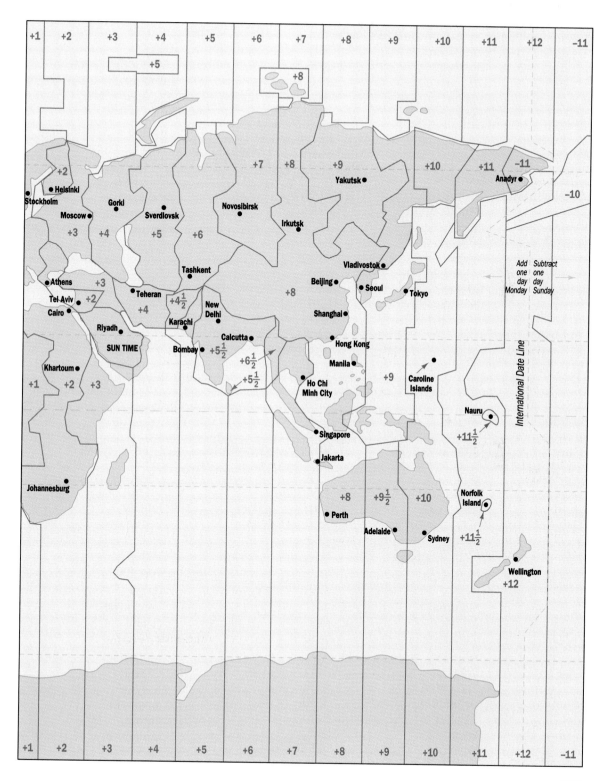

DATA BANK

Flying Distances

Within Canada		km
Calgary	Edmonton	248
	Montreal	3003
	Ottawa	2877
	Regina	661
	Saskatoon	520
	Toronto	2686
	Vancouver	685
	Victoria	725
	Winnipeg	1191
Charlottetown	Ottawa	976
	Toronto	1326
Edmonton	Calgary	248
	Ottawa	2848
	Regina	698
	Saskatoon	484
	Toronto	2687
	Vancouver	826
	Winnipeg	1187
Halifax	Montreal	803
	Ottawa	958
	Saint John	192
	St. John's	880
	Toronto	1287
Montreal	Calgary	3003
	Halifax	803
	Ottawa	151
	Saint John	614
	St. John's	1618
	Toronto	508
	Vancouver	3679
	Winnipeg	1816
Ottawa	Calgary	2877
	Charlottetown	976
	Edmonton	2848
	Halifax	958
	Montreal	151
	Toronto	363
	Vancouver	3550
	Winnipeg	1687
Regina	Calgary	661
	Edmonton	698
	Saskatoon	239
	Toronto	2026
	Vancouver	1330
	Winnipeg	533

Within Canada		km
Saint John	Halifax	192
	Montreal	614
	Toronto	1103
Saskatoon	Calgary	520
	Edmonton	484
	Regina	239
	Toronto	2207
	Vancouver	1203
	Winnipeg	707
St. John's	Halifax	880
	Montreal	1618
	Toronto	2122
Toronto	Calgary	2686
	Charlottetown	1326
	Edmonton	2687
	Halifax	1287
	Montreal	508
	Ottawa	363
	Regina	2026
	Saint John	1103
	Saskatoon	2207
	St. John's	2122
	Vancouver	3342
	Winnipeg	1502
Vancouver	Calgary	685
	Edmonton	826
	Montreal	3679
	Ottawa	3550
	Regina	1330
	Saskatoon	1203
	Toronto	3342
	Victoria	62
	Winnipeg	1862
Victoria	Calgary	725
	Vancouver	62
Winnipeg	Calgary	1191
	Edmonton	1187
	Montreal	1816
	Ottawa	1687
	Regina	533
	Saskatoon	707
	Toronto	1502
	Vancouver	1862

From Canada to USA		km
Calgary	Chicago	2222
	Los Angeles	1942
	New York	3245
	San Francisco	1637
Halifax	Boston	659
	New York	961
	Orlando	2409
	Tampa	2522
Montreal	Miami	2262
	New York	522
	Orlando	2006
	Tampa	2095
Ottawa	Orlando	1941

From Canada to USA		km
Toronto	Boston	717
	Chicago	700
	Honolulu	7465
	Los Angeles	3492
	Miami	1988
	New York	574
	Orlando	1702
	San Francisco	3625
	Tampa	1706
Vancouver	Honolulu	4350
Winnipeg	Chicago	1137
	Honolulu	6124
	Orlando	2740
	Tampa	2739

From Canada to Europe		km
Calgary	Glasgow	6473
	London	7012
Edmonton	London	6782
Halifax	Glasgow	4218
	London	4597
Montreal	Athens	7622
	Lisbon	5248
	London	5217
	Paris	5526

From Canada to Europe		km
St. John's	London	3741
Toronto	Dublin	5262
	Frankfurt	6340
	Glasgow	5280
	Lisbon	5737
	London	5735
	Paris	6015
	Rome	7110
	Zurich	6488

From Canada to Asia/Australia		km
Toronto	Hong Kong	13 048
	Sydney	16 753
	Tokyo	10 116
	Wellington	15 052

From Canada to Asia/Australia		km
Vancouver	Hong Kong	10 253
	Singapore	12 810
	Sydney	13 628
	Tokyo	7533
	Wellington	11 927

From Canada to Mexico/Caribbean		km
Calgary	Puerto Vallarta	3467
Montreal	Acapulco	4003
	Barbados	3875
	Cancun	2935
	Cuba	2590
	Guadeloupe	3469
	Haiti	3016
	Martinique	3665
	Montego Bay	3042
	Nassau	2313
	Puerto Plata	2899
	Puerto Vallarta	3979
	St. Lucia	3755

From Canada to Mexico/Caribbean		km
Toronto	Acapulco	3541
	Antigua	3386
	Barbados	3906
	Cancun	2605
	Cuba	2289
	Guadeloupe	3492
	Martinique	3685
	Montego Bay	2801
	Nassau	2080
	Puerto Plata	2790
	Puerto Vallarta	3484
Vancouver	Cancun	4472
	Puerto Vallarta	3543
Winnipeg	Montego Bay	3891

Driving Distances Between Cities

(All distances
are in kilometres.)

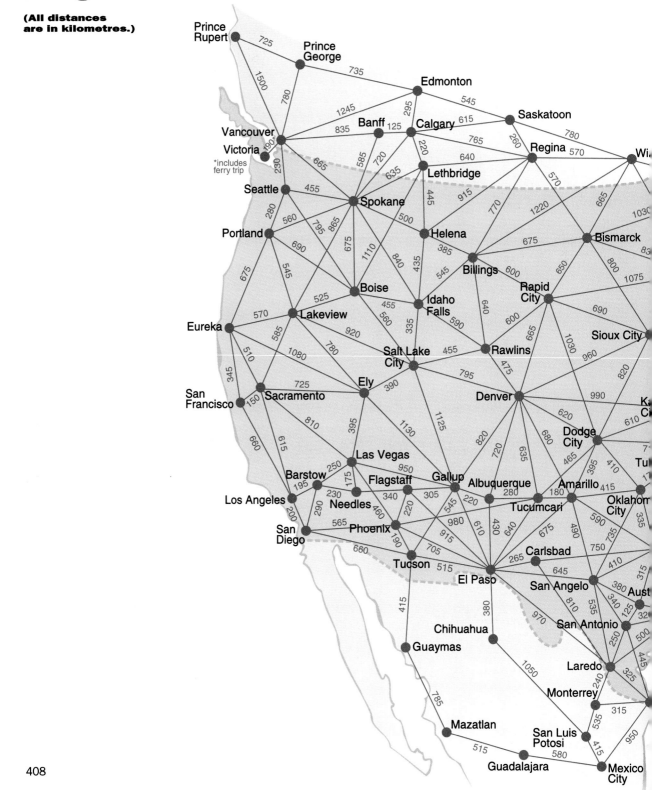

Prince Rupert
725
Prince George
735
Edmonton
545
Saskatoon
1500
780
1245
835
Banff 125
Calgary 615
765
780
Regina 570
Wi...
Vancouver
295
220
260
Victoria
*includes ferry trip
190
230
665
585
720
640
570
Lethbridge
635
Seattle 455
Spokane
445
915
770
1220
665
1030
280
560
795
865
675
500
445
Helena
675
Bismarck
83...
Portland
690
1110
840
435
385
545
Billings 600
650
800
1075
675
545
Boise
455
Idaho Falls
590
640
Rapid City
600
695
690
525
335
Eureka
570
Lakeview
920
560
Salt Lake City
455
Rawlins
600
1030
960
Sioux City
585
780
San Francisco
510
1080
795
475
820
345
150
Sacramento
725
Ely
390
Denver
990
K... C...
810
395
1125
620
610
660
615
1130
820
680
Dodge City
465
395
410
Tu...
Las Vegas
950
Gallup
635
Amarillo 415
1...
Barstow
250
175
Flagstaff
305
Albuquerque
180
Oklahom... City
335
195
230
340
220
280
Tucumcari
590
Los Angeles
290
Needles
460
545
220
430
675
490
750
410
315
200
565
Phoenix
980
915
610
640
Carlsbad
265
380
Aust...
San Diego
190
705
San Angelo
340
125
32...
660
Tucson 515
El Paso
645
410
250
500
415
San Antonio
445
380
970
810
535
Chihuahua
Laredo
325
Guaymas
240
Monterrey
1050
315
785
535
415
950
Mazatlan
San Luis Potosi
515
580
Guadalajara
Mexico City

St. John's

690* *includes ferry trip

Sydney

Rivière du Loup

205

515

445

Quebec

235

385

485

Halifax

280

260

Montreal

200

635

535

365

610

Saint John

nder

710

Sault Ste. Marie

425

North Bay

Ottawa

350

365

305

415

400

Bangor

875

90

655

680

635

Mackinaw City

470

415

Toronto

170

400

250

Syracuse

440

210

690

250

265

Albany

Boston

polis

on

230

Detroit

450

285

Buffalo

305

360

675

645

435

355

160

Boston

Chicago

580

615

405

Cleveland

250

Pittsburgh

370

515

235

New York

Indianapolis

325

300

Cincinnati

480

415

Philadelphia

ld

340

175

130

600

405

Washington, D.C.

155

675

Lexington

285

Charleston

is

330

345

630

555

785

565

405

Raleigh

Nashville

525

470

460

Asheville

475

Memphis

340

395

330

630

550

220

605

510

280

410

Montgomery

585

Savannah

ttle ock

345

390

440

560

225

ckson

735

510

345

480

Jacksonville

290

710

Tallahassee

260

New Orleans

440

Tampa

325

580

280

Miami

DATA BANK

The Solar System

Planet/Star	Diameter (km)	Relative Mass (Earth = 1)	Period of Orbit About the Sun	Period of Rotation on Axis (days)
Sun	1 392 000	332 830	—	25.38
Mercury	4 878	0.06	88.0 days	58.60
Venus	12 104	0.8	224.7 days	243.00
Earth	12 756	1.0	365.3 days	0.99
Mars	6 787	0.1	1.88 years	1.02
Jupiter	142 800	317.8	11.86 years	0.41
Saturn	120 000	95.2	29.63 years	0.42
Uranus	51 200	14.5	83.97 years	0.45
Neptune	48 680	17.2	164.80 years	0.67
Pluto	2 300	0.002	248.63 years	6.38

Wind Chill Temperatures

Wind chill temperatures are represented by the curved lines on the graph. At wind speeds below 8 km/h, wind chill temperatures cannot be found accurately. Therefore, the curved lines are dashed for wind speeds below this value.

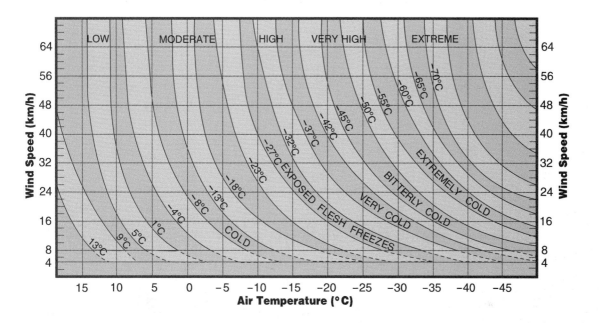

Suppose you want to find the wind chill temperature for a wind speed of 25 km/h and an air temperature of –10°C. First, find the horizontal line that represents a wind speed of 25 km/h and the vertical line that represents a temperature of –10°C. Then, find the point of intersection of these two lines. The point of intersection lies between the curves that represent wind chill temperatures of –23°C and –18°C. Estimation gives a wind chill temperature of about –22°C.

Humidex Readings

On hot days, a humidex reading gives an estimate of how hot the weather feels to us. In hot weather, our bodies are cooled by evaporation from our skin. On humid days, the air contains a lot of moisture, and evaporation does not cool us effectively. Therefore, we feel that the temperature is higher than the actual air temperature.

The table shows that, on a day with an air temperature of 24°C and a relative humidity of 80%, the humidex reading is 32°C. The weather feels as hot as it would feel at 32°C with no humidity.

Relative Humidity (%)

Air Temperature (°C)	100	95	90	85	80	75	70	65	60	55	50	45	40	35	30	25	20
43													56	54	51	49	47
42												56	54	52	50	48	46
41											56	54	52	50	48	46	44
40										57	54	52	51	49	47	44	43
39									56	54	53	51	49	47	45	43	41
38							57	56	54	52	51	49	47	46	43	42	40
37					58	57	55	53	51	50	49	47	45	43	42	40	
36			58	57	56	54	53	51	50	48	47	45	43	42	40	38	
35		58	57	56	54	52	51	49	48	47	45	43	42	41	38	37	
34	58	57	55	53	52	51	49	48	47	45	43	42	41	39	37	36	
33	55	54	52	51	50	48	47	46	44	43	42	40	38	37	36	34	
32	52	51	50	49	47	46	45	43	42	41	39	38	37	36	34	33	
31	50	49	48	46	45	44	43	41	40	39	38	36	35	34	33	31	
30	48	47	46	44	43	42	41	40	38	37	36	35	34	33	31	31	
29	46	45	44	43	42	41	39	38	37	36	34	33	32	31	30		
28	43	42	41	41	39	38	37	36	35	34	33	32	31	29	28		
27	41	40	39	38	37	36	35	34	33	32	31	30	29	28	28		
26	39	38	37	36	35	34	33	32	31	31	29	28	28	27			
25	37	36	35	34	33	33	32	31	30	29	28	27	27	26			
24	35	34	33	33	32	31	30	29	28	28	27	26	26	25			
23	33	32	32	31	30	29	28	27	27	26	25	24	23				
22	31	29	29	28	28	27	26	26	24	24	23	23					
21	29	29	28	27	27	26	26	24	24	23	23	22					

Average Annual Hours of Sunshine

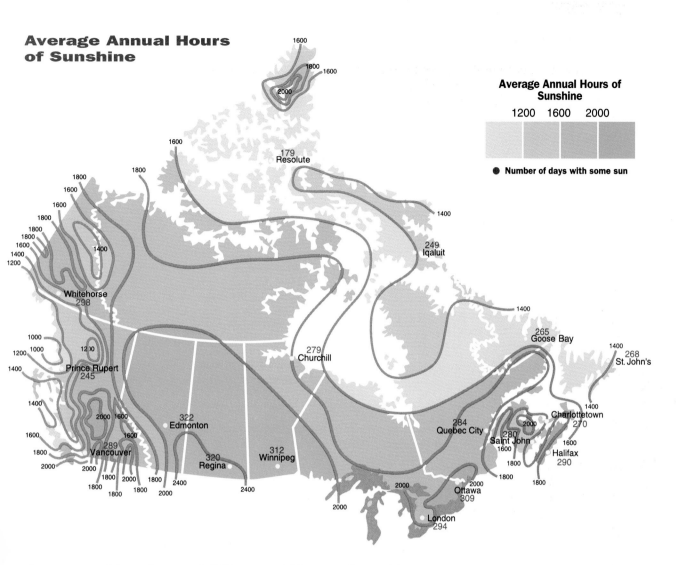

1600

1800 1600

2000

1600

179
Resolute

1600

1400

1800

1600

1600

249
Iqaluit

1800
1800
1600
1400
1200

1400

1400
1400
268
St. John's

1400

265
Goose Bay

1400

Whitehorse
298

1000
1200 1000

1200

279
Churchill

1400

Charlottetown
270

2000

Prince Rupert
245

1400

2000 1600

284
Quebec City

280
Saint John
1600

1600
1800

Edmonton
322

1600

1600

Halifax
290

2000

1800

Vancouver
289

1800 1800

2000

1800

312
Winnipeg

1800

320
Regina

2000 1800 2400

1800 1800 2000

2400

2000

2000

1800

Ottawa
309

London
294

Average Number of Blizzard Hours in a Year

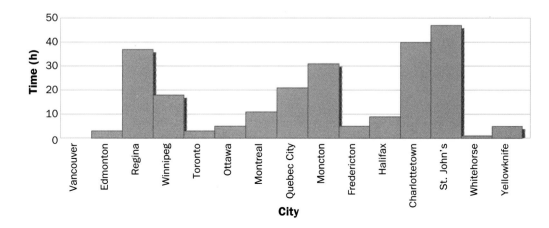

ANSWERS

Math Standards

Mathematics as Problem Solving p. xvi
1 Solving Problems **1.** 3 **2.** 11 **3.** 3 **4.** RS = 4, ST = 3
2 Comparing Strategies **1. a)** 16 **b)** 48 **c)** 8

Mathematics as Communication p. xvii
1 Writing Instructions **1.** Answers may vary.
2. Answers may vary. **3.** Answers may vary.
4. Answers may vary.
2 Describing Real Objects **1.** Answers may vary.
2. Answers may vary.

Mathematics as Reasoning p. xviii
2 Arrangement of Sticks **1.** 2, 3, 6, 7, 8, 9, 11, 12
2. 7 **3.** 1: other end; 4: numbered end; 5: numbered
end; 10: other end **4.** no **5.** yes **6. a)** 5 **b)** 6, 8, 9, 12
7. a) 2 **b)** 1, 4, 5, 10 **8.** Answers may vary. 1, 10, 4, 5,
7, 11, 9, 12, 3, 6, 8, 2

Mathematical Connections p. xix
1 Transforming Music **1. a)** translation
c) transposition **2. a)** reflection **c)** inversion
2 Water Sprinklers **2.** 294 km

Algebra p. xx
1 Solving Equations **1.** -7 **2.** 5 **3.** -6 **4.** 4 **5.** -18
6. $\dfrac{1}{3}$ **7.** 6 **8.** -5 **9.** 4 h **10.** 24 days
2 Using Patterns **1. a)** 40 **b)** 54 **2. a)** $n(n + 3)$ **b)** 2650
3. a) 15 **b)** 21 **4. a)** $\dfrac{n(n+1)}{2}$ **b)** 1275
3 Simplifying Expressions **1.** $6t^2 - t$ **2.** $y^2 + 2y$ **3.** $-20x^2$
4. $-12x^4y^3$ **5.** $-4st^2$ **6.** $-4yz^2$

Functions p. xxi
1 T-Patterns **1.** Number of Squares: 8, 14, 20, 26, 32;
Perimeter: 18, 30, 42, 54, 66 **2. a)** $6n + 2$ **b)** 452; 620
c) 28 **3. a)** $12n + 6$ **b)** 822; 1254 **c)** 22 **4. a)** $2s + 2$
b) 186; 630 **c)** 152; 428
2 E-Patterns **1.** Number of Squares: 8, 13, 18, 23, 28;
Perimeter: 18, 28, 38, 48, 58 **2.** $5n + 3$ **3.** $10n + 8$
4. a) $2s + 2$ **b)** the same

Geometry from a Synthetic Perspective p. xxii
1 Geometry in Art **1.** Square 1: 196 square units;
Square 4: 64 square units; Square 5: 1 square unit;
Square 6: 49 square units; Square 7: 16 square units;
Square 8: 324 square units; Square 9: 225 square
units **2.** Squares 5 and 8 **3.** Squares 4 and 7; Squares
1 and 6; Squares 8 and 3 **4.** Squares 3, 4, 5, 7 **5.** No;
its dimensions are 33 by 32.
2 Fractal Geometry **1.** 1, 8, 64, 512 **2.** Each entry is
8 times the previous entry. **3.** 8^n **4.** 32 768, 262 144

Geometry from an Algebraic Perspective p. xxiii
1 Describing Flight **1.** reflects in a vertical axis; while
translating, rotates 180° nose to tail; while translating,
rotates 180° side to side; while translating, rotates
360° top to bottom to top and 180° side to side; while
translating, rotates 180° side to side and then another
180° side to side; and then translates
3 Transforming Linear Relations **2. b)** $x + y = 6$
3. b) $x - y = 2$ **4. b)** $y - x = 2$

Trigonometry p. xxiv
1 Finding Sides and Angles **1. a)** 4.5 m **b)** 6.3 m
c) 16.4 m **2. a)** 51.3° **b)** 30° **c)** 36.9°
2 Problem Solving **1.** 128 m **2.** 190 m

Statistics p. xxv
1 Goals Against Average **1.** 1.948 **2.** 0.921
2 Comparing Averages **1. a)** Sharon 2.857; Danica
3.000 **b)** Sharon **2. a)** Sharon 2.609; Danica 2.721
b) Sharon **3. a)** Sharon 2.833; Danica 2.780
b) Danica **4.** Answers may vary.

Probability p. xxvi
1 Rolling Dice **1.** 0, 1, 2, 3, 4, 5; 1, 0, 1, 2, 3, 4; 2, 1,
0, 1, 2, 3; 3, 2, 1, 0, 1, 2; 4, 3, 2, 1, 0, 1; 5, 4, 3, 2, 1,
0
2. a) $\dfrac{1}{6}$ **b)** $\dfrac{5}{18}$ **c)** $\dfrac{2}{9}$ **d)** $\dfrac{1}{6}$ **e)** $\dfrac{1}{9}$ **f)** $\dfrac{1}{18}$ **g)** 0 **h)** $\dfrac{7}{18}$ **i)** $\dfrac{1}{6}$
j) $\dfrac{2}{3}$ **k)** 1
2 Selecting Cubes **1.** $\dfrac{3}{4}$ **2. a)** $\dfrac{3}{4}$ **b)** $\dfrac{2}{3}$

Discrete Mathematics p. xxvii
1 Finding Possibilities **1.** 6 **2.** 11 **3. a)** 8
2 Seating Arrangements **1.** 6 **2. a)** Person A has
person B on the left and person C on the right. **b)** 6
3 Making Words **1. a)** 3 **b)** 9 **c)** 27 **d)** 81 **2.** 120

Investigating Limits p. xxviii
1 Volume of a Pyramid **1.** 8 cm³; 72 cm³; 200 cm³;
392 cm³; 648 cm³; 968 cm³ **2.** 2288 cm³ **3.** 1 cm³;
9 cm³; 25 cm³; 49 cm³; 81 cm³; 121 cm³; 169 cm³;
225 cm³; 289 cm³; 361 cm³; 441 cm³; 529 cm³
4. 2300 cm³ **5.** 2304 cm³ **6.** The estimates get better
as the rectangular prisms get thinner. **7.** yes
2 Other Limits **2.** Divide the cone into a stack of
cylinders of equal height but decreasing radius.

Mathematical Structure p. xxix
1 Geoboard Properties **1.** A, B, C, D, E, F; B, A, D,
C, F, E; C, E, A, F, B, D; D, F, B, E, A, C; E, C, F,
A, D, B; F, D, E, B, C, A **2.** no **3.** A **4.** yes

Chapter 1

Real Numbers p. 1
1. 5 000 000 000 000 000 000 000
2. 160 000 000 000 000 years **3.** 3000
4. 1 700 000 000 000 000 000 **5.** 300

Getting Started pp. 2–3
Using the Pythagorean Theorem
1. a) $a = \sqrt{2}$; $b = \sqrt{3}$; $c = 2$; $d = \sqrt{5}$; $e = \sqrt{6}$; $f = \sqrt{7}$;
$g = \sqrt{8}$ **b)** $k = \sqrt{12}$ **2. a)** 8.6 cm **b)** 12.7 m **c)** 18.3 m
d) 5.1 cm

Warm Up
1. $3a + 12$ **2.** $x^2 - 3x$ **3.** $-5y + 10$ **4.** $-3t - 18$
5. $2x^2 + 10x$ **6.** $8a - 4a^2$ **7.** $-2m^2 - 12m$ **8.** $8y^2 + 4y$
9. $-15t^2 + 40t$ **10.** $-a^2 + a$ **11.** $6x^2 + 30x + 6$
12. $6x^3 - 2x^2 - 4x$ **13.** $6x + 26$ **14.** $2m - 22$
15. $-2t^2 - 37t$ **16.** $4a^2 - 5a - 5$ **17.** $-x^2 - 5x$
18. $-2y^2 + 13y$ **19.** $2t^2 + 30t$ **20.** $x^2 + 7x + 12$
21. $a^2 + 3a - 10$ **22.** $m^2 - 9m + 18$ **23.** $y^2 - 6y - 7$
24. $b^2 - 8b + 16$ **25.** $c^2 - 36$ **26.** $x^2 + 10x + 25$
27. $y^2 - 6y + 9$ **28.** $2x^2 + 5x + 2$ **29.** $6a^2 + 19a + 10$
30. $4m^2 - 15m - 4$ **31.** $8y^2 - 22y - 21$
32. $8b^2 - 30b + 27$ **33.** $9x^2 - 25$ **34.** $4x^2 + 4x + 1$
35. $4x^2 - 4x + 1$ **36.** 4.5×10^4 **37.** 2.3×10^{-4}
38. 8×10^6 **39.** 6×10^{-7} **40.** 570 000 000
41. 1 250 000 000 000 **42.** 0.000 000 38
43. 0.000 000 000 0408 **44.** 5.6×10^8 **45.** 7.81×10^{-4}
46. 1.23×10^{10} **47.** 3.4×10^{-7} **48.** 8.2×10^7
49. 4.3×10^{-8} **50.** 7.06×10^9 **51.** 9.84×10^{-10}
52. 7×10^7 **53.** 1.6×10^{10} **54.** 1×10^7 **55.** 5×10^7
56. 3×10^{-7} **57.** 7×10^{-9} **58.** 9×10^5 **59.** 9×10^{-3}
60. 6×10^{-6} **61.** 6×10^4

Mental Math
Evaluating Powers and Square Roots 1. 16 **2.** 64 **3.** 16
4. -25 **5.** 36 **6.** 2 **7.** 64 **8.** 17 **9.** 5 **10.** 1 **11.** 1 **12.** 13
13. 21 **14.** 12 **15.** 20 **16.** 44 **17.** 3 **18.** 5 **19.** 3 **20.** 4
Multiplying by Multiples of 5 1. 130 **2.** 270 **3.** 365
4. 510 **5.** 4490 **6.** 2175 **7.** 26 **8.** 64 **9.** 182 **10.** 2800
11. 3050 **12.** 6700 **13.** 12 800 **14.** 155 **15.** 510
16. Multiplying by 0.5 is the same as dividing by 2.
17. 44 **18.** 28.5 **19.** 66 **20.** 225 **21.** 4.8 **22.** 7.2
23. Multiply by $\frac{10}{2} = 5$.

Technology p. 4
1 One Digit Repeating 1. a) $0.\overline{7}$ **b)** $0.0\overline{7}$ **c)** $0.00\overline{7}$
2. a) $\frac{5}{9}$ **b)** $\frac{2}{3}$ **c)** $\frac{4}{45}$ **d)** $\frac{1}{450}$ **e)** $\frac{1}{900}$ **f)** $1\frac{1}{3}$ **g)** $2\frac{2}{45}$
h) $7\frac{1}{450}$ **3. a)** $0.\overline{1} = \frac{1}{9}, 0.\overline{2} = \frac{2}{9}, \ldots$, therefore $0.\overline{9} = \frac{9}{9}$
b) $3 \times \frac{3}{9} = \frac{9}{9} = 1$, yes

2 Two Digits Repeating 1. a) $0.\overline{47}$ **b)** $0.0\overline{47}$ **c)** $0.00\overline{47}$
2. a) $\frac{32}{99}$ **b)** $\frac{7}{11}$ **c)** $\frac{1}{22}$ **d)** $\frac{14}{2475}$ **e)** $3\frac{5}{11}$ **f)** $6\frac{2}{3}$
g) $11\frac{7}{198}$ **h)** $22\frac{13}{1650}$

3 Non-Repeating Parts 1. a) $\frac{31}{90}$ **b)** $\frac{17}{30}$ **c)** $\frac{61}{495}$ **d)** $\frac{211}{900}$
e) $4\frac{23}{90}$

Section 1.1 pp. 8–9
1. rational: repeating decimal **2.** irrational: non-repeating, non-terminating decimal **3.** irrational: non-repeating, non-terminating decimal **4.** rational: fraction **5.** rational: terminating decimal **6.** rational: repeating decimal **7.** irrational: 35 not a perfect square **8.** rational: 225 a perfect square **9.** irrational: 0.29 not a perfect square **10.** rational: 25 and 49 are perfect squares **11.** irrational: 0.1 not a perfect square **12.** rational: repeating decimal **13.** rational: 16 and 4 are perfect squares **14.** rational: fraction **15.** irrational: non-repeating, non-terminating decimal **16.** natural, whole, integer, rational, real **17.** rational, real **18.** irrational, real **19.** rational, real **31.** 6 **32.** 2 **33.** 4
34. $\frac{2}{3}$ **35.** $\frac{1}{4}$ **36.** 0.3 **37.** 4 **38.** 3 **39.** -1 **45.** $\frac{1}{9}$ **46.** $\frac{2}{9}$
47. $\frac{23}{99}$ **48.** $\overline{Q}$: 1.101 001..., $-\sqrt{10}$; Q: $\sqrt{36}$, $-\sqrt{100}$
$-\frac{3}{4}$, 0.06, $4.\overline{6}$, $\sqrt[3]{8}$, $\sqrt[4]{1}$; I: $\sqrt{36}$, $-\sqrt{100}$
$\sqrt[3]{8}$, $\sqrt[4]{1}$; W: $\sqrt{36}$, $\sqrt[3]{8}$, $\sqrt[4]{1}$; N: $\sqrt{36}$, $\sqrt[3]{8}$, $\sqrt[4]{1}$
49. $x > -3$ **50.** $x \leq 2$ **51.** $-1 \leq x \leq 5$ **52.** $-3 < x < 0$
53. $-5 \leq x < -1$
Applications and Problem Solving 54. 2.35, $2.3\overline{5}$, $2.3\overline{5}$
56. Answers may vary. **57.** 0, since the distance from 0 to 0 on a number line is 0 **58.** $\frac{223}{71}, \frac{355}{113}, \frac{22}{7}$
59. Venus, Sirius, Canopus, Vega, Altair, Antares, Spica, Polaris **60.** 1 000 000 km (using 5 digits per cm)
61. 4 **62. a)** sometimes true **b)** always true **c)** never true
d) never true **e)** sometimes true **f)** never true
63. a) No. The calculator is rounding the number
b) 3.000 000 001; The square of the rounded square root of 3 will only approximate 3. **64. a)** $\frac{D}{A} + \frac{C}{B}$
b) $\frac{B}{D} + \frac{A}{C}$ **65. a)** whole numbers **b)** natural numbers
c) whole numbers **d)** positive real numbers and zero
66. a) sometimes true **b)** always true **c)** never true
d) sometimes true **e)** always true **f)** sometimes true

Section 1.2 p. 11
Applications and Problem Solving 3. 16 to 18 h
5. 5959 m **6.** Mary: 12; Sasha: 17; Paula: 19; Amandi: 21; Heather: 24 **7.** 118, 119 **8. A:** Clockwise from top: 1, 4, 5, 2, 3, 6 **B:** Clockwise from top left: 4, 5, 2, 3, 6, 1 **C:** Clockwise from top: 5, 3, 4, 2, 6, 1 **9.** 9
10. $x = 6, y = 3, z = 1$; $x = 6, y = 1, z = 3$ **11.** 2, 6, 10, 14, 18 **12.** 2, 9, 4, 3; 5, 7, 6; 10, 8; 1 **13.** 37, 82
14. 33, 41, 26 **15. a)** $\dfrac{1}{2} + \dfrac{1}{3} + \dfrac{1}{6}$ **b)** No; since $\dfrac{1}{2}$ is the largest unit fraction and two of these are required to sum to 1.

Section 1.3 pp. 14–16
Practice 1. less than; greater than; less than; greater than **2.** 7; 6.63 **3.** 9; 9.33 **4.** 30; 28.14 **5.** 80; 81.73 **6.** 100; 154.31 **7.** 9; 9.09 **8.** 2; 2.25 **9.** 0.5; 0.47 **10.** 0.2; 0.21 **11.** 0.06; 0.06 **12.** 0.02; 0.02 **13.** 1; 1.42 **14.** 13; 13.9 **15.** 2; 2.0 **16.** 12; 13.1 **17.** 126; 123.7 **18.** 13; 13.4 **19.** 11; 13.0 **20.** 42; 42.6 **21.** 120; 126.5 **22.** 2.5; 2.0 **23.** 2.4; 2.4 **24.** 1.6; 1.5 **25.** 7; 6.1 **26.** 1.8; 1.8 **27.** 4; 5.6 **28.** 3.4641; 3.4641; equal **29. a)** 5 **b)** 7 **c)** 4 **d)** 2 **e)** 3 **f)** 1 **g)** 15 **h)** 15 **i)** 20 **j)** 20 **30. a)** > **b)** = **c)** < **d)** =
31. $\sqrt{17}, 3\sqrt{2}, 2\sqrt{5}$ **32.** $2\sqrt{15}, \sqrt{62}, 3\sqrt{7}, 8$
33. $6\sqrt{2}, \sqrt{74}, 5\sqrt{3}, 4\sqrt{5}$ **34.** 5 **35.** 2 **36.** 2 **37.** -12
38. 5 **39.** 3.27 **40.** 4.48 **41.** 5.60 **42.** -12.43 **43.** 8.27 **44.** 27.20
Applications and Problem Solving 45. a) 4, 5 **b)** 5, 6 **c)** 14, 15 **d)** $-4, -3$ **46. a)** estimate: $2 + 3 + 3 = 8$ **b)** 8.2
47. a) 2 **b)** 3 **c)** 2.83 **d)** 0.5 **48. a)** 68.9 km **b)** 77.8 km
49. 598 km **50. a)** 500 cm^2 **b)** 22.4 cm **c)** 11.3 cm
51. a) 5.7 h **b)** 64.5 km **52. a)** 2.8 s **b)** 9.5 s **c)** 0.25 m
d) 140 m **53. a)** 2.4 m **b)** 2.2 m; midpoint **c)** 6.7 m
54. a) 27.6 m **56. a)** no **b)** yes

Section 1.4 pp. 19–20
1. $2\sqrt{3}$ **2.** $2\sqrt{5}$ **3.** $3\sqrt{5}$ **4.** $5\sqrt{2}$ **5.** $2\sqrt{6}$ **6.** $3\sqrt{7}$
7. $10\sqrt{2}$ **8.** $4\sqrt{2}$ **9.** $2\sqrt{11}$ **10.** $2\sqrt{15}$ **11.** $3\sqrt{2}$ **12.** $3\sqrt{6}$
13. $8\sqrt{2}$ **14.** $3\sqrt{10}$ **15.** $5\sqrt{5}$ **16.** $\sqrt{2}$ **17.** $\sqrt{5}$ **18.** $2\sqrt{5}$
19. $2\sqrt{2}$ **20.** $\sqrt{11}$ **21.** $\dfrac{\sqrt{7}}{2}$ **22.** 6 **23.** $9\sqrt{3}$ **24.** 15
25. $\sqrt{12}$ **26.** $\sqrt{32}$ **27.** $\sqrt{90}$ **28.** $\sqrt{45}$ **29.** $\sqrt{28}$
30. $\sqrt{288}$ **31.** $2\sqrt{11}, 3\sqrt{5}, 4\sqrt{3}, 5\sqrt{2}$
32. $2\sqrt{15}, 3\sqrt{7}, 8, 6\sqrt{2}$ **33.** $4\sqrt{7}, 2\sqrt{30}, 5\sqrt{5}, 3\sqrt{14}$
34. $2\sqrt{5}$ **35.** $3\sqrt{2}$ **36.** $5\sqrt{3}$ **37.** $\sqrt{77}$ **38.** $4\sqrt{21}$ **39.** 54
40. $12\sqrt{3}$ **41.** $30\sqrt{2}$ **42.** $36\sqrt{5}$ **43.** $56\sqrt{2}$ **44.** 6

45. 42 **46.** $\dfrac{\sqrt{3}}{3}$ **47.** $\dfrac{\sqrt{21}}{7}$ **48.** $\dfrac{\sqrt{30}}{6}$ **49.** $\dfrac{5\sqrt{15}}{6}$ **50.** $\dfrac{2}{3}$
51. $\dfrac{2\sqrt{3}}{3}$ **52.** 2 **53.** $\sqrt{2}$ **54.** $\dfrac{3\sqrt{15}}{20}$
Applications and Problem Solving 55. a) 5 **b)** 5 **c)** 5
d) -5 **e)** -5 **f)** -5 **56.** $\dfrac{\sqrt{15}}{2}$ **57.** $15\sqrt{3}$ **58. a)** $\sqrt{5}, 2\sqrt{5}, 3\sqrt{5}$
b) The length of the diagonal is equal to $\sqrt{5}$ times the width of the flag **c)** $75\sqrt{5}$ cm **59. a)** $2\sqrt[3]{2}$
b) $2\sqrt[3]{4}$ **c)** $3\sqrt[3]{2}$ **d)** $3\sqrt[3]{3}$ **e)** $2\sqrt[4]{2}$ **f)** $2\sqrt[4]{4}$ **60. a)** 100
b) 144 **61.** $\dfrac{1}{\sqrt[3]{2}} = \dfrac{1}{\sqrt[3]{2}} \times \dfrac{\sqrt[3]{2}}{\sqrt[3]{2}} \times \dfrac{\sqrt[3]{2}}{\sqrt[3]{2}} = \dfrac{\sqrt[3]{4}}{2}$ **62. a)** $\sqrt{7}$
b) $\sqrt{2}$ **c)** $3\sqrt{2}$ **d)** $\sqrt{6}$ **63. a)** The square root of a negative number is not defined **b)** $b \neq 0$, since the denominator of the fraction would then be 0.

Section 1.5 pp. 23–24
Practice 1. $11\sqrt{5}$ **2.** $5\sqrt{3}$ **3.** $9\sqrt{2}$ **4.** $6\sqrt{7}$ **5.** $-\sqrt{10}$
6. 0 **7.** $4\sqrt{5}$ **8.** $8\sqrt{3} + 2\sqrt{6}$ **9.** $4\sqrt{5} + 4\sqrt{7}$
10. $7\sqrt{2} - \sqrt{10}$ **11.** $8\sqrt{6} - 5\sqrt{13}$ **12.** $6\sqrt{11} + 3\sqrt{14}$
13. $9\sqrt{7} + 13$ **14.** $-1 - 2\sqrt{11}$ **15.** $5\sqrt{3}$ **16.** $5\sqrt{5}$ **17.** $\sqrt{2}$
18. $11\sqrt{2}$ **19.** $12\sqrt{3}$ **20.** $5\sqrt{6} + 2\sqrt{2}$ **21.** $5\sqrt{7} + 7\sqrt{3}$
22. $12\sqrt{7}$ **23.** $7\sqrt{2}$ **24.** $31\sqrt{3}$ **25.** $12\sqrt{2}$ **26.** $\sqrt{5}$ **27.** $9\sqrt{5}$
28. $24\sqrt{3} - 20\sqrt{2}$ **29.** $2\sqrt{5} + 4\sqrt{2}$ **30.** $3\sqrt{2} - \sqrt{3}$
31. $2\sqrt{3} + 6$ **32.** $12\sqrt{3} - 2\sqrt{6}$ **33.** $\sqrt{6} + 4\sqrt{2}$
34. $12\sqrt{3} + 6\sqrt{5}$ **35.** $23 + 4\sqrt{30}$ **36.** $16 + \sqrt{3}$
37. $26 + 14\sqrt{14}$ **38.** $28 + 6\sqrt{3}$ **39.** $13 - 4\sqrt{10}$
40. 1 **41.** 4 **42.** -17 **43.** $\dfrac{2 - \sqrt{2}}{2}$ **44.** $\dfrac{3\sqrt{5} + 3}{4}$
45. $\dfrac{2\sqrt{3} + 3\sqrt{2}}{-3}$ **46.** $\dfrac{2\sqrt{6} - 2\sqrt{3}}{3}$ **47.** $\sqrt{5} + \sqrt{2}$
48. $3 - \sqrt{6}$ **49.** $\dfrac{24 - 2\sqrt{6}}{23}$ **50.** $3 - 2\sqrt{2}$
51. $\dfrac{2\sqrt{3} + 2\sqrt{5} + \sqrt{30} + 5\sqrt{2}}{-4}$ **52.** $\dfrac{52 + 7\sqrt{35}}{43}$
Applications and Problem Solving 53. $5\sqrt[3]{2}$ **54.** $5\sqrt[3]{3}$
55. $19\sqrt[3]{4}$ **56.** $13\sqrt[3]{2}$ **57.** $-\sqrt[3]{2}$ **58.** $\sqrt[3]{4}$ **59.** $3\sqrt[3]{5}$
60. $4\sqrt[3]{6}$ **61.** $(\sqrt{3} + 1)^2$, $\sqrt{3}(\sqrt{3} + 1)$, $(\sqrt{3} + 1)(\sqrt{3} - 1)$, $(1 - \sqrt{3})^2$ **62. a)** $6\sqrt{8} + \sqrt{8} - 5\sqrt{8}$ **b)** It is twice the others.

416

63. 1 **64. a)** $8\sqrt{6} - 6$ **b)** $8\sqrt{2} + 2\sqrt{3}$ **65.** $13 - 4\sqrt{10}$
66. $10\sqrt{5}$ **67.** $38\sqrt{15} - 38\sqrt{2}$ **68.** $2\sqrt{7} + 2\sqrt{5}$
69. $97 + 56\sqrt{3}$ **70. a)** $4\sqrt{13} + 8$ **b)** $3\sqrt{2} + 2\sqrt{5}$
71. It is only true if $a = 0$, $b = 0$, or $a = b = 0$.

Technology p. 25
1 Simplifying Radicals 1. $5\sqrt{17}$ **2.** $7\sqrt{6}$ **3.** $13\sqrt{3}$
4. $4\sqrt{5}$ **5.** $14\sqrt{15}$ **6.** $120\sqrt{3}$ **7.** $\frac{\sqrt{3}}{3}$ **8.** $\frac{5\sqrt{7}}{28}$ **9.** $-\frac{\sqrt{2}}{2}$
10. $4\sqrt{2}$ **11.** $\frac{5\sqrt{6}}{18}$ **12.** $\frac{2\sqrt{6}}{3}$

2 Operations With Radical Expressions 1. $5\sqrt{5}$ **2.** $24\sqrt{2}$
3. $\sqrt{7}$ **4.** $-\sqrt{6}$ **5.** $21\sqrt{3}$ **6.** $9\sqrt{10}$ **7.** $2\sqrt{5}$ **8.** $-5\sqrt{11}$
9. $5\sqrt{2} + 5\sqrt{3}$ **10.** $6\sqrt{3} - 3\sqrt{2}$ **11.** $18 + 60\sqrt{2}$
12. $84\sqrt{3} - 56\sqrt{2}$ **13.** -7 **14.** $226 + 40\sqrt{22}$
15. $304 - 60\sqrt{15}$ **16.** $26\sqrt{6} - 54$ **17.** $2\sqrt{3} - 2\sqrt{2}$
18. $\sqrt{2} + 1$ **19.** $\frac{6\sqrt{2} + 3\sqrt{6}}{2}$ **20.** $\frac{8 + \sqrt{10}}{9}$ **21.** $-3 - 2\sqrt{2}$
22. $\frac{30 + 5\sqrt{14}}{-10}$

3 Problem Solving 1. $\frac{48 - 12\sqrt{2}}{7}$
2. a) $528 + 128\sqrt{2}$ **b)** 709 cm²

Investigating Math pp. 26–27
1 The 45°–45°–90° Triangle 1. a) $2\sqrt{2}$ **b)** $3\sqrt{2}$ **c)** $5\sqrt{2}$
2. The length of the hypotenuse is $\sqrt{2}$ times the length of a leg. **3. a)** $k = 7$, $h = 7\sqrt{2}$ **b)** $k = 10$, $h = 10\sqrt{2}$
c) $a = b = 6$ **d)** $c = d = 4\sqrt{2}$ **4.** $\sqrt{2}$ **5.** $s\sqrt{2}$

2 The 30°–60°–90° Triangle 1. a) $a = 2$, $b = 2\sqrt{3}$
b) $x = 5\sqrt{3}$, $y = 5$ **c)** $d = 4\sqrt{3}$, $e = 4$ **d)** $g = 1$, $k = \sqrt{3}$
2. The length of the side opposite the 30° angle is half the length of the hypotenuse. **3.** The length of the side opposite the 60° angle is $\sqrt{3}$ times the length of the side opposite the 30° angle. **4. a)** $h = 12$, $a = 6\sqrt{3}$
b) $h = 16$, $k = 8\sqrt{3}$ **c)** $a = 3$, $h = 6$ **5.** $\frac{1}{2}, \frac{\sqrt{3}}{2}$ **6.** $\frac{s}{2}, \frac{s\sqrt{3}}{2}$

3 The Equilateral Triangle 1. a) $2\sqrt{3}$ **b)** $4\sqrt{3}$ **2. a)** $\frac{\sqrt{3}}{2}$
b) $\frac{\sqrt{3}}{4}$ **3. a)** $\frac{s\sqrt{3}}{2}$ **b)** $\frac{s^2\sqrt{3}}{4}$ **4. a)** $9\sqrt{3}$ **b)** $25\sqrt{3}$ **c)** $\frac{49\sqrt{3}}{4}$

4 Problem Solving 1. 9.1 cm, 7.9 cm **2.** 15.2 cm

3. 4.8 cm, 8.3 cm, 9.6 cm **4.** $48\sqrt{3}$ **5. a)** 16 cm²
b) 32 cm²

Section 1.7 p. 33
Practice 1. 2^7 **2.** 2^4 **3.** 2^{12} **4.** 2^8 **5.** 2^{3+m} **6.** 2^{7-y}
7. 2^{x-4} **8.** 2^{xy} **9.** 2 **10.** 2^3 **11.** 2^{-3} **12.** 2^{-4} **13.** $\frac{1}{9}$ **14.** 1
15. $\frac{1}{8}$ **16.** $\frac{1}{16}$ **17.** $\frac{1}{4}$ **18.** -1 **19.** 25 **20.** -4 **21.** $-\frac{1}{64}$
22. a^7 **23.** m^8 **24.** b^{12} **25.** a^5b^2 **26.** x^8y^5 **27.** x^{-2} **28.** m^{-9}
29. y^{-2} **30.** a^5 **31.** $a^{-1}b^{-2}$ **32.** x^3 **33.** m^6 **34.** t^6 **35.** y^{-2}
36. m^4 **37.** t^5 **38.** x^6 **39.** a^8b^{12} **40.** x^{-2} **41.** 1 **42.** a^2b^{-4}
43. $x^{-6}y^{-9}$ **44.** $\frac{x^3}{8}$ **45.** $\frac{a^4}{b^4}$ **46.** $\frac{x^{10}}{y^{15}}$ **47.** $\frac{3}{x}$ **48.** 1
49. $\frac{a^4}{b^6}$ **50.** $15m^6$ **51.** $-20a^4b^6$ **52.** $30a^2b^5$ **53.** $24m^4n^7$
54. 42 **55.** $-6y^{-1}$ **56.** $6a^{-4}b^{-2}$ **57.** $5x^3$ **58.** $5ab^2$ **59.** $2m^4$
60. $2a^3b^{-1}$ **61.** $\frac{4a^2}{b^5}$ **62.** $7x^8$ **63.** $-3m^{-3}n^8$ **64.** $9a^7b^{-4}$
65. $4x^{-8}y^2$ **66.** $4m^6$ **67.** $-64x^6$ **68.** $9m^6n^4$ **69.** $\frac{c^6}{25d^6}$
70. $\frac{a^9b^6}{8}$ **71.** $\frac{y^8}{81x^{12}}$ **72.** $\frac{16x^2}{9y^2}$ **73.** $-\frac{8a^6}{27y^9}$
74. $\frac{81a^4}{b^{16}}$ **75.** $\frac{n^6}{4m^4}$ **76.** $27b^6$ **77.** $\frac{4x^{10}}{y^{12}}$ **78.** 3 **79.** $\frac{3}{8}$
80. 12 **81.** 3
Applications and Problem Solving 82. 12 000 years
83. a) 1024 s **b)** $\frac{1}{32}$ s **84.** 20^{100} **85. a)** $\frac{37}{35}$ **b)** $\frac{12}{65}$ **c)** -1
d) 0 **86. a)** 2 **b)** 4 **c)** 3 **d)** 5 **87.** all values but $x = 0$

Section 1.8 pp. 37–38
Practice 1. $\sqrt[3]{2}$ **2.** $(\sqrt{37})^3$ **3.** $\sqrt{x}$ **4.** $\sqrt[5]{a^3}$ **5.** $(\sqrt[3]{6})^4$
6. $(\sqrt[4]{6})^3$ **7.** $\frac{1}{\sqrt{7}}$ **8.** $\frac{1}{\sqrt[3]{9}}$ **9.** $\frac{1}{\sqrt[7]{x^3}}$ **10.** $\frac{1}{\sqrt[5]{b^6}}$ **11.** $\sqrt{3x}$
12. $3\sqrt{x}$ **13.** $7^{\frac{1}{2}}$ **14.** $34^{\frac{1}{2}}$ **15.** $(-11)^{\frac{1}{3}}$ **16.** $a^{\frac{2}{5}}$ **17.** $6^{\frac{4}{3}}$
18. $b^{\frac{4}{3}}$ **19.** $x^{-\frac{1}{2}}$ **20.** $a^{-\frac{1}{3}}$ **21.** $x^{-\frac{4}{5}}$ **22.** $2^{\frac{1}{3}}b$ **23.** $3^{\frac{1}{2}}x^{\frac{5}{2}}$
24. $5^{\frac{1}{4}}t^{\frac{3}{4}}$ **25.** 2 **26.** 5 **27.** $\frac{1}{2}$ **28.** -2 **29.** 5 **30.** $-\frac{1}{3}$
31. $\frac{1}{2}$ **32.** 0.2 **33.** 3 **34.** 0.1 **35.** $\frac{2}{3}$ **36.** $\frac{3}{2}$ **37.** 4 **38.** 8
39. 243 **40.** 27 **41.** $\frac{1}{8}$ **42.** 4 **43.** $-\frac{1}{32}$ **44.** $\frac{1}{9}$ **45.** 1
46. 1 **47.** $\frac{1000}{27}$ **48.** $\frac{4}{9}$ **49.** not possible **50.** 1000
51. $\frac{9}{4}$ **52.** 3 **53.** -3 **54.** 4 **55.** -32 **56.** 4 **57.** not possible

58. 5 **59.** $\dfrac{11}{6}$ **60.** 27 **61.** not possible **62.** $\dfrac{50}{9}$ **63.** 5

64. 3 **65.** 5 **66.** 2 **67.** 3 **68.** 12 **69.** x **70.** x **71.** $3^{\frac{1}{4}}x^2$

72. $2^{\frac{1}{2}}x^{\frac{7}{6}}$ **73.** $3x^2$ **74.** x^2y **75.** a^4b^3 **76.** $-3x^{\frac{1}{3}}$ **77.** $3a^2b$

78. $9x^4y^{-6}$ **79.** $x^{\frac{11}{6}}$ **80.** $x^{\frac{17}{12}}$ **81.** $x^{\frac{19}{15}}$ **82.** $a^{\frac{4}{3}}b^{\frac{8}{3}}$ **83.** $a^{\frac{3}{8}}b^{\frac{5}{8}}$

84. 2.05 **85.** 21.67 **86.** 0.19 **87.** 1.71 **88.** 0.31 **89.** 2.68

Applications and Problem Solving **90.** 2.7 m/s

91. 278, 294, 312, 330, 350, 371, 393, 416, 441, 467, 495, 524 **92. a)** 5 **b)** 3 **c)** 2, 4, 6, ... **d)** 4 **e)** 2

f) 1, 3, 5, ... **93. a)** 4; 5 **c)** $2^{\frac{1}{2}}$, $5^{\frac{1}{2}}$ **d)** $2^{\frac{1}{2}}$, $5^{\frac{1}{2}}$, $8^{\frac{1}{2}}$, $10^{\frac{1}{2}}$

e) 11 different squares; $2^{\frac{1}{2}}$, $5^{\frac{1}{2}}$, $8^{\frac{1}{2}}$, $10^{\frac{1}{2}}$, $13^{\frac{1}{2}}$, $17^{\frac{1}{2}}$

Investigating Math p. 39

1 Perfect Squares and Prime Factors **1. a)** 7; 1
b) 3×3; 2 **c)** 2×5; 2 **d)** $2 \times 3 \times 3$; 3 **e)** $2 \times 2 \times 2$; 3
f) $2 \times 2 \times 2 \times 2$; 4 **2. a)** 7×7; 2 **b)** $3 \times 3 \times 3 \times 3$; 4
c) $2 \times 2 \times 5 \times 5$; 4 **d)** $2 \times 2 \times 3 \times 3 \times 3 \times 3$; 6
e) $2 \times 2 \times 2 \times 2 \times 2 \times 2$; 6
f) $2 \times 2 \times 2 \times 2 \times 2 \times 2 \times 2 \times 2$; 8 **3.** even **4. a)** $2n$
b) $2n$ is an even number
2 Proof by Contradiction
* All rational numbers can be written in the form $\dfrac{a}{b}$, where a and b are natural numbers.
* The square of $\sqrt{2}$ is 2. The square of $\dfrac{a}{b}$ is $\dfrac{a^2}{b^2}$.

* Multiply both sides of the equation by b^2.
* The square of a natural number has an even number of numbers in its prime factorization.
* The square of a natural number has an even number of numbers in its prime factorization. $2b^2$ can be broken up into 2 (one number in its prime factorization) and b^2 (an even number of numbers in its prime factorization). An even number plus one is an odd number. Therefore, $2b^2$ has an odd number of numbers in its prime factorization.
* There is only one prime factorization for each number. Thus, a number with an even number of numbers in its prime factorization cannot equal a number with an odd number of numbers in its prime factorization.

Section 1.9 p. 41

Applications and Problem Solving **1. a)** 33 h 29 min
b) 174 km/h **2. a)** 5685 km **3. a)** $-37°C$ **b)** $-23°C$
4. 40°C **5.** 08:20 Wednesday **9.** Uranus
10. Saskatchewan

Computer Data Bank p. 42

1 Ski Trails **1.** 111 **2.** 88, $\dfrac{88}{111}$

2 Movie Costs **3.** multiply the cost in millions of dollars by 1 000 000 and divide by the length in minutes **4.** *True Lies, Easy Rider*
3 Winning Times **3.** 11.08 s, the 8th and 9th times are both 11.08 s **4.** 11.50 s, 3 times **5.** 11.28 s
4 Food Graphs **2.** divide protein (or carbohydrates or fat) in grams by mass in grams and multiply by 100
5 Collision Coverage Premiums **2.** 34 821 vehicles, $19 376 251 **3.** $556

Career Connection p. 43

1 Comparing Distances **1. a)** 1.5×10^6 km
b) 9×10^6 km **c)** 9×10^7 km **2. a)** 9.5×10^{12} km
b) 1.6×10^{-5} light-years **c)** 6.65×10^{20} km
d) 4.4×10^{12} AU **e)** 7×10^7 years
2 Comparing Sizes **1.** 110 **2. a)** 638 **b)** 70 000

Connecting Math and Physiology pp. 44–45

1 Calculating Speeds **1.** 11.5 m/s **2.** 1.9 m/s
3. 2.9 m/s **4. a)** 12.0 m/s **b)** no
2 Calculating Masses **1.** 30 727 kg **2.** 3943 kg

Review pp. 46–47

1. natural, whole, integer, rational, real **2.** rational, real **3.** irrational, real **4.** rational, real **5.** 7 **6.** $\dfrac{1}{10}$

7. 1 **11.** 7; 6.68 **12.** 200; 204.89 **13.** 0.03; 0.03
14. 2; 2.726 **15.** 38; 42.451 **16.** 3.5; 3.347 **17.** 2; 1.851
18. 8.603 **19.** 2.520 **20. a)** 2.605 **b)** 7.318 km, 4.534 km
c) 1.614 **d)** part a) is the square of part c)
21. $3\sqrt{2}$ **22.** $4\sqrt{2}$ **23.** $10\sqrt{5}$ **24.** $\sqrt{5}$ **25.** $\sqrt{7}$ **26.** $6\sqrt{6}$

27. 10 **28.** $\sqrt{44}$ **29.** $\sqrt{112}$ **30.** $\sqrt{108}$ **31.** $2\sqrt{15}$

32. $30\sqrt{2}$ **33.** $\dfrac{\sqrt{5}}{5}$ **34.** $\dfrac{\sqrt{6}}{3}$ **35.** $\dfrac{2\sqrt{2}}{3}$ **36.** $\dfrac{\sqrt{30}}{40}$

37. $2\sqrt{10}$, $3\sqrt{5}$, $4\sqrt{3}$, $5\sqrt{2}$ **38.** $30\sqrt{3}$ **39.** $5\sqrt{2}$

40. $4\sqrt{3}+3\sqrt{6}$ **41.** $7\sqrt{5}$ **42.** $-\sqrt{3}$ **43.** $2\sqrt{2}$ **44.** $\sqrt{5}$

45. $\sqrt{6}+5\sqrt{3}$ **46.** $2\sqrt{5}-2\sqrt{3}$ **47.** $-7-11\sqrt{10}$

48. $17+4\sqrt{15}$ **49.** 4 **50.** $\sqrt{3}+1$ **51.** $\dfrac{4\sqrt{5}-4\sqrt{2}}{3}$

52. $\dfrac{2\sqrt{6}+10\sqrt{3}}{-23}$ **53.** $\dfrac{48-7\sqrt{21}}{51}$ **54.** $-\sqrt[3]{2}$ **55.** $19\sqrt[3]{3}$

56. $21-8\sqrt{5}$ **57.** $\dfrac{1}{25}$ **58.** 1 **59.** $\dfrac{1}{27}$ **60.** $\dfrac{1}{81}$ **61.** $\dfrac{1}{25}$

62. -3 **63.** m^7 **64.** y^{-5} **65.** t^3 **66.** m^{-5} **67.** x^8y^{12} **68.** 1

69. x^4y^{-6} **70.** $\dfrac{m^{12}}{n^8}$ **71.** $\dfrac{x^6}{y^4}$ **72.** $10x^5y^7$ **73.** $9ab$ **74.** $12m^4$

75. $-5x$ **76.** $4a^{10}b^6$ **77.** $-\dfrac{m^9n^3}{27}$ **78.** $\dfrac{27m^6}{8n^9}$ **79.** $\dfrac{9x^6}{4y^8}$

80. $-2x^3y^3$ **81.** $\dfrac{6b^2}{5a}$ **82.** $-\dfrac{5t^3}{2s^7}$ **83.** $\dfrac{a^8b^4}{9}$ **84.** $\sqrt{6}$ **85.** $\dfrac{1}{\sqrt{5}}$

86. $(\sqrt[5]{7})^3$ **87.** $(\sqrt[3]{10})^{-4}$ **88.** $(-8)^{\frac{1}{3}}$ **89.** $m^{\frac{5}{3}}$ **90.** $x^{\frac{2}{3}}$

91. $2^{\frac{1}{5}}a^{\frac{2}{5}}$ **92.** 5 **93.** $\dfrac{1}{3}$ **94.** $\dfrac{1}{7}$ **95.** 1 **96.** 0.3 **97.** $-\dfrac{1}{2}$

98. 5 **99.** 9 **100.** $-\dfrac{1}{8}$ **101.** $\dfrac{243}{32}$ **102.** $\dfrac{1}{243}$ **103.** $\dfrac{25}{9}$

104. 16 **105.** -2 **106.** 2 **107.** $y^{\frac{2}{3}}$ **108.** $3m^2$ **109.** $-2x^{\frac{1}{3}}$

110. x^2 **111.** $-4x$ **112.** $-4x^{\frac{1}{3}}$ **113.** 4

Exploring Math p. 47

1. Drive 100 km, drop off 30 L of fuel, drive back to A, refuel, drive 100 km, pick up 10 L, drive another 100 km, drop off 30 L, drive back 100 km, pick up 10 L, drive to A, refuel, drive 100 km, pick up 10 L, drive another 100 km, pick up 10 L, drive another 100 km, drop off 30 L, drive back 100 km, pick up 20 L, return to A, refuel, drive 300 km, pick up 30 L, drive to B.

2. The drop-off points are C: $\dfrac{500}{15}$ km,

D: $\dfrac{500}{15}+\dfrac{500}{13}$ km, E: $\dfrac{500}{15}+\dfrac{500}{13}+\dfrac{500}{11}$ km,

F: $\dfrac{500}{15}+\dfrac{500}{13}+\dfrac{500}{11}+\dfrac{500}{9}$ km,

G: $\dfrac{500}{15}+\dfrac{500}{13}+\dfrac{500}{11}+\dfrac{500}{9}+\dfrac{500}{7}$ km,

H: $\dfrac{500}{15}+\dfrac{500}{13}+\dfrac{500}{11}+\dfrac{500}{9}+\dfrac{500}{7}+\dfrac{500}{5}$ km, and

I: $\dfrac{500}{15}+\dfrac{500}{13}+\dfrac{500}{11}+\dfrac{500}{9}+\dfrac{500}{7}+\dfrac{500}{5}+\dfrac{500}{3}$ km.

Drop off the amount of fuel shown below at each point, and then return to A, refuelling completely at A. Each time you pass a drop-off point on the way out, pick up enough fuel to refill the tank. The

drop-offs at each point are C: $\dfrac{13}{15}\times50$ L; D: $\dfrac{11}{13}\times50$ L;

E: $\dfrac{9}{11}\times50$ L; F: $\dfrac{7}{9}\times50$ L; G: $\dfrac{5}{7}\times50$ L; D: $\dfrac{3}{5}\times50$ L;

D: $\dfrac{1}{3}\times50$ L

1. $|2-5|,\ |2|,\ |-1|,\ \left|\dfrac{3}{2}-\dfrac{9}{4}\right|,\ |-3|-|-4|$

6. $5\sqrt{2}$ **7.** $2\sqrt{11}$ **8.** $4\sqrt{5}$ **9.** $\sqrt{50}$ **10.** $\sqrt{45}$

11. $\sqrt{1600}$ **12.** $\sqrt{35}$ **13.** $6\sqrt{2}$ **14.** $\sqrt{5}$ **15.** $30\sqrt{5}$

16. $4\sqrt{3},\ 5\sqrt{2},\ 3\sqrt{6},\ 2\sqrt{15}$

17. $9;\ 8.66$ **18.** $80;\ 77.65$ **19.** $0.9;\ 0.88$ **20.** $0.07;\ 0.07$

21. $14;\ 13.50$ **22.** $2;\ 1.83$ **23.** $3\sqrt{10},\ \sqrt{93},\ 7\sqrt{2},\ 6\sqrt{3}$

24. $3\sqrt{3}$ **25.** $7\sqrt{7}$ **26.** $14\sqrt{3}$ **27.** $5\sqrt{3}-7$ **28.** $22-12\sqrt{2}$

29. 11 **30.** $\dfrac{2\sqrt{7}}{7}$ **31.** $\dfrac{3\sqrt{3}+12}{-13}$ **32.** $\dfrac{5\sqrt{6}-5\sqrt{3}}{3}$

33. $7\sqrt[3]{3}$ **34.** $3\sqrt[3]{2}$ **35.** 1 **36.** -8 **37.** $\dfrac{1}{36}$ **38.** $-\dfrac{1}{16}$ **39.** 6

40. $\dfrac{1}{25}$ **41.** y^4 **42.** a^{10} **43.** y^{12} **44.** $\dfrac{x^4}{25y^8}$ **45.** $\dfrac{m^5}{n^5}$ **46.** s^6t^9

47. $12a^5$ **48.** $-6a^3b$ **49.** $-\dfrac{6}{m}$ **50.** $-\dfrac{4}{x^3}$ **51.** $9a^4b^{10}$

52. $\dfrac{y^6}{8x^{12}}$ **53.** $-5n^5$ **54.** $\dfrac{t^4}{4}$ **55.** $\sqrt{13}$ **56.** $(\sqrt[3]{6})^2$

57. $(\sqrt{3})^{-5}$ **58.** $5^{\frac{2}{3}}$ **59.** $x^{\frac{5}{2}}$ **60.** $3^{\frac{1}{8}}t^{\frac{7}{8}}$ **61.** 6 **62.** $\dfrac{1}{2}$

63. $-\dfrac{1}{1000}$ **64.** 4 **65.** $\dfrac{1}{27}$ **66.** 3 **67.** $\dfrac{2}{3}$ **68.** $\dfrac{9}{4}$ **69.** $a^{\frac{2}{3}}$

70. $2x^{\frac{1}{2}}$ **71.** $-3a^2$ **72.** 7 **73.** 625

Using the Strategies p. 49

1. 9 **3.** Austin: First Officer, Puerto Plata; Greenberg: Captain, Miami; Hill: Chef, Nassau; Lo: Navigator, San Juan; Pearson: Engineer, Havana **4.** $\dfrac{1}{57}$ **6. a)** 8 **b)** 8

Data Bank 1. a) Chicago, St. Louis, Houston, New Orleans, Atlanta, Cincinnati, Pittsburgh, Philadelphia, New York, Montreal; 5440 km **b)** 8393 km **c)** 72 h
2. a) +22 °C **b)** +9 °C

Chapter 2

Number Patterns p. 51

1. first place: 5; second place: 3; third place: 1
2. Answers may vary; first place: 5, second place: 4, third place: 1; first place: 6, second place: 5, third place: 1 **4.** 28

Getting Started pp. 52–53

1 Patterns in Numbers and Letters 1. a) The next term equals the previous term plus 3; 16, 19, 22
b) The next term equals twice the previous term;

48, 96, 192 **c)** The next term equals the previous term plus the term number of the previous term; 22, 29, 37 **d)** Successive terms are obtained by alternately multiplying by 3 and dividing by 2; 27, 13.5, 40.5 **e)** Successive terms are obtained by alternately adding 1 and multiplying by 3; 31, 93, 94 **f)** The next term is the sum of the previous 2 terms; 47, 76, 123 **g)** The next term equals the previous term plus 18; 81, 99, 117 **2. a)** every third letter, beginning with A; M, P, S **b)** Beginning with A, alternately go forward 3 letters, then back 2 letters; G, E, H **c)** Beginning with A, alternately go forward 2 letters, then forward 1 letter; L, M, O **d)** Letters and numbers alternate. Beginning with number 1, the next number is obtained so that the difference between successive numbers is 3, 4, 5, … Beginning with Z, go backward in the alphabet; W, 19, V. **4.** The product of the left and right numbers divided by the sum of the top and bottom numbers is 2.

2 Percent **1.** $150 874.06 **2. a)** $4147 **b)** $290 **c)** $219.24 **3. a)** $213.79 **b)** $75.11 **c)** $25.71 **4. a)** $420 **b)** $67.20 **c)** $102.80 **d)** Answers may vary.

Mental Math

Evaluating Expressions and Percents **1.** 10 **2.** 1 **3.** 9 **4.** 35 **5.** 3 **6.** −22 **7.** 28 **8.** 44 **9.** −6 **10.** −26 **11.** 13 **12.** 2 **13.** −1 **14.** 28 **15.** −13 **16.** 49 **17.** $10 **18.** $25 **19.** $75 **20.** $30 **21.** $60 **22.** $2

Dividing by Multiples of 5 **1.** 17 **2.** 21 **3.** 27 **4.** 2.8 **5.** 7.8 **6.** 8.4 **7.** 12.2 **8.** 24.6 **9.** 20.2 **10.** 3.2 **11.** 4.2 **12.** 7.4 **13.** 19.6 **14.** 2.7 **15.** 5.7 **16.** 23 **17.** 0.9 **18.** 1.3 **19.** The first step would be to divide by 1, a step that can be omitted. **20.** 52 **21.** 106 **22.** 236 **23.** 1300 **24.** 19 **25.** 6.4 **26.** Dividing by 5 is the same as dividing by 10 and then multiplying by 2.

Section 2.1 pp. 57–59

Practice **1.** $7503.65 **2.** $9751.96 **3.** $28 615.38 **4.** $6762.96 **5.** $339.25 **6.** $3988.86 **7.** $23.31 **8.** $199.36

Applications and Problem Solving **9. a)** no; the data in row 2 are not related to the data in row 1 **b)** $g + a$ **c)** Gordie Howe: 1850; Marcel Dionne: 1771; Phil Esposito: 1590; Stan Mikita: 1467; Bryan Trottier: 1410; John Bucyk: 1369; Guy Lafleur: 1353; Bobby Clarke: 1210 **d)** $2g + a$ **e)** They would not change. **f)** Marcel Dionne: 1.314; Phil Esposito: 1.240; Guy Lafleur: 1.202; Bryan Trottier: 1.139; Bobby Clarke: 1.058; Stan Mikita: 1.052; Gordie Howe: 1.047; John Bucyk: 0.889 **10.** Answers may vary. **a)** gold: 5; silver: 4; bronze: 3 **b)** gold: 3, silver: 2, bronze: 1 **c)** not possible assuming positive numbers are used for points **11. a)** 7% **b)** $0.75 **c)** 5.85; 95.85; 122.88

12. a) $5049.91 **b)** $5145.64 **c)** No, since $4000(1.06)^2(1.07)^2 = 4000(1.07)^2(1.06)^2$ **13. a)** 325, 650, 1300, 2600, 5200 **b)** 3120, 1872, 1123 **14. a)** 7% **b)** 10.7% or 10% **c)** $390, $27.30, $31.20, $448.50; $160, $11.20, $12.80; $184 **d)** $390, $27.30, $25.04, $442.34; $160, $11.20, $10.27, $181.47 **15. a)** $1123.60 **b)** $1125.51 **c)** $0.65 **16. a)** 6.5 **b)** 1.7 **c)** Asia **d)** 80% **e)** 50% **f)** Northern Hemisphere **17. a)** 1 year **b)** monthly payment − interest; opening balance − closing balance **c)** $16 232.83 **d)** $103 832.10 **e)** $98 255.27 **f)** $28.68 **g)** rounding **h)** $33 326.75 **i)** $983.43 more

Section 2.2 p. 61

Applications and Problem Solving **1.** Tia: carpenter, turtle; Jane: lawyer, dog; Fran: police officer, parrot; Marta: teacher, cat **2.** 10 **3.** 25, 85 **4.** 8 **5.** Bonnie Zimmer: rock; Cari Lennon: rhythm and blues; Deborah Falco: country; Alonzo Jones: classical; Eric Schultz: jazz **6.** 9

Computer Data Bank p. 62

1 Favourite Type of Movie **3.** action, 45, 6.7%; comedy, 67, 11.9%; drama, 90, 31.1%; family, 25, 8.0%; horror, 3, 33.3%; musical, 31, 32.3%; science fiction, 11, 45.5%; western, 8, 37.5% **4.** science fiction, action

2 Academy Awards **2.** Ben-Hur, 11

3 Comparing Profits **2.** subtract the cost in millions of dollars from the income in millions of dollars **3.** *Star Wars, Jurassic Park, Return of the Jedi, The Empire Strikes Back, Jaws* **4.** yes, the last 5 of the sorted records have negative profits **6.** *Gone with the Wind, Star Wars, Fantasia, Pinocchio, Jaws; Star Wars* and *Jaws* in both lists

Investigating Math p. 63

1 Trapezoids **1.** 17; 21 **2.** Each diagram has 4 more toothpicks than the previous diagram. **3.** $4n + 1$ **4.** 201; 321

2 I-Shapes **1.** 26; 32 **2.** Each diagram has 6 more asterisks than the previous diagram. **3.** $6n + 2$ **4.** 392; 602

3 Triangles **1.** 64; 256 **2.** Each diagram has 4 times as many triangles as the previous diagram. **3.** 262 144 **4.** 4^{n-1}

4 Generating a Sequence **1.** 1, 4, 2 **2.** 1, 4, 2 **3.** yes

Section 2.3 pp. 66–68

Practice **1.** 3, 6, 9, 12, 15 **2.** 6, 8, 10, 12, 14 **3.** 3, 1, −1, −3, −5 **4.** 9, 8, 7, 6, 5 **5.** 2, 4, 8, 16, 32 **6.** 0, 3, 8, 15, 24 **7.** $t_n = 5n$; 25, 30, 35 **8.** $t_n = n + 1$; 6, 7, 8 **9.** $t_n = 7 − n$; 2, 1, 0 **10.** $t_n = n^2$; 25, 36, 49

11. $t_n = 2n$; 12, 14, 16 **12.** $t_n = -3n$; $-15, -18, -21$
13. $t_n = n - 2$; 4, 5, 6 **14.** $t_n = nx$; $5x, 6x, 7x$
15. $t_n = 1 + (n - 1)d$; $1 + 4d$; $1 + 5d$; $1 + 6d$ **16.** 0, 3, 6, 9
17. 0, 1, 4, 9 **18.** $1, \dfrac{1}{2}, \dfrac{1}{3}, \dfrac{1}{4}$ **19.** $2, \dfrac{3}{2}, \dfrac{4}{3}, \dfrac{5}{4}$
20. 0, 3, 15 **21.** $-1, 1, -1, 1$ **22.** 1, 2, 4, 8 **23.** 1, 3, 7, 15
24. $0, \dfrac{1}{3}, \dfrac{1}{2}, \dfrac{3}{5}$ **25.** 1, -1, 1, -1 **26.** $\dfrac{1}{3}, \dfrac{1}{9}, \dfrac{1}{27}, \dfrac{1}{81}$
27. $\dfrac{3}{2}, \dfrac{5}{4}, \dfrac{9}{8}, \dfrac{17}{16}$ **28.** 19, 37 **29.** 67, 83 **30.** 27, 62
31. $-7, -23$ **32.** 13, 53 **33.** 1, 5 **34.** 1, 144 **35.** 4, 7, 10,
13, 16 **36.** $-1, 3, 11, 27, 59$ **37.** $3, 1, -1, -3, -5$
38. 6, 10, 16, 24, 34 **39.** 5, 4, 3, 2, 1 **40.** 1, 1, 2, 3, 5

Applications and Problem Solving **41.** 0, 1, 2, 3;
$t_n = n - 1$ **42.** $-3, -1, 1, 3$; $t_n = 2n - 5$ **43.** 3, 6, 11, 18;
$t_n = n^2 + 2$ **44.** 1, 3, 9, 27; $t_n = 3^{n-1}$ **45.** 156; 176
46. a) 169.5 t; 174 t **b)** 246 t; 291 t **47. a)** 10.25 cm;
10.5 cm; 11.5 cm **b)** 20 weeks **48. a)** 1020 kJ; 1100 kJ;
1300 kJ **b)** $t_n = 1000 + 20n$ **49. a)** 30, 32, 34, 36, 38
b) $28 + 2n$ **c)** 148 **50.** 1, 1, 2, 3, 5, 8, 13, 21
51. b) $\dfrac{48+4}{10}, \dfrac{96+4}{10}, \dfrac{192+4}{10}, \dfrac{384+4}{10}, \dfrac{768+4}{10}$
c) 0.4 AU, 0.7 AU, 1 AU, 1.6 AU, 2.8 AU, 5.2 AU,
10 AU, 19.6 AU, 38.8 AU, 77.2 AU **d)** Neptune
53. a) \$48 000; \$38 400 **b)** $60\ 000(0.8)^n$ **c)** 8 years
54. No. $t_n = 2^{n-1}$ and $t_n = t_{n-1} + n - 1$ both have these
as the first 3 terms.

Section 2.4 pp. 70–71
Applications and Problem Solving **1.** D **2.** 6889
3. 149 **4.** 26 **5. a)** 1 m by 72 m, 2 m by 36 m,
3 m by 24 m, 4 m by 18 m, 6 m by 12 m, 8 m by 9 m
b) 1 m by 72 m **c)** 8 m by 9 m **6.** 1, 240; 3, 80; 5, 48;
15, 16 **7.** 4 **8. a)** 2 **b)** 4 **c)** 30 **9.** 3, 1, 1, 1, 3; 3, 1, 1, 1,
3; 3, 2, 1, 0, 4 or 3, 1, 0, 2, 4; 3, 0, 1, 2, 2 **10.** \$26
11. 3 **12.** 7 **13.** 36 **14.** 24 **15. a)** 6 km **b)** 36 min **16.** 8

Section 2.5 pp. 74–76
1. 15, 19, 23 **2.** 15, 9, 3 **3.** $-8, -3, 2$ **4.** 4, -3, -10
5. 10, 11.4, 12.8 **6.** $\dfrac{9}{4}, \dfrac{11}{4}, \dfrac{13}{4}$ **7.** 5, 4
8. not arithmetic **9.** not arithmetic **10.** $-1, -3$
11. not arithmetic **12.** not arithmetic **13.** x, x **14.** $c, 2d$
15. 7, 9, 11, 13, 15 **16.** 3, 7, 11, 15, 19 **17.** $-4, 2, 8,$
14, 20 **18.** $2, -1, -4, -7, -10$ **19.** $-5, -13, -21, -29,$
-37 **20.** $8, 8 + x, 8 + 2x, 8 + 3x, 8 + 4x$ **21.** $6, 7 + y,$
$8 + 2y, 9 + 3y, 10 + 4y$ **22.** $3m, 2m + 1, m + 2, 3, 4 - m$
23. 24, 72 **24.** 80, 172 **25.** 65, 702 **26.** 20, 53
27. $-89, -329$ **28.** $11 - 6n; -73$ **29.** $4 + 3n; 94$
30. $12 - 2n; -32$ **31.** $x + 4(n - 1); x + 52$ **32.** 49
33. 41 **34.** 36 **35.** 42 **36.** 35 **37.** 44 **38.** 13 **39.** 20, 26
40. $-22, -41$ **41.** 0, 1.5, 3 **42.** 23, 44, 65, 86

43. $m + 31, m + 22, m + 13$ **44. a)** 38, 46, 54, 62
45. a) $8, 5, 2, -1, -4$
Applications and Problem Solving **46. a)** 5, 20, 35,
50, 65 **b)** 740, 2990 **47. a)** 2, 8, 14, 20, 26 **b)** 10, 3, -4,
$-11, -18$ **48. a)** $a = 1896, d = 4$ **b)** 1916, 1940, 1944
c) t_6, t_{12}, t_{13} **49.** 88 **50.** 7 **51. a)** 132 km **b)** 212 km
c) $52 + 80t$ km **52.** 2016, 2023, 2030 **53. a)** \$45
b) \$285 **54.** \$39 200; \$39 400; \$39 600 **55. a)** 45
b) $t_n = 47 - 2n$ **c)** 23 **56. a)** 8, 2 **b)** $t_n = 6 + 2n$ **c)** 50;
2000 m **57. a)** 16 **b)** $t_n = 4 + 3n$ **c)** 79 **d)** 45
58. a) 160 kPa; 220 kPa **b)** 110 kPa **59. a)** 14
b) $t_n = 9 + 5n$ **60.** 3.8 **61.** 6, 8, 10 **62.** 5, 11, 17
63. a) equal **b)** Answers may vary. **64.** 319 **65.** $5x + y$
66. a) 5 **b)** 4 **c)** -1 **d)** 8 **e)** -2

Career Connection p. 77
1 Precipitation in Canada **2. a)** 1.98 trillion tonnes
b) 0.002 mg
2 Greatest Precipitation **1.** 20.4 mm **2. a)** 20.4, 40.8,
61.2 **b)** equal **c)** 367.2 **3.** all 20.4 less; 0, 20.4, 40.8;
346.8

Section 2.6 p. 79
Applications and Problem Solving **1.** 32 **2.** 22.5
3. 44.5 **4.** 70 **5.** 100 **6. a)** 20 100 **b)** 500 500
c) 500 000 500 000 **d)** 400 **7.** 301 **8.** 401 **9.** 756

Section 2.7 pp. 82–83
Practice **1.** 19 900 **2.** 15 050 **3.** -8900 **4.** -9900
5. 110 **6.** 1150 **7.** 2550 **8.** 414 **9.** 780 **10.** 10 098
11. 28 920 **12.** 400 **13.** 3275 **14.** -3960 **15.** -3738
16. 51 **17.** 84 **18.** -40 **19.** -550 **20.** 132 **21.** 2015
Applications and Problem Solving **22. a)** 1275
b) 10 000 **c)** 8550 **23.** 790 **24. a)** \$1400 **b)** \$53 850
c) 4th year **d)** \$414 000 **25.** 6, 10, 14; 7, 10, 13; 8, 10,
12; 9, 10, 11; 10, 10, 10 **26.** 122.5 m **27.** 312
28. a) 5 weeks **b)** \$60 **29.** 8, 11, 14 **30.** 126°, 122°,
118°, 114°, 110° **31.** 10, 304 **32.** 18 **33. a)** $130x$
b) $10x + 55y$

Investigating Math pp. 84–85
1 Writing Terms **1.** 3; 81, 243, 729 **2.** 2; 80, 160, 320
3. -4; 512, -2048, 8192 **4.** -1; 7, -7, 7 **5.** $\dfrac{1}{2}$; 4, 2, 1
6. $\dfrac{1}{10}$; 0.9, 0.09, 0.009 **7.** $-\dfrac{1}{2}$; 50, -25, 12.5
8. 4, 12, 36, 108, 324 **9.** 8, 20, 50, 125, 312.5
10. 1000, 200, 40, 8, 1.6 **11.** $-3, -9, -27, -81, -243$
12. $2, -8, 32, -128, 512$ **13.** $-80, 40, -20, 10, -5$
2 Finding Geometric Means **1.** 9, 27 **2.** 4, 16 **3.** 12,
24 **4.** $-125, -25$ **5.** $20, -20$ **6.** $30, -30$ **7.** $-8, -16,$
-32; 8, -16, 32 **8.** 40, 20, 10; -40, 20, -10 **9.** 36, 6
10. 6, 12, 24, 48

421

3 Applications 1. $12 000, $14 400, $17 280
2. $72 000, $64 800, $58 320, $52 488 **3. a)** $1000
b) $1562.50 **4.** 4
4 Critical Thinking 1. The square of the geometric
mean equals the product of the two terms.
2. geometric **3. a)** 36, 67, 98, 129 **b)** 10, 20, 40, 80

Technology p. 86

1 Interest Calculated Annually 1. a) Opening
Balance ($): 80 000.00, 72 202.58, 63 859.34,
54 932.07, 45 379.89, 35 159.06, 24 222.77,
12 520.94; Interest ($): 5600, 5054.18, 4470.15,
3845.24, 3176.59, 2461.13, 1695.59, 876.47; Closing
Balance ($): 72 202.58, 63 859.34, 54 932.07,
45 379.89, 35 159.06, 24 222.77, 12 520.94, 0
2. Opening Balance ($): 80 000.00, 72 061.02,
63 606.01, 54 601.42, 45 011.53, 34 798.30,
23 921.21, 12 337.11; Interest ($): 5200, 4683.97,
4134.39, 3549.09, 2925.75, 2261.89, 1554.88, 801.91;
Closing Balance ($): 72 061.02, 63 606.01, 54 601.42,
45 011.53, 34 798.30, 23 921.21, 12 337.11, 0
2 Interest Calculated Monthly Month: January,
February, March, April, May, June, July, August,
September, October, November, December;
Opening Balance ($): 10 000.00, 9189.34, 8374.63,
7555.84, 6732.96, 5905.96, 5074.83, 4239.54,
3400.08, 2556.42, 1708.54, 856.42; Annual Interest
Rate (%): 6, 6, 6, 6, 6, 6, 6, 6, 6, 6, 6, 6; Interest ($):
50.00, 45.95, 41.87, 37.78, 33.66, 29.53, 25.37, 21.20,
17.00, 12.78, 8.54, 4.28; Monthly Payment ($):
860.66, 860.66, 860.66, 860.66, 860.66, 860.66,
860.66, 860.66, 860.66, 860.66, 860.66, 860.70;
Closing Balance ($): 9189.34, 8374.63, 7555.84,
6732.96, 5905.96, 5074.83, 4239.54, 3400.08,
2556.42, 1708.54, 856.42, 0

Connecting Math and English pp. 87–89

1 Measuring Readability Grade Level: 3.86,
Reading Ease score: 88.29

Review pp. 90–91

1. $12 624.77 **2.** $7029.96 **3. a)** $1158.84 **b)** $1165.13
4. a) $321.94 **b)** $323.51 **5. a)** the sum of rainfall and
snowfall, ignoring the units **b)** 1 cm **6. a)** $33.44
b) $501.65 **c)** yes **7.** 3, 5, 7, 9, 11 **8.** −2, 1, 6, 13, 22
9. 5, 3, 1, −1, −3 **10.** 2, 8, 26, 80, 242 **11.** 4 **12.** 195
13. −5, −8, −11, −14, −17 **14.** 2, 0, −3, −7, −12 **15.** 2,
2, 6, 10, 22 **16. a)** 15.6 mm **b)** 16.2 mm **c)** 17.4 mm
17. 27, 33, 39 **18.** −9, −14, −19 **19.** 3, 8, 13, 18
20. −5, −3, −1, 1 **21.** 17 **22.** −98 **23.** $7n - 3$ **24.** 34
25. 32 **26. a)** 8, 17 **27.** 23, 37, 51 **28. a)** 14; 16
b) $2n + 6$ **c)** 56 **d)** 43 **29.** 1010 **30.** 100 **31.** 847

32. −260 **33.** 1425 **34.** −190 **35.** −1275 **36.** 8 cm,
11 cm, 14 cm **37.** 368 **38.** 231

Exploring Math p. 91

1. a) 4, 7 **b)** 7 **2. a)** 1, 2, 4, 5, 8, 11 **b)** 11 **3. a)** 1, 2, 3,
5, 6, 9, 10, 13, 17 **b)** 17 **4. a)** 1, 2, 3, 5, 7, 9, 11, 13, 15,
17 **b)** no; odd totals are impossible **5. a)** 1, 2, 4, 5, 7,
8, 10, 11, 13, 14 **b)** no; only totals that are multiples
of 3 are possible

Chapter Check p. 92

1. $10 122.55 **2. a)** $806.20 **b)** $810.58 **3.** −1, 1, 3, 5, 7
4. 4, 7, 12, 19, 28 **5.** 40 **6.** 169 **7.** 7, 4, 1, −2 **8.** −2, 0,
3, 7 **9.** 4, −1, −6 **10.** 86 **11.** $1 - 6n$ **12.** 46 **13.** −33,
−61 **14.** 795 **15.** 243 **16.** 800 **17.** Answers may vary
a) gold: 7; silver: 6; bronze: 5 **b)** gold: 3; silver: 2;
bronze: 1 **c)** gold: 4; silver: 3; bronze: 2 **d)** gold: 4;
silver: 2; bronze: 1 **18.** 11:30, 13:00 **19. a)** $3685.76
b) $4983.06 **c)** at the end of the 3rd year

Using the Strategies p. 93

1. a) 2 **2.** 8 **3.** Smith: Vancouver farmer; Wong:
London accountant; Bevan: Toronto baker; Lee:
Paris writer; Kostash: Melbourne biologist
5. front face, clockwise from lower right: 4, 7, 2, 5;
rear face, clockwise from lower right: 6, 1, 8, 3
6. 14 m^2 **7. a)** 28 **b)** 168 **8.** $3 \times 54 = 162$ **9.** 15
Data Bank 1. about 45 km/h **2.** about 5 000 000 000

Chapter 3

Polynomials

Getting Started p. 97

Algebra Skills 1. 15 **2.** 30 **3.** 30 **4.** 3 **5.** 0 **6.** 31
7. 93 **8.** 36 **9.** −2 **10.** −8 **11.** −7 **12.** −22 **13.** 48
14. 11 **15.** 17 **16.** −18 **17.** 144 **18.** 6 **19.** 36 **20.** −39
21. 2^7 **22.** 3^7 **23.** a^5 **24.** y^9 **25.** m^7 **26.** $8y^5$ **27.** $-8a^7$
28. n^8 **29.** $8x^6$ **30.** $12m^7$ **31.** 2^2 **32.** 3^0 **33.** m^3 **34.** a^3
35. y^6 **36.** $4a^2$ **37.** $6x$ **38.** $-3a + 5b$ **39.** $8a - 2b + 2$
40. $7x^2 + x - 11$ **41.** $9m - 9mn + n$ **42.** $-2x^2 + 4y^2$
43. $2x - 10$ **44.** $8a + 12b - 8c$ **45.** $4 - 8x + 20y$
46. $-6x + 14$ **47.** $-3x^2 + 6x + 3$ **48.** $-2a + 4b + c$
49. $3x^2 + 6x$ **50.** $4x^3 - 12x^2 + 4x$ **51.** $-2y^3 - 2y^2 + 6y$
52. $-2a^3 + 3a^2 + 5a$ **53.** $5x - 1$ **54.** $5y + 7$
55. $5m^2 + 3m + 1$ **56.** $x^2 + 9x + 10$ **57.** $7n^2 - 13n - 9$
58. $3x^2 - 19x$ **59.** $-2t^3 + 6t^2 - 5t$ **60.** $17z^2 - 17z - 12$
Mental Math
Integers 1. 2, 2 **2.** 2, 3 **3.** 1, 5 **4.** 2, 4 **5.** 3, 4 **6.** 2, 6
7. 4, 5 **8.** 3, 8 **9.** 4, 6 **10.** 2, 15 **11.** −1, −1 **12.** −2, 1
13. −5, 1 **14.** −3, 3 **15.** −2, 3 **16.** −2, 5 **17.** −7, −1
18. −4, −4 **19.** −8, −2 **20.** −5, 3

Multiplying Two Numbers That Differ by 2   1. 143
2. 168 **3.** 399 **4.** 899 **5.** 624 **6.** 1599 **7.** 9999 **8.** 3599
9. 999 999 **10.** 1.68 **11.** 1.95 **12.** 62 400 **13.** 89 900
14. 99.99 **15.** 1430 **16.** 399 **17.** 15.99 **18.** 24 990
19. 288 **20.** 80 990 **21.** 0.4899 **22.** Square their
average, then subtract 4 **a)** 165 **b)** 396 **c)** 221 **d)** 896
e) 621 **f)** 9996 **g)** 25.2 **h)** 159 600 **i)** 2496

Investigating Math pp. 98–99
1 Representing Expressions With Tiles   1. a) $2x^2 + 3x + 4$
b) $y^2 + 2y - 3$ **c)** $x^2 + xy + 3y$ **d)** $2x^2 - 2x + 3y$ **2. a)** 48
b) 5 **c)** 30 **d)** 30
2 The Zero Principle   1. a) 1 $-x$-tile, 3 1-tiles,
1 $-x^2$-tile **b)** 3 x^2-tiles, 2 $-y$-tiles **c)** 5 1-tiles, 2 x-tiles,
3 $-y$-tiles **d)** 1 $-xy$-tile, 1 y^2-tile
3 Adding Expressions   1. $-x^2 - x - 2$; -8 **2.** $3y^2 - 2y$; 21
3. $2xy - x + 3$; 13 **4.** $-3xy + 2x + 2y$; -8 **5.** $3xy + x + 3$; 23
4 Designing Algebra Tiles   1. cube-shaped

Section 3.1 pp. 102–103
Practice   1. linear, binomial **2.** quadratic, binomial
3. cubic, trinomial **4.** cubic, polynomial **5.** linear,
monomial **6.** quartic, trinomial **7.** cubic, binomial
8. constant, monomial **9.** 0 **10.** 14 **11.** 18
12. $4x^3 + 6x^2 - 5x - 7$ **13.** $-5x^5 + 7x^4 - 3x^3 - 8x^2 + 1$
14. $2x^5 + 9x^4y - 8x^3y^2 + 4xy^4 - y^5$
15. $11 + 2x - 6x^2 - 5x^3$ **16.** $-4 - 2x^2 + 3x^3 - 5x^4 + x^5$
17. $2 - 2xy^5 + x^2y^4 + x^3y^3 + x^5y - x^6$ **18.** $6x + 17$
19. $12a + 4b$ **20.** $13x^2 - x - 5$ **21.** $10x^2 - 3xy + 5$
22. $4x + 4$ **23.** $a - 4b$ **24.** $-y^2 + 8y - 12$
25. $-2m^2n - 2mn + n$ **26.** $8x + 6$ **27.** $9a + 3$ **28.** $9 + 3yz$
29. $6m - n$ **30.** $3x + 6$ **31.** $2a - 3b$ **32.** $2d$ **33.** $-m - 5n$
34. $7x^2 + x + 17$ **35.** $15x^2y + 2xy - 3y^2$ **36.** $a^2 + 4a + 4$
37. $-t^2 - 5t - 19$ **38.** -6 **39.** $x^2 - 6y^2$ **40.** $2\frac{1}{3}x^3$ **41.** $-12a^2b^3$
42. $42m^3n^5$ **43.** $8x^4y^3z^3$ **44.** $32r^3s^5t^2$ **45.** $18x^5y^4$ **46.** $4x^3$
47. $2a$ **48.** $-9x$ **49.** $9ac^2$ **50.** $-5yz^2$ **51.** $3t^3$
Applications and Problem Solving   52. Answers will
vary. **a)** $2x^3$ **b)** $x^4 + 1$ **c)** $x^2 + x + 1$ **d)** $x^5 + x^4 + x^3 + x^2$
53. 176.4 m **54. a)** $x^2 + 7x - 9$ **b)** $-x^2 - 7x + 9$
c) opposite **55. a)** x **b)** $2x^2 - 4x + 3$ **56. a)** $18x^2y$
b) $30xy + 12x^2$ **57. a)** $\dfrac{\pi d^2}{4}$ **b)** d^2 **c)** $\dfrac{\pi}{4}$ **d)** 79% **58.** 3, 5
59. 30 cm by 20 cm

Section 3.2 p. 105
Applications and Problem Solving   1. 90 **2.** 160
3. 204 **4.** 385 **5.** $y = x + 4$; 13, 16, 25 **6.** $y = 2x + 1$; 17,
12, 101 **7.** $y = 9 - x$; 3, 8, 0 **8.** $y = x - 7$; 5, 2, 7
9. $y = x^2 + 1$; 6, 401, 10 001 **10.** $y = 4x - 1$; 39, 16, 119
11. third letter; N, H **12.** number of consonants; 4, 3
13. second last letter; I, E **14.** number of consonants
minus number of vowels; 1, 0 **15. a)** 6 **b)** 3 **16. a)** 156
b) 1260 **17.** 41, 82, 81 **18.** 42, 68, 110 **19.** 37, 50, 65

20. $A = p - q$, -11, 5, 3; $B = q - p$, 11, -5, -3; $C = 2p + q$,
-4, -2, -12; $D = p^2 - q^2$, -11, -15, -27; $E = pq + 1$,
-29, -3, 19 **21. a)** 146 **b)** $146^2 = 21\ 316$ **22.** 6 **23.** 4

Section 3.3 pp. 107–109
Practice   1. $6x^2 - 8x$ **2.** $12a^2 - 9ab$ **3.** $-20st + 4t^2$
4. $-2x^3 + 7x^2$ **5.** $6y^3 - 2y^2$ **6.** $-3m^4 + 18m^3$ **7.** $6x - 2$
8. $20a^2 + 15a$ **9.** $-3 + 18y$ **10.** $-4x^3 + 12x^2$ **11.** $7x + 7$
12. $-3m + 33$ **13.** $-12x - 73$ **14.** $11t - 36$ **15.** $-20x - 9$
16. $-10 - 8y$ **17.** $18a - 24b - 54$ **18.** $-12t^2 + 4t + 4$
19. $-4m^2 + m + 7$ **20.** $63y^2 - 21y - 49$
21. $-10x^2 - 15xy + 5y^2$ **22.** $8y^3 + 12y^2 - 4y$
23. $-18x^4 + 36x^3 + 54x^2$ **24.** $8a^3b^3 - 2a^2b^3 + 6a^2b^4$
25. $3a^2bc + 4ab^2c - 2abc^2$ **26.** $-15x$
27. $4a^2 + 16ab - 2b^2 - 6a + 6b$ **28.** $-10x^3 - 3x^2 - 23x$
29. $-3y^3 - 3y^2 + 9y$ **30.** $6s^3 - s^2t - 2st^2 - 3t^3$
31. b) $x^2 + 9x + 20$ **32.** $t^2 + 9t + 20$ **33.** $m^2 + 5m + 4$
34. $x^2 - 7x + 12$ **35.** $w^2 - 15w + 56$ **36.** $x^2 - 8x + 16$
37. $y^2 - 2y - 63$ **38.** $s^2 - 3s - 4$ **39.** $a^2 - 13a + 36$
40. $28 + 3x - x^2$ **41.** $6 - 5y + y^2$ **42.** $x^2 + 14x + 49$
43. $b^2 - 64$ **44. b)** $6x^2 + 7xy + 2y^2$ **45.** $3x^2 + 10x + 3$
46. $3a^2 + 17a + 20$ **47.** $4y^2 - 7y - 15$ **48.** $5m^2 - 22m + 8$
49. $9x^2 - 24x + 16$ **50.** $4 - 19t - 30t^2$ **51.** $9a^2 - 25$
52. $3x^2 + 13xy + 4y^2$ **53.** $8a^2 - 22ab + 5b^2$
54. $20m^2 - 7mn - 6n^2$ **55.** $20s^2 - 39st + 18t^2$
56. $7a^2 + ab - 8b^2$ **57.** $2x^4 - 7x^3y + 3x^2y^2$
58. $-6a^2 - ab + 12b^2$ **59.** $2x^2 + 15x + 30$ **60.** 15
61. $14x^2 + 26x - 11$ **62.** $3b^2 - 39b + 126$
63. $2m^2 + 24m + 42$ **64.** $3x^2 - 11x - 34$
65. $30t^2 - 61t + 25$ **66.** $6x^2 + 36x + 5$
67. $-22y^2 - 20y - 5$ **68. b)** $2x^2 + 7x + 4xy + 2y + 3$
69. $x^3 + 5x^2 + 10x + 12$ **70.** $y^3 - 3y^2 - 3y + 10$
71. $6m^3 + 13m^2 - 6m - 8$ **72.** $2t^3 - 9t^2 - 19t - 7$
73. $x^4 + x^3 - 7x^2 - 7x + 4$ **74.** $y^4 - 4y^3 + 7y^2 - 7y + 2$
75. $3a^4 - 7a^3 - 9a^2 + 18a - 10$ **76.** $3x^6 - 14x^3 - 49$
77. $2x^4 - 5x^3 + 12x^2 - 11x + 3$
Applications and Problem Solving   78. $12x - 69$
79. $22t + 6$ **80.** $-16y - 38$ **81.** $3x^2 - 8x$
82. $-3y^3 + 10y^2 - y - 2$ **83.** $6x^3 - 4x^2 - 13x + 5$
84. $12x^3 - 12x^2 - 15x - 3$ **85. a)** $7x^2 - 72x + 20$
b) 29 280 cm²; 2.928 m² **86. a)** $x^2 + 10x - 200$;
$2x^2 + 22x + 56$ **b)** 5400 m²; 11 396 m² **87. a)** $x^2 + x - 2$
b) $x^2 + 3xy + 2y^2 + 3x - 3y$ **88. a)** $14x^2 + 17x - 3$
b) $2xy - y + x - 2$ **89. a)** $x^3 - 7x - 6$ **b)** $x^3 - 7x - 6$
c) $x^3 - 7x - 6$ **d)** no **90. a)** 42 cm² **b)** $(n + 2)(n + 3)$
c) $n^2 + 5n + 6$ **d)** 930 cm² **91. a)** $10x^2 + 10x - 10$
b) $2x^3 + 3x^2 - 11x - 6$ **c)** 290 cm² **d)** 264 cm³
92. a) The product of 3 consecutive numbers plus the
middle number. **b)** 8, 27, 64, 125 **c)** The answer
equals the cube of the middle number.
d) $(x - 1)(x)(x + 1) + x = x^3$ **e)** The simplified
expression equals the cube of the middle number.

93. a) Of 4 consecutive numbers, subtract the product of the outer 2 from the product of the inner 2 **b)** 2, 2, 2, 2 **c)** $(x + 1)(x + 2) - x(x + 3) = 2$ **d)** The simplified expression equals 2. **94.** no

Section 3.4 pp. 112–113
Practice **1.** $x^2 + 10x + 25$ **2.** $y^2 + 2y + 1$
3. $x^2 - 12x + 36$ **4.** $m^2 - 6m + 9$ **5.** $x^2 - 9$ **6.** $y^2 - 36$
7. $m^2 - 49$ **8.** $t^2 - 64$ **9.** $9x^2 + 12x + 4$
10. $25x^2 - 10x + 1$ **11.** $4x^2 - 9$ **12.** $4m^2 + 28m + 49$
13. $9y^2 - 4$ **14.** $16y^2 - 24y + 9$ **15.** $1 - 25m^2$
16. $4 - 12t + 9t^2$ **17.** $4x^2 - 9y^2$ **18.** $4x^2 + 12xy + 9y^2$
19. $9a^2 - b^2$ **20.** $16t^2 - 40st + 25s^2$ **21.** $16m^2 - 25n^2$
22. $9c^2 + 42cd + 49d^2$ **23.** $y^2 - 36x^2$ **24.** $a^2 - 16ab + 64b^2$
25. $2x^2 + 4x + 20$ **26.** $2y^2 + 14y + 13$ **27.** $-16m + 65$
28. $5a^2 + 12a - 6$ **29.** $-2x^2 + 100x - 94$
30. $-19t^2 - 30t + 105$ **31.** $-x^2 - 26x - 107$
32. $-8x^2 - 23x + 14$ **33.** $-7m^2 - 33m + 24$
34. $19t^2 + 12t - 14$ **35.** $-21y^2 + 13y + 28$
36. $-6t^3 + 36t^2 + 2t - 4$ **37.** $100s^2 - 22t^2 + 6t$
38. $12m^2 - 12mn - 3m + 2n^2 + 45$ **39.** $5x^2 + 4xy - 3y^2$
40. $-13a^2 - 28ab + 8b^2$ **41.** $50x^3 + 125x^2 - 18x - 45$
42. $18m^3 + 39m^2 + 8m - 16$ **43.** $x^3 + 6x^2 + 12x + 8$
44. $y^3 - 3y^2 + 3y - 1$ **45.** $64x^3 + 240x^2 + 300x + 125$
46. $27 - 54t + 36t^2 - 8t^3$ **47.** $8y^3 - 12xy^2 + 6x^2y - x^3$
48. $27a^3 + 54a^2b + 36ab^2 + 8b^3$
Applications and Problem Solving **49. a)** $x^2 - 2x + 1$; $4x^2 - 16x + 16$; $4x^2 + 40x + 100$ **b)** 11 881 m²; 46 656 m²; 52 900 m² **50. a)** $y^2 - 6y + 11$ **b)** $3y^2 - 7y - 7$
51. a) $x^3 - 3x^2 - 9x + 27$ **b)** 275 m³
52. a) $343x^3 - 294x^2 + 84x - 8$ **b)** 64 000 m³
53. a) $24x^2 - 24xy + 6y^2$ **b)** 96 cm²
c) $8x^3 - 12x^2y + 6xy^2 - y^3$ **d)** 64 cm³ **54. a)** $x^4 + 2x^2 + 1$
b) $y^4 - 2y^2 + 1$ **c)** $x^4 + 2x^2y^2 + y^4$ **d)** $x^4 - 2x^2y^2 + y^4$
e) $4x^4 + 12x^2 + 9$ **f)** $9y^4 - 24y^2 + 16$ **g)** $x^4 - 4x^2y^2 + 4y^4$
h) $16x^4 + 24x^2y^2 + 9y^4$ **55. a)** $x^4 - 1$ **b)** $y^4 - 4$ **c)** $x^4 - y^4$
d) $64a^4 - 9$ **e)** $9x^4 - 4y^4$ **f)** $16 - 9c^4$
56. a) $36y^4 - 72y^3 + 35y^2 + 2y - 1$
b) $m^4 + 2m^3 - 11m^2 - 12m + 36$ **57.** square, by 9 cm²
58. a) $a^2 + b^2 + c^2 + 2ab + 2ac + 2bc$
b) $4x^2 + 9y^2 + 1 + 12xy + 4x + 6y$ **59. a)** $a^2 + 2ab + b^2$; equal **b)** $a^3 + 3a^2b + 3ab^2 + b^3$; equal
c) $a^4 + 4a^3b + 6a^2b^2 + 4ab^3 + b^4$ **d)** $x^3 + 9x^2y + 27xy^2 + 27y^3$
e) $x^4 + 8x^3 + 24x^2 + 32x + 16$ **60. a)** $a^3 - 3a^2b + 3ab^2 - b^3$
b) $27 - 54y + 36y^2 - 8y^3$; $64y^3 - 144y^2x + 108yx^2 - 27x^3$
61. a) 24, 25; 41, 40 and 15, 12; 37, 35 and 20, 16 and 15, 9 and 13, 5

Section 3.5 p. 115
Applications and Problem Solving **1.** 4 time periods; A and D: 13:00 to 13:40; B: 13:40 to 14:20; C and F: 14:20 to 15:00; E: 15:00 to 15:40 **2. a)** Alliston to

Bevan to Gaston **b)** Clearwater to Flagstaff to Dunstan **c)** Gaston to Flagstaff to Clearwater to Bevan **d)** Flagstaff to Clearwater to Bevan to Alliston **3. a)** 20 **b)** 35 **4.** 276 m² **5.** 60 **6.** 300 **7.** 12 L **8.** 4 blocks east, 7 blocks south

Technology pp. 116–117
1 Multiplying Polynomials **1.** $8x^2 + 42x + 27$
2. $6x^2 - 5x - 25$ **3.** $20y^2 - 52y + 33$ **4.** $32x^2 - 4xy - 21y^2$
5. $20x^2 + 7xy - 6y^2$ **6.** $12x^4 + x^2 - 35$ **7.** $3x^4 + x^2y - 10y^2$
8. $10x^6 - 13x^3y + 4y^2$ **9.** $x^3 - x^2 + 12$
10. $4x^3 + 4x^2 - 13x + 3$ **11.** $5m^3 - 34m^2 - 33m + 18$
12. $x^4 + 5x^3 + 12x^2 + 17x + 5$ **13.** $2y^4 - 6y^3 + 5y^2 + y - 3$
14. $20x^3 - x^2 - 14x - 3$ **15.** $6x^4 + 10x^2y - 3x^2 - y^2$
16. $x^3 + 7x^2 - 11x + 3$
2 Special Products **1.** $x^2 + 30x + 225$ **2.** $t^2 - 10t + 25$
3. $81 - 18y + y^2$ **4.** $9x^2 - 48x + 64$ **5.** $16m^4 + 56m^2 + 49$
6. $36 - 60r + 25r^2$ **7.** $64x^2 + 48xy + 9y^2$
8. $25x^2 - 20xy + 4y^2$ **9.** $27x^3 + 54x^2 + 36x + 8$
10. $64x^3 + 240x^2y + 300xy^2 + 125y^3$
11. $8y^3 - 84xy^2 + 294x^2y - 343x^3$
12. $4x^2 + 12xy + 4xz + 9y^2 + 6yz + z^2$ **13.** $4x^2 - 121$
14. $16 - 25x^2$ **15.** $9y^2 - 25x^2$ **16.** $36t^4 - 25s^4$ **17.** $4x$
18. $-4x - 13$ **19.** $44y^2 + 68y + 26$
20. $-40m^2 + 48m + 15$
21. $4x^4 + 24x^3 + 11x^2 - 150x - 225$
22. $3y^4 + 18y^3 - 33y^2 - 180y + 300$
3 Multiplying and Evaluating **1.** -1 **2.** 33 **3.** 617 **4.** 79
5. 214 **6.** 1976 **7.** 40.25 **8.** 33.52 **9. a)** $13x^2 - 27x + 2$
b) 450 m² **c)** $46y^2 - 21y - 1$ **d)** 350 m²
10. a) $25x^2 - 30x + 9$ **b)** 288 m²

Section 3.6 p. 120
Practice **1.** $5(x + 5)$ **2.** not possible **3.** $8(x + 1)$
4. $9(y - 1)$ **5.** $3(x - 5y)$ **6.** $5x(5x + 2)$
7. $2a(2x + 4y - 3z)$ **8.** $pq(5r - s - 10t)$ **9.** $2(x^2 - x - 3)$
10. not possible **11.** $9(a^3 + 3b^2)$ **12.** $3x(x^2 - 1)^2$
13. $4y(3 - 2y + 6y^2)$ **14.** $6w^3(2w + 1)(2w - 1)$
15. not possible **16.** $11b(3a + 2c - b)$ **17.** $8xy(3y + 2x)$
18. $5y(7x - 2y)$ **19.** not possible **20.** $12xy(2y - 1 + 3x)$
21. $9a^2b^2(3b + 1 - 2a)$ **22.** $6mn^2(m^2 + 3mn - 2 + 4n)$
23. $(a + b)(5x + 3)$ **24.** $(x - 1)(3m + 5)$ **25.** not possible
26. $(p + q)(4y - x)$ **27.** $(m + 7)(4t + 1)$ **28.** not possible
29. $(m - n)(8x + 6)$ **30.** $(w + z)(x + y)$ **31.** $(x + 3)(4 + y)$
32. $(x + 1)(x - y)$ **33.** $(m + 4)(m - n)$ **34.** $(x + 2)(2x + 3y)$
35. $(t - 2)(5m^2 + t)$ **36.** $(a - 3)(3a - 2b^2)$
Applications and Problem Solving **37. a)** $20t - 5t^2$
b) 0 m, 15 m, 20 m, 15 m, 0 m, −25 m **c)** 20 m
d) Negative distances are impossible. **e)** 0 s and 4 s
f) $5t(4 - t)$ **g)** It is easy to see when the expression equals 0. **38. a)** $4\pi x^2 - x^2$ **b)** $x^2(4\pi - 1)$ **39. a)** $6xy - 3xz$
b) $3x(2y - z)$ **40. a)** $\pi r^2 - 2r^2$ **b)** $r^2(\pi - 2)$

41. a) $10x + 10y + 100$ **b)** $10(x + y + 10)$
42. a) $6a + 3b + 3c + 6d + 36$ **b)** $3(2a + b + c + 2d + 12)$
43. a) $4a + 4b - 16$ **b)** $4(a + b - 4)$ **44.** k is divisible by 2
45., 46. Answers will vary.

Career Connection p. 121
1 Cardio-respiratory Assessment 1. a) $2a + 2b + 2c$
b) the three heartbeat measurements for one minute
2. $\dfrac{50d}{a+b+c}$ **3. a)** 57.3, low average **b)** 82.2, good
c) 76.9, average
2 Burning Energy 1. a) 579.4 kJ **b)** 895.0 kJ
c) 1849.8 kJ **2.** walking, 189 kJ

Section 3.7 pp. 123–124
Applications and Problem Solving 1. a) 26 min;
north, north, east, east, north, east **b)** 51 min; south,
west, south, south, south, west, west **c)** 50 min; west,
west, north, west, west, north, north, north **2.** 36
3. 05:20 **4. a)** west on Hwy 363, south on Hwy 58 to
Gravelbourg, east on Hwy 43 to Vantage, south on
Hwy 2 to Assiniboia, west on Hwy 13 into Limerick
b) east to Assiniboia on Hwy 13, north on Hwy 2 to
Vantage, west on Hwy 43 to Gravelbourg, north on
Hwy 58 until you meet Hwy 363, east on Hwy 363
into Courval **5. a)** 4000 **6.** 256 **7.** $\dfrac{3}{4}$

Section 3.8 pp. 126–127
Practice 1. 3, 5 **2.** −4, −3 **3.** −3, 10 **4.** −5, 4 **5.** 2, 5
6. −15, 1 **7.** −9, −4 **8.** −6, −4 **9.** $(x + 1)(x + 4)$
10. $(x + 3)(x + 5)$ **11.** $(m + 2)(m + 5)$ **12.** not possible
13. $(r - 6)(r - 7)$ **14.** $(n + 5)(n + 6)$ **15.** $(r - 2)(r - 5)$
16. $(w - 2)(w - 8)$ **17.** not possible **18.** $(m - 4)(m - 6)$
19. $(y + 4)(y - 5)$ **20.** $(x + 9)(x - 2)$ **21.** not possible
22. $(x + 7)(x - 2)$ **23.** $(n + 2)(n - 12)$ **24.** $(m + 3)(m - 7)$
25. not possible **26.** $(x + 2)(x - 10)$ **27.** $(m + 8)(m + 10)$
28. $(m + 4)(m - 3)$ **29.** not possible **30.** $(r - 3)(r - 14)$
31. $(y - 8)(y - 9)$ **32.** $(x + 2)(x - 8)$ **33.** $(t - 2)(t - 13)$
34. not possible **35.** not possible **36.** $(n + 11)(n - 4)$
37. $(x - 3)(x - 7)$ **38.** $(w + 2)(w + 10)$ **39.** $(r + 5)(r - 6)$
40. $(y - 2)(y - 18)$ **41.** not possible **42.** $(x + 7y)(x + 5y)$
43. $(a + 7b)(a - 11b)$ **44.** $(c + d)(c - 2d)$
45. $(x + 9y)(x - 4y)$ **46.** not possible **47.** $(xy + 1)(xy + 2)$
48. $(ab - 2)(ab - 3)$ **49.** $(8 - y)(1 + y)$ **50.** $(8 + x)(2 - x)$
51. $2(x - 1)(x - 2)$ **52.** $3(x + 1)(x + 3)$ **53.** $5(y + 2)(y + 6)$
54. $4(t + 3)(t - 5)$ **55.** $6(x + 4)(x - 1)$ **56.** $a(x + 12)(x - 2)$
57. $x(x + 6)(x + 12)$ **58.** $2(x - 4)(x - 7)$ **59.** $5(w + 6)(w - 2)$
60. $x(3 + x)(1 - x)$
Applications and Problem Solving 61. a) $(x + 1)(x - 4)$
b) 18 m by 13 m **64. a)** −4, −24, 4, −28, 24, −3, −6,
15, 3, −20 **b)** MACK SENNETT; KEYSTONE
KOPS **65.** Answers may vary. **a)** 1, −24, −15

b) 6, −6, −50 **66. a)** −5, 5, −7, 7 **b)** −11, 11, −4, 4, −1, 1
c) −25, 25, −14, 14, −11, 11, −10, 10 **d)** −17, 17, −7, 7,
−3, 3 **67. a)** $(x^2 + 1)^2$ **b)** $(x^2 + 3)(x^2 - 2)$
c) $(x^2 + 2)(x^2 - 5)$ **d)** $(x^2 + 9y)(x^2 + y)$
68. a) $(x + a + 1)(x + a + 2)$ **b)** $(x - b + 5)(x - b - 1)$
69. a) $(x + 5)(x - 7); (t + 8)(t - 5)$ **b)** one factor or the
other must be zero

Section 3.9 pp. 130–131
Practice 1. $(y + 3)(2y + 3)$ **2.** $(m + 3)(3m + 1)$
3. $(t + 1)(5t + 2)$ **4.** not possible **5.** $(x + 2)(2x + 7)$
6. $(2x + 3)(3x + 1)$ **7.** $(x - 1)(2x - 3)$ **8.** $(3x - 2)(x - 1)$
9. $(t - 2)(3t - 4)$ **10.** $(m - 2)(5m - 1)$
11. $(3m - 2)(2m - 3)$ **12.** not possible **13.** $(x - 2)(2x + 3)$
14. $(3x - 4)(2x + 1)$ **15.** $(t + 5)(2t - 1)$
16. $(3n + 1)(5n - 2)$ **17.** $(x - 1)(3x + 4)$
18. $(4x + 3)(x - 1)$ **19.** $(5y + 1)(y - 3)$
20. $(4x + 1)(2x - 3)$ **21.** not possible **22.** $(5t - 2)(2t + 3)$
23. $(2t + 3)(2t + 1)$ **24.** $(5x - 1)(2x - 3)$ **25.** not
possible **26.** $(2y + 5)(y + 3)$ **27.** $2(y - 2)(4y - 3)$
28. not possible **29.** $3(r + 1)(2r + 3)$ **30.** $(3y - 2)(4y - 1)$
31. $2(x - 5)(2x + 1)$ **32.** $m(m + 6)(2m - 5)$
33. $t(t + 4)(2t + 1)$ **34.** $(2s - 1)(9s + 1)$
35. $3(r + 1)(4r + 5)$ **36.** $s(r - 1)(5r - 2)$
37. $(2 - y)(3 + 4y)$ **38.** $(2 - m)(1 - 3m)$
39. $2(6 + 9t + 4t^2)$ **40.** $(3 - 2y)(2 + 3y)$
41. $(2m - n)(3m + 2n)$ **42.** $(3x + y)(x + 2y)$
43. $(2a - b)(5a + b)$ **44.** $(2x - y)(x - 5y)$
45. $(6c + d)(c + 2d)$ **46.** $(m - n)(4m - n)$
47. $3(x - y)(2x - y)$ **48.** $2(m + n)(m - 3n)$
49. $4(y + 2x)(y - x)$ **50.** $2(3a - 2b)(a - 3b)$
Applications and Problem Solving
51. a) $(10x + 3)(x - 1)$ **b)** 503 m by 49 m **53. a)** 2, 5,
−2, 4, −3, −4, 3, −6, 6, −1, 7, 8, −9, 10, −5, 1, −7
b) JONI MITCHELL - folk singer; NED
HANLAN - rower; MARSHALL MCLUHAN -
author; EMILY STOWE - first female doctor in
Canada **54. a)** −16, 16, −8, 8 **b)** −41, 41, −22, 22, −14,
14, −13, 13 **c)** −5, 5, −1, 1 **d)** −35, 35, −16, 16, −9, 9,
−5, 5, 0 **55.** Answers will vary. **a)** 1, −2, −5 **b)** 4, 5, −3
56. a) $(x^2 + 1)(2x^2 + 1)$ **b)** $(x^2 + 3)(2x^2 - 1)$
c) $(3x^2 - 4)(x^2 + 1)$ **d)** $(2x^2 - 3)(3x^2 - 2)$
e) $(x^2 + 2y)(2x^2 + y)$ **f)** $(3x^2 - y)(x^2 + 4y)$

Section 3.10 pp. 133–134
Practice 1. $(x + 3)(x - 3)$ **2.** $(y + 4)(y - 4)$
3. not possible **4.** $(5a - 6)(5a + 6)$ **5.** $(1 - 8t)(1 + 8t)$
6. $(6 - 7a)(6 + 7a)$ **7.** not possible **8.** $(5x - 8y)(5x + 8y)$
9. $(2t - 3s)(2t + 3s)$ **10.** $(10p + 11q)(10p - 11q)$
11. $(16 - 9y)(16 + 9y)$ **12.** $(15b - a)(15b + a)$
13. $(x + 3)^2$ **14.** $(y - 5)^2$ **15.** not possible **16.** $(2t + 1)^2$
17. $(m - 10)^2$ **18.** $(4t + 3)^2$ **19.** $(7 + x)^2$ **20.** $(1 - 8t)^2$

21. $(3x - 4)^2$ **22.** $(2 + 7r)^2$ **23.** not possible
24. $(6m + 5n)$ **25.** $(11m - 1)^2$ **26.** $(3a + 2b)^2$
27. $(y - 12)(y + 12)$ **28.** not possible **29.** $(3a - 4)^2$
30. $2(x - 4)(x + 4)$ **31.** not possible **32.** $3(x + 1)^2$
33. $(m - 7)^2$ **34.** $(2p + 5q)^2$ **35.** $(7x - 11y)(7x + 11y)$
36. $5(4a + 3b)(4a - 3b)$ **37.** not possible
38. $y(y - 6)(y + 6)$ **39.** $y(y - 9)^2$ **40.** $4(9x^2 + 25y^2)$
41. $3x(x + 4)(x - 4)$ **42.** $5m(m - 4)^2$
43. $(9x - 12)(9x + 12)$ **44.** $3(b - 10)(b + 10)$
Applications and Problem Solving **46.** $(x + 5)(x - 1)$
47. $(7 - y)(1 + y)$ **48.** $-2m - 3$ **49.** $(x^2 + 11)^2$
50. $(t^3 - 9)^2$ **51.** $\left(\dfrac{x}{2} - \dfrac{1}{3}\right)\left(\dfrac{x}{2} + \dfrac{1}{3}\right)$ **52.** $(5x^2 + 9)(5x^2 - 9)$
53. $8xy$ **54. a)** 600 **b)** 800 **c)** $640\,000$ **55. a)** $-8, 8$
b) $-42, 42$ **c)** 4 **d)** 9 **e)** 25 **f)** 16 **56. a)** $2(x - 1)^2$
b) $L = 2(x - 1)$, $W = x - 1$ **c)** 18 m by 9 m
57. a) $2x(x - 6)^2$ **b)** $2x, x - 6, x - 6$ or $x, 2(x - 6), x - 6$
c) 16 cm by 2 cm by 2 cm or 8 cm by 4 cm by 2 cm
d) No; dimensions must be greater than zero.
58. $a = 5, b = 2$; or $a = 11, b = 10$ **59. a)** $6\pi(5.6^2 - 4.4^2)$
b) 72π **c)** 226 cm^3 **60. b)** $(s - 1)^2$ **c)** $121; 8100$ **d)** 24
e) $2s - 1$ **f)** 105 **61.** 4 cm, 12 cm, 20 cm
62. a) $(x + 3 - y)(x + 3 + y)$ **b)** $(x - 2 - 3y)(x - 2 + 3y)$
c) $(2x + 3y - 2z)(2x + 3y + 2z)$ **d)** $(x^2 - y - z)(x^2 - y + z)$
63. 16 cm

Investigating Math p. 135
1 Factoring $x^2 + bx + c$ **1. a)** $x^2 + 5x + 4$ **b)** $(x + 4)$,
$(x + 1)$ **2. b)** no
2 Factoring $ax^2 + bx + c$, $a \neq 1$ **1. a)** $2x^2 + 7x + 3$
b) $(2x + 1), (x + 3)$ **2. b)** no
3 Factoring Special Trinomials **1. a)** $x^2 + 4x + 4$
b) $(x + 2), (x + 2)$ **2. b)** no

Technology p. 136
1 Factoring Polynomials **1.** $3(2x^2 + 5x - 4)$
2. $7(2y^2 - 6y + 3)$ **3.** $4(5 + 2m^2 + 3m)$ **4.** $5(4x - 3x^2 + 2)$
5. $2xy(2x + 3 - 4y)$ **6.** $3pq(p^2 + 6pq + 2q^2)$
7. $2b^2(6a^3 + 2a^2b + 4ab^2 - 3b^3)$
8. $5x^2y^2(2x^3 + x^2y - 3xy^2 + 5y^3)$ **9.** $(x + 2)(x + 17)$
10. $(x + 6)(x - 12)$ **11.** not possible **12.** $(5 - t)(3 - t)$
13. $(x + 2y)(x + 5y)$ **14.** $(2y - x)(14y - x)$
15. $(n + 1)(4n + 9)$ **16.** not possible **17.** $(x - 4)(5x + 3)$
18. $(3y - 2)(5y + 7)$ **19.** $(2 + 3z)(10 + 3z)$
20. $(x^2 + 5)(x^2 - 3)$ **21.** $(3x^2 - 5)(4x^2 + 5)$
22. $(x - 4y)(3x - 2y)$ **23.** $(3a - 2b)(5a + 3b)$
24. $(7x - 4y)(2x + 9y)$ **25.** $(x + a + 4)(x + a + 2)$
26. $(x - y - 2)(x - y - 3)$ **27.** $3(x - 1)(x - 9)$
28. $2(2x - 3)(x + 4)$ **29.** $7(x + 6)(3x - 1)$
30. $5(3y + 8)(5y + 1)$ **31.** $2(u - v)(u - 2v)$
32. $6(3x - y)(2x + 3y)$ **33.** $x(x + 1)(x + 2)$
34. $2t(t - 7)(2t + 1)$ **35.** $3(5x^2 + 2)(2x^2 + 5)$

36. $8(3x^2 + 1)(x - 1)(x + 1)$ **37.** $4(x^2 - 5y^2)(2x^2 - y^2)$
38. $11(9a^2 - 2b^2)(a^2 + 3b^2)$
2 Factoring Special Products **1.** $(5x + 6)^2$ **2.** $(3y - 5)^2$
3. $(2m + 9n)^2$ **4.** $(7x - 4y)^2$ **5.** $(2a^2 + 5b^2)^2$
6. $(3x - 8)(3x + 8)$ **7.** $(5 - 13x)(5 + 13x)$
8. $(2x^2 - 3y)(2x^2 + 3y)$ **9.** $(x^2 + 2 - 2y^2)(x^2 + 2 + 2y^2)$
10. $(6x - y^2 + 5)(6x + y^2 - 5)$ **11.** $16(m - 2)(m + 2)$
12. $4(3 - 2x)(3 + 2x)$ **13.** $5(5x^2 + 4)(5x^2 - 4)$
14. $2(6x + 7y^2)(6x - 7y^2)$ **15.** $2(x - 7)^2$
16. $3(2x + 5)^2$ **17.** $8w(2w - 5)^2$ **18.** $12(5 - 2x^2)(5 + 2x^2)$
19. $x^3(2x - 3y)^2$ **20.** $4(3y^2 + 5x^2)^2$
3 Egyptian Pyramids **1. a)** $(5x - 3), (4x + 7)$
b) 122 m by 107 m **2. a)** $6x - 4$ **b)** 188 m

Career Connection p. 137
1 Inflation **3. b)** 1.5% **c)** 1981, 12.4% **4. a)** $1979.96
b) $2180.82 **c)** $2717.39

Connecting Math and Architecture
pp. 138–139
1 Writing Expressions for Areas **1.** πr^2 **2.** $r; r + 1$
3. a) $\pi(r + 1)^2 - \pi r^2$ **b)** $\pi(2r + 1)$ **4. a)** $\pi(r + 2)^2 - \pi(r + 1)^2$;
$\pi(2r + 3)$ **b)** $\pi(r + 3)^2 - \pi(r + 2)^2$; $\pi(2r + 5)$
c) $\pi(r + 4)^2 - \pi(r + 3)^2$; $\pi(2r + 7)$
d) $\pi(r + 5)^2 - \pi(r + 4)^2$; $\pi(2r + 9)$ **5. b)** $\pi(2r + 2n - 1)$
c) $\pi(2r + 15)$ **6. a)** 44 m^2 **b)** 57 m^2 **c)** 75 m^2 **d)** 88 m^2
7. a) $\pi(26r + 169)$ **b)** 740 m^2
2 Writing Expressions for Circumference **1.** $2\pi r$
2. a) $2\pi(r + 1)$ **b)** $2\pi(r + 2)$ **c)** $2\pi(r + 5)$ **d)** $2\pi(r + 12)$
3. a) $26\pi(r + 6)$ **b)** 690 m
3 Estimating Seating Capacities Estimates may vary.
1. At 0.75 m per seat **a)** 21 **b)** 54 **c)** 80 **d)** 926
2. $1\,047\,200$

Review pp. 140–141
1. $5x - 3y$ **2.** $8x^2 - 4x + 3$ **3.** $-a^2 - 6a - 8$
4. $m^2 + 3mn + n^2$ **5.** $-12x^4y^4$ **6.** $24r^2s^4t^6$ **7.** $-4a$ **8.** $4n^3p$
9. $8x + 18$ **10.** $4a + 28$ **11.** $8t^2 - 3t$ **12.** $y^3 - 7y$
13. a) The digits are reversed. **14.** $x^2 + 2x - 8$
15. $a^2 - a - 30$ **16.** $6y^2 - y - 12$ **17.** $3x^2 - 11xy - 4y^2$
18. $2y^2 - 4y - 6$ **19.** $-7x^2 - 12x + 6$ **20.** $8a^2 + 12a + 19$
21. $x^3 - 6x^2 + 7x + 6$ **22.** $6t^3 + t^2 - 3t - 1$ **23.** $x^2 + 8x + 16$
24. $y^2 - 16$ **25.** $a^2 - 10a + 25$ **26.** $9t^2 - 1$
27. $4x^2 - 12xy + 9y^2$ **28.** $25a^2 - 9b^2$ **29.** $18m^2 + 12m + 2$
30. $1 - 6x + 12x^2 - 8x^3$ **31.** $2m^2 - 8m + 7$ **32.** $-12x + 61$
33. $30t^2 + 12t + 1$ **34.** $-9x^2 + 18xy - 11y^2$ **35. b)** Square
their average and subtract 9. **c)** $(x + 3)(x - 3) = x^2 - 9$
36. $5(t - 7)$ **37.** $5y(y - 4)$ **38.** $2xy(1 - 4y)$ **39.** not
possible **40.** $(m - 3)(x + 4)$ **41.** not possible
42. $(2x - 1)(m + n)$ **43.** $(m + 2)(x + y)$ **44.** $(x - 1)(x - y)$
45. $(m - 3)(2m + t)$ **46. a)** πr^2 **b)** $\pi r^2 + 6\pi r + 9\pi$
c) $3\pi(2r + 3)$ **d)** 122.5 cm^2 **47.** $(x + 3)(x - 4)$

48. $(y + 6)(y - 3)$ **49.** $(m + 3)(m + 8)$ **50.** $(t - 3)(t - 5)$
51. not possible **52.** $(n - 5)(n - 8)$ **53.** $(w + 5)(w - 6)$
54. $(7 - m)(2 + m)$ **55.** $(x + y)(x + 8y)$ **56.** $(c - 2d)(c - 8d)$
57. $2(x + 4)(x - 5)$ **58.** $a(y + 14)(y - 2)$
59. $3(x + 6)(x - 2)$ **60.** $5(x - 1)(x - 2)$ **63.** $(5m + 2)(m + 3)$
64. $(3x + 2)(2x + 1)$ **65.** $(2x - 5)(x - 1)$
66. $(t - 2)(3t + 10)$ **67.** not possible **68.** $(2y + 1)(3y - 1)$
69. $(3x + 1)(2x - 1)$ **70.** $(3z - 1)(3z - 2)$
71. $(x + 5y)(2x + y)$ **72.** $(4p - 7q)(p + q)$
73. $2(2x + 1)(x + 1)$ **74.** $3(3t - 2)(t + 1)$
75. $4(m - 1)(5m + 3)$ **76.** $y(3y - 1)(y - 2)$
77. a) $(11x - 7)$, $(7x + 1)$ **b)** 180 cm by 120 cm
78. $(x - 5)(x + 5)$ **79.** $(1 + 7m)(1 - 7m)$ **80.** not
possible **81.** $(7t + 9s)(7t - 9s)$ **82.** $4(a - 2b)(a + 2b)$
83. $(12p - q)(12p + q)$ **84.** $(x + 5)^2$ **85.** $(y - 6)^2$
86. $(3t - 1)^2$ **87.** not possible **88.** $(2x + 3)^2$ **89.** $(5r - 2s)^2$
90. $5(x + 1)(x - 1)$ **91.** $4(2m + 3n)(2m - 3n)$
92. $2(3y + 5)^2$ **93.** $5(x - 2y)^2$ **94. a)** $3x - 2$ **b)** 220 m

Exploring Math p. 141
1. 3, 5, 8, 13, 21 **2.** For 3 or more panels, each number
is the sum of the previous two numbers. **3. a)** 89 **b)** 987

Chapter Check p. 142
1. $8y^2 - 3y - 3$ **2.** $-7x + 2$ **3.** $10a^4b^5$ **4.** $5xz^2$ **5.** $-6m - 14$
6. $5x^2 - 11x$ **7.** $-4a^3 - 7a^2 + 4a$ **8.** $y^2 + 2y - 15$
9. $2a^2 - 5ab - 3b^2$ **10.** $m^2 - 12m - 3$ **11.** $6x^2 + 2x + 19$
12. $3x^3 - 10x^2 - 17x - 6$ **13.** $x^2 - 2x + 1$ **14.** $4y^2 - 9$
15. $x^3 - 6x^2y + 12xy^2 - 8y^3$ **16.** $30x^2 + 12x - 1$
17. $4m(m - 7)$ **18.** $3ab(1 - 3b + 2a)$ **19.** $(x - 4)(m - 2)$
20. $(y + x)(y + 2)$ **21.** $(x - 3)(x - 4)$ **22.** $(a + 7)(a - 3)$
23. $(y + 4)(y + 5)$ **24.** $(t + 3)(t - 9)$ **25.** $(x + 5y)(x + y)$
26. $(m - 2n)(m - 7n)$ **27.** $3(x + 1)(x - 2)$
28. $4(t - 2)(t - 5)$ **29.** $a(y + 4)(y - 3)$ **30.** $(3t + 5)(t + 1)$
31. $(2m - 1)(m - 4)$ **32.** $(2x + 1)(3x - 2)$
33. $(2y + 3)(2y - 1)$ **34.** $(5r - s)(r - 2s)$
35. $(2x + y)(2x - 5y)$ **36.** $(x + 2)(x - 2)$
37. $(1 - 6m)(1 + 6m)$ **38.** $(6t + 7s)(6t - 7s)$
39. $(11a + b)(11a - b)$ **40.** $(x + 4)^2$ **41.** $(y - 3)^2$
42. $(2t - 1)^2$ **43.** $(2m + 5)^2$ **44.** $3(y + 3)(y - 3)$
45. $t(3 - 2t)(3 + 2t)$ **46.** $2(2x + 3)^2$ **47.** $x(x - 2)^2$
48. a) $25x^2 + 30x + 9$ **b)** 1444 cm² **49. a)** $(5x - 8)$,
$(2x + 5)$ **b)** 152 mm by 69 mm

Using the Strategies p. 143
1. 7.5 L/min **2.** 32 **3.** Pedro: 50; Lucia: 30 **4.** 120
5. 9 **6.** $\dfrac{4}{3} + \dfrac{5}{2}$ **7. a)** 5 **b)** 1 **8.** 12 **9.** 18, 20, 24
10. 4 cm by 8 cm by 18 cm **11.** 28
Data Bank **1.** 09:00 Saturday **2. a)** Answers may
vary. Vancouver, Edmonton, Toronto **b)** St. John's,
Charlottetown

Chapter 4

Rational Expressions and Equations p. 145
a) days in a leap year **b)** minutes in an hour **c)** angles
in a triangle **d)** freezing point of water in degrees
Celsius **e)** half a dozen **f)** sides on a quadrilateral
g) equal sides on an equilateral triangle **h)** trunk on an
elephant **i)** wheels on a tricycle **j)** legs on a spider
k) holes on a golf course **l)** lives of a cat **m)** keys on a
piano **n)** provinces in Canada **o)** tires on a car
p) Olympic rings **q)** faces on a cube **r)** strings on a
violin **s)** time at noon **t)** toes on a three-toed sloth
u) sides on a stop sign **v)** sun in our solar system
w) stars on the Stars and Stripes

Getting Started pp. 146–147
Developing Formulas From Patterns **1. a)** 8, 12, 16,
20, 24 **b)** $M = 4c + 4$ **c)** 44 **d)** 12, 20, 28, 36, 44
e) $T = 8c + 4$ **f)** 84 **2. a)** 10, 14, 18, 22, 26 **b)** $P = 2n + 6$
c) 50 **3. a)** 2, 5, 8, 11, 14 **b)** $T = 3s - 1$ **c)** 44
Using Unit Rates **1.** 84 L **2.** 7 **3.** 100 L **4.** 12.5 s
5. 7.2 h **6.** 12 500 kJ **7.** 4 365 000 km² **8.** 1.36 s
Mental Math
Solving Equations **1.** 1 **2.** 5 **3.** 4 **4.** 7 **5.** 2 **6.** 2 **7.** 2
8. 3 **9.** 3 **10.** 2 **11.** 3 **12.** 4 **13.** 2 **14.** 6 **15.** 15 **16.** 16
Adding Using Compatible Numbers **1.** 81 **2.** 122
3. 112 **4.** 196 **5.** 281 **6.** 383 **7.** 4.1 **8.** 5.3 **9.** 12.3
10. 12.3 **11.** 19.2 **12.** 310 **13.** 420 **14.** 420 **15.** 1110
16. c) 10

Section 4.1 pp. 152–153
Practice **1.** $5y$, $x \neq 0$ **2.** $15m$, $m, n \neq 0$
3. $-4x^2y^2$, $x, y \neq 0$ **4.** $-8ac^2$, $b, c \neq 0$ **5.** $5b^3$, $a, b, c \neq 0$
6. -1, $x, y \neq 0$ **7.** $2x$, $x \neq 0$ **8.** $-a$, $a \neq 0$ **9.** $4m^2n$,
$m, n \neq 0$ **10.** $-4a^2b^2$, $a, b \neq 0$ **11.** $x + 9$ **12.** $2a - 3b$
13. $-x^2 + 3x + 4$ **14.** $2a + b - 1$ **15.** $y - 4$, $y \neq 0$
16. $-a + 4b$, $a, b \neq 0$ **17.** $-4y + 5x^2 + 6x^3y^2$, $x, y \neq 0$
18. $x + 5$, $x \neq -3$ **19.** $a - 2$, $a \neq 5$ **20.** $y + 3$, $y \neq 4$
21. $t - 2$, $t \neq -2$ **22.** $x^2 + x + 2$, $x \neq -1$
23. $t^2 - t - 1$, $t \neq -4$ **24.** $a^2 - 1$, $a \neq 3$
25. $x^2 + 5x - 8$, $x \neq 1$ **26.** $m^2 + m - 2$, $m \neq -2$
27. $y^2 - 2$, $y \neq 4$ **28.** $n + 2$, $n \neq \pm 1$ **29.** $x - 3$
30. $y + 5$ **31.** $a - 1$, $a \neq \pm 2$
32. $x + 3$, $x \neq -\dfrac{5}{2}$ **33.** $y + 3$, $y \neq \dfrac{1}{3}$ **34.** $r - 6$, $r \neq \dfrac{1}{5}$
35. $2t - 3$, $t \neq -\dfrac{1}{2}$ **36.** $2r - 7$, $r \neq \dfrac{2}{3}$ **37.** $2s + 3$, $s \neq -\dfrac{3}{5}$
38. $2x + 5$, $x \neq \dfrac{3}{4}$ **39.** $2x^2 + 3$, $x \neq 1$ **40.** $3z^2 + 5$, $z \neq -2$
41. $2m^2 - 3$, $m \neq -\dfrac{1}{2}$ **42.** $3n^2 - 4$, $n \neq \dfrac{3}{2}$ **43.** $2d + 3$

44. $4g - 3$ **45.** $4s + 1$, $s \neq \pm\sqrt{\dfrac{5}{3}}$ **46.** $3t - 4$, $t \neq \pm\sqrt{\dfrac{2}{7}}$

47. $\dfrac{t^2 + 4t + 2}{t + 4} = t + \dfrac{2}{t + 4}$,

$t^2 + 4t + 2 = (t + 4)t + 2$, $t \neq -4$

48. $\dfrac{p^2 - 3p - 8}{p - 3} = p - \dfrac{8}{p - 3}$,

$p^2 - 3p - 8 = (p - 3)p - 8$, $p \neq 3$

49. $\dfrac{y^2 - 7y + 10}{y - 4} = y - 3 - \dfrac{2}{y - 4}$,

$y^2 - 7y + 10 = (y - 4)(y - 3) - 2$, $y \neq 4$

50. $\dfrac{2w^2 + w - 3}{w + 2} = 2w - 3 + \dfrac{3}{w + 2}$,

$2w^2 + w - 3 = (w + 2)(2w - 3) + 3$, $w \neq -2$

51. $\dfrac{4a^2 - 7a - 7}{a - 2} = 4a + 1 - \dfrac{5}{a - 2}$,

$4a^2 - 7a - 7 = (a - 2)(4a + 1) - 5$, $a \neq 2$

52. $\dfrac{x^2 - 8}{x - 3} = x + 3 + \dfrac{1}{x - 3}$,

$x^2 - 8 = (x - 3)(x + 3) + 1$, $x \neq 3$

53. $\dfrac{5c^2 + 14c + 11}{5c + 4} = c + 2 + \dfrac{3}{5c + 4}$,

$5c^2 + 14c + 11 = (5c + 4)(c + 2) + 3$, $c \neq -\dfrac{4}{5}$

54. $\dfrac{6m^2 - 5m - 5}{3m - 4} = 2m + 1 - \dfrac{1}{3m - 4}$,

$6m^2 - 5m - 5 = (3m - 4)(2m + 1) - 1$, $m \neq \dfrac{4}{3}$

55. $\dfrac{8n^2 - 18n + 13}{2n - 3} = 4n - 3 + \dfrac{4}{2n - 3}$,

$8n^2 - 18n + 13 = (2n - 3)(4n - 3) + 4$, $n \neq \dfrac{3}{2}$

56. $\dfrac{4y^2 - 29}{2y - 5} = 2y + 5 - \dfrac{4}{2y - 5}$,

$4y^2 - 29 = (2y - 5)(2y + 5) - 4$, $y \neq \dfrac{5}{2}$

57. $\dfrac{6x^3 + 4x^2 + 9x + 8}{3x + 2} = 2x^2 + 3 + \dfrac{2}{3x + 2}$,

$6x^3 + 4x^2 + 9x + 8 = (3x + 2)(2x^2 + 3) + 2$, $x \neq -\dfrac{2}{3}$

58. $\dfrac{8y^3 - 4y^2 + 6y - 9}{2y - 1} = 4y^2 + 3 - \dfrac{6}{2y - 1}$,

$8y^3 - 4y^2 + 6y - 9 = (2y - 1)(4y^2 + 3) - 6$, $y \neq \dfrac{1}{2}$

59. $\dfrac{12f^3 - 10f^2 - 18f + 25}{6f - 5} = 2f^2 - 3 + \dfrac{10}{6f - 5}$,

$12f^3 - 10f^2 - 18f + 25 = (6f - 5)(2f^2 - 3) + 10$,

$f \neq \dfrac{5}{6}$

60. $\dfrac{6m^3 + 2m^2 - 5}{3m + 1} = 2m^2 - \dfrac{5}{3m + 1}$,

$6m^3 + 2m^2 - 5 = (3m + 1)(2m^2) - 5$, $m \neq -\dfrac{1}{3}$

61. $\dfrac{6n^3 + 14n^2 + 9n + 12}{2n^2 + 3} = 3n + 7 - \dfrac{9}{2n^2 + 3}$,

$6n^3 + 14n^2 + 9n + 12 = (2n^2 + 3)(3n + 7) - 9$

62. $\dfrac{10x^3 + 2x^2 - 5x - 2}{2x^2 - 1} = 5x + 1 - \dfrac{1}{2x^2 - 1}$,

$10x^3 + 2x^2 - 5x - 2 = (2x^2 - 1)(5x + 1) - 1$,

$x \neq \pm\dfrac{1}{\sqrt{2}}$

63. $\dfrac{8m^3 - 4m^2 + 2m}{4m^2 + 1} = 2m - 1 + \dfrac{1}{4m^2 + 1}$,

$8m^3 - 4m^2 + 2m = (4m^2 + 1)(2m - 1) + 1$

64. $\dfrac{9z^3 + 24z^2 - 12z - 10}{3z^2 - 4} = 3z + 8 + \dfrac{22}{3z^2 - 4}$,

$9z^3 + 24z^2 - 12z - 10 = (3z^2 - 4)(3z + 8) + 22$,

$z \neq \sqrt{\dfrac{4}{3}}$

Applications and Problem Solving   65. a) dividend: 156; divisor: 12; quotient: 13; remainder: 0 **b)** dividend: 350; divisor: 9; quotient: 38; remainder: 8 **c)** dividend: $a^2 - 5a + 6$; divisor: $a - 3$; quotient: $a - 2$; remainder: 0 **d)** dividend: $2t^2 + 5t - 2$; divisor: $2t - 1$; quotient: $t + 3$; remainder: 1 **e)** dividend: $x^3 + x^2 + x - 3$; divisor: $x^2 + 1$; quotient: $x + 1$; remainder: 4 **f)** dividend: $6x^3 + 3x^2 - x$; divisor: $3x^2 - 2$; quotient: $2x + 1$; remainder: $3x + 2$
66. a) dividend: 255; divisor: 11 or 23; quotient: 11 or 23; remainder: 2 **b)** dividend: $8y^3 + 6y^2 - 4y - 5$; divisor: $4y + 3$ or $2y^2 - 1$; quotient: $4y + 3$ or $2y^2 - 1$; remainder: -2 **c)** dividend: $x^2 + x + 3$; divisor: x or $x + 1$; quotient: x or $x + 1$; remainder: 3   **67. a)** $x + 4$, $x \neq -1$ **b)** $3x + 1$, $x \neq -2$ **c)** $x + 3$, $x \neq -\dfrac{1}{2}$

68. a) $(2x^2 + 18x - 72) \div (2x - 6)$ **b)** $x + 12$ **c)** 274 cm by 152 cm **69. a)** $x^2 + x + 1$; $x^2 + 2x + 4$; $x^2 + 3x + 9$ **b)** 1^0, 1^1, 1^2; 2^0, 2^1, 2^2; 3^0, 3^1, 3^2 **c)** $x^2 + 4x + 16$; $x^2 + 5x + 25$ **70. a)** $x^4 + x^3 + x^2 + x + 1$ **b)** $x^4 + 2x^3 + 4x^2 + 8x + 16$ **71. a)** $x^2 + 3x + 1$

428

b) $y^2 + 3$, $y \neq \pm 1$ c) $2z^2 + 4$, $z \neq \pm 2$ d) $2a^2 - 1$, $a \neq \pm\sqrt{\frac{7}{3}}$

e) $3t^2 + 5t - 3$ **72. a)** $3x + 2 + \frac{x+1}{2x^2+1}$; yes

b) $y - 3 - \frac{y}{3y^2+1}$; yes c) $t + 2 - \frac{11}{2t^2-4}$, $t \neq \pm\sqrt{2}$; no

d) $3d^2 - 2 + \frac{d-2}{2d^2-3}$, $d \neq \pm\sqrt{\frac{3}{2}}$; yes **73. a)** 12 **b)** 2 **c)** 8

d) 4 **74.** $4m + 2$ **75.** $3n - 2$ **76.** The rational expression $\frac{5x^2+14x-3}{x+3}$ is not defined when $x = -3$, but the quotient $5x - 1$ is defined when $x = -3$.

Investigating Math pp. 154–155
1 Dividing Polynomials by x – b **1.** $x + 5$ **2.** $y + 6$
3. $n - 3$ **4.** $2t + 4$ **5.** $5n - 3$ **6.** $4r^2 - 2r - 1$ **7.** $6a^2 + 1$
8. $3c^2 - c - 5$ **9.** $x^3 + 1$ **10.** $w^2 + 2w + 7 + \frac{6}{w-2}$

11. $3x + 7 - \frac{3}{x+3}$ **12.** $4y^2 + 1 + \frac{1}{y-1}$

13. $2z^2 - z - 1 - \frac{5}{z+10}$ **14.** $8g^2 + g + 3 - \frac{1}{g-3}$

15. $q^3 + 2q^2 + 2 + \frac{7}{q-4}$ **16.** $10x^3 - 1 + \frac{8}{x+8}$

17. $9y^3 - y + 2 + \frac{3}{y+2}$

2 Dividing Polynomials by ax – b **1.** $x + 1$ **2.** $3t - 4$
3. $2a + 1$ **4.** $4y^2 + 2y + 3$ **5.** $3c^2 - 1$ **6.** $5m^2 - 4m + 6$
7. $8u^3 + 1$ **8.** $6t^3 - 8t^2 + 2t - 5$ **9.** $3x + 7$ R3
10. $5e^2 + 6$ R–8 **11.** $3f^2 + 5f - 2$ R5 **12.** $6n^2 - 5$ R–10
13. $2r^3 + 3r + 5$ R1
3 Finding a Pattern **1. a)** 2 **b)** 2; equal **2. a)** −5 **b)** −5; equal **3. a)** −4 **b)** −4; equal **4. a)** The remainder when a polynomial $P(x)$ is divided by $x - b$ is equal to $P(b)$. **b)** The remainder when a polynomial $P(x)$ is divided by $ax - b$ is equal to $P\left(\frac{b}{a}\right)$. **5. a)** 1 **b)** −2 **c)** 0 **d)** 0

Section 4.2 pp. 158–159
Practice

1. $\frac{y^2}{3x}$, $x, y \neq 0$ **2.** $\frac{4}{bc}$, $a, b, c \neq 0$ **3.** $\frac{-x}{5y}$, $x, y, z \neq 0$

4. $\frac{x}{x+4}$, $x \neq -4$ **5.** $\frac{3(m-4)}{m}$, $m \neq 0$

6. $\frac{2t(t+5)}{t-5}$, $t \neq 0, 5$ **7.** $\frac{1}{2x}$, $x \neq 0, 3$

8. $\frac{m+2}{m+4}$, $m \neq 1, -4$ **9.** $\frac{x}{x+4}$, $x \neq -4$

10. $\frac{y}{y+2}$, $y \neq 0, -2$ **11.** $\frac{2}{x-3}$, $x \neq 0, 3$

12. $\frac{m-2n}{n}$, $m, n \neq 0$ **13.** $\frac{2a+3}{4a}$, $a \neq 0$

14. $\frac{1}{2x-4y}$, $x, y \neq 0$, $x \neq 2y$

15. $t^2 + 2t - 5$, $t \neq 0$ **16.** $\frac{1}{4x^2-3}$, $x \neq 0$, $\pm\frac{\sqrt{3}}{2}$

17. $\frac{7n^3 - 2n^2 + 3n + 4}{n}$, $n \neq 0$ **18.** $2y^3 + y^2 - 3y$, $y \neq 0$

19. $\frac{4}{5}$, $x \neq -y$ **20.** $\frac{a+2}{a-3}$, $a \neq 0, 3$ **21.** 6, $t \neq 6$

22. $\frac{m+6}{2m-6}$, $m \neq 3$ **23.** $\frac{5}{3}$, $x \neq 2$ **24.** $\frac{4}{3}$, $x \neq 0, -\frac{1}{2}$

25. $\frac{x-1}{x+1}$, $x \neq 0, -1$ **26.** $\frac{2a+4b}{3a-3}$, $a \neq 0, 1$

27. $\frac{5x}{2y}$, $y \neq 0, -2$ **28.** $\frac{1}{m-3}$, $m \neq 2, 3$

29. $y + 5$, $y \neq -5$ **30.** $\frac{2}{x-9}$, $x \neq 9, -3$

31. $\frac{r-2}{5}$, $r \neq -2$ **32.** $\frac{a}{a+1}$, $a \neq -1$

33. $\frac{x+3}{2xy}$, $x \neq 0, 3$, $y \neq 0$ **34.** $\frac{2}{2w+1}$, $w \neq -1, -\frac{1}{2}$

35. $\frac{t-2}{2t}$, $t \neq 0$, $\frac{2}{3}$ **36.** $\frac{2z}{3z-4}$, $z \neq \pm\frac{4}{3}$

37. $\frac{5x-2y}{3x}$, $x \neq 0, -y$ **38.** $\frac{x+2}{x+3}$, $x \neq -2, -3$

39. $\frac{a+3}{a-5}$, $a \neq 4, 5$ **40.** $\frac{m-2}{m+5}$, $m \neq 3, -5$

41. $\frac{y-3}{y+5}$, $y \neq \pm 5$ **42.** $\frac{x-4}{x-6}$, $x \neq 6$

43. $\frac{n+1}{n+3}$, $n \neq 2, -3$ **44.** $\frac{p+4}{p-4}$, $p \neq \pm 4$

45. $\frac{2t+1}{t-2}$, $t \neq 1, 2$ **46.** $\frac{3v+1}{2v+1}$, $v \neq -\frac{1}{2}, -\frac{3}{2}$

47. $\frac{3x-2}{4x+3}$, $x \neq -\frac{3}{4}, \frac{3}{2}$ **48.** $\frac{z-2}{3z-1}$, $z \neq \frac{1}{3}$

49. $\frac{m-n}{2m-3n}$, $m \neq -\frac{1}{2}n, \frac{3}{2}n$

Applications and Problem Solving 50. a) −1, $x \neq 1$
b) not possible **c)** not possible **d)** 1, $t \neq \frac{7}{3}$

e) $\frac{t-s}{s+t}$, $s \neq -t$ **f)** $\frac{x}{2}$ **51. a)** $x = y$ **b)** $x = -\frac{y}{3}$

429

c) $x = 0$ d) $x = 2$ e) $x = \pm 1$ f) $x = \pm\dfrac{3y}{2}$

52. a) no b) yes c) yes d) no e) no f) no

53. a) $\dfrac{x^2 + 3x + 2}{x + 1}$ b) $x + 2$ c) 3:2 54. $\dfrac{(x+1)^3}{6(x+1)^2} = \dfrac{x+1}{6}$

55. a) $\dfrac{x^2 + 1}{x^2 - 1}$, $x \neq \pm 1$ b) $\dfrac{3a^2 - 4}{a^2 - 1}$, $a \neq \pm 1$

c) $\dfrac{2y^2 - 3z^2}{2y^2 + 3z^2}$ d) $\dfrac{2m - 3n}{4m^2 + 9n^2}$, $m \neq \pm\dfrac{3}{2}n$

e) $\dfrac{x + y}{(x^2 + y^2)(x - y)}$, $x \neq y$ 56. a) $n + 1$

b) $(n + 1)(n + 3)$ c) $n + 3$ d) 13 e) 440 f) 35

57. Answers will vary.

Section 4.3 pp. 163–164

Practice 1. $\dfrac{8y}{3}$, $y \neq 0$ 2. $-\dfrac{x}{4}$, $x \neq 0$ 3. $\dfrac{1}{9n^3}$, $n \neq 0$

4. $-\dfrac{8m}{3}$ 5. $\dfrac{x}{4}$, $x \neq 0$ 6. $-\dfrac{y}{2}$, $y \neq 0$

7. $-\dfrac{9m^2}{4}$, $m \neq 0$ 8. $\dfrac{20t^2}{9}$, $t \neq 0$ 9. $-\dfrac{32}{9x^3}$, $x \neq 0$

10. $-\dfrac{2}{3r}$, $r \neq 0$ 11. $\dfrac{4x^2 y}{3}$, $x, y \neq 0$

12. $\dfrac{16m}{5n}$, $m, n \neq 0$ 13. $\dfrac{9}{xt^2}$, $x, y, t \neq 0$

14. $\dfrac{1}{2a^3 b^3}$, $a, b \neq 0$ 15. $\dfrac{9}{2mt}$, $m, t \neq 0$

16. $-\dfrac{45a}{32c^2}$, $a, b, c \neq 0$ 17. $\dfrac{2xy^2}{3ab}$, $a, b, x, y \neq 0$

18. $\dfrac{4nx}{3}$, $m, n, x, y \neq 0$ 19. $\dfrac{3x}{4}$, $x, y \neq 0$

20. $16a^2$, $a, b \neq 0$ 21. $-9x^2 y^3$, $x, y \neq 0$

22. $-\dfrac{2ab}{9c}$, $a, b, c \neq 0$ 23. $\dfrac{1}{2}$, $x \neq 4$

24. $\dfrac{2m + 4}{y + 1}$, $y \neq -1$ 25. $\dfrac{y - 2}{2}$, $y \neq -1$ 26. 2, $x \neq -1, 2$

27. $-\dfrac{a}{6b}$, $a \neq -b$, $a, b \neq 0$ 28. $\dfrac{9m^2}{5}$, $m \neq 0, -4$

29. $\dfrac{8}{5}$, $x \neq \pm 1$ 30. $\dfrac{15}{4}$, $m \neq 0, -3$

31. $-\dfrac{1}{a}$, $a \neq 0, -2$ 32. $\dfrac{3(x + 2)}{4}$, $x \neq 2, -3$

33. $\dfrac{2}{y(y - 3)}$, $y \neq 0, \pm 3$ 34. $\dfrac{2(m + 5)}{(m - 4)}$, $m \neq 5, \pm 4$

35. $\dfrac{1}{3x}$, $x \neq 0, \dfrac{3}{2}$, $y \neq 0$ 36. $x + 2$, $x \neq -2$, $\dfrac{1}{2}$

37. $\dfrac{x + 2}{x - 1}$, $x \neq -6, -3, 1, 5$ 38. $\dfrac{a + 4}{a + 2}$, $a \neq 2, \pm 3$

39. $\dfrac{m + 1}{m}$, $m \neq -5, 0, 3, 4$

40. $\dfrac{x + 2y}{x - 2y}$, $x \neq -4y, \pm 2y, 3y, 5y$

41. $\dfrac{x}{x + 6y}$, $x \neq -6y, \pm 3y, 7y$

42. $\dfrac{a + 7b}{a - 9b}$, $a \neq -8b, -6b, 2b, 9b$

43. $\dfrac{4a - 5}{2a + 3}$, $a \neq \dfrac{1}{3}, \pm\dfrac{3}{2}$ 44. 1, $x \neq -\dfrac{1}{2}, \dfrac{5}{2}, 3$

45. $\dfrac{(4w + 1)(4w + 5)}{(4w + 7)(4w + 3)}$, $w \neq -\dfrac{7}{4}, -\dfrac{5}{4}, -\dfrac{3}{4}, \dfrac{2}{3}, \dfrac{3}{2}$

46. $\dfrac{3s + 5t}{5s + t}$, $s \neq -\dfrac{5}{3}t, -\dfrac{1}{5}t, \dfrac{6}{5}t, \dfrac{5}{4}t$

47. $\dfrac{(x + 5)(2x - 1)}{2(x + 1)}$, $x \neq -1, -\dfrac{3}{4}, 0, \dfrac{1}{3}, \dfrac{1}{2}$

48. -1, $x \neq \pm\dfrac{3}{2}y$ 49. 10, $x \neq 0$ 50. $-\dfrac{y^2}{4}$, $x, y \neq 0$

51. $\dfrac{2t - 1}{2(3t + 5)}$, $t \neq -\dfrac{5}{3}, \dfrac{1}{2}, 2$

52. $\dfrac{s(s - 1)}{(s + 1)(s - 2)}$, $s \neq -3, -1, 0, 2$

53. $\dfrac{3(n - 1)}{2}$, $n \neq -2, \pm 1$

54. $\dfrac{5x + 3}{2x + 3}$, $x \neq -\dfrac{3}{2}, -\dfrac{3}{5}, \dfrac{1}{6}, 3, 6$

Applications and Problem Solving 55. a) $\dfrac{10x^2}{27}$

b) $\dfrac{22x^2}{9}$ c) $\dfrac{33}{5}$ d) No; The answer to c) is independent of x. 56. $(x - 3)(x - 2)$ 57. $3y + 2$

58. a) $\dfrac{(6x - 9)(2x + 4)}{2}$ b) $\dfrac{(2x - 3)(3x + 6)}{2}$ c) 2

59. $\dfrac{x - 2}{3(x - 3)}$ 60. Division by zero is not defined.

61. $\dfrac{1}{x - 2}$, $x \neq \pm 2, 3$ 62. a) decreases
b) stays the same 63. Answers may vary.
64. Answers may vary.

Section 4.4 pp. 167–168

Practice **1.** $\dfrac{1}{y}$, $y \neq 0$ **2.** $\dfrac{8}{x^2}$, $x \neq 0$ **3.** $\dfrac{9}{x+3}$, $x \neq -3$

4. $\dfrac{x-y}{x-2}$, $x \neq 2$ **5.** $\dfrac{2x+11}{2}$ **6.** $\dfrac{5y-7}{3}$ **7.** $\dfrac{-a-3}{a}$, $a \neq 0$

8. $\dfrac{x-2y}{3x}$, $x \neq 0$ **9.** $\dfrac{3x^2+4}{x+1}$, $x \neq -1$ **10.** $\dfrac{11t-11}{7}$

11. 60 **12.** 36 **13.** 120 **14.** 60 **15.** $\dfrac{4x}{3}$ **16.** $\dfrac{11a}{12}$

17. $\dfrac{2x-5y+7}{10}$ **18.** $-\dfrac{11m}{24}$ **19.** $\dfrac{20m+29}{14}$ **20.** $\dfrac{16x-1}{12}$

21. $\dfrac{-4y-1}{12}$ **22.** $\dfrac{-16x+11y}{10}$ **23.** $\dfrac{3t+1}{2}$ **24.** $\dfrac{-12a+19b}{18}$

25. $\dfrac{10x+39}{30}$

Applications and Problem Solving **26. a)** $2x-1$

b) $\dfrac{13x-5}{3}$ **c)** 23 cm by 15 cm; 23 cm by 33 cm

27. a) $\dfrac{1191}{s}$ **b)** $\dfrac{685}{s}$ **c)** $\dfrac{1876}{s}$ **d)** 2.68 h

28. a) $\dfrac{\pi d^2}{4}$ **b)** $\dfrac{\pi(d+1)^2}{4}$ **c)** $\dfrac{\pi(2d+1)}{4}$ **d)** 16.5 cm²

29. a) $\dfrac{13x+1}{3}$ **b)** 2, 5, 8 **30. a)** $\dfrac{(2x+1)(x-3)}{16}$

b) $\dfrac{(x-1)(x-3)}{4}$ **c)** $\dfrac{3(2x-1)(x-3)}{16}$ **d)** $\dfrac{4x-1}{4}$

e) $\dfrac{3(2x-1)(x-3)}{16}$ **f)** equal **31. a)** $\dfrac{n(n+1)}{2}$

b) 15, 21, 28, 36, 45 **c)** square **d)** $\dfrac{(n+1)(n+2)}{2}$

e) $(n+1)^2$ **f)** It is a square.

Section 4.5 pp. 172–173

Practice **1.** $\dfrac{24xy}{12x^2y^2}$, $x, y \neq 0$ **2.** $\dfrac{12x^3y}{12x^2y^2}$, $x, y \neq 0$

3. $\dfrac{20x}{12x^2y^2}$, $x, y \neq 0$ **4.** $-\dfrac{2y^3}{12x^2y^2}$, $x, y \neq 0$

5. $20a^2b^3$ **6.** $6m^2n^2$ **7.** $12x^3y^2$ **8.** $60s^2t^2$

9. $\dfrac{23}{10x}$, $x \neq 0$ **10.** $\dfrac{1}{y}$, $y \neq 0$ **11.** $\dfrac{x+2x^2-4}{2x^3}$, $x \neq 0$

12. $\dfrac{15n^2-10+8mn^2}{10m^2n^3}$, $m, n \neq 0$ **13.** $\dfrac{x^2+5x-2}{x}$, $x \neq 0$

14. $\dfrac{3m-2mn-n+4}{mn}$, $m, n \neq 0$

15. $\dfrac{-30x^2-10x+1}{15x^2}$, $x \neq 0$

16. $\dfrac{-3x^2-2y^2+5xy-4x+y}{xy}$, $x, y \neq 0$

17. $6(m+2)$ **18.** $15(y-1)(y+2)$ **19.** $12(m-2)(m-3)$

20. $20(2x-3)$ **21.** $\dfrac{21}{4(x+3)}$, $x \neq -3$

22. $-\dfrac{5}{3(y-5)}$, $y \neq 5$ **23.** $\dfrac{t}{3(t-4)}$, $t \neq 4$

24. $\dfrac{8}{3(m+1)}$, $m \neq -1$ **25.** $\dfrac{1}{12(y-2)}$, $y \neq 2$

26. $\dfrac{11}{6(2a+1)}$, $a \neq -\dfrac{1}{2}$ **27.** $\dfrac{5x+7}{(x+1)(x+2)}$, $x \neq -1, -2$

28. $\dfrac{m^2-3m+15}{(m-3)(m+2)}$, $m \neq -2, 3$ **29.** $\dfrac{8x-3}{x(x-1)}$, $x \neq 0, 1$

30. $\dfrac{11t-1}{5(t-1)}$, $t \neq 1$ **31.** $\dfrac{10x-x^2}{(x-2)(x+2)}$, $x \neq \pm 2$

32. $\dfrac{15-n}{(3n-1)(2n+3)}$, $n \neq -\dfrac{3}{2}, \dfrac{1}{3}$

33. $\dfrac{5x-7}{4(x-1)(x-2)}$, $x \neq 1, 2$

34. $\dfrac{2t^2-9t-5}{6(t+5)(t-4)}$, $t \neq -5, 4$

35. $\dfrac{-s^2+16s-10}{5(s-6)(s-1)}$, $s \neq 1, 6$

36. $\dfrac{11m^2-31m}{12(m-5)(m-2)}$, $m \neq 2, 5$

37. $(x+2)^2$ **38.** $(y-2)(y+2)(y+4)$

39. $(t+3)(t-4)(t+1)$ **40.** $2(x-2)(x+1)(x-4)$

41. $(m+3)^2(m-5)$ **42.** $\dfrac{2x+7}{(x+3)(x+2)}$, $x \neq -3, -2$

43. $\dfrac{-3y+16}{(y-4)(y+4)}$, $y \neq \pm 4$ **44.** $\dfrac{3x^2+5x}{(x-5)(x+1)}$, $x \neq -1, 5$

45. $\dfrac{9a-2a^2}{(a-3)(a-4)}$, $a \neq 3, 4$

46. $\dfrac{6+2x}{(2x+1)(x+1)}$, $x \neq -1, -\dfrac{1}{2}$

47. $\dfrac{18n-9}{(2n-1)(3n-1)}$, $n \neq \dfrac{1}{3}, \dfrac{1}{2}$

48. $\dfrac{3}{(m+1)(m+4)}$, $m \neq -4, -3, -1$

49. $\dfrac{-2x-8}{(x+2)^2(x-2)}$, $x \neq \pm 2$

431

50. $\dfrac{a^2-6a-10}{(a-5)(a-5)(a-4)}$, $a \neq \pm 5,\ 4$

51. $\dfrac{6m^2-26m}{(m-3)(m-6)(m-5)}$, $m \neq 3,\ 5,\ 6$

52. $\dfrac{7x-3}{(3x+1)(x+1)(x-1)}$, $x \neq -\dfrac{1}{3},\ \pm 1$

53. $\dfrac{2y^2-15y}{(2y-3)^2(2y+3)}$, $y \neq \pm \dfrac{3}{2}$

54. $\dfrac{t^2-3t-2}{(t-1)(t-4)}$, $t \neq 1,\ 4$

55. $\dfrac{y^2+4y+1}{(y-1)(y+2)}$, $y \neq -2,\ 1$

56. $\dfrac{-2x^2-2x-3}{(x+3)(x+1)}$, $x \neq -3,\ -1$

57. $\dfrac{4n+7}{3(n+4)(n-4)}$, $n \neq \pm 4$

58. $-\dfrac{4}{(m+3)(m-4)(m-1)}$, $m \neq -3,\ 1,\ 4$

59. $\dfrac{5a+1}{(a-1)(a+1)^2}$, $a \neq \pm 1$

60. $\dfrac{5w^2-11w+5}{(w+1)(w+4)(w-2)}$, $w \neq -4,\ -1,\ 2$

61. $\dfrac{10x^2+3x}{(2x+1)(x+1)(3x+1)}$, $x \neq -1,\ -\dfrac{1}{2},\ -\dfrac{1}{3}$

62. $\dfrac{-20z-26}{(2z-5)(2z+5)(2z+1)}$, $z \neq \pm \dfrac{5}{2},\ -\dfrac{1}{2}$

Applications and Problem Solving 63. a) 96, 24, 4, 96; $135y^3$, $45y^2$, $3y$, $135y^3$; a^2-1, a^2-1, 1, a^2-1; $6(t-1)^2$, $6(t-1)$, $t-1$, $6(t-1)^2$ **64. a)** $\dfrac{45}{s}$ **b)** $\dfrac{45}{2s}$

c) $\dfrac{135}{2s}$ **d)** 6.75 h **65. a)** $\dfrac{100}{x}$ **b)** $\dfrac{100}{x+4}$ **c)** $\dfrac{400}{x(x+4)}$

d) 1 s **66. a)** $\dfrac{2x+1}{2x-1}$, $x \neq 0,\ \dfrac{1}{2}$ **b)** $\dfrac{4y+1}{2(2y-1)}$, $y \neq \dfrac{1}{2}$

c) $\dfrac{m+1}{2m}$, $m \neq 0,\ 1$ **d)** $\dfrac{2t^2}{t+2}$, $t \neq -2,\ 0$

e) $\dfrac{n+3}{2(n-3)}$, $n \neq \pm 3$ **f)** $\dfrac{3(2x+1)}{2(3x+1)}$, $x \neq -\dfrac{1}{3},\ 3$

67. Answers may vary.

Technology pp. 174–175
1 Simplifying Rational Expressions

1. $\dfrac{1}{2x^3y^2}$, $x,\ y \neq 0$ **2.** $-\dfrac{3a^5}{c^2}$, $a,\ b,\ c \neq 0$

3. $-\dfrac{5p^5}{2qrs^4}$, $p,\ q,\ r,\ s \neq 0$ **4.** $\dfrac{2x^2-3x+4}{x}$, $x \neq 0$

5. $\dfrac{t}{4t^2+2t-1}$, $t \neq 0$ **6.** $\dfrac{m+5}{3-m}$, $m \neq 0,\ 3$

7. $\dfrac{4x^2+3x-2}{5x^2+x+2}$, $x \neq 0$ **8.** $\dfrac{x+5}{x-5}$, $x \neq -2,\ 5$

9. $\dfrac{x+y}{x-y}$, $x \neq \pm y$ **10.** $\dfrac{3n-1}{2n+5}$, $n \neq -\dfrac{5}{2},\ \dfrac{3}{4}$

11. $\dfrac{5m-6}{4m-5}$, $m \neq \dfrac{5}{4},\ \dfrac{1}{2}$ **12.** $\dfrac{x^2+2}{x^2-2}$, $x \neq \pm \sqrt{2}$

13. $\dfrac{a(a+b)}{3a+2b}$, $a \neq \pm \dfrac{2b}{3}$

2 Dividing Polynomials 1. $y-5$, $y \neq 6$ **2.** x^2+2x+3, $x \neq -2$ **3.** $4t-1$, $t \neq -\dfrac{3}{2}$ **4.** $v+4$, $v \neq \pm \dfrac{\sqrt{6}}{3}$

5. $x+12-\dfrac{10}{x+3}$, $x \neq -3$ **6.** $n^2-3n-6+\dfrac{2}{n-2}$, $n \neq 2$

7. $3d+5-\dfrac{4}{6d-5}$, $d \neq \dfrac{5}{6}$ **8.** $7m-6+\dfrac{7}{2m^2+7}$

3 Operations on Rational Expressions

1. $-\dfrac{2n}{5m}$, $m,\ n \neq 0$ **2.** $\dfrac{2b}{5x}$, $a,\ b,\ x,\ y \neq 0$

3. $\dfrac{1}{6a}$, $a \neq 0,\ 3$ **4.** $\dfrac{x+1}{x-2}$, $x \neq -1,\ 3$

5. $\dfrac{3x-2}{4x+3}$, $x \neq \pm \dfrac{3}{2},\ -\dfrac{3}{4},\ -\dfrac{2}{3}$

6. $\dfrac{(4p-1)(2p-1)}{(3p+1)(3p+2)}$, $p \neq -\dfrac{2}{3},\ -\dfrac{3}{5},\ -\dfrac{1}{3},\ \dfrac{5}{2}$

7. $\dfrac{s}{6t}$, $s,\ t \neq 0$ **8.** $-\dfrac{9my^2z}{2n}$, $m,\ n,\ y,\ z \neq 0$

9. $\dfrac{4}{9}$, $x \neq -2,\ 3$

10. $\dfrac{(y-6)(y+5)}{(y-2)(y+4)}$, $y \neq -5,\ -4,\ -3,\ 2,\ 7$

11. $\dfrac{3m-1}{2(3m-2)}$, $m \neq -\dfrac{5}{2},\ \pm \dfrac{2}{3},\ \dfrac{1}{2}$

12. $\dfrac{(4x+1)(3x+1)}{(3x+2)(5x+3)}$, $x \neq -\dfrac{2}{3},\ -\dfrac{3}{5},\ -\dfrac{1}{3},\ \dfrac{5}{6}$

13. a) $\dfrac{16q+1}{12}$ **14.** $\dfrac{24x-13}{10}$ **15.** $\dfrac{2y^2+5y+3}{y^3}$, $y \neq 0$

16. $\dfrac{25}{12(n-3)}$, $n \neq 3$ **17.** $\dfrac{2t^2+t}{2(t-1)(t+1)}$, $t \neq \pm 1$

18. $\dfrac{5n^2+7n-2}{(n+2)^2(n-2)}$, $n\neq\pm2$

19. $\dfrac{3c^2-2c+3}{2(2c+1)(c-1)(c+1)}$, $c\neq-\dfrac{1}{2}$, ±1

20. a) $\dfrac{4q-5}{12}$ **21.** $\dfrac{z+13}{30}$ **22.** $\dfrac{9x-2}{6x^2}$, $x\neq0$

23. $\dfrac{4}{3(2r+3)}$, $r\neq-\dfrac{3}{2}$ **24.** $\dfrac{x^2-17x}{(3x+4)(2x-1)}$, $x\neq-\dfrac{4}{3}$, $\dfrac{1}{2}$

25. $-\dfrac{y^2}{(y-2)(y-1)(y-3)}$, $y\neq1$, 2, 3

26. $\dfrac{t^2-21t+21}{(3t+1)^2(2t-7)}$, $t\neq-\dfrac{1}{3}$, $\dfrac{7}{2}$

4 Problem Solving 1. a) $\dfrac{(x+3)(x-1)}{12}$ **b)** $\dfrac{7x+5}{6}$

c) $\dfrac{2(7x+5)}{(x+3)(x-1)}$ **d)** No; the area is 0 when $x=1$.

2. a) $\dfrac{y-2}{2}$ **b)** $\dfrac{y-2}{12}$

Investigating Math pp. 176–177
1 Representing Equations 1. a) $x-1=3$ **b)** $2x+1=5$
2 Solving Equations by Addition and Subtraction
1. a) $x-2=1$ **b)** 2 **d)** 3 **2. a)** 3 **b)** 2 **c)** 0 **d)** -5 **e)** -4
f) -5 **3. a)** 2 **b)** 5 **c)** 1 **d)** 2 **e)** 0 **f)** 4
3 Solving Equations by Division 1. a) 2 **b)** 3 **c)** 2
2. a) -1 **b)** -2 **c)** -2
4 Solving Equations in More Than One Step
1. a) $2x-1=3$ **b)** 1 **e)** 2 **2. a)** 3 **b)** 2 **c)** 1 **d)** 1 **e)** 2 **f)** 2
5 Solving Equations With Variables on Both Sides
1. b) 2 **d)** 1 **2. a)** 2 **b)** -2 **c)** -3 **d)** 2 **e)** 1 **f)** -3 **g)** 2 **h)** 1 **i)** -1

Section 4.6 pp. 179–180
Practice 1. 2 **2.** 10 **3.** 5 **4.** -2 **5.** -3 **6.** 7 **7.** -9 **8.** -4
9. 2 **10.** 1 **11.** -4 **12.** 4 **13.** -2 **14.** 2 **15.** -3 **16.** -2
17. 4 **18.** 3 **19.** -3 **20.** -1 **21.** 2 **22.** 4 **23.** 4 **24.** -1
25. 1 **26.** -1 **27.** 1 **28.** 8 **29.** -3 **30.** -1 **31.** 0 **32.** 2
33. 5 **34.** 3 **35.** -7 **36.** 1 **37.** 5 **38.** -2
Applications and Problem Solving 39. 1903
40. 70 cm **41.** 10 h **42.** 4 days **43.** 65 **44.** 832 m
45. a) -1 **b)** 4 by 3 **46.** 14 m by 26 m **47. a)** 12 by 4;
6 by 8 **b)** 48 **48. a)** ±4 **b)** ±3 **49.–51.** Answers may vary.

Career Connection p. 181
1 Shutter Speed 1. high **2. a)** high **b)** low
2 F-stops 1. a) geometric **b)** To get the next term,
multiply the previous term by $\sqrt{2}$. **c)** 8, 8$\sqrt{2}$, 16,
16$\sqrt{2}$, 32 **d)** 1.4, 2, 2.8, 4, 5.6, 8, 11.3, 16, 22.6, 32

2. a) $A=\dfrac{\pi d^2}{4}$ **b)** $A=\dfrac{\pi F^2}{4f^2}$ **c)** 1963 mm^2, 962 mm^2,
491 mm^2 **d)** decreases, roughly in half with each
f-stop **3. a)** high **b)** low

Section 4.7 pp. 185–186
Practice 1. 6 **2.** -8 **3.** $-\dfrac{5}{2}$ **4.** 6 **5.** 8 **6.** 12 **7.** -5

8. $\dfrac{1}{4}$ **9.** $-\dfrac{5}{2}$ **10.** $\dfrac{4}{3}$ **11.** 1.3 **12.** -4.3 **13.** 3.6

14. -1.5 **15.** 0.8 **16.** -6 **17.** 0.6 **18.** -0.3 **19.** 3.4 **20.** -5

21. 0.36 **22.** -0.18 **23.** 1.05 **24.** $-\dfrac{1}{3}$ **25.** $\dfrac{3}{2}$ **26.** -5

27. $-\dfrac{5}{9}$ **28.** $\dfrac{1}{2}$ **29.** $-\dfrac{4}{3}$ **30.** 2 **31.** $\dfrac{7}{2}$ **32.** 2 **33.** $\dfrac{5}{3}$

34. -1 **35.** $-\dfrac{1}{3}$ **36.** $\dfrac{2}{3}$ **37.** 12 **38.** $\dfrac{3}{2}$, $x\neq0$

39. -2, $n\neq0$ **40.** $\dfrac{1}{2}$, $a\neq0$ **41.** 2, $x\neq0$ **42.** $-\dfrac{1}{6}$, $m\neq0$

43. 10, $x\neq0$ **44.** 1, $y\neq0$ **45.** $\dfrac{1}{2}$, $x\neq0$ **46.** $\dfrac{1}{3}$, $x\neq0$

47. -1, $d\neq0$ **48.** 8, $s\neq0$ **49.** $-\dfrac{9}{2}$, $x\neq0$ **50.** -2, $x\neq0$

51. 2, $t\neq0$ **52.** $\dfrac{2}{3}$, $y\neq0$ **53.** $-\dfrac{5}{3}$, $x\neq0$ **54.** 4, $z\neq0$

55. $\dfrac{1}{3}$, $x\neq0$ **56.** $\dfrac{3}{2}$, $x\neq0$ **57.** -2, $y\neq0$ **58.** 1, $y\neq-1$, 0

59. -6, $k\neq0$, 3 **60.** $-\dfrac{3}{2}$, $x\neq0$, 1 **61.** $\dfrac{3}{2}$, $y\neq0$, 2

62. 3, $x\neq\pm1$ **63.** -3, $x\neq-5$, 1 **64.** 2, $x\neq-1$, $\dfrac{1}{3}$

65. $\dfrac{7}{2}$, $x\neq2$, $\dfrac{1}{2}$ **66.** -3, $w\neq-\dfrac{1}{3}$ **67.** $-\dfrac{1}{2}$, $w\neq-\dfrac{3}{2}$

68. $\dfrac{5}{3}$, $t\neq1$ **69.** -1, $m\neq\dfrac{1}{2}$

Applications and Problem Solving 70. 300 g
71. 464°C **72.** 22 h **73.** -44°C **74.** 40.5 cm
75. 80 km/h **76.** -9 **77.** 5 **78. a)** Each side of the
equation represents the width of one rectangle.
b) 10 cm; 15 cm **c)** 2 cm **79. a)** 4, $x\neq0$ **b)** -1, $x\neq0$
c) -4, $x\neq0$, ±2 **d)** 0, $x\neq\pm1$, -2 **e)** -7, $x\neq\pm4$

f) $\dfrac{1}{2}$, $x\neq1$ **80. b)** No; the equation is not defined
when $x=1$. **81.** Answers will vary.

Section 4.8 pp. 190–191

Practice **1.** $m-2$ **2.** $6h+1$ **3.** $\frac{3}{5}p-2$ **4.** $n+2n$

5. $n+n+1$ **6.** $0.05n$ **7.** $2w+1=10$ **8.** $0.50p=4.5$

9. $16=\frac{1}{3}a$ **10.** $40-4t=16$ **11.** $5+\frac{1}{2}m+m=35$

Applications and Problem Solving **12. a)** $\frac{1}{2}g$

b) $\frac{1}{2}g-8$ **c)** $\frac{1}{2}g-8=92$ **d)** 200 cm **13.** 36 weeks
14. 395 km **15. a)** 12, 13, 14 **b)** 26, 27 **16.** 687 days
17. 4 nickels, 3 dimes, 5 quarters **18.** Lake Winnipeg:
24 400 km²; Lake Winnipegosis: 5400 km² **19.** 3 kg
20. 85 km/h **21. a)** 4.8 km **b)** 24 min **22.** hydrofoil:
30 km/h; boat: 20 km/h **23.** 300 km **24.** 9 dimes,
5 quarters **25.** 14:40 **26. a)** 13:00 **b)** 360 km
27. 3000 km **28.** 700 km/h **29.** plane: 840 km/h;
train: 105 km/h **30.** 11 **31.** commuter plane:
400 km/h; passenger jet: 800 km/h **32.** Answers may
vary.

Section 4.9 p. 193

Practice **1.** $\frac{b}{2a}$, $a\neq0$ **2.** $\frac{3-by}{a}$, $a\neq0$

3. $\frac{4+b}{a-c}$, $a\neq c$ **4.** $2c-10y$ **5.** $-b-6$

6. $\frac{2}{3y}+1$, $y\neq0$, $x\neq1$ **7.** $s=\frac{d}{t}$, $t\neq0$ **8.** $r=\frac{C}{2\pi}$

9. $y=\frac{-C-Ax}{B}$, $B\neq0$ **10.** $x=\frac{y-b}{m}$, $m\neq0$

11. $a=\frac{2A}{h}-b$, $h\neq0$

12. $R_1=\frac{RR_2}{R_2-R}$, $R,R_1,R_2\neq0$, $R_2\neq R$

Applications and Problem Solving **13. a)** $V=\frac{m}{D}$

b) 6 cm³ **14. a)** $l=\frac{P}{2}-w$ **b)** 10 cm **15.** 45
16. a) 38.5 g **b)** 24 g **17.** Answers may vary.

Technology pp. 194–195

1 Solving Equations **1. a)** 3 **b)** 5.75 **c)** $-\frac{1}{4}$ **d)** $-\frac{1}{2}$

e) -3 **f)** 4 **g)** $-\frac{8}{5}$ **h)** $\frac{4}{5}$ **i)** 6 **j)** 15 **2. a)** 1.5 **b)** -1.6

c) $\frac{3}{2}$ **d)** -4 **e)** 13 **f)** $\frac{1}{2}$ **g)** $-\frac{1}{2}$ **h)** $-\frac{1}{4}$ **i)** $\frac{13}{8}$ **j)** $-\frac{5}{6}$

3. a) -5.83 **b)** 3.67 **c)** 0.79 **d)** 0.22 **4. a)** ERROR

2 Solving Formulas **1. a)** $w=\frac{A}{l}$ **b)** $l=\frac{V}{wh}$

c) $x=\frac{y-b}{m}$ **d)** $w=\frac{P-2l}{2}$ **e)** $r=\frac{pq}{p+q}$ **f)** $q=\frac{pr}{p-r}$

2. a) $l\neq0$ **b)** $w,h\neq0$ **c)** $m\neq0$ **d)** no restrictions
e) $p,q,r\neq0$, $p\neq-q$ **f)** $p,q,r\neq0$, $p\neq r$
3 Applications **1.** 1502 km **2.** 88 beats/min

3. a) $r=\frac{s+25}{1.25}$ **b)** 100; 84; 60; 20 **c)** No; since a

person who gets no right answers still gets 20% on

the test. **4. a)** $T=\frac{60(n+0.5)}{7}$ **b)** 13°C; 21°C

5. a) Robyn: 70 km/h; Karl: 90 km/h **b)** 5 h

Connecting Math and Transportation
pp. 196–197
1 Writing Directions **1.** Drive N on the local road to
Bonnechere, then NE on Highway 62 to Pembroke,
then S on Highway 41 to Eganville, then SW on
Highway 512 through Foymount to Latchford
Bridge. **2.** Drive E on Highway 60 to Barry's Bay,
then S on Highway 62 to Combermere, then SE
on Highway 515 to Latchford Bridge, then S on
Highway 514 to Hardwood Lake, then SW on
Highway 28 to McArthur Mills, then NE on
Highway 28 to Hardwood Lake, then SE on
Highway 28 to Denbigh, then NE on Highway 41
to Dacre.
2 Comparing Routes **1. a)** 130 km **b)** 161 km
c) impossible **2. a)** 130 km; 2 h 3 min **b)** 161 km;
2 h 24 min **3. a)** 162 km **b)** 185 km **c)** impossible
4. a) 185 km; 2 h 49 min **b)** 162 km; 3 h

Computer Data Bank pp. 198–199
1 Fat and Fibre **2.** 8 g, 12 g **3. a)** 68 **c)** Except for
chili con carne with beans and some nuts, the fibre
content is 0 g **4.** With a few exceptions like chili con
carne with beans and some nuts, the fat content is 0 g,
1 g, or 2 g **5. a)** 7g of fat, 0 g of fibre **6. a)** 14 g of fat,
0 g of fibre **b)** 0 g of fat, 1.6 g of fibre **c)** 5 g of fat, 0 g
of fibre **d)** 7 g of fat, 0 g of fibre
2 Energy from Fat **2.** rounding errors caused by the
factor, 38, the energy, and the mass of fat being given
to the ones place; if the factor, 38, and the mass of fat
were rounded up, and the energy was rounded down,
a greater result occurs than would occur if they were
given to one or more decimal places

Review pp. 200–201

1. $-2x+3$, $x \neq 0$ **2.** $4a^2 - 2a + 5$, $a \neq 0$ **3.** $n + 8$, $n \neq -2$

4. $y^2 + 2y + 2$, $y \neq 1$ **5.** $2w - 1 + \dfrac{2}{2w+3}$, $w \neq -\dfrac{3}{2}$

6. $t^2 + 3 - 2t - \dfrac{18}{t+3}$, $t \neq -3$ **7.** $3x - 2 - \dfrac{1}{x^2+4}$

8. a) $(28y^2 - 15y + 2) \div (7y - 2)$ **b)** $4y - 1$, $y \neq \dfrac{2}{7}$

c) 89 mm by 51 mm **9.** $\dfrac{x}{x+3}$, $x \neq -3$

10. $\dfrac{4y - 5x}{2}$, $y \neq 0$ **11.** $\dfrac{5}{7}$, $x \neq y$ **12.** $\dfrac{3}{4}$, $m \neq 0$, 1

13. $\dfrac{1}{t-1}$, $t \neq 1$, 2 **14.** $2a + 3$, $a \neq 5$

15. $\dfrac{y+3}{y+4}$, $y \neq -4$, 3 **16.** $\dfrac{2n-3}{4n+1}$, $n \neq -\dfrac{1}{4}$, $-\dfrac{1}{3}$

17. a) $2x + 2$, $x \neq -1$ **b)** 2:1 **18.** $\dfrac{4x}{3}$, $x, y \neq 0$

19. $-2a^2b$, $a, b \neq 0$ **20.** $\dfrac{5ax}{8b}$, $a, b, x, y \neq 0$

21. $\dfrac{b}{6x^2}$, $b, x, y \neq 0$ **22.** $\dfrac{5}{4}$, $x \neq \pm 1$

23. $\dfrac{14(m+2)}{3(m+3)}$, $m \neq -3$, $n \neq 1$

24. $\dfrac{(t+2)(t-3)}{3}$, $t \neq \pm 2$, 3

25. $\dfrac{(2x+1)(x+2)}{(x-2)(x-1)}$, $x \neq \dfrac{1}{2}$, 1, 2, 3

26. $\dfrac{(2y-1)(y+1)}{(3y+2)(y-1)}$, $y \neq -\dfrac{2}{3}$, $\dfrac{1}{4}$, $\dfrac{1}{3}$, ± 1

27. a) $\dfrac{2t^2 - 3t + 1}{2t - 1}$ **b)** $\dfrac{3t^2 - 2t - 1}{3t + 1}$ **c)** $(t-1)^2$, $t \neq \dfrac{1}{2}$, $-\dfrac{1}{3}$

d) square **28.** $-\dfrac{2}{x}$, $x \neq 0$ **29.** $\dfrac{5m-4}{m-2}$, $m \neq 2$

30. $\dfrac{z-2}{z^2}$, $z \neq 0$ **31.** $\dfrac{t}{12}$ **32.** $\dfrac{26x-1}{20}$ **33.** $-\dfrac{5a}{12}$

34. $\dfrac{3x+1}{4}$ **35.** $\dfrac{y+4}{y^2}$, $y \neq 0$

36. $\dfrac{4y^2 - 5xy + 2x^2}{x^2 y^2}$, $x, y \neq 0$ **37.** $\dfrac{3a+4}{6(a-1)}$, $a \neq 1$

38. $\dfrac{-2x - 10}{(x+3)(x+1)}$, $x \neq -3$, -1

39. $\dfrac{t-3}{(t+1)(t+2)(t-1)}$, $t \neq -2$, ± 1

40. $\dfrac{3(x-1)}{(3x+1)(x-2)}$, $x \neq -1$, $-\dfrac{1}{3}$, 2

41. -2 **42.** 4 **43.** -9 **44.** 1 **45.** 1070 km **46.** -9

47. $-\dfrac{1}{2}$ **48.** -1.5 **49.** 0.5 **50.** -3 **51.** 4 **52.** 0

53. Banff: 6433 km²; Jasper: 11 067 km² **54.** 260 km

55. $A = \dfrac{F}{P}$, $P \neq 0$ **56.** $h = \dfrac{3V}{b}$, $b \neq 0$ **57.** $a = \dfrac{P-b}{2}$

58. $b = 2a - c$

Exploring Math p. 201

1. $\dfrac{6-6}{6} + \dfrac{6}{6} = 1$, $\dfrac{6+6}{6} + 6 - 6 = 2$, $\dfrac{6+6}{6} + \dfrac{6}{6} = 3$,

$\dfrac{6+6+6+6}{6} = 4$, $6 - \dfrac{6}{6} + 6 - 6 = 5$, $6 + \dfrac{6}{6} - \dfrac{6}{6} = 6$,

$6 + \dfrac{6}{6} + 6 - 6 = 7$, $6 + \dfrac{6}{6} + \dfrac{6}{6} = 8$, $\dfrac{6+6+6}{6} + 6 = 9$

2. $\dfrac{4}{4} + 4 - 4 = 1$, $\dfrac{4}{4} + \dfrac{4}{4} = 2$, $\dfrac{4+4+4}{4} = 3$,

$\dfrac{4-4}{4} + 4 = 4$, $4 + 4^{4-4} = 5$, $\dfrac{4+4}{4} + 4 = 6$,

$\dfrac{44}{4} - 4 = 7$, $4\left(\dfrac{4+4}{4}\right) = 8$, $\dfrac{4}{4} + 4 + 4 = 9$

3. $\dfrac{5-5}{5} + \dfrac{5}{5} = 1$, $\dfrac{5+5}{5} + 5 - 5 = 2$, $\dfrac{5+5}{5} + \dfrac{5}{5} = 3$,

$\dfrac{5+5+5+5}{5} = 4$, $5 - \dfrac{5}{5} + \dfrac{5}{5} = 5$, $5 + \dfrac{5}{5} + 5 - 5 = 6$,

$5 + \dfrac{5}{5} + \dfrac{5}{5} = 7$, $5 + 5 - \dfrac{5+5}{5} = 8$, $5 + 5 - 5^{5-5} = 9$

4. $\dfrac{n}{n} + \dfrac{n}{n} + \dfrac{n-n}{n} = 2$, $\dfrac{n+n+n}{n} + \dfrac{n-n}{n} = 3$,

$\dfrac{n+n+n+n}{n} + n - n = 4$, $\dfrac{n+n+n+n}{n} + \dfrac{n}{n} = 5$,

$\dfrac{n+n+n+n+n+n}{n} = 6$

Chapter Check p. 202

1. $3y^2 + 2y - 4$, $y \neq 0$ **2.** $4t - 3$, $t \neq -\dfrac{1}{3}$

3. $x^2 + 3x + 1 + \dfrac{4}{x-2}$, $x \neq 2$ **4.** $n - 1 + \dfrac{1}{2n^2 + 1}$

5. $\dfrac{3}{5}$, $x \neq y$ **6.** $\dfrac{2}{3}$, $y \neq -2$, 0 **7.** $\dfrac{t+4}{t+3}$, $t \neq -3$, 4

8. $\dfrac{2m+3}{3m+5}$, $m \neq -\dfrac{5}{3}$, 1 **9.** $\dfrac{6x}{5}$, $x, y \neq 0$

10. $-2xy^2$, $x, y \neq 0$ **11.** $\dfrac{3(y-3)}{y+3}$, $y \neq -3$, 4

435

Exploring Math p. 357

1. yes **2.** no, the lengths are not all equal **3.** no, the faces are not all congruent equilateral triangles
4. skeleton

Chapter Check p. 358

1. a) 302 cm² **b)** 302 cm³ **2. a)** 4430 m² **b)** 22 619 m³
3. a) 468 m² **b)** 951 m³ **4. a)** 516 cm² **b)** 849 cm³
5. a) 1.4 cm **b)** 1.2 cm **6.** 54.5 m **7.** 0.9910 **8.** 0.5075
9. 0.9403 **10.** −0.9348 **11.** no; an angle with a negative cosine is always obtuse **12.** yes; every sine corresponds to two angles between 0° and 180°— an acute angle and an obtuse angle **13.** ∠P = 68.4°, m = 14.3 cm, n = 10.6 cm **14.** d = 2.6 m, ∠B = 63.0°, ∠C = 83.9° **15.** 68 m² **16.** 44 cm² **17. a)** 408 cm²
b) 776 cm³ **18.** 1200 km **19.** 470 m

Using the Strategies p. 359

1. Fill the glass until it is obviously more than half full. Tilt the glass to one side, and pour out the water until the water goes from one side of the bottom of the glass to the opposite side of the top of the glass.
2. hour hand: 88 cm; minute hand: 1508 cm
3. interchange G and B **4.** 30 **5. a)** 10 **b)** 10 **6.** Weigh 3 coins against another 3. If the scales balance, then the real coin is among the 2 not weighed and can be determined with a second weighing. If the scales do not balance, weigh 2 of the three coins that tipped the scales most. If one side tips more, the real coin is found. If the coins balance, the real coin was the other of the three. **7.** 3 **8.** 512 **9.** left to right: G, P, W, R **10.** 60°, 75°, 105°, 120° **11.** 68
Data Bank 1. The surface area of the moon is about 2.3 times as great, assuming the moon and Pluto are spheres. **2.** The volume of the sun is about 550 times as great.

Chapter 8

Statistics and Probability p. 361

1. a) 3 153 600 000 **b)** 8 640 000 **2. a)** 63 072 000
b) 172 800 **3. a)** Alberta: 4 162 752; British Columbia: 5 991 840; Manitoba: 4 099 680; New Brunswick: 441 504; Newfoundland: 2 585 952; Nova Scotia: 378 432; Ontario: 6 748 704; Prince Edward Island: 63 072; Quebec: 9 776 160; Saskatchewan: 4 099 680
b) Alberta: 11 405; British Columbia: 16 416; Manitoba: 11 232; New Brunswick: 1210; Newfoundland: 7085; Nova Scotia: 1037; Ontario: 18 490; Prince Edward Island: 173; Quebec: 26 784; Saskatchewan: 11 232

Getting Started pp. 362–363

1 Scrabble® 1. a) $\frac{9}{100}$; $\frac{1}{100}$; $\frac{1}{50}$ **b)** B, C, F, H, M, P, V, W, Y **c)** D, L, S, U **2. a)** $\frac{1}{100}$ **b)** $\frac{21}{50}$ **c)** $\frac{14}{25}$
d) $\frac{11}{25}$ **e)** $\frac{17}{25}$ **f)** $\frac{7}{100}$ **g)** $\frac{1}{50}$ **h)** 0 **i)** $\frac{1}{20}$ **j)** $\frac{53}{100}$ **k)** 1
3. a) $\frac{1}{9}$ **b)** $\frac{2}{33}$ **c)** $\frac{41}{99}$ **d)** $\frac{58}{99}$ **e)** $\frac{56}{99}$ **f)** $\frac{67}{99}$ **g)** $\frac{8}{99}$
h) $\frac{4}{99}$
2 Stone, Scissors, Paper 2. stone, scissors; scissors, scissors; scissors, paper; scissors, stone; paper, paper; paper, stone; paper, scissors; 9

3. a) $\frac{1}{3}$ **b)** $\frac{1}{3}$ **c)** $\frac{1}{3}$ **4.** Yes. Each player has the same probability of winning.
Mental Math
Order of Operations With Rational Numbers 1. 2 **2.** 2
3. 8 **4.** −1 **5.** −4 **6.** 5 **7.** 55 **8.** 7 **9.** 2 **10.** 5.4
Multiplying by Multiples of 9 1. 171 **2.** 252 **3.** 495
4. 792 **5.** 990 **6.** 1125 **7.** 40.5 **8.** 23.4 **9.** 51.3
10. 1170 **11.** 2430 **12.** 3690 **13.** 13 500 **14.** 33 300
15. 31 500 **16.** 441 **17.** 576 **18.** 945 **19.** 14.4 **20.** 34.2
21. 37.8 **22.** Let the number be x. $10x − x = 9x$.
23. Multiply the number by 10. Then add the number itself. **24.** 396 **25.** 528 **26.** 58.3 **27.** 93.5
28. 2310 **29.** 7260

Section 8.1 pp. 368–369

Applications and Problem Solving 17. a) simple random sampling; consumers of that brand of mineral water **b)** systematic sampling; people who mailed in the coupon for the mineral water **c)** simple random sampling; people in the province with listed telephone numbers **d)** stratified sampling; mailing list **e)** convenience sampling; people drinking mineral water at the park **18. a)** convenience sampling **b)** mall shoppers **19. a)** convenience sampling **20. a)** sampling of volunteers **b)** Answers may vary; people who spend a lot of time outdoors **23. a)** none **b)** no **24. a)** yes **b)** no **25. a)** no **c)** no **d)** yes **e)** no **26.** Answers may vary. **a)** Assign each book a number, and generate 30 random numbers. **b)** Randomly choose the appropriate percent of books from each subject area. **c)** Choose the first 30 books on a shelf. **d)** If there are 3000 books in the library, choose every 100th book. **e)** Randomly choose 30 books from the fiction section.

Section 8.2 pp. 372–373

Practice **1.** selection bias **2.** non-response bias **3.** response bias **4.** selection bias **5.** response bias **6.** selection bias **7.** Answers may vary. Which tastes better: Cola A or Cola B? **8.** Answers may vary. Do small dogs make good pets? **9.** Answers may vary. Should logging be restricted in Canada? **10.** Answers may vary. What makes a painting art? **11.** Answers may vary. What types of projects make school work more interesting? **12.** Answers may vary. Which type of power is better for the environment: hydro-electric or nuclear? **13.** Answers may vary. What is Canada's favourite winter sport?

Applications and Problem Solving **14. a)** Answers may vary; education ministries, school boards, teachers' organizations, parents **d)** Answers may vary. Survey the same students after they graduate to see if they have found employment. **15. a)** Use a telephone survey instead of a mail-in questionnaire. **16. a)** Banks are now treating women and men business owners equally. **17.** selection bias, non-response bias, or people may phone in more than once if they feel very strongly about the subject **18.** non-response bias **19.** selection bias **20. a)** selection bias, non-response bias **b)** Answers may vary. Survey a random sample of students in the school. **21. a)** yes **b)** no **c)** no, except for government census where replying is mandatory **22. b)** Their headache tablets are preferred by 75% of doctors.

Investigating Math pp. 374–375

1 Theoretical Probability **1.** $\frac{1}{10}$ **2.** $\frac{1}{2}$ **3.** $\frac{1}{10}$ **4.** $\frac{2}{5}$ **5.** $\frac{1}{5}$ **6.** $\frac{3}{10}$ **7.** 0 **8.** 1 **9.** $\frac{3}{5}$ **10.** $\frac{3}{10}$ **11.** $\frac{1}{5}$ **12.** $\frac{1}{2}$

2 Independent Events — Tossing Coins **2. a)** $\frac{1}{8}$ **b)** $\frac{3}{8}$ **c)** $\frac{1}{4}$ **3. a)** $\frac{1}{16}$ **b)** $\frac{1}{4}$ **c)** $\frac{3}{8}$ **d)** $\frac{1}{2}$ **e)** $\frac{3}{8}$ **f)** $\frac{15}{16}$

3 Independent Events — Rolling Dice **2. a)** $\frac{1}{9}$ **b)** $\frac{1}{6}$ **c)** $\frac{1}{12}$ **d)** $\frac{1}{4}$ **e)** $\frac{1}{36}$ **f)** $\frac{1}{9}$

Technology pp. 376–377

1 Generating Random Numbers **1.** The numbers for both calculators are between 0 and 1. The numbers for the graphing calculator are rounded to 10 decimal places and the numbers for the scientific calculator are rounded to 3 decimal places. **3. b)** $0 < 2r < 2$ **4. b)** $1 < 2r + 1 < 3$ **5.** 1 and 2

2 Simulating Coin Tosses and Rolling Dice **1. a)** 1 or 2 **b)** Let 1 represent heads and 2 represent tails. **2. a)** int(6rand)+1 **b)** Assign each number to its corresponding outcome on the die.

3 Sampling **1. a)** int(750rand)+1 **b)** Generate another random number. **2. a)** Generate 23 random numbers using int(460rand)+1. Generate 21 random numbers using int(420rand)+1. **3.** There are 670 students altogether, so 10% of the students will be sampled. Select 10% from each grade: 25 from grade 10; 22 from grade 11; and 20 from grade 12. Generate 25 random numbers using int(250rand)+1. Generate 22 random numbers using int(220rand)+1. Generate 20 random numbers using int(200rand)+1.

Section 8.3 pp. 382–383

Practice **1.** 4 **2.** $-\frac{1}{3}$ **3.** 7 **4.** 0.5 **5.** -1 **6.** 1

Applications and Problem Solving **7. b)** Yes. The mathematical expectation of both players is $\frac{1}{2}$.

8. a) 0.5 **b)** yes **9. a)** -0.5 **b)** 10 **10. a)** 20% **b)** $11 **11. a)** 3.5 **b)** 7 **c)** 7 **12. b)** 5 **c)** 20 **13.** $18 **14. a)** 4.94 **b)** There are 13 cards and 4 suits. If the longest suit had 3 cards, there would be 10 cards to divide among the other 3 suits. So one of the remaining suits must have at least 4 cards, which makes it the longest suit. This is a contradiction. **15.** 7 **16.** 10 **17.** 4 **18.** -3 **19. a)** $2.50 **b)** no **c)** $-$3 **d)** no **e)** $1.11 **21. a)** cube: 3.5; tetrahedron: 2.5; octahedron: 4.5; dodecahedron: 6.5; icosahedron: 10.5 **b)** Answers may vary. 1, -2, 3, 4, 5, 6, 7, -8

Section 8.4 pp. 387–389

Applications and Problem Solving **1.** 0.000 0005; 0.000 001; 0.000 025; 0.0004; 0.005 **2.** $4420 **3.** $140 000 **4. a)** 710 **b)** $5680 **c)** 9 **d)** $990 **e)** $4390 **5. a)** $0.30 **b)** $70 **6.** week 7 **7. a)** $150 **b)** $202.50 **c)** $-$24.30 **8. a)** 38; you will pass **b)** 26; you will barely pass **c)** 35; you will pass **9.** The expected loss is $0.40, so it is worthwhile to spend $0.25 on tape protection. **10.** novice, expert **11.** 4 months **12. a)** 0.000 015, 0.000 05, 0.000 125, 0.000 25, 0.005 **b)** $0.60 **c)** $3.09 **13. a)** 0.94 **14.** $0.50 **15. a)** $168 **b)** $-$168

Connecting Math and English pp. 390–391

1 Comparing Keyboards 1. a) 585 **b)** 4.5 **c)** 306
2. a) 32.2% **b)** 70.8% **3.** They are more difficult to get to. **4. a)** 50% **b)** 100%

2 Popular Words 1. $\frac{2}{3}$ **2.** 333 **3. a)** 12% **b)** 40

4. a) 64% **b)** 213 **5. a)** 47.8% **b)** 69.6%
3 Keyboard Trivia 1. DFGHJKL
4 Letter Game 1. A: 0.082; B: 0.014; C: 0.028;
D: 0.038; E: 0.13; F: 0.03; G: 0.02; H: 0.053; I: 0.065;
J: 0.001; K: 0.004; L: 0.034; M: 0.025; N: 0.07;
O: 0.08; P: 0.02; Q: 0.001; R: 0.068; S: 0.06;
T: 0.105; U: 0.025; V: 0.009; W: 0.015; X: 0.002;
Y: 0.02; Z: 0.0005 **2.** 1.382 **3.** 22 **4. a)** yes **b)** Yes; it would take over 200 turns.

Career Connection p. 392

1 Car Liability Insurance 1. 0.0502, 0.0446, 0.0434,
0.0355 **2.** $262, $220, $206, $171 **3.** Answers may vary. $352, $330, $274, using a 60% mark up

Computer Data Bank p. 393

2 Loss Ratio 1. greater than 1 **3.** collision coverage in Alberta: females, with driver training, licensed less than 3 years, 46–55 years old; third party liability coverage in Alberta: males, with driver training, licensed less than 3 years, 56–65 years old; collision coverage in Ontario: females, with driver training, licensed less than 3 years, 66+ years old; third party liability coverage in Ontario: females, with driver training, licensed less than 3 years, 66+ years old
3 Expected Value 1. a) low **b)** a loss to insurance companies. **3.** collision coverage in Alberta: females, with driver training, licensed less than 3 years, 46–55 years old; third party liability coverage in Alberta: males, with driver training, licensed less than 3 years, 56–65 years old; collision coverage in Ontario: females, with driver training, licensed less than 3 years, 66+ years old; third party liability coverage in Ontario: females, with driver training, licensed less than 3 years, 66+ years old
4 Paying Premiums 1. collision coverage in Alberta: females, with driver training, licensed less than 3 years, 46–55 years old; third party liability coverage in Alberta: males, with driver training, licensed less than 3 years, 56–65 years old; collision coverage in Ontario: females, with driver training, licensed less than 3 years, 66+ years old; third party liability coverage in Ontario: females, with driver training, licensed less than 3 years, 66+ years old **2.** same

results for both types of coverage in Ontario; all four results are with driver training and licensed less than 3 years, three of four are females, three different age groups, but all over 45 years

Review pp. 394–395

1. a) school population; stratified sampling by grade
b) households in the town; simple random sampling
c) amateur musicians; simple random sampling
6. non-response bias, selection bias **7.** response bias
8. non-response bias **9. a)** people who are at home during the day **b)** selection bias: survey excludes people who are not at home during the day
10. b) students in your school **d)** the bias that occurs when people do not fill in the questionnaire
11. a) Their cola is better. **b)** no **12.** 2 **13.** –0.5 **14.** 5.5
15. a) 20 **b)** Answers may vary. **16. a)** $110 200
b) $72.45 **c)** $96.60; $10 200 **d)** 2898; $100
17. a) $15.40 **b)** $11.40 **18.** 5 days

Exploring Math p. 395

1. a) 24 **b)** 24, 120; $4 \times 3 \times 2 \times 1$; $5 \times 4 \times 3 \times 2 \times 1$
2. 4!; 5! **3. a)** 10! **b)** 3 628 800 **4.** 106 298 years

Chapter Check p. 396

5. a) stratification by age or grade, and/or gender
6. response bias **7.** selection bias **8.** 1 **9. a)** $\frac{2}{3}$ **b)** yes
c) 15 **10. a)** $30 **b)** $350 000 **c)** $37.50 **d)** 1500
11. a) $110 **b)** $154

Using the Strategies p. 397

1. 48.5 square units **2.** 2, 7, 31 **3.** 321 cm² **4.** 25
5. $10^2 - 2^2$; $11^2 - 5^2$; $14^2 - 10^2$ **6. a)** 16 m by 32 m
b) 1600 m² **7.** 865 013 + 75 013 = 940 026 **8.** 14
Data Bank 1. 0.21

Cumulative Review, Chapters 5–8 pp. 398–399
Chapter 5

1. a) domain {–1, 0, 4, 5}, range {–2} **c)** For every value in the domain, the corresponding value in the range is –2. **2. a)** domain {1, 2, 4, 11}, range {4, 8, 16, 44}
c) Each value in the range is 4 times the corresponding value in the domain. **3. a)** domain {–9, 3, 6, 12}, range {–3, 1, 2, 4} **c)** Each value in the domain is 3 times the corresponding value in the range. **8.** –2, –1, –1, 2, 2, 7, 7 **9.** 5, 6, 6, 9, 9, 14, 14
10. function **11.** not a function **12.** function **13. a)** 4
b) 25 **c)** 1 **d)** –2 **e)** –11 **f)** –0.8 **g)** –2.3 **h)** 28 **i)** 298
14. Range is R. **15.** 45 **16. a)** $E = 500 + 0.04S$ **b)** $720
c) $11 875

Chapter 6

1. $2\sqrt{5}$ **2.** $\sqrt{89}$ **3.** $\sqrt{89}$ **4.** $10\sqrt{2}$ **5.** (3, 4) **6.** (5, 5)
7. (2.5, −2.5) **8.** (3.4, 0.7) **9.** −1 **10.** 1 **11.** 4 **12.** 0.2
13. a) \$3 800 000/year **b)** \$72.1 million

17. $2x + y - 7 = 0$ **18.** $x + 2y - 2 = 0$; $\dfrac{1}{2}$; Answers may

vary. (0, 1), (4, −1) **19.** 5; $-\dfrac{1}{5}$ **20.** $3x - y - 4 = 0$

Chapter 7

1. 137 cm³ **2.** 42 cm² **3. a)** 6.25 : 1 **b)** 15.625 : 1
4. 11.6 km **5.** 155° **6.** $\angle C = 40.3°$, $a = 48.2$ cm,
$c = 31.3$ cm **7.** $\angle S = 43.1°$, $r = 32.2$ cm, $t = 34.0$ cm
8. $e = 20.4$ m, $\angle F = 64.7°$, $\angle D = 44.9°$ **9.** $\angle X = 56.7°$,
$\angle Y = 73.4°$, $\angle Z = 49.9°$ **10.** 6511 m² **11.** 401 km

Chapter 8

1. simple random sampling, stratified sampling, clustered sampling, systematic sampling
2. convenience sampling **4. a)** response bias, selection bias **b)** Reword the question to "What do you think the age for obtaining a driver's learning permit should be?" and ask people other than just high school students. **5. a)** 12.5 **b)** 8 **6. a)** \$500 000 **b)** \$4
7. 5th day

Cumulative Review, Chapters 1–8 pp. 400–403

1. rational, real **2.** irrational, real **3.** natural, whole, integer, rational, real **4.** rational, real **5.** 7 **6.** 5 **7.** −1

8. $2\sqrt{7}$ **9.** $\sqrt{5}$ **10.** $2\sqrt{6}$ **11.** $6\sqrt{15}$ **12.** $11\sqrt{2}$ **13.** $5\sqrt{5}$
14. $\sqrt{5} + 2\sqrt{10}$ **15.** $2\sqrt{32}$, $3\sqrt{16}$, $7\sqrt{3}$, $9\sqrt{2}$
16. $6\sqrt{2} - 12\sqrt{5}$ **17.** $3\sqrt{6} - 2$ **18.** −4 **19.** $22 - 4\sqrt{10}$

20. $2\sqrt{5}$ **21.** $\dfrac{2\sqrt{6} + 2\sqrt{3}}{3}$ **22.** $8x^9y^{-3}$ **23.** $-12x^2y^7$

24. $3a^4b$ **25.** $3y^7$ **26.** 9 **27.** $\dfrac{2}{3}$ **28.** $\dfrac{1}{4}$ **29.** 0.09 **30.** 32

31. 2 **32.** $10\sqrt{5} - 10\sqrt{3}$ **33.** \$16 244 **34.** \$10 145.93
35. a) \$3242.20 **b)** \$3259.81 **36.** 5, 8, 11, 14, 17; 32
37. 1, 7, 17, 31, 49; 199 **38.** 7, 5, 3, 1, −1 **39.** −15, −11, −7, −3, 1 **40.** 19, 94 **41.** −70, −125 **42.** 67 **43.** 12
44. 25 **45.** 21, 34, 47 **46.** 555 **47.** 400 **48.** −15 **49.** 88
50. 230 **51.** $-x - 3y$ **52.** $5a^2 + 4ab + b^2$ **53.** $-42x^3y^2z^4$
54. $-4ab^2$ **55.** $14u + 17$ **56.** $-10t^3 - 20t^2 + 109t$
57. $2y^2 + 9y - 3$ **58.** $3a^3 + 5a^2 - a - 2$ **59.** $4x^2 - 4x + 1$
60. $36q^2 - 49$ **61.** $3g^2 + 6g + 3$ **62.** $125c^3 - 150c^2 + 60c - 8$
63. $7(d + 7)$ **64.** $(t + 3)(t + 4)$ **65.** $(x + 3)(x - 2)$
66. not possible **67.** $(n + 3)(n - 3)$ **68.** not possible
69. $3(x + 1)^2$ **70.** not possible **71.** $(2m + 1)^2$
72. $(3t + 1)(3t - 1)$ **73.** $3(x - 2)(x + 2)$
74. $2(2x - 3y)(2x + 3y)$ **75.** $2(x - 3y)^2$
76. $2(d + 1)(2d - 3)$ **77.** $3(4k - 1)(k + 3)$

78. $3x^2 - x + 2$, $x \neq 0$ **79.** $-6xy^3 + 4$, $x, y \neq 0$
80. $x + 6$, $x \neq -6$ **81.** $4x + 1$, $x \neq 1$

82. $5r - 2$, $r \neq -\dfrac{1}{2}$ **83.** $x^2 + 7x + 13 = (x + 4)(x + 3) + 1$

84. $6y^2 + y - 3 = (2y - 1)(3y + 2) - 1$
85. $12m^3 + 17m^2 + 3m = (3m + 2)(4m^2 + 3m - 1) + 2$
86. $8n^3 - 10n^2 - 12n + 12 = (2n^2 - 3)(4n - 5) - 3$

87. $\dfrac{m-2}{m+3}$, $m \neq -3$, 1 **88.** $\dfrac{n}{3n+1}$, $n \neq -\dfrac{1}{3}$

89. $b + 1$, $b \neq \dfrac{2}{5}$ **90.** $\dfrac{2t+1}{3t-5}$, $t \neq \dfrac{1}{2}$, $\dfrac{5}{3}$

91. $2m^2p$, $m, n, p, q \neq 0$ **92.** $\dfrac{y^4}{10x^5}$, $x, y \neq 0$

93. $\dfrac{(b-5)(b-2)}{b+5}$, $b \neq -7$, -5

94. $\dfrac{2a+3}{a+5}$, $a \neq -5$, $-\dfrac{8}{3}$, 4 **95.** $\dfrac{19a+2}{6}$ **96.** $\dfrac{3-7x}{3}$

97. $\dfrac{2t-15}{6t-12}$, $t \neq 2$ **98.** $\dfrac{4n-4}{(n+2)^2(n-2)}$, $n \neq \pm 2$

99. 3.5 **100.** $-\dfrac{1}{2}$ **101.** −8 **102.** 2.2

103. Banff: 6600 km²; Kluane: 22 000 km²
104. a) domain: {0, 1, 3, 4}, range: {−2, −1, 1, 2}
b) The sum of the element in the range and the corresponding element in the domain is 2.
105. a) domain: {1, 3, 8, 10}, range: {−2, 0, 5, 7}
b) Each element in the range is 3 less than the corresponding element in the domain.
106. a) domain: {−2, 2, 3, 6}, range: {−4, 4, 6, 12}
b) Each element in the range is 2 times the corresponding element in the domain. **111.** −1.9; −4.9
112. −19.6; 7.1 **113.** −2.9; −1.6 **114.** 10.8; −2.6
115. range: $y \geq -3$ **116.** range: $y \geq -1$ **117.** not a function **118.** function **119.** function **120. a)** 7 **b)** 1
c) −8 **d)** 2.2 **e)** −0.5 **f)** 40 **121.** 35 **122. a)** 2, 4, 6, 8, 10; directly **b)** 4, 5, 6, 7, 8; partially **c)** −5, −10, −15, −20, −25; directly **123. a)** $d = 100 + 90t$ **b)** 325 km **c)** 3.6 h
124. 5 **125.** $\sqrt{2}$ **126.** $3\sqrt{5}$ **127.** $2\sqrt{5}$ **128.** (1, 5)
129. (−3, 3) **130.** (3, −1) **131. a)** 26 units

b) 36 square units **132.** undefined **133.** $\dfrac{1}{7}$ **134.** 2

135. a) 500 000/year **b)** 39.3 million
139. $3x - y - 15 = 0$ **140.** $4x + 7y + 20 = 0$; Answers

may vary. $\left(0, -\dfrac{20}{7}\right)$, $\left(1, -\dfrac{24}{7}\right)$ **141.** $-\dfrac{1}{2}$; 2

142. $2x - y - 21 = 0$ **144. a)** 133 cm² **b)** 144 cm³
145. 938 m³ **146.** 3375 cm³ **147.** 140 m
148. 72 m **149.** 60°, 120° **150.** $\angle A = 49.2°$, $b = 11.3$ m, $c = 13.5$ m **151.** $\angle D = 26.4°$, $d = 16.0$ cm, $e = 31.6$ cm
152. $f = 8.2$ m, $\angle G = 75.8°$, $\angle H = 51.9°$
153. $\angle R = 53.7°$, $\angle S = 57.9°$, $\angle T = 68.4°$
154. 128° **155.** 3117 m² **157. a)** selection bias, non-response bias **b)** Answers may vary. **158.** –1
159. b) 0.25 **c)** 20 turns **160.** $8.50

GLOSSARY

A

absolute value   The distance of a real number, a, from zero on a real number line, denoted by $|a|$.

$$|-5| = 5 \text{ and } |12| = 12.$$

acute angle   An angle whose measure is between $0°$ and $90°$.

acute triangle   A triangle with all angles acute.

additive inverse   The inverse, $-a$, of a real number, a, such that $a + (-a) = -a + a = 0$.

adjacent angles   Two angles with a common vertex, a common side, and no interior points in common.

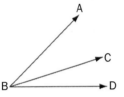

$\angle ABC$ and $\angle CBD$ are adjacent.

alternate angles   Two angles formed on opposite sides of a transversal.

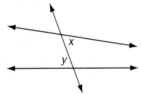

$\angle x$ and $\angle y$ are alternate angles.

altitude of a triangle   The perpendicular distance from one vertex to the opposite side.

angle   A figure formed by two rays with a common endpoint called the vertex.

angle bisector   A ray that divides an angle into two angles having the same measure.

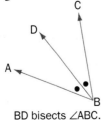

BD bisects $\angle ABC$.

angle of depression   The angle, measured downwards, between the horizontal and the line of sight from an observer to an object.

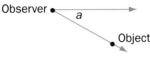

$\angle a$ is an angle of depression.

angle of elevation   The angle, measured upwards, between the horizontal and the line of sight from an observer to an object.

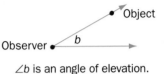

$\angle b$ is an angle of elevation.

area   The number of unit squares contained in a region.

arithmetic mean(s)   The term(s) between two non-consecutive terms of an arithmetic sequence.

The three arithmetic means between 8 and 16 are 10, 12, and 14.

arithmetic sequence   A sequence where the difference between consecutive terms is a constant, called the common difference.

1, 4, 7, 10, 13, 16, 19, 22, … is an arithmetic sequence.

arithmetic series   The sum of the terms of an arithmetic sequence.

arrow diagram   A diagram used to show how each element in the domain of a relation is paired with some element(s) in the range.

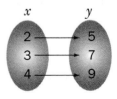

ascending order   From least to greatest. In a polynomial, an arrangement of the terms so that the powers of one variable are in order, beginning with the least exponent.

average   The mean of a set of n numbers, found by dividing the sum of the numbers by n.

The average of 4, 5, 8, and 10 is 6.75.

axis   A number line used for reference in locating points on a coordinate plane.

axis of symmetry   A line that is invariant under a reflection.

B

bar graph   A graph using bars to represent data.

base design   For an object built from linking cubes, the top view of the object with the number of cubes in each column recorded.

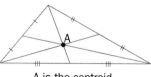

bias   An unwanted influence that prevents a sample from being truly representative of the population from which it is selected.

binomial   A polynomial consisting of two terms.

$x^2 - 1$ is a binomial.

broken-line graph   A graph using line segments to represent data.

C

Cartesian coordinate system   The system developed by René Descartes to plot points on a grid defined by two perpendicular number lines.

census   A survey in which data are collected from every member of a population.

centroid   The point of intersection of the three medians of a triangle.

A is the centroid.

chord of a circle   A line segment having its endpoints on the circumference.

circle   The set of all points in the plane that are equidistant from a fixed point called the centre.

circle graph   A graph using sectors of a circle to represent data.

circumference   The perimeter of a circle.

clustered sampling   Choosing a random sample from one group within a population.

collinear points   Points that lie in the same straight line.

$(-3, -2)$, $(0, 1)$, and $(4, 5)$ are collinear points. They lie on the line $y = x + 1$.

common difference   The difference between two consecutive terms of an arithmetic sequence.

The common difference of the sequence 1, 4, 7, 10, ... is 3.

common ratio   The ratio of consecutive terms in a geometric sequence.

The common ratio of the sequence 2, 6, 18, 54, 162, ... is 3.

complementary angles   Two angles whose sum is $90°$.

complex fraction   A rational expression that contains a fraction in its numerator or denominator.

$\dfrac{\frac{x}{5}}{8}$ is a complex fraction.

compound interest   Interest that is paid at regular intervals and added to the principal for the next interest period.

concentric circles   Circles having the same centre.

cone   A three-dimensional object with one circular face (in a circular cone) and one curved face.

congruent angles   Angles with the same measure.

congruent figures   Figures having the same size and shape.

conjugates   Two binomials of the form $a\sqrt{b} + c\sqrt{d}$ and $a\sqrt{b} - c\sqrt{d}$.

consecutive numbers Numbers obtained by counting by ones from any given number.

constant of variation In a direct variation, the ratio of corresponding values of the variables.

continuous graph A line graph in which the line is unbroken.

convenience sampling Choosing a sample from any group of the population that happens to be handy.

coordinate A real number paired with a point on a number line.

coordinate plane A one-to-one pairing of all ordered pairs of real numbers with all points of a plane. Also called the Cartesian coordinate plane.

cosine ratio In a right triangle, for acute angle θ, the ratio of the length of the side adjacent to $\angle\theta$ and the length of the hypotenuse.

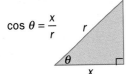

$$\cos\theta = \frac{x}{r}$$

coterminal angles Angles, in standard position, that have the same initial and terminal rays (arms).

$30°$, $390°$, and $-330°$ are coterminal angles.

cube A polyhedron with six congruent faces.

cubic equation An equation of degree three.

$y = x^3 - 1$ is a cubic equation.

cubic polynomial A polynomial of the form $ax^3 + bx^2 + cx + d$, where $a \neq 0$.

$x^3 + 2x^2 + 5$ is a cubic polynomial.

cylinder A three-dimensional object with two parallel, congruent circular faces (in a circular cylinder).

D

degree A unit of angle measure equal to $\frac{1}{360}$ of a rotation.

degree of a monomial The sum of the exponents of the variables.

The degree of $7ab^3c^2$ is 6.

degree of a polynomial The degree of the monomial with the greatest degree.

The degree of $7x^4y + x^2y + 3xy^3$ is 5.

dependent variable A variable whose value is determined by another variable.

descending order From greatest to least. In a polynomial, an arrangement of the terms so that the powers of one variable are in order, beginning with the greatest exponent.

$7x^2y + 3xy^2$ is in descending order for the variable x.

diagonal A line segment with endpoints on two non-adjacent vertices of a polygon.

diameter of a circle A chord that contains the centre of the circle.

difference of squares A binomial that is the difference of two perfect square terms.

$25a^2 - 16$ is a difference of squares.

dilatation A transformation that maps each point of a figure to an image point so that, for a centre C and a point P, $CP' = k(CP)$, where k is the scale factor.

direct variation A function defined by an equation of the form $y = kx$.

discrete graph A graph that is a series of separate points.

distance between two points The length of the line segment joining the points. For points (x_1, y_1) and (x_2, y_2), $\sqrt{(x_2 - x_1)^2 + (y_2 - y_1)^2}$.

distance from a point to a line The length of the perpendicular segment drawn from the point to the line.

distributive property The property defined by $a(b + c) = ab + ac$.

domain   The set of numbers for which a relation is defined. The set of all first coordinates of the ordered pairs in the relation.

domain of a variable   The set of numbers that can be used as replacements for a variable.

E

enlargement   A dilatation for which the image is larger than the original.

equation   An open sentence formed by two expressions related by an equal sign.

$$2x - 7 = 3(4x + 1) \text{ is an equation.}$$

equilateral triangle   A triangle with all sides equal.

equivalent equations   Equations that have the same solution over a given domain.

event   Any possible outcome of an experiment in probability.

expected value   The theoretically calculated outcome for a set of observations based on probability. The expected value for 100 tosses of a coin is 50 heads, because the theoretical probability of heads for one toss of a coin is $\frac{1}{2}$.

experimental probability   The probability of an outcome occurring, determined by conducting an experiment.

exponent   The number of times the base occurs in a power.

$$\text{In } 3x^4, \text{ the exponent is 4.}$$

exponent laws   The rules governing the behaviour of the exponents when multiplying and dividing powers.

exterior angle of a polygon   An angle contained between one side of a polygon and the extension of the adjacent side.

extrapolation   The use of a graph to find values outside the given ordered pairs.

F

factor   A number that is multiplied by another number to give a product.

$$3 \text{ is a factor of 18.}$$

factoring   Finding the factors of a number or expression.

family of lines   Lines that share a common characteristic.

Fermi problems   Problems that involve large numbers and have approximate answers.

Fibonacci sequence   The number sequence $\{1, 1, 2, 3, 5, 8, \ldots\}$ in which each term, except the first two, is the sum of the two preceding terms.

formula   An equation that states the relationship among quantities that can be represented by variables.

frequency of an event   The number of times an event has taken place.

function   A relation in which for each element in the domain there is a single corresponding element in the range.

function notation   A method of writing a relation that is a function. f names a function and $f(x)$ is another name for y.

G

general term, t_n   The formula by which any term in an arithmetic or geometric sequence may be found.

The sequence 0, 3, 8, 15, … has the general term $t_n = n^2 - 1$.

geometric mean(s)   The term(s) between two given terms of a geometric sequence.

The two geometric means between 5 and 40 are 10 and 20.

geometric sequence   A sequence in which the ratio of every pair of successive terms is constant.

2, 6, 18, 54, 162, … is a geometric sequence.

glide reflection   The composition of a translation and a line reflection.

golden ratio   The ratio of the length to the width of a golden rectangle, approximately 1.6:1.

golden rectangle   A rectangle whose sides are in a ratio that is pleasing to the eye.

graph   A diagram that shows a relation between two or more quantities.

greatest common factor   The monomial, with the greatest numerical coefficient and greatest degree, that is a factor of two or more terms.

The greatest common factor of $3x^3 + 6x^2 + 3x$ is $3x$.

H

Heron's formula   The equation for calculating the area of a triangle based on the lengths of the sides, a, b, and c, and half the perimeter, s.

$$A = \sqrt{s(s-a)(s-b)(s-c)}$$

hexagon   A polygon with six sides.

histogram   A bar graph used to display a large set of data that is grouped in consecutive classes.

hypotenuse   The side opposite the right angle in a right triangle.

I

identity   An equation whose sides are equivalent expressions. The equation is true for every value of the variable.

identity elements   The identity element for addition is 0, since $a + 0 = a$. The identity element for multiplication is 1, since $a \times 1 = a$.

image   The representation of a point, line, or lines following a transformation.

independent events   Events whose outcomes do not influence each other.

independent variable   In a relation, the variable whose value may be freely chosen and upon which the value(s) of the other variable depend.

indirect reasoning   Assuming the opposite of what is to be proved and showing that this leads to a contradiction.

inequality   Two expressions related by an inequality symbol (>, ≥, < or ≤).

$3x \le 12$ is an inequality.

initial arm   For an angle in standard position on a coordinate grid, the arm fixed on the positive x-axis.

inscribed polygon   A polygon with its vertices on a circle.

integer   A member of the set $\{..., -3, -2, -1, 0, 1, 2, 3, ...\}$.

interest period   The time, in days, months, or years, over which interest on the principal is calculated.

interpolation   The use of a graph to find values between the given ordered pairs.

inverse variation   A function defined by an equation of the form $xy = k$, $k \ne 0$.

irrational number   A real number that cannot be expressed as a terminating or repeating decimal.

isometric view   A two-dimensional representation of an object, showing the corner view.

isosceles triangle   A triangle with two sides equal.

L

lateral area   The sum of the areas of the faces of a three-dimensional object, other than the base(s).

latitude   The position north or south of the equator, measured in degrees.

law of cosines   The relationship between the lengths of the three sides and the cosine of an angle in any triangle.

$$a^2 = b^2 + c^2 - 2bc \cos A$$

law of sines   The relation stating the lengths of the sides of any triangle are in the same proportion as the sines of the angles opposite the sides.

$$\frac{a}{\sin A} = \frac{b}{\sin B} = \frac{c}{\sin C}$$

least common multiple   The monomial, with the least positive numerical coefficient and least degree, that is a multiple of several monomials.

The least common multiple of $3xy$ and $5x^2y$ is $15x^2y$.

length of a line segment For a line segment with endpoints (x_1, y_1) and (x_2, y_2), $\sqrt{(x_2 - x_1)^2 + (y_2 - y_1)^2}$.

like terms   Terms that have the same variable factors, such as $3ab$ and $-5ab$.

line (curve) of best fit   A line drawn as close as possible to the most points of a scatter plot.

line segment   Two points on a line and the points between them.

line symmetry   A property of a figure in which there is a line such that the figure coincides with its reflection image over the line.

linear equation   An equation in which each term is either a constant or has degree 1.

$y = 7x - 2$ is a linear equation.

linear function   A function of the form $f(x) = mx + b$.

linear relation   The relation between two values whose graph is or approximates a straight line.

literal coefficient   The variable part of a monomial.

The literal coefficient of $3m^2n$ is m^2n.

locus   A set of points that satisfy a given condition.

longitude   The position east or west of the meridian through Greenwich, England, measured in degrees.

lowest common denominator (LCD)   The least common multiple of the denominators of two or more rational expressions.

The lowest common denominator of $\frac{4}{5x}$ and $\frac{3}{10y}$ is $10xy$.

M

mapping   A pairing of each element in the domain of a function with some element(s) in the range. A correspondence of points between an object and its image.

mass   The amount of matter in an object, measured in grams or a related unit.

mean   The sum of a set of values divided by the number of values.

The mean of 4, 5, 8, and 10 is 6.75.

median   The middle number of a set of values arranged in order from least to greatest.

The median of 4, 5, 8, and 10 is 6.5.

median of a triangle   The line segment joining a vertex to the midpoint of the opposite side.

midpoint   The point that divides a line segment into two equal parts.

mixed number   A number that is part whole number and part fraction, such as $4\frac{2}{5}$.

mode   The number that occurs most often in a set of values.

The mode of 3, 8, 11, 11, 15, 19, and 22 is 11.

monomial   A number, a variable, or a product of numbers and variables.

8, x, $-y^2$, and $3mn$ are monomials.

multiplicative identity   The number 1. The product of any number and 1 is identical to the original number.

multiplicative inverses   Two numbers whose product is 1.

N

natural numbers   The set of numbers {1, 2, 3, 4, 5, 6, ...}.

net   A pattern for constructing a three-dimensional object.

nonagon   A polygon with nine sides.

non-linear relation   A relation whose graph is not a straight line.

non-response bias   Bias arising from a lack of response from many of the people surveyed.

number line   A line on which real numbers may be represented as distances from zero.

numeral   A symbol that represents a number.

numerical coefficient   The number part of a monomial.

The numerical coefficient of $3m^2n$ is 3.

O

oblique triangle   A triangle that is not right-angled.

obtuse angle   An angle whose measure is greater than 90° but less than 180°.

obtuse triangle   A triangle with one obtuse angle.

octagon   A polygon with eight sides.

octahedron   A polyhedron with eight faces.

order of operations   The rules to be followed when simplifying expressions. Sometimes referred to as BEDMAS.

ordered pair   A pair of numbers used to name a point on a graph, such as (−5, 3).

origin   The intersection of the horizontal axis and the vertical axis on a Cartesian coordinate grid. Described by the ordered pair (0, 0).

orthocentre   The point where the altitudes of a triangle intersect.

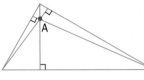

A is the orthocentre.

orthographic view   A two-dimensional representation of an object, showing how it appears from the top, front, and right side.

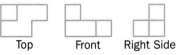

Top   Front   Right Side

outcome   The result of an experiment or a trial.

P

palindrome   A number that reads the same forwards as backwards, such as 232.

parallel lines   Two lines in the same plane that never meet.

parallelogram   A quadrilateral with opposite sides parallel.

partial variation   A relation between two variables that involves a fixed amount plus a variable amount, such as $C = 3d + 15$.

Pascal's triangle   The triangular arrangement of numbers with 1 in the first row, and 1 1 in the second row. Each number in succeeding rows is the sum of the two numbers above it in the preceding row.

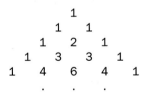

payoff   The amount won or lost for each outcome of a game.

pentagon   A polygon with five sides.

percent   A fraction (or ratio) in which the denominator is 100.

perfect square trinomial   A trinomial that can be factored as the square of a binomial.

$$a^2x^2 + 2abx + b^2 = (ax + b)^2, \text{ where } a, b \neq 0.$$

perimeter   The distance around a polygon.

period   In calculating interest, the unit time for which interest is charged.

period of a decimal   The digits that repeat in a periodic or repeating decimal.

The period of $3.5\overline{64}$ is 64.

periodic decimal   A repeating decimal, such as $0.\overline{4}$.

perpendicular bisector   The line that intersects a line segment at right angles, dividing it into two equal parts.

perpendicular lines   Lines that intersect at right angles.

pi (π)   The ratio of the circumference of a circle to its diameter.

pictograph   A graph using pictures to represent data.

point-slope form of a linear equation   For a line with a point, (x_1, y_1), and slope, m, the equation $y - y_1 = m(x - x_1)$.

polygon   A closed figure formed by line segments.

polyhedron   A three-dimensional object having polygons as faces.

polynomial   A monomial or the sum of monomials.

population   The entire set of items from which data are taken.

power   A product obtained by using a base as a factor one or more times.

5^3 is the power that equals 125.

prime factorization   The factoring of a number into the product of factors that are prime numbers.

$$24 = 2 \times 2 \times 2 \times 3$$

prime number   A number with exactly two factors — itself and 1.

2, 5, and 7 are prime numbers.

principal square root   The positive square root.

prism   A polyhedron with two parallel and congruent bases in the shape of polygons.

probability   The ratio of the number of favorable outcomes to the number of possible outcomes.

probability sampling   The random selection of units from a population.

proportion   An equation that states that two ratios are equal.

pyramid   A polyhedron with three or more triangular faces and the base in the shape of a polygon.

Pythagorean theorem   The relation that expresses the area of the square drawn on the hypotenuse of a right triangle as equal to the sum of the areas of the squares drawn on the other two sides.

Q

quadrant   One of the four regions formed by the intersection of the x-axis and the y-axis.

quadratic equation   An equation of degree two.

$y = x^2 + 3$ is a quadratic equation.

quadrilateral   A polygon with four sides.

quotient   The result of a division.

R

radical   A number in the form $\sqrt{2}, \sqrt{13}, \sqrt{26}$. $\sqrt{75}$ is an entire radical. $5\sqrt{3}$ is a mixed radical.

radical sign   The symbol $\sqrt{}$.

radius   The length of the line segment that joins the centre of a circle and a point on the circumference.

random sample   A sample in which each member of the population has the same chance of being selected.

range   The set of all second coordinates of the ordered pairs of a relation.

rate   A ratio of two measurements having different units.

$$100 \text{ km/h is a rate.}$$

ratio   A comparison of two numbers.

$$7:5 \text{ is a ratio.}$$

rational equation   An equation that contains one or more rational expressions.

$$\frac{x+1}{x^2-4} = 3 \text{ is a rational equation.}$$

rational expression   An algebraic fraction whose numerator and denominator are polynomials.

$$\frac{x}{x-5} \text{ is a rational expression.}$$

rational function   A function of the form $\frac{P}{Q}$, where P and Q are polynomials.

rationalizing the denominator   The process of eliminating a radical from the denominator of a fraction.

$$\frac{3}{2\sqrt{2}} = \frac{3}{2\sqrt{2}} \times \frac{\sqrt{2}}{\sqrt{2}} = \frac{3\sqrt{2}}{4}$$

rational number   A number that can be expressed as the ratio of two integers.

$$0.75, \frac{3}{8}, \text{ and } -2 \text{ are rational numbers.}$$

ray   Part of a line extending in one direction without end.

real numbers   The set of all terminating decimals, all repeating decimals, and all non-terminating, non-repeating decimals.

$$0, -8, \frac{5}{6}, \text{ and } 0.\overline{4} \text{ are real numbers.}$$

reciprocals   Two numbers that have a product of 1.

3 and $\frac{1}{3}$ are reciprocals, and $-\frac{2}{5}$ and $-\frac{5}{2}$ are reciprocal.

rectangle   A parallelogram with four right angles.

recursion formula   The formula by which a term of a sequence may be determined from the previous term.

The sequence 6, 10, 14, 18, 22 has the recursion formula $t_1 = 6$, $t_n = t_{n-1} + 4$.

reduction   A dilatation for which the image is smaller that the original.

reference angle   For an angle θ in standard position, the acute angle formed by the terminal arm and the x-axis.

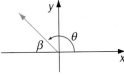

$\angle\beta$ is a reference angle.

reflection   A transformation that maps an object onto an image by a reflection in a line.

reflex angle   An angle whose measure is greater than 180° and less than 360°.

regular polygon (polyhedron)   A polygon (polyhedron) in which all sides and all angles are equal.

relation   A set of ordered pairs.

repeating decimal   A decimal in which one or more digits repeat without end.

response bias   Bias arising from the phrasing of the questions in a survey.

rhombus   A parallelogram in which all sides are equal.

right angle   An angle whose measure is 90°.

right cone   A three-dimensional figure generated by rotating a right triangle about one of its legs.

right cylinder   A cylinder in which the lateral surface is perpendicular to the bases.

right prism   A prism in which the lateral edges are perpendicular to the bases.

right triangle   A triangle with one right angle.

root of an equation   A solution of the equation.

The root of $3x - 1 = 20$ is 7.

rotation   A transformation that maps an object onto its image by a rotation about a point.

rotational symmetry   A property of a figure that maps onto itself after a turn.

rounding   A process of replacing a number by an approximate number.

6.1784 rounded to the nearest tenth is 6.2.

S

sample   A selection from a population.

sample space   The set of all possible outcomes of an experiment or investigation.

scale drawing   A drawing in which all distances are reduced or enlarged by a fixed factor.

scale factor   The multiplication factor used in dilatations (enlargements and reductions) and in scale drawings.

scalene triangle   A triangle with no two sides equal.

scatter plot   The result of plotting data that can be represented as ordered pairs on a graph.

scientific notation   Expressing a number as the product of a number, n, where $1 \le n < 10$, and a power of ten.

$$2700 = 2.7 \times 10^3$$

sector of a circle   A region bounded by two radii and an arc.

segment of a circle   A region bounded by a chord and an arc.

selection bias   Bias arising from excluding one or more groups of the population from a sample.

sequence   An ordered list of numbers.

series   The sum of the terms of a sequence.

set   A collection of objects.

shell   A three-dimensional object whose interior is empty.

similar figures   Figures having corresponding angles equal and corresponding lengths proportional.

simple random sampling   The selection of items at random from a population, so that every item has the same probability of being selected.

simplest form   The form of a rational expression in which the numerator and denominator have no common factor other than 1.

The simplest form of $\dfrac{x^2 + 2x + 1}{x^2 - 1}$ is $\dfrac{x+1}{x-1}$.

simulation   Modelling a situation, often with the help of a computer or calculator.

sine ratio   In a right triangle, for acute angle θ, the ratio of the length of the side opposite $\angle \theta$ and the length of the hypotenuse.

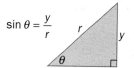

skeleton   A representation of the edges of a polyhedron.

slope and y-intercept form of a linear equation   A linear equation written in the form $y = mx + b$, where m is the slope and b is the y-intercept.

slope of a line   The ratio $\dfrac{\text{rise}}{\text{run}}$. For a non-vertical line containing two distinct points (x_1, y_1) and (x_2, y_2), $\dfrac{y_2 - y_1}{x_2 - x_1}$.

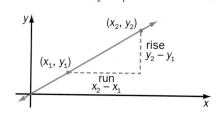

solid   A three-dimensional object whose interior is completely filled.

solution set   Replacement(s) for a variable that result in a true sentence.

solve a triangle   Finding the lengths of the three sides and the measures of the three angles.

sphere   The set of all points in space that are a given distance from a given point.

spreadsheet   A table that is used to manage data, including performing calculations with the data.

square   A quadrilateral with four congruent sides and four right angles.

square root of a number   A number that, when multiplied by itself, gives the original number.

standard form of a linear equation   A linear equation written in the form $Ax + By + C = 0$.

standard position   The position of an angle with its vertex at the origin of a coordinate grid and its initial arm fixed on the positive x-axis.

statistics   The science of collecting and analyzing numerical information.

stem-and-leaf plot   A graph using digits of numbers to display data.

step function   A function that has different constant values in the range corresponding to successive intervals in the domain.

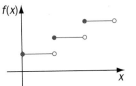

straight angle   An angle whose measure is $180°$.

stratified sampling   Dividing a population into groups and selecting items at random from the groups.

supplementary angles   Two angles whose sum is $180°$.

surface area   The number of unit squares covering the surface of an object.

survey   A method of collecting data by asking questions or counting.

synthetic division   A method for dividing polynomials that uses the numerical coefficients of the terms.

systematic sampling   When items of a population are selected from a list using a selection interval, k.

T

table of values   A method of organizing values of a relation.

tangent ratio   In a right triangle, for acute angle θ, the ratio of the length of the side opposite $\angle\theta$ and the length of the adjacent side.

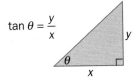

tangent to a circle   A line that lies in the plane of a circle and intersects the circle in exactly one point.

term of a polynomial   A monomial that is one of the addends in a polynomial.

$3x^2 - 7x + 1$ has terms $3x^2$, $-7x$, and 1.

terminal arm   For an angle in standard position, the ray that rotates about the origin.

terminating decimal   A decimal whose digits terminate.

tessellation   A repeated pattern of geometric figures that completely cover a surface.

tetrahedron   A polyhedron with four triangular faces.

theoretical probability   The probability of an outcome determined mathematically, without doing an experiment.

transformation   A mapping of points on a plane onto points on the same plane.

translation   A transformation that maps an object onto its image so that each point in the object is moved the same distance in the same direction.

transversal   A line that intersects two lines in the same plane in two distinct points.

trapezoid   A quadrilateral with one pair of parallel sides.

tree diagram   A diagram illustrating the possible outcomes of an event.

triangle   A polygon with three sides.

trigonometry   The branch of mathematics concerned with the measurement properties of triangles.

trinomial   A polynomial with three terms, such as $x^2 + 2xy + y^2$.

V

value of a function   For an element of the domain of a function, the corresponding element in the range.

The value of the function $f(x) = x + 2$ when $x = -3$ is -1.

variable   A letter or symbol used to represent a number, such as x.

vertex of an angle   The common endpoint of two rays.

vertex of a polygon   The point where two adjacent sides meet.

vertical line test   A test for determining whether a given graph represents a function. If any vertical line intersects the graph in more than one point, then the relation is not a function.

volume   The number of cubic units contained in a solid.

W

whole numbers   Numbers in the set $\{0, 1, 2, 3, 4, 5, \ldots\}$.

X

x-axis   The horizontal line used as a scale for the independent variable in the Cartesian coordinate system.

x-coordinate   The first value in an ordered pair.

x-intercept   The x-coordinate of the point where a line or curve crosses the x-axis.

Y

y-axis   The vertical line used as a scale for the dependent variable in the Cartesian coordinate system.

y-coordinate   The second value in an ordered pair.

y-intercept   The y-coordinate of the point where a line or curve crosses the y-axis.

Z

zero-product property   If $ab = 0$, then $a = 0$ or $b = 0$.

APPLICATIONS INDEX

TECHNOLOGY INDEX

Text Credits

88–89 From *Deadly Appearances* by Gail Bowen, © 1990, published by Douglas & McIntyre. Reprinted with permission of the publisher. From *Deadly Appearances: A Joanne Kilbourn Mystery* by Gail Bowen. Courtesy of Bella Pomer Agency Inc; **404–405** From *MATHPOWER 9, Western Edition*, by George Knill. Reprinted by permission of McGraw-Hill Ryerson Limited; **410** From 1997 *Canadian Global Almanac*. Reprinted by permission of Macmillan Canada; **411** From 1997 *Canadian Global Almanac*. Reprinted by permission of Macmillan Canada; **412** From 1997 *Canadian Global Almanac*. Reprinted by permission of Macmillan Canada; **413 top and bottom** Environment Canada, *Climates of Canada*. Reproduced with the permission of the Minister of Public Works and Government Services Canada, 1996.

Photo Credits

v Ian Crysler; **vi** Ian Crysler; **vii** Courtesy of NASA; **ix** Ian Crysler; **x** Courtesy of Imperial Oil Limited; **xii top** Courtesy of Mary Russel; **centre top** Charles Michael Murray/First Light; **centre bottom** House of Commons photo; **bottom** Canapress Photo Service/Andrew Vaughan; **xiii top** First Light; **centre top** David Stoecklein/First Light; **centre bottom** Ian Crysler; **xv** Courtesy of the Royal Ontario Museum, Toronto, Canada; **xvii left** Courtesy Department of Tourism, Culture, and Recreation. Government of Newfoundland and Labrador; **right** Courtesy of Canada Post Corporation; **xix** Canapress Photo Service/ABC Robert Freeman; **xx** Ron Watts/First Light; **xxii** Courtesy of Mary Russel; **xxiv** Thomas Kitchin/First Light; **xxv** Canadian Sport Images/Claus Andersen; **xxvi** Courtesy of Harold Hosein and City TV. Taffi Rosen/Photographer; **xxvii** Ian Crysler; **xxix** Ian Crysler; **xxx–1** Courtesy of NASA; **2** Charles Michael Murray/First Light; **5** Ione Rice/Earthwatch photo archives; **10** Canapress Photo Service/Al Behrman; **12** Bridgeman/Art Resource, New York; **21** Canada's Radarsat Satellite Supplied by Spar Aerospace Ltd; **30** Canapress Photo Service; **34–35** National Research Council Canada; **39** Ian Crysler; **40** Canapress Photo Service; **42** Ian Crysler; **43** Courtesy of NASA; **44** Canapress Photo

Service/Andrew Vaughan; **50–51** Toronto Blue Jays Baseball Club; **54** Bowers and Merena Galleries Inc./Photographer Doug Plasencia; **62** Copyright © 1997 by Universal City Studios, Inc. Courtesy of Universal Studios Publishing Rights. All Rights Reserved; **64–65** Hale Telescope, Palomar Observatory/California Institute of Technology; **69** CIDA Photo: Benoit Aquin; **72** Corel Corporation/#177087; **77** G. Kopelow/First Light; **78** D. Baswick/First Light; **80** Dick Hemingway; **85** Stamp reproduced courtesy of Canada Post Corporation; **86** Dick Hemingway; **87** Ian Crysler; **88** Courtesy of Gail Bowen; **89** Ian Crysler; **100** Comstock Photofile; **104** Courtesy of the Royal Ontario Museum, Toronto, Canada; **106** Simon Fraser University, Instructional Media Centre; **110** Courtesy of the Royal British Columbia Museum, Victoria, B.C., neg. no. PN 10982; **116** Texas Instruments; **117** Comstock/George Hunter; **118** Toyota Canada Inc; **121** Comstock; **122** Don Ford; **125** Photo of Jumbotron at Skydome in Toronto, Ontario. Skydome is North America's premier sports and entertainment facility; **128** Canadian Olympic Association/Claus Andersen; **132** Courtesy of NASA; **135** Ian Crysler; **136** Comstock/Frank Viola; **137** Dan Bosler/Tony Stone Images; **138** Cornwall Tourist Board; **145** Canadian Olympic Association; **148** Peter Vadnai/First Light; **156** British Columbia Protocol Office; **160** Canadian Sport Images/F. Scott Grant; **165** Canapress Photo Service/Dan Bailey; **176** Ian Crysler; **178** Canadian Olympic Association; **181** Yousuf Karsh/Comstock; **182–183** First Light; **187 top** Robert McCaw; **centre** Robert Lankinen/First Light; **bottom** Ed Cesar; **194** Robert McCaw; **206–207** Photo courtesy of West Edmonton Mall; **210–211** Ian Crysler; **212** Roger Ressmeyer/© 1990 Corbis; **213** 1/50th scale model of the Gemini Facility built by Bob Rice, Tucson, AZ. Photo by Mark Hanna and Glynn Pickens, NOAO; **216** Courtesy of NASA; **217** Courtesy of Bentall Property Management; **220** Canada's Sports Hall of Fame; **221** First Light; **222** House of Commons photo; **223** Canadian Sport Images; **224** First Light/Miwako Ikeda; **228** Canadian Sport Images/Claus Andersen; **232** Richard Sears/MICS Photo; **236** Tourism Newfoundland and Labrador;